Extremophiles for a Sustainable Future

Masahiro Ito • Issay Narumi • Noriyuki Doukyu
Editors

Extremophiles for a Sustainable Future

Springer

Editors
Masahiro Ito
Faculty of Life Sciences,
and Bio-Resilience
Research Project (BRRP)
Toyo University
Asaka, Saitama, Japan

Issay Narumi
Faculty of Life Sciences,
and Bio-Resilience
Research Project (BRRP)
Toyo University
Asaka, Saitama, Japan

Noriyuki Doukyu
Faculty of Life Sciences,
and Bio-Resilience
Research Project (BRRP)
Toyo University
Asaka, Saitama, Japan

ISBN 978-981-95-6969-4 ISBN 978-981-95-6970-0 (eBook)
https://doi.org/10.1007/978-981-95-6970-0

This Springer imprint is published by the registered company Springer Nature Singapore Pte Ltd.
The registered company address is: 152 Beach Road, #21-01/04 Gateway East, Singapore 189721, Singapore

Preface

Earth is a mosaic of environments that defy conventional limits—heat, salinity, acidity, and alkalinity, radiation, desiccation, and pressure. Extremophiles do more than survive at these "edges." Through their metabolisms, enzymes, and cellular machinery, they offer practical routes to a more sustainable society. This volume connects the frontiers of extremophile microbiology to social and industrial implementation, for researchers, engineers, students, and policy practitioners alike.

The book is designed to carry readers seamlessly from fundamentals to applications and future outlooks. Part I (General Aspects) surveys the basic characteristics and diversity of extremophiles (Chaps. 1 and 2). Part II focuses on enzymes and catalysts that operate under extreme conditions and their industrial value, including thermostability engineering and biocatalysis under heat, salt, and other stresses (Chaps. 3, 4, 5, 6, and 7). Part III then highlights the production of valuable compounds and synthetic-biology platforms in extremophiles, covering genome editing in hyperthermophilic archaea, bio-based production by extreme halophiles, Antarctic-derived fungi, ectoine production, and solvent-tolerant enzymes (Chaps. 8, 9, 10, 11, and 12). Part IV addresses fermentation and biomass applications for renewable energy, including sustainable composting and methanation (Chaps. 13 and 14). Part V is dedicated to biohydrogen production technologies using extremophiles, particularly hyperthermophilic archaeal systems (Chaps. 15 and 16). Finally, Part VI presents next-generation biotechnologies for a sustainable future, spanning bioleaching, microbial fuel cells, radiation-resistant bacteria, AI-enabled approaches, and an integrated view toward sustainable development (Chaps. 17, 18, 19, 20, and 21). This structure balances clear learning pathways for newcomers with efficient navigation for implementation-oriented readers.

The contributors are leading researchers from universities, public institutes, and industry. From thermostable enzymes and salt-tolerant catalysts to the use of hyperthermophilic archaea, lignin-degradation metabolism, methanation using methanogenic archaea, marine-strain-based resource recovery, and topics tied directly to environmental remediation and resource circulation, the chapters present up-to-date perspectives with an international outlook. Examples include advances in thermostable enzymes that undergird PCR, industrial applications of halophilic enzymes, genome editing and enzyme production platforms in extremophile archaea, and the challenges and approaches for implementing biohydrogen and biomethane. Each chapter examines these themes from both theoretical and practicable angles.

First, interdisciplinary. We highlight connections among microbiology, enzyme engineering, chemical engineering, materials and resource circulation, energy systems, and policy design for practical implementation; where appropriate, chapter endings include sections on "implementation issues" and "scale-up challenges," organized editorially.

Second, to ensure broad readership, we aim to serve experts with an up-to-date review and provide students or readers from other fields with a self-study primer. To achieve this, we offer a coherent pathway using figures and tables, glossaries, and references.

Third, future outlook. We weave in perspectives that connect research outcomes to social value: data-driven approaches including AI, sustainability assessment such as Life Cycle Assessment (LCA), regulation and standardization, and frameworks for international collaboration. We have explicitly asked authors to reflect these especially in Chaps. 20 and 21.

Research on extremophiles is not merely a theoretical inquiry into "adaptation to harshness." It calls for design thinking that turns constraints into advantages: creating bio-based materials, converting recalcitrant resources, valorizing wastes, and engineering biocatalysts that function under salt, pH, and temperature stress. Each chapter aims to illustrate this turning point—where constraints expand the design space—through concrete examples. Cases in which salt or heat enhances process stability or contamination resistance; cases where the solvent tolerance or thermostability of extremozymes enables precision synthesis or continuous production; and flows that connect fermentation by-products or hard-to-treat residues to resource recovery should offer immediately applicable insights.

This project grew from sustained dialogue with colleagues, collaborators, and industry partners in Japan and abroad. In shaping the table of contents, we mapped the achievements and blind spots of existing books and strengthened the focus on applications and real-world implementation. During revision, we reorganized the contents and added a new chapter on AI and biotechnology, refining the material in step with advances in scholarship and evolving societal needs. This flexibility safeguards both the timeliness and practicality of the first edition at publication.

We extend our sincere thanks to all chapter authors, to the reviewers for their insightful comments, and to the editorial and production staff who supported this work. The daily efforts of students and early-career researchers pushing forward in research, education, and societal engagement also animate many parts of this volume. We hope readers will draw from the knowledge and methods compiled here to take their next steps—in deepening research, implementing technologies, and enriching teaching and training.

Finally, the "Sustainable Future" invoked by this book is not a distant ideal. Designing ways to harness extremophiles in real-world settings is the shortest path to it. May this volume serve as a practical bridge between laboratory and society and accompany your thinking and practice.

Asaka, Saitama, Japan Masahiro Ito
Asaka, Saitama, Japan Issay Narumi
Asaka, Saitama, Japan Noriyuki Doukyu
October 15, 2025

Contents

Part I
General Aspects

Chapter 1
General Aspects: Overview of the Basic Characteristics of Extremophiles and Their Role in Sustainability

Masahiro Ito

Abstract Extremophiles are microorganisms that thrive under physical and chemical extremes that were once considered incompatible with life. This chapter outlines the physiological boundaries of life across temperature, pH, salinity, pressure, radiation, and desiccation and distinguishes between growth and tolerance. Core molecular adaptations such as protein stabilization, archaeal membranes, DNA repair, and compatible solutes have been highlighted as strategies that enable survival. We also summarize the key industrial and ecological applications, including extremozymes, bioremediation, and bioenergy production, and relate these contributions to the UN Sustainable Development Goals. Finally, we note the remaining challenges in cultivation, genetic manipulation, and scale-up and highlight future opportunities in synthetic biology and space biotechnology.

Keywords Extremophiles · Physiological limits · Extremozymes · Bioremediation · Sustainable Development Goals

1.1 Introduction

Extremophiles are microorganisms that flourish in hostile environments (Rothschild & Mancinelli, 2001; Cavicchioli et al., 2011). These organisms have been discovered in environments characterized by extreme temperature, salinity, pH, pressure, radiation, and desiccation (Oren, 2011; Horikoshi & Bull, 2011). The study of extremophiles has reshaped our understanding of fundamental life limitations, offering insights into the resilience and adaptability of microbial systems (Rampelotto, 2013).

M. Ito (✉)
Faculty of Life Sciences, and Bio-Resilience Research Project (BRRP), Toyo University, Asaka, Saitama, Japan
e-mail: masahiro.ito@toyo.jp

M. Ito et al. (eds.), *Extremophiles for a Sustainable Future*,
https://doi.org/10.1007/978-981-95-6970-0_1

Rather than providing a taxonomic overview or detailed habitat-specific examples, this chapter focuses on the overarching characteristics that make extremophiles significant to both science and society. Emphasis has been placed on their physiological and molecular adaptations to extreme environments and their practical applications in sustainable technologies and industries (Cavicchioli, 2019). The diversity and ecological distribution of these organisms, which are critical for understanding their evolutionary success and biogeographic relevance, are examined in detail in Chap. 2.

This chapter serves as a conceptual foundation for those that follow, providing context for how extremophiles contribute to our understanding of life at its limits and to global efforts toward sustainable development (Harrison et al., 2013).

Cross-Reference A comprehensive treatment of the biodiversity and biogeography of extremophiles follows in Chap. 2.

1.2 Physiological Definitions and Limits of Life

Life on Earth is characterized by remarkable adaptability; however, every organism operates within physiological boundaries defined by its biochemistry and cellular architecture. Extremophiles challenge and expand these boundaries by surviving and even thriving in conditions once deemed incompatible with life. This section outlines the key environmental parameters that constrain biological activity, and summarizes the currently known physiological limits of life.

Temperature is a fundamental factor affecting life as it influences enzymatic reactions, membrane fluidity, and metabolic rate. Thermophiles grow optimally at temperatures >60 °C, and the hyperthermophilic methanogenic archaeon *Methanopyrus kandleri* survives at 122 °C under 40 MPa (400 atm) pressure (Takai et al., 2008). In contrast, psychrophiles grow in environments <15 °C, with some capable of metabolic activity at −20 °C, such as in permafrost or deep-sea ice (Mykytczuk et al., 2013). The existence of these organisms compels the reevaluation of the thermal stability of biological macromolecules.

Most organisms prefer near-neutral pH. However, some microorganisms inhabit highly acidic or alkaline environments. The acidophilic archaeon *Picrophilus oshimae* can grow at the extreme acidity of pH −0.06 (equivalent to 1.2 M sulfuric acid), representing the lowest pH at which growth has been confirmed. In contrast, *Serpentinomonas* sp. B1, an alkaliphilic bacterium isolated in 2014 by Suzuki et al. from the serpentine environment "The Cedars" in California, grows at pH 12.5 (Suzuki et al., 2014). As of 2025, it holds the record for growth at the highest pH. These microorganisms maintain intracellular pH through proton pumps and specialized membrane structures, enabling adaptation to extreme pH environments.

Salinity poses another major challenge for cellular systems, particularly regarding protein solubility and osmotic balance. Halophiles, such as *Halobacterium salinarum*, inhabit hypersaline environments with sodium chloride concentrations of up to 5 M (Oren, 2002). Their survival strategies include the accumulation of compatible solutes such as ectoine (Galinski et al., 1985) and structural adaptations that allow enzymes to function efficiently under highly ionic conditions (Mevarech et al., 2000).

Pressure also influences cellular processes, particularly in deep-sea environments where hydrostatic pressure increases by approximately 0.1 MPa per 10 m of depth (Yayanos, 1995). Piezophiles grow optimally at elevated pressures (~10–60 MPa); hyperpiezophiles prefer ≥70 MPa. Some taxa remain active at ≥100 MPa, such as the obligate hyperpiezophile *Colmarinimaniae* isolated from the Mariana Trench (optimum ~120 MPa; growth range 80–140 MPa) (Bartlett, 2002; Kusube et al., 2017).

In addition to these well-defined parameters, some extremophiles exhibit tolerance to ionizing radiation, desiccation, and heavy metal toxicity. For example, *Deinococcus radiodurans* can withstand radiation doses thousands of times higher than those that are lethal to humans because of its highly efficient DNA repair system (Battista, 1997; Slade & Radman, 2011). Cyanobacteria such as *Chroococcidiopsis* spp. persist in arid desert crusts (Billi et al., 2000), whereas certain acidophiles tolerate extreme metal concentrations (Johnson & Hallberg, 2003). Although these conditions may not always support growth, they highlight the extraordinary resilience of microbial life.

Understanding the physiological limits of life refines our definitions of habitability on Earth and broadens the scope of astrobiological exploration and industrial applications (Rothschild & Mancinelli, 2001). The ability of extremophiles to persist under such conditions lays the groundwork for further discussions on their diversity and ecological distribution, which are addressed in Chap. 2.

Table 1.1 summarizes the major categories of extremophiles, their physiological limitations, representative taxa, example species, notable applications, and key references.

Cross-Reference Detailed discussions on the specific adaptations of radiation-resistant bacteria are presented in Chap. 19.

1.3 Molecular and Cellular Adaptation Mechanisms

Salinity, pH, water activity (aw), and chaotropicity are fundamental constraints for life (Williams & Hallsworth, 2009; Cray et al., 2015). Low-aw conditions and chaotropic salts can destabilize macromolecules, even when the ionic strength is not extreme, shaping ecological limits and biotechnological opportunities (Williams & Hallsworth, 2009). Survival in extreme environments requires structural resilience and intricate molecular adaptations that preserve the functionality of biological systems under stress. Extremophiles have evolved various strategies at the molecular and cellular levels to cope with harsh environmental conditions (Cavicchioli, 2002).

Table 1.1 Major categories of extremophiles

Extreme factor	Definition/growth limit	Representative taxa	Example species	Notable applications	References
Thermophiles	Optimal growth ≥60 °C	Bacteria, Archaea	*Thermus aquaticus*, *Methanopyrus kandleri*	PCR enzymes (Taq polymerase), biofuel production	Brock Thomas and Freeze (1969), Takai et al. (2008)
Hyperthermophiles	Optimal growth ≥80 °C	Archaea	*Pyrolobus fumarii*, *Thermococcus gammatolerans*	High-temperature biocatalysis, industrial enzymes	Stetter (1999)
Psychrophiles	Optimal growth ≤15 °C, active at subzero	Bacteria, Archaea	*Colwellia psychrerythraea*, *Planococcus halocryophilus*	Cold-active enzymes for food, bioremediation in polar regions	Deming, (2002), Mykytczuk et al. (2013)
Acidophiles	Optimal growth pH ≤3	Bacteria, Archaea	*Acidithiobacillus ferrooxidans*, *Picrophilus oshimae*	Bioleaching, acid mine drainage remediation	Schleper et al. (1995), Johnson and Hallberg (2003)
Alkaliphiles	Optimal growth pH ≥9	Bacteria, Archaea	*Serpentinomonas* sp., *Alkaliphilus transvaalensis*	Enzymes for detergents, waste treatment	Takai et al. (2001), Suzuki et al. (2014)
Halophiles	Optimal growth at ≥3 M NaCl	Bacteria, Archaea	*Ectothiorhodospira halochloris*, *Halobacterium salinarum*	Compatible solute production (ectoine), fermentation	Galinski et al. (1985), Oren (2002)
Piezophiles	Optimal growth at 10–60 MPa	Bacteria, Archaea	*Shewanella benthica*, *Photobacterium profundum*	Deep-sea bioprocessing, novel biomaterials	Yayanos (1995), Bartlett (2002)
Hyperpiezophiles	Optimal growth ≥70 MPa	Bacteria	*C. marinimaniae*	High-pressure industrial processes	Kusube et al. (2017)
Radiotolerant microbes	Survive ≥5 kGy ionizing radiation	Bacteria	*Deinococcus radiodurans*, *Rubrobacter radiotolerans*	Bioremediation in radioactive sites, DNA repair studies	Battista (1997), Slade and Radman, (2011)
Xerophiles	Thrive in low water activity ($a_w \leq 0.85$)	Bacteria, Fungi	*Chroococcidiopsis* sp., *Aspergillus penicillioides*	Food preservation, astrobiology studies	Billi et al. (2000), Stevenson et al. (2015)

The stability of extremophile proteins under extreme temperature, pH, or salinity is an important feature. These proteins often exhibit enhanced hydrophobic core packing, increased electrostatic interactions, and rigid secondary structures that prevent denaturation (Sterner & Liebl, 2001). Membrane adaptations are equally important; for example, archaeal lipids often comprise ether-linked isoprenoids forming monolayers, which confer superior thermal and chemical stability compared to bacterial bilayers (Koga & Morii, 2007).

DNA stability is critical, particularly in environments with high radiation or high temperatures. Extremophiles employ robust DNA repair systems, including efficient homologous recombination, base excision repair, and unique radiation resistance pathways (Daly, 2009; Slade & Radman, 2011). Molecular chaperones such as heat shock proteins assist in protein folding and prevent aggregation under thermal or oxidative stress (Feder & Hofmann, 1999). Additionally, extremophiles often accumulate compatible solutes such as trehalose or ectoine, which stabilize proteins and membranes without interfering with their metabolism (Galinski, 1995; Empadinhas & da Costa, 2008).

These adaptations illustrate how extremophiles fine-tune their cellular machinery to survive and offer valuable insights into protein engineering and synthetic biology (Elleuche et al., 2014).

Cross-Reference For applications, see Chaps. 3 and 11; see also Chaps. 6 and 9 for broader enzyme use and metabolite valorization.

1.4 Industrial and Environmental Relevance of Extremophiles

Although extremophiles exhibit remarkable resilience, their true value has emerged in practical applications. Their unique biochemical traits make them indispensable tools in industrial biotechnology and environmental engineering, particularly under conditions in which conventional organisms or enzymes fail (van den Burg, 2003; Elleuche et al., 2014).

Extremozyme production is an important research topic. Enzymes derived from extremophiles, such as thermostable DNA polymerases, proteases, lipases, and cellulases, retain their activity in high-temperature, high-salinity, and alkaline environments (Madigan & Marrs, 1997; Littlechild, 2015). These enzymes have been successfully used for PCR (Saiki et al., 1988), biofuel production (Turner et al., 2007), textile processing (Singh et al., 2016), and as detergents (Beg & Gupta, 2003). Their robustness reduces the need for expensive temperature control and chemical additives.

In bioremediation, extremophiles play a vital role in detoxifying polluted environments including acidic mine drainage, saline soils, and radioactive waste sites (Rawlings, 2002; Dopson et al., 2004). Some strains can transform or immobilize toxic metals (Gadd, 2010), whereas others can degrade organic pollutants under extreme pH or temperature (Mougin et al., 2009).

Extremophiles also contribute to renewable energy, particularly through the methanogenic archaea used in anaerobic digestion (Zinder, 1993) and hydrogen-producing microbes in biohydrogen generation systems (Mishra & Das, 2014). Their ability to operate under non-sterile and harsh industrial conditions makes extremophiles attractive for large-scale sustainable bioprocesses (Feller, 2010).

These applications underscore the potential of extremophiles as agents of green technology and not just scientific curiosities.

Cross-Reference Thermostable and haloalkaline enzymes for PCR, detergents, and biomass processing are covered in Chaps. 3, 4, 5, 6, and 7 (incl. lignin/biomass in Chap. 5); value-added metabolites such as ectoine are discussed in Chap. 11; biomethane (methanation) in Chap. 14; biohydrogen systems in Chaps. 15 and 16; bioelectrochemical approaches in Chap. 18; and environmental applications, including biomining and radiation-related processes, in Chaps. 17 and 19.

1.5 Roles of Extremophiles in Sustainable Development Goals (SDGs)

The United Nations SDGs provide a global framework for addressing environmental, economic, and social challenges (United, 2015). Extremophiles offer innovative biological solutions that align with many of these goals, particularly clean energy, responsible production, and ecosystem protection (Cavicchioli et al., 2019).

For example, biohydrogen and biomethane production by thermophilic or halophilic microbes contributes to SDG 7 (Affordable and Clean Energy) (Kumar & Das, 2000; McGenity & Sorokin, 2018). The use of extremophilic enzymes in industry promotes SDG 12 (Responsible Consumption and Production) by reducing resource use and waste (Elleuche et al., 2014; Littlechild, 2015). Microbial processes based on extremophiles also support SDGs 13, 14, and 15, which are related to climate action, marine conservation, and terrestrial ecosystem restoration, respectively (Sorokin et al., 2014; Ando et al., 2021).

Furthermore, the potential role of extremophiles in space biotechnology, including closed-loop life support systems and biomining on other planets, aligns with SDG 9 (Industry, Innovation, and Infrastructure) (Olsson-Francis & Cockell, 2010; Verseux et al., 2016). Their ability to function under extraterrestrial conditions also informs planetary protection policies and astrobiological exploration (Horneck et al., 2010).

By bridging the gap between microbiology and sustainability, extremophiles represent key biological tools for addressing the critical issues of this century.

Cross-Reference Concrete case studies aligned with SDG 7 (clean energy: methanation, biohydrogen, microbial fuel cells) appear in Chaps.14, 15, 16 and 18; SDG

9 (industry, innovation: extremozymes and genome editing) in Chaps. 3, 4, 5, 6, 7 and 8 (see also Chap. 19); SDG 12 (responsible production: enzyme-enabled processing, ectoine, biomass valorization) in Chaps. 3, 4, 5, 6, 7, 11 and 12; and SDGs 13–15 (climate, oceans, land: biomining, composting, radiation-resilience applications) in Chaps. 13, 16, and 19.

1.6 Challenges and Future Perspectives

Despite the promising applications of extremophiles, several challenges hinder the realization of their full potential. One major limitation is the cultivation barrier; most extremophiles remain uncultured and are known only from metagenomic data (Rinke et al., 2013; Steen et al., 2019). Unlocking their genetic and metabolic capabilities requires advances in single-cell genomics, high-throughput cultures, and in situ monitoring technologies (Nichols et al., 2010; Stepanauskas, 2012).

However, challenges related to scalability and implementation exist. Although extremozymes are commercially successful (Elleuche et al., 2014), whole-cell applications of extremophiles are often limited by their growth rates, nutrient requirements, and incompatibility with standard bioprocess equipment (van den Burg, 2003). Moreover, genetic tools for extremophiles, particularly archaea, are still under development, restricting metabolic engineering efforts (Leigh et al., 2011).

Synthetic biology and genome editing hold promise for customizing extremophilic traits (Atomi et al., 2012), but raise ethical, biosafety, and biosecurity considerations, particularly when considering environmental release or space applications. These concerns have been discussed in the broader context of synthetic biology governance (Douglas & Stemerding, 2013; Lai et al., 2019).

Finally, the understanding of the limits of life through extremophiles informs astrobiology and the search for life beyond Earth (Horneck et al., 2010). The taxonomic and ecological diversity underlying these limitations, such as specific organism distributions, is addressed in Chap. 2.

Cross-Reference Data-driven design and AI-enabled strain/process engineering are highlighted in Chap. 20, and forward-looking roadmaps are synthesized in Chap. 21. For foundational context on taxonomic and ecological diversity, see Chap. 2.

1.7 Conclusions

Extremophiles represent a unique frontier of microbiology. Their ability to persist at the margins of habitability renders them scientifically intriguing and practically valuable (Rothschild & Mancinelli, 2001; Cavicchioli et al., 2019). In this chapter, we explore the general physiological characteristics and adaptation mechanisms that define extremophiles along with their potential roles in industrial innovation and sustainability initiatives (Cavicchioli et al., 2019).

The next section discusses the remarkable biodiversity of extremophiles across different extreme environments. By examining their ecological distribution and phylogenetic variety, Chap. 2 expands on the foundations laid here, providing a clearer picture of how these organisms have adapted to various habitats on Earth and potentially beyond (Horneck et al., 2010).

Cross-Reference Readers seeking ecological context may proceed to Chap. 2, whereas those interested in industrial applications can refer to Chaps. 3, 4, 5, 6 and 7 (enzymes and catalysts), 11, 12 and 14 (value-added products and methanation), and 15, 16, 17, 18 and 19 (biohydrogen, bioelectrochemical systems, biomining, and radiation-related applications).

References

Ando, N., Barquera, B., Bartlett, D. H., Boyd, E., Burnim, A. A., Byer, A. S., et al. (2021). The molecular basis for life in extreme environments. *Annual Review of Biophysics, 50*, 343–372. https://doi.org/10.1146/annurev-biophys-100120-072804

Atomi, H., Imanaka, T., & Fukui, T. (2012). Overview of the genetic tools in the Archaea. *Frontiers in Microbiology, 3*. https://doi.org/10.3389/fmicb.2012.00337

Bartlett, D. H. (2002). Pressure effects on in vivo microbial processes. *Biochimica et Biophysica Acta (BBA) – Protein Structure and Molecular Enzymology, 1595*(1), 367–381. https://doi.org/10.1016/S0167-4838(01)00357-0

Battista, J. R. (1997). Against all odds: The survival strategies of Deinococcus radiodurans. *Annual Review of Microbiology, 51*, 203–224. https://doi.org/10.1146/annurev.micro.51.1.203

Beg, Q. K., & Gupta, R. (2003). Purification and characterization of an oxidation-stable, thiol-dependent serine alkaline protease from Bacillus mojavensis. *Enzyme and Microbial Technology, 32*(2), 294–304. https://doi.org/10.1016/S0141-0229(02)00293-4

Billi, D., Friedmann, E. I., Hofer Kurt, G., Caiola Maria, G., & Ocampo-Friedmann, R. (2000). Ionizing-radiation resistance in the desiccation-tolerant Cyanobacterium *Chroococcidiopsis*. *Applied and Environmental Microbiology, 66*(4), 1489–1492. https://doi.org/10.1128/AEM.66.4.1489-1492.2000

Brock Thomas, D., & Freeze, H. (1969). Thermus aquaticus gen. n. and sp. n., a nonsporulating extreme thermophile. *Journal of Bacteriology, 98*(1), 289–297. https://doi.org/10.1128/jb.98.1.289-297.1969

Cavicchioli, R. (2002). Extremophiles and the search for extraterrestrial life. *Astrobiology, 2*(3), 281–292. https://doi.org/10.1089/153110702762027862

Cavicchioli, R., Amils, R., Wagner, D., & McGenity, T. (2011). Life and applications of extremophiles. *Environmental Microbiology 13*(8), 1903–1907. https://doi.org/10.1111/j.1462-2920.2011.02512.x, https://doi.org/10.1111/emi.2011.13.issue-8

Cavicchioli, R. (2019). A vision for a 'microbcentric' future. *Microbial Biotechnology 12*(1), 26–29. https://doi.org/10.1111/mbt2.2019.12.issue-1, https://doi.org/10.1111/1751-7915.13262

Cavicchioli, R., Ripple, W. J., Timmis, K. N., Azam, F., Bakken, L. R., Baylis, M., et al. (2019). Scientists' warning to humanity: Microorganisms and climate change. *Nature Reviews Microbiology, 17*(9), 569–586. https://doi.org/10.1038/s41579-019-0222-5

Cray, J. A., Stevenson, A., Ball, P., Bankar, S. B., Eleutherio, E. C. A., Ezeji, T. C., et al. (2015). Chaotropicity: A key factor in product tolerance of biofuel-producing microorganisms. *Current Opinion in Biotechnology, 33*, 228–259. https://doi.org/10.1016/j.copbio.2015.02.010

Daly, M. J. (2009). A new perspective on radiation resistance based on Deinococcus radiodurans. *Nature Reviews Microbiology, 7*(3), 237–245. https://doi.org/10.1038/nrmicro2073

Deming, J. W. (2002). Psychrophiles and polar regions. *Current Opinion in Microbiology, 5*(3), 301–309. https://doi.org/10.1016/S1369-5274(02)00329-6

Dopson, M., Baker-Austin, C., Hind, A., Bowman John, P., & Bond Philip, L. (2004). Characterization of Ferroplasma isolates and Ferroplasma acidarmanus sp. nov., extreme acidophiles from acid mine drainage and industrial bioleaching environments. *Applied and Environmental Microbiology, 70*(4), 2079–2088. https://doi.org/10.1128/AEM.70.4.2079-2088.2004

Douglas, C. M. W., & Stemerding, D. (2013). Governing synthetic biology for global health through responsible research and innovation. *Systems and Synthetic Biology, 7*(3), 139–150. https://doi.org/10.1007/s11693-013-9119-1

Elleuche, S., Schröder, C., Sahm, K., & Antranikian, G. (2014). Extremozymes—Biocatalysts with unique properties from extremophilic microorganisms. *Current Opinion in Biotechnology, 29*, 116–123. https://doi.org/10.1016/j.copbio.2014.04.003

Empadinhas, N., & da Costa, M. S. (2008). Osmoadaptation mechanisms in prokaryotes: Distribution of compatible solutes. *International Microbiology, 11*(3), 151–161.

Feder, M. E., & Hofmann, G. E. (1999). Heat-shock proteins, molecular chaperones, and the stress response: Evolutionary and ecological physiology. *Annual Review of Physiology, 61*, 243–282. https://doi.org/10.1146/annurev.physiol.61.1.243

Feller, G. (2010). Protein stability and enzyme activity at extreme biological temperatures. *Journal of Physics. Condensed Matter, 22*(32), 323101. https://doi.org/10.1088/0953-8984/22/32/323101

Gadd, G. M. (2010). Metals, minerals and microbes: Geomicrobiology and bioremediation. *Microbiology, 156*(3), 609–643. https://doi.org/10.1099/mic.0.037143-0

Galinski, E. A. (1995). Osmoadaptation in bacteria. In R. K. Poole (Ed.), *Advances in microbial physiology* (pp. 273–328). Academic Press.

Galinski, E. A., Pfeiffer, H. P., & Truper, H. G. (1985). 1,4,5,6-Tetrahydro-2-methyl-4--pyrimidinecarboxylic acid. A novel cyclic amino acid from halophilic phototrophic bacteria of the genus Ectothiorhodospira. *European Journal of Biochemistry, 149*(1), 135–139. https://doi.org/10.1111/j.1432-1033.1985.tb08903.x

Harrison, J. P., Gheeraert, N., Tsigelnitskiy, D., & Cockell, C. S. (2013). The limits for life under multiple extremes. *Trends in Microbiology, 21*(4), 204–212. https://doi.org/10.1016/j.tim.2013.01.006

Horikoshi, K., & Bull, A. T. (2011). Prologue: definition, categories, distribution, origin and evolution, pioneering studies, and emerging fields of extremophiles. *In Extremophiles handbook*, 3–15. Springer, Tokyo

Horneck, G., Klaus David, M., & Mancinelli Rocco, L. (2010). Space microbiology. *Microbiology and Molecular Biology Reviews, 74*(1), 121–156. https://doi.org/10.1128/mmbr.00016-09

Johnson, D. B., & Hallberg, K. B. (2003). The microbiology of acidic mine waters. *Research in Microbiology, 154*(7), 466–473. https://doi.org/10.1016/S0923-2508(03)00114-1

Koga, Y., & Morii, H. (2007). Biosynthesis of ether-type polar lipids in archaea and evolutionary considerations. *Microbiology and Molecular Biology Reviews, 71*(1), 97–120. https://doi.org/10.1128/mmbr.00033-06

Kumar, N., & Das, D. (2000). Enhancement of hydrogen production by Enterobacter cloacae IIT-BT 08. *Process Biochemistry, 35*(6), 589–593. https://doi.org/10.1016/S0032-9592(99)00109-0

Kusube, M., Kyaw, T. S., Tanikawa, K., Chastain, R. A., Hardy, K. M., Cameron, J., et al. (2017). Colwellia marinimaniae sp. nov., a hyperpiezophilic species isolated from an amphipod within the challenger deep, Mariana trench. *International Journal of Systematic and Evolutionary Microbiology, 67*(4), 824–831. https://doi.org/10.1099/ijsem.0.001671

Lai, H.-E., Canavan, C., Cameron, L., Moore, S., Danchenko, M., Kuiken, T., et al. (2019). Synthetic biology and the United Nations. *Trends in Biotechnology, 37*(11), 1146–1151. https://doi.org/10.1016/j.tibtech.2019.05.011

Leigh, J. A., Albers, S.-V., Atomi, H., & Allers, T. (2011). Model organisms for genetics in the domain archaea: Methanogens, halophiles, thermococcales and sulfolobales. *FEMS Microbiology Reviews, 35*(4), 577–608. https://doi.org/10.1111/j.1574-6976.2011.00265.x

Littlechild, J. A. (2015). Enzymes from extreme environments and their industrial applications. *Frontiers in Bioengineering and Biotechnology, 3*, 161. https://doi.org/10.3389/fbioe.2015.00161

Madigan, M. T., & Marrs, B. L. (1997). Extremophiles. *Scientific American, 276*(4), 82–87. https://doi.org/10.1038/scientificamerican0497-82

McGenity, T. J., & Sorokin, D. Y. (2018). Methanogens and methanogenesis in hypersaline environments. In A. J. M. Stams & D. Sousa (Eds.), *Biogenesis of hydrocarbons* (pp. 1–27). Springer International Publishing.

Mevarech, M., Frolow, F., & Gloss, L. M. (2000). Halophilic enzymes: Proteins with a grain of salt. *Biophysical Chemistry, 86*(2), 155–164. https://doi.org/10.1016/S0301-4622(00)00126-5

Mishra, P., & Das, D. (2014). Biohydrogen production from Enterobacter cloacae IIT-BT 08 using distillery effluent. *International Journal of Hydrogen Energy, 39*(14), 7496–7507. https://doi.org/10.1016/j.ijhydene.2013.08.100

Mougin, C., Boukcim, H., & Jolivalt, C. (2009). Soil bioremediation strategies based on the use of fungal enzymes. In A. Singh, R. C. Kuhad, & O. P. Ward (Eds.), *Advances in applied bioremediation* (pp. 123–149). Springer Berlin Heidelberg.

Mykytczuk, N. C. S., Foote, S. J., Omelon, C. R., Southam, G., Greer, C. W., & Whyte, L. G. (2013). Bacterial growth at −15 °C; molecular insights from the permafrost bacterium Planococcus halocryophilus Or1. *The ISME Journal, 7*(6), 1211–1226. https://doi.org/10.1038/ismej.2013.8

Nichols, D., Cahoon, N., Trakhtenberg, E. M., Pham, L., Mehta, A., Belanger, A., et al. (2010). Use of Ichip for high-throughput in situ cultivation of "uncultivable" microbial species. *Applied and Environmental Microbiology, 76*(8), 2445–2450. https://doi.org/10.1128/AEM.01754-09

Olsson-Francis, K., & Cockell, C. S. (2010). Experimental methods for studying microbial survival in extraterrestrial environments. *Journal of Microbiological Methods, 80*(1), 1–13. https://doi.org/10.1016/j.mimet.2009.10.004

Oren, A. (2002). Diversity of halophilic microorganisms: Environments, phylogeny, physiology, and applications. *Journal of Industrial Microbiology and Biotechnology, 28*(1), 56–63. https://doi.org/10.1038/sj/jim/7000176

Oren, A. (2011). Thermodynamic limits to microbial life at high salt concentrations. *Environmental Microbiology, 13*(8), 1908–1923. https://doi.org/10.1111/j.1462-2920.2010.02365.x, https://doi.org/10.1111/emi.2011.13.issue-8

Rampelotto, P. (2013). Extremophiles and extreme environments. *Life, 3*(3), 482–485. https://doi.org/10.3390/life3030482

Rawlings, D. E. (2002). Heavy metal mining using microbes. *Annual Review of Microbiology, 56*, 65–91. https://doi.org/10.1146/annurev.micro.56.012302.161052

Rinke, C., Schwientek, P., Sczyrba, A., Ivanova, N. N., Anderson, I. J., Cheng, J.-F., et al. (2013). Insights into the phylogeny and coding potential of microbial dark matter. *Nature, 499*(7459), 431–437. https://doi.org/10.1038/nature12352

Rothschild, L. J., & Mancinelli, R. L. (2001). Life in extreme environments. *Nature, 409*(6823), 1092–1101. https://doi.org/10.1038/35059215

Saiki, R. K., Gelfand, D. H., Stoffel, S., Scharf, S. J., Higuchi, R., Horn, G. T., et al. (1988). Primer-directed enzymatic amplification of DNA with a thermostable DNA polymerase. *Science, 239*(4839), 487–491. https://doi.org/10.1126/science.2448875

Schleper, C., Puehler, G., Holz, I., Gambacorta, A., Janekovic, D., Santarius, U., et al. (1995). Picrophilus gen. nov., fam. nov.: A novel aerobic, heterotrophic, thermoacidophilic genus and family comprising archaea capable of growth around pH 0. *Journal of Bacteriology, 177*(24), 7050–7059. https://doi.org/10.1128/jb.177.24.7050-7059.1995

Singh, R., Kumar, M., Mittal, A., & Mehta, P. K. (2016). Microbial enzymes: industrial progress in 21st century. *3 Biotech, 6*(2), 174. https://doi.org/10.1007/s13205-016-0485-8

Slade, D., & Radman, M. (2011). Oxidative stress resistance in Deinococcus radiodurans. *Microbiology and Molecular Biology Reviews, 75*(1), 133–191. https://doi.org/10.1128/mmbr.00015-10

Sorokin, D. Y., Berben, T., Melton, E. D., Overmars, L., Vavourakis, C. D., & Muyzer, G. (2014). Microbial diversity and biogeochemical cycling in soda lakes. *Extremophiles, 18*(5), 791–809. https://doi.org/10.1007/s00792-014-0670-9

Steen, A. D., Crits-Christoph, A., Carini, P., DeAngelis, K. M., Fierer, N., Lloyd, K. G., et al. (2019). High proportions of bacteria and archaea across most biomes remain uncultured. *The ISME Journal, 13*(12), 3126–3130. https://doi.org/10.1038/s41396-019-0484-y

Stepanauskas, R. (2012). Single cell genomics: An individual look at microbes. *Current Opinion in Microbiology, 15*(5), 613–620. https://doi.org/10.1016/j.mib.2012.09.001

Sterner, R. H., & Liebl, W. (2001). Thermophilic adaptation of proteins. *Critical Reviews in Biochemistry and Molecular Biology, 36*(1), 39–106. https://doi.org/10.1080/20014091074174

Stetter, K. O. (1999). Extremophiles and their adaptation to hot environments. *FEBS Letters, 452*(1–2), 22–25. https://doi.org/10.1016/s0014-5793(99)00663-8

Stevenson, A., Cray, J. A., Williams, J. P., Santos, R., Sahay, R., Neuenkirchen, N., et al. (2015). Is there a common water-activity limit for the three domains of life? *The ISME Journal, 9*(6), 1333–1351. https://doi.org/10.1038/ismej.2014.219

Suzuki, S., Kuenen, J. G., Schipper, K., van der Velde, S., Ishii, S. I., Wu, A., et al. (2014). Physiological and genomic features of highly alkaliphilic hydrogen-utilizing betaproteobacteria from a continental serpentinizing site. *Nature Communications, 5*(1), 3900. https://doi.org/10.1038/ncomms4900

Takai, K., Moser, D. P., Onstott, T. C., Spoelstra, N., Pfiffner, S. M., Dohnalkova, A., et al. (2001). Alkaliphilus transvaalensis gen. nov., sp. nov., an extremely alkaliphilic bacterium isolated from a deep south African gold mine. *International Journal of Systematic and Evolutionary Microbiology, 51*(Pt 4), 1245–1256. https://doi.org/10.1099/00207713-51-4-1245

Takai, K., Nakamura, K., Toki, T., Tsunogai, U., Miyazaki, M., Miyazaki, J., et al. (2008). Cell proliferation at 122°C and isotopically heavy CH_4 production by a hyperthermophilic methanogen under high-pressure cultivation. *Proceedings of the National Academy of Sciences, 105*(31), 10949–10954. https://doi.org/10.1073/pnas.0712334105

Turner, P., Mamo, G., & Karlsson, E. N. (2007). Potential and utilization of thermophiles and thermostable enzymes in biorefining. *Microbial Cell Factories, 6*(1), 9. https://doi.org/10.1186/1475-2859-6-9

United Nations. (2015). *Transforming our world: The 2030 agenda for sustainable development*. United Nations.

van den Burg, B. (2003). Extremophiles as a source for novel enzymes. *Current Opinion in Microbiology, 6*(3), 213–218. https://doi.org/10.1016/S1369-5274(03)00060-2

Verseux, C., Baqué, M., Lehto, K., de Vera, J.-P. P., Rothschild, L. J., & Billi, D. (2016). Sustainable life support on Mars—The potential roles of cyanobacteria. *International Journal of Astrobiology, 15*(1), 65–92. https://doi.org/10.1017/S147355041500021X

Williams, J. P., & Hallsworth, J. E. (2009). Limits of life in hostile environments: No barriers to biosphere function? *Environmental Microbiology, 11*(12), 3292–3308. https://doi.org/10.1111/j.1462-2920.2009.02079.x

Yayanos, A. A. (1995). Microbiology to 10,500 meters in the deep sea. *Annual Review of Microbiology, 49*, 777–805. https://doi.org/10.1146/annurev.mi.49.100195.004021

Zinder, S. H. (1993). Physiological ecology of methanogens. In J. G. Ferry (Ed.), *Methanogenesis: Ecology, physiology, biochemistry & genetics* (pp. 128–206). Springer US

Chapter 2
Diversity and Distribution of Extremophiles

Issay Narumi

Abstract Extremophilic microorganisms on Earth are groups of organisms that thrive in harsh environments and serve as important models for understanding evolutionary adaptation and genetic diversification of life. Many thermophiles have been isolated from composts and hot springs. Thermophiles have also been isolated from desert sand and sediments of deep-sea hydrothermal vents. Psychrophiles are found in polar regions, permafrost, and glaciers. Deep-sea environments are home to the psychrophilic piezophiles. Halophiles have been found in salt lakes, rock salts, and deep-sea high-salinity anaerobic basins. Acidophiles are found in acidic environments such as acidic hot springs and mine drainage, whereas haloalkaliphilic bacteria are isolated from hypersaline alkaline lakes. Bacteria of the genera *Chroococcidiopsis* and *Deinococcus*, which are resistant to desiccation, ultraviolet rays, and ionizing radiation, have been isolated from desert environments. Extremophilic microorganisms play a significant role in a wide range of research fields, including the study of the evolution of life, the possibility of extraterrestrial life, and applications in biotechnology.

Keywords Biodiversity · Distribution · Extremophiles · Extreme environments · Ecosystem

---

I. Narumi (✉)
Faculty of Life Sciences, and Bio-Resilience Research Project (BRRP), Toyo University, Asaka, Saitama, Japan
e-mail: narumi@toyo.jp

M. Ito et al. (eds.), *Extremophiles for a Sustainable Future*,
https://doi.org/10.1007/978-981-95-6970-0_2

2.1 Introduction

Most living organisms on Earth inhabit temperate environments, specifically those with an average temperature of approximately 15 °C, an atmospheric pressure of 1 atm, river water with a pH near 7, surface seawater with a pH of approximately 8.1, a salinity of approximately 3%, an oxygen concentration of approximately 20%, and moderate humidity and nutrients. Such conditions form environments that are conducive to human survival. In contrast, there are regions on Earth with extremely harsh conditions that are nearly uninhabitable by humans and the most common plants and animals. These regions are collectively referred to as "extreme environments" and are defined as habitats that would be intolerable or even lethal for most other life forms on Earth. Thus, the term "extreme" is defined from an anthropocentric perspective. From a somewhat more objective viewpoint, however, these extreme environments are characterized by certain physical and chemical parameters reaching extreme values, such as high or low temperatures, salinity, pH, high pressure, radiation, or dryness. The distribution of extreme environments on Earth has changed over geological time, and even today, much of Earth's surface is occupied by such extreme environments. However, microbial activity is not limited to specific environments or conditions. This fact is directly connected to the evolutionary development of life's ability to adapt to a diverse array of conditions as it spreads across Earth.

Extremophiles are a group of organisms that prefer and thrive in environments that were once thought to be uninhabitable by humans. Their ability to live in extreme environments provides important models for understanding the mechanisms of adaptive evolution and strategies for genetic diversification. Research on extremophiles not only advances our fundamental understanding, such as the study of stress response networks and mechanisms underlying the acquisition of new traits, but also leads to industrial applications through the exploitation of extremozymes and extremolytes produced by these microorganisms. In this chapter, I define a variety of extreme environments on Earth and explain the major classifications and geographical distributions of extremophiles that have been isolated from such environments. Although some extremophiles exist within the Eukarya, this chapter focuses solely on bacterial and archaeal extremophiles (Fig. 2.1).

2.2 High-Temperature Environments: Compost, Hot Springs, Deserts, Hydrothermal Vents

Compost is organic matter that has been decomposed by microorganisms and is a fertile soil amendment that promotes plant growth. It is widely used in agriculture and horticulture to maintain soil health and to increase crop productivity. The rise in temperature of compost is an important phenomenon caused by the activity of aerobic microorganisms. When organic matter is broken down, microorganisms

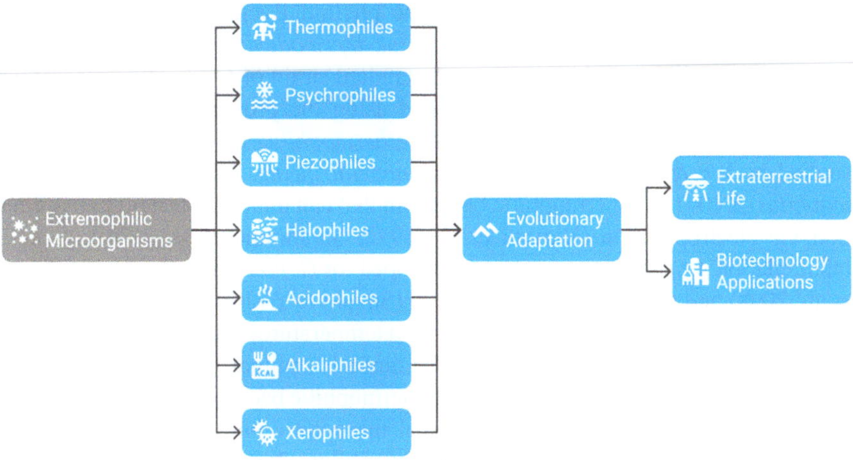

Fig. 2.1 Classification of extremophiles and directions for research and development

generate heat through respiration, which accumulates in the composting material and increases the temperature of the compost. This increase in temperature is essential for compost maturation. Thermophilic bacteria isolated from the compost play an important role in the decomposition of organic matter. Below are some examples of the investigations of thermophilic microbial communities living in compost.

Strom (1985) studied the microbial communities in samples collected from an experimental artificial compost made from a mixture of food waste and shredded newspaper, a composting facility in Pennsylvania, USA, and a rotary drum composting accelerator and two windrow piles with different maturation periods in Leicester, UK. Of the 652 colonies, 87% were identified as belonging to the genus *Bacillus*. The isolated species included *Bacillus circulans* complex, *B. stearothermophilus*, *B. coagulans* types A and B, *B. licheniformis*, *B. brevis*, *B. sphaericus*, and *B. subtilis*. Additionally, two unidentified genera of non-spore-forming bacteria as well as actinomycetes, including *Streptomyces* sp. and *Thermoactinomyces* sp., were isolated. Blanc et al. (1999) investigated thermophilic bacterial communities in high-temperature compost composed of 45% grass, 25–35% kitchen and garden waste, and 10–25% shredded wood using methods such as Polymerase Chain Reaction-Restriction Fragment Length Polymorphism (PCR-RFLP). Among the 200 clones, 69 were identified as *Thermus thermophilus* or thermophilic *Bacillus* species (*B. thermodenitrificans*, *B. pallidus*, and *B. thermoglucosidasius*). *Saccharococcus thermophilus* and *Rhodothermus marinus* have also been identified. Charbonneau et al. (2012) isolated thermophilic bacterial strains that produce heat-resistant hydrolytic enzymes from compost made from livestock manure in Quebec, Canada. Based on phylogenetic analysis of 16S rRNA gene sequences and biochemical property evaluations, they identified *Geobacillus thermodenitrificans*, *B. smithii*, *Ureibacillus suwonensis*, and *Aneurinibacillus thermoaerophilus*. Papale et al. (2021) investigated the microbial diversity of high-temperature compost made from coffee grounds. They found that the phylum Bacillota was the dominant microbial group, and identified *Geobacillus thermodenitrificans* subsp. *calidus*, *Aeribacillus pallidus*, and *U. terrenus*. A standard strain of *Geobacillus stearothermophilus* (ATCC 12980) was transformed using the protoplast technique established by Imanaka et al. (1982). *G. stearothermophilus* NUB36 was extensively studied by Wu & Welker (1989). *G. thermodenitrificans* (formerly *B. stearothermophilus*) K1041 is a thermophilic bacterium that can undergo plasmid transformation via electroporation (Narumi et al., 1992; Studholme et al., 1999). Genetic manipulation systems have also been developed for *G. thermoglucosidasius* (Taylor et al., 2008) and *G. kaustophilus* HTA-426 (Suzuki & Yoshida, 2012).

Hot spring water generally maintains high temperatures (above 40 °C, and sometimes close to 100 °C). This temperature range is optimal for the habitation and growth of both the thermophilic and hyperthermophilic bacteria. A high-temperature environment inhibits the growth of typical mesophilic microorganisms, which is advantageous for efficient isolation of thermophiles. Depending on the geological origin of the hot spring and the pathways of underground water, there are significant variations in pH, salinity, and the types and concentrations of dissolved minerals. These diverse chemical conditions foster the adaptation of a variety of thermophiles to unique environments. Hot springs can be found in volcanic regions worldwide, each potentially harboring different microbial diversity. *B. caldolyticus*, *B. caldovelox*, and *B. caldotenax* are highly thermophilic bacteria isolated from hot springs in the United States, with an optimal growth temperature of 70 °C (Heinen & Heinen, 1972). Benammar et al. (2020) investigated the diversity of thermophilic bacteria in eight hot springs in northeastern Algeria and isolated genera such as *Aeribacillus*, *Aneurinibacillus*, *Anoxybacillus*, *Bacillus*, *Brevibacillus*, *Geobacillus*, *Laceyella*, *Meiothermus*, *Saccharomonospora*, *Thermoactinomyces*, *Thermobifida*, and *Thermus*. Ortega-Villar et al. (2024) studied the diversity of thermophilic bacteria in calcium sulfate hot springs in Mexico. Of the 18 strains isolated, 17 formed spores and were classified as *Anoxybacillus*, *Bacillus*, *Anerunibacillus*, or *Paenibacillus*. The remaining one strain was identified as *Deinococcus sahariens*. The standard strain of *D. sahariens* (DSM 18496) is a thermophilic bacterium isolated from a hot spring in the Sahara Desert of Tunisia (Bouraoui et al., 2012).

Bacteria of the genus *Deinococcus* belong to the phylum Deinococcota and are non-spore-forming, exhibiting resistance to ionizing radiation, ultraviolet light, and desiccation. *D. geothermalis* is a radiation-resistant thermophilic bacterium isolated from the Agnano hot spring in Italy (Ferreira et al., 1997). A radiation-resistant thermophilic bacterium, *D. murrayi*, has also been isolated from the São Pedro do Sul hot spring in Portugal (Ferreira et al., 1997). *Rubrobacter radiotolerans* (formerly *Arthrobacter radiotolerans*) is a radiation-resistant thermophilic bacterium with an optimum growth temperature of 45 °C isolated from the Misasa hot spring (a radon hot spring) in Japan (Yoshinaka et al., 1973). *R. radiotolerans* belongs to the phylum *Actinomycetota* and is the most resistant to ionizing radiation among the prokaryotes known to date. A radiation-resistant thermophilic bacterium, *D. misasensis*, was isolated from the Misasa hot spring (Asker et al., 2008). In a hot spring on São Miguel Island in the Azores (a territory of Portugal), a radiation-resistant thermophilic bacterium belonging to the phyla Deinococcota, *Truepera radiovictrix* (Albuquerque et al., 2005), which grows optimally at 50 °C, and two ultraviolet-resistant bacteria, *Rubrobacter calidifluminis* and *Rubrobacter naiadicus*, both with an optimum growth temperature of 60 °C, were isolated (Albuquerque et al., 2014). A radiation-resistant, halophilic, thermophilic bacterium, *Rubrobacter taiwanensis*, with an optimum growth temperature of 60 °C, has been isolated from the Lushan Hot Spring in Taiwan (Chen et al., 2004). *T. aquaticus*, isolated from a hot spring in Yellowstone National Park in the United States, belongs to the phylum *Deinococcota* and is a hyperthermophilic bacterium with an optimal growth temperature of 70 °C (Brock & Freeze, 1969). *T. thermophilus* (formerly *Flavobacterium thermophilus*), isolated from the Mine Hot Spring on the Izu Peninsula in Japan, is a hyperthermophilic bacterium with an optimum growth temperature of 70 °C (Oshima & Imahori, 1971; Oshima & Imahori, 1975). *T. thermophilus* has an established genetic manipulation system and is advancing in molecular biological research and development (Cava et al., 2009). *Acidothermus cellulolyticus*, an acidophilic thermophilic bacterium belonging to the phylum *Actinomycetota* with an optimum growth temperature of 55 °C, was obtained from a hot spring in the Yellowstone National Park (Mohagheghi et al., 1986). *Thermoanaerobacter ethanolicus*, belonging to the class *Clostridia* in the phylum *Bacillota*, is a non-spore-forming, gram-positive, anaerobic thermophilic rod bacterium isolated from a hot spring in the Yellowstone National Park. While its optimum growth temperature is approximately 70 °C, it is notable for its broad growth temperature range of 37–77 °C. This bacterium can break down many sugars comprising cellulose and produce ethanol. *Anaerocellum bescii* (formerly *Caldicellulosiruptor bescii*), belonging to the order Caldicellulosiruptorales in the class *Clostridia*, is an anaerobic thermophilic bacterium isolated from hot springs in Kamchatka, Russia. Its optimum growth temperature is approximately 78–80 °C; however, it can grow even at 90 °C (Yang et al., 2010; Ford et al., 2025). With the development of genetic manipulation techniques, this bacterium can be used to produce industrial chemicals and biofuels such as ethanol and acetone (Lipscomb et al., 2016; Cha et al., 2024).

Thermotoga thermarum, isolated from a terrestrial solfataric spring in Djibouti, is an obligate anaerobic hyperthermophilic bacterium with an optimal growth temperature of 70 °C (Windberger et al., 1989). In addition to terrestrial hot springs (Mori et al., 2014; Itoh et al., 2016), thermophiles belonging to the phylum Thermotogae have also been isolated from marine thermal environments and oil reservoirs (L'Haridon et al., 2001; Purwasena et al., 2014). *Thermotoga maritima*, isolated from marine thermal environments, is an obligate anaerobic hyperthermophilic bacterium with an optimal growth temperature of 80 °C (Huber et al., 1986). *Saccharolobus solfataricus* (formerly *Sulfolobus solfataricus*), which belongs to the class Thermoprotei in the phylum Thermoproteota, is an obligate aerobic hyperthermophilic archaeon with an optimal growth temperature of 80 °C isolated from a solfatara in the Campi Flegrei caldera in Italy (de Rosa & Gambacorta, 1975; Zillig et al., 1980). *Sulfolobus. acidocaldarius* is an obligate aerobic hyperthermophilic archaeon isolated from hot springs in the Yellowstone National Park, USA (Brock et al., 1972). Both species are acidophilic with an optimal growth pH of approximately 2–4. *S. solfataricus*, *S. acidocaldarius*, and *T. maritima* have established genetic manipulation systems that allow for molecular biological research and development (Cannio et al., 1998; Yu et al., 2001; Berkner et al., 2007).

Deserts are harsh environments characterized by extremely high temperatures, dryness, ultraviolet radiation exposure, nutrient scarcity, and large temperature variations between day and night. Aanniz et al. (2015) explored thermophilic bacteria in hot springs, salt marshes, and desert soils in Morocco, and reported that *B. licheniformis*, *B. aerius*, *B. sonorensis*, *B. subtilis*, and *B. amyloliquefaciens* were commonly isolated from all three sources, with the highest proportion of thermophiles (45%) found in desert soils. Valenzuela et al. (2024) explored thermophiles from a hypersaline lagoon in the Atacama Desert, northern Chile, and identified *A. pallidus*, *Geobacillus thermoleovorans*, and *Parageobacillus thermoglucosidasius*.

Sediments near deep-sea hydrothermal vents form unique ecosystems despite the extreme conditions of high temperature, high pressure, and total darkness. In these environments, the dominant microorganisms are "chemoautotrophs" that fix carbon by obtaining energy from the oxidation of chemicals such as sulfides, hydrogen, methane, and iron, since sunlight does not reach these depths. Among the thermophilic bacteria isolated from this biosphere, in addition to those from the genus *Thermotoga*, bacteria from the genus *Aquifex* belonging to the phylum *Aquificota* are known. *Aquifex pyrophilus*, isolated from the Kolbeinsey Ridge, located north off the coast of Iceland, is a hyperthermophilic bacterium with an optimal growth temperature of 85 °C. Its growth temperature range is 67–95 °C, which makes it one of the bacterial species capable of thriving at the highest temperatures. The optimal NaCl concentration is 0.6 M (Huber et al., 1992). *A. pyrophilus* exhibits resistance to ionizing radiation (Beblo et al., 2011). *Desulfurobacterium thermolithotrophum*, a member of the phylum Aquificota, was isolated from a deep-sea hydrothermal chimney sample collected from the Mid-Atlantic Ridge. This thermophilic sulfur-reducing bacterium has an optimal growth temperature of 70 °C and an optimal NaCl concentration of 0.6 M (L'Haridon et al., 1998). *Hydrogenothermus marinus*, belonging to the phylum Aquificota, was isolated from the marine hydrothermal area of the Vulcano Island, Italy. This thermophilic bacterium has an optimal growth temperature of 65 °C and an optimal NaCl concentration of 0.3–0.5 M (Stöhr et al., 2001). This bacterium is known to exhibit resistance to desiccation, high NaCl concentrations, and sodium perchlorate (Beblo et al., 2009; Beblo-Vranesevic et al., 2017a, b).

A heterotrophic hyperthermophilic archaeon, *Thermococcus celer*, belonging to the order Thermococci in the phylum Methanobacteriota, was isolated from a sulfur-rich shallow volcanic crater on the coast of Vulcano Island (Zillig et al., 1983). This archaeon is an obligate anaerobe and a sulfate reducer, with an optimal growth temperature of 88 °C and an optimal NaCl concentration of 0.7 M. It grew best at a pH of 5.8, making it slightly acidophilic. A heterotrophic hyperthermophilic archaeon, *Pyrococcus furiosus*, belonging to the order Thermococci, has been isolated from geothermally heated marine sediments on Vulcano Island (Fiala & Stetter, 1986). The optimal growth temperature was 100 °C, optimal NaCl concentration was 0.34 M, and optimal pH was 7.0. *P. furiosus*, *P. abyssi*, *T. stetteri*,

T. gammatolerans, *T. marinus*, and *T. radiotolerans* exhibit resistance to ionizing radiation (DiRuggiero et al., 1997; Jolivet et al., 2003; Jolivet et al., 2004). Another species isolated from Vulcano Island, *T. alcaliphilus*, has an optimal growth temperature of 86 °C, is alkaliphilic, and has an optimal pH of 9.0 (Keller et al., 1995). *T. kodakarensis* (formerly *Pyrococcus* sp.) was isolated from sulfur vents on the coast of Kotakara Island, Japan (Morikawa et al., 1994). This obligate anaerobic hyperthermophilic archaeon has an optimal growth temperature of 95 °C and genetic manipulation systems have been established (Sato et al., 2003; Santangelo et al., 2008). It is the most extensively studied strain of *Thermococcus* spp. *P. yayanosii* is a hyperthermophilic, piezophilic archaeon isolated from a hydrothermal vent named Ashadze, located at a depth of 4,100 meters on the Mid-Atlantic Ridge (Birrien et al., 2011). The optimal growth temperature and pressure were 98 °C and 52 MPa, respectively.

Thermoproteus tenax, belonging to the class Thermoprotei in the phylum Thermoproteota, is an obligate anaerobic, hyperthermophilic, acidophilic archaeon isolated from a sulfur vent in an Icelandic hot spring with an optimal growth temperature of 86 °C and an optimal pH of 5.6 (Zillig et al., 1981). It is a rod-shaped archaeon, with a diameter of approximately 0.4 μm, and its length ranges from less than 1 μm to sometimes reaching up to 100 μm. *Pyrodictium occultum*, *P. brockii*, and *P. abyssi*, all of which belong to the class Thermoprotei, are obligate anaerobic hyperthermophilic archaea with optimal growth temperatures of 97–105 °C and can grow even at 110 °C (Stetter, 1982). *Pyrolobus fumarii*, also classified as Thermoprotei, was isolated from a hydrothermal vent on the Mid-Atlantic Ridge, has an optimal growth temperature of 106 °C, and can grow even at 113 °C (Blöchl et al., 1997). *Geogemma barossii*, another hyperthermophilic iron-reducing archaeon within the Thermoprotei, was isolated from a submarine hydrothermal vent in the northeastern Pacific Ocean (Kashefi & Lovley, 2003). The optimal growth temperature was 105 °C and it was capable of growth at 121 °C. *Pyrobaculum aerophilum*, belonging to the class Thermoprotei, is an aerobic hyperthermophilic archaeon isolated from shallow hydrothermal vents on Ischia Island, Italy, with an optimal growth temperature of 100 °C (Völkl et al., 1993). The *Methanopyrus kandleri* AV19 strain, which belongs to the Methanopyri class of the Methanobacteriota phylum, is a methanogenic hyperthermophilic archaeon isolated from hydrothermally heated deep-sea sediments in the Gulf of California. Its optimal growth temperature is 98 °C and it can grow at 110 °C (Kurr et al., 1991). Additionally, *M. kandleri* strain 116, isolated from a hydrothermal vent in the Central Indian Ridge, has an optimal growth temperature of 100 °C and can grow even at 116 °C. It has been shown that under a hydrostatic pressure of 40 MPa, strain 116 can grow at temperatures of up to 122 °C (Takai et al., 2008). *Archaeoglobus fulgidus*, which belongs to the class Archaeoglobi in the phylum Methanobacteriota, is an obligate anaerobic hyperthermophilic archaeon (Stetter, 1988). Although originally isolated from the North Sea oilfields, it also inhabits coastal hydrothermal vent ecosystems. The optimal growth temperature was 83 °C and the optimal pH was 6.5. They are also involved in the acidification of petroleum in oil facilities. Symbionts of hyperthermophilic archaea are known. *Nanoarchaeum equitans*, belonging to the class

Nanoarchaeia in the phylum Nanobdellota and characterized by its nano-sized cells, lives in symbiosis with *Ignicoccus hospitalis*, which belongs to the class Thermoprotei in the phylum Thermoproteota (Huber et al., 2002).

2.3 Low-Temperature Environments: Polar Regions, Permafrost, and Glaciers

Psychrophilic bacteria inhabit environments below 10 °C, such as seawater in polar regions (Arctic and Antarctic), permafrost, glaciers, and deep sea. *Psychrobacter giacincola*, a salt-tolerant psychrophilic bacterium belonging to the class Gammaproteobacteria in the phylum Pseudomonadota, was isolated from Antarctic seawater. Its optimal growth temperature is 13–15 °C, and it can grow at temperatures ranging from −18 to 22 °C. Its optimal NaCl concentration is 0.3–0.5 M, but it can also grow in the presence of 2 M NaCl (Bowman et al., 1997). *Psychromonas antarcticus*, also belonging to the class Gammaproteobacteria, is an aerotolerant anaerobic psychrophilic bacterium isolated from a hypersaline lake on the McMurdo Ice Shelf in Antarctica, with optimal conditions of 12 °C, pH 6.5, and 0.5 M NaCl (Mountfort et al., 1998). *Polaribacter filamentus*, in the class Flavobacteriia, the phylum Bacteroidota, is a psychrophilic bacterium isolated from surface seawater of the Arctic Ocean off the coast of Alaska, characterized by the presence of gas vesicles within its cells (Gosink et al., 1998). *Polaribacter* species containing intracellular gas vesicles have also been isolated from Antarctica. *Maribacter arcticus*, in the class Flavobacteriia, the phylum Bacteroidota, is a psychrophilic bacterium isolated from marine sediments in the Arctic Ocean off the coast of Norway, with an optimal growth temperature of 20–22 °C, an optimal pH of 6, and an optimal NaCl concentration 0.3 of 0.5 M (Cho et al., 2008). *Colwellia demingiae*, *C. hornerae*, *C. rossensis*, and *C. psychrotropica*, all belonging to the class Gammaproteobacteria, are psychrophilic bacteria isolated from Antarctic sea ice (Bowman et al., 1998a). These species are capable of synthesizing polyunsaturated fatty acids, such as docosahexaenoic acid. *Glaciecola punicea* and *G. pallidula*, also in the class Gammaproteobacteria, are psychrophilic bacteria isolated from Antarctic sea ice (Bowman et al., 1998b). *Marinobacter psychrophilus*, belonging to the class Alphaproteobacteria, the phylum Pseudomonadota, is a psychrophilic bacterium isolated from sea ice in the Canada Basin. Its optimal growth temperature is 16–18 °C, optimal pH is 6–9, and it does not require NaCl for growth (Zhang et al., 2008a). *Phaeobacter arcticus*, in the class *Alphaproteobacteria*, is a psychrophilic bacterium isolated from marine sediments in the Arctic Ocean, with an optimal growth temperature of 19–20 °C, optimal pH of 6–9, and requiring 0.3–1.5 M NaCl for growth (Zhang et al. 2008b). *Halorubrum lacusprofundi*, belonging to the class Halobacteria in the phylum Methanobacteriota, is a cold-adapted halophilic archaeon isolated from the Deep Lake in Antarctica. It grows at 1.5 M up to saturated concentrations of NaCl, with an optimal growth temperature of 35–37 °C, but it is also able to grow at 4 °C (Franzmann et al., 1988). Genetic engineering systems have been established for this microorganism and research and development are ongoing (Liao et al., 2016).

In permafrost, soil, sediment, or rock is maintained at or below 0 °C for at least 2 consecutive years. *Psychrobacter cryohalolentis* and *Psychrobacter arcticus* are psychrophilic bacteria isolated from the Arctic permafrost (Bakermans et al., 2006). The optimal growth temperature for *P. cryohalolentis* was 22 °C, and it could grow from −10 to 30 °C. Although growth does not require NaCl, it can tolerate up to 1.8 M NaCl. *P. arcticus* has an optimal growth temperature of 22 °C and can grow from −10 to 28 °C. It requires trace amounts of NaCl for growth, and can withstand up to 1.25 M NaCl. Low-acyl lipopolysaccharides isolated from these two bacteria have been shown to induce inflammatory responses in macrophages, which may lead to reduced immune function (Korneev et al., 2014).

Glaciers are massive blocks of ice that deform plastically under weight pressure. *Sphingomonas glacialis*, in the class Alphaproteobacteria, the phylum Pseudomonadota, is a psychrophilic bacterium isolated from cryoconite on the Stubai Glacier in Austria and grows well between 1 and 25 °C (Zhang et al., 2011). *Chryseobacterium glaciei*, belonging to the class Flavobacteria in the phylum Bacteroidota, is a psychrophilic bacterium isolated from the surface soil on glaciers in the Himalayas and grows well between 10 °C and 8 °C (Pal et al., 2018). Psychrophilic bacteria have also been isolated from China's Urumqi Glacier No.1, including *Cryobacterium psychrotolerans* (phylum Actinomycetota, class Actinomycetes) (Zhang et al., 2007), *Salinibacterium xinjiangense* (phylum Actinomycetota, class Actinomycetes) (Zhang et al., 2008c), and *Planomicrobium glaciei* (phylum Bacillota, class Bacilli) (Zhang et al., 2009).

2.4 Low-Temperature, High-Pressure Environments: The Deep Sea

The deep sea has an average temperature of 1–3 °C and is simultaneously subjected to high-pressure water. Consequently, many psychrophiles isolated from the deep sea exhibit piezophilic characteristics, making them psychropiezophiles. *C. hasaliensis*, which belongs to the class Gammaproteobacteria, is a piezophilic bacterium isolated from particle-rich seawater at a depth of 7410 meters in the Puerto Rico Trench, with an optimum pressure of 75 MPa at 2 °C (Deming et al., 1988). *Shewanella violacea* DSS12 and *Photobacterium profundum* DSJ4, both in the class Gammaproteobacteria, are psychropiezophiles isolated from marine sediments at a depth of 5110 meters in the Ryukyu Trench (Nogi et al., 1998a, b). The former has an optimum NaCl concentration of approximately 0.5 M, an optimum growth temperature of 8 °C, and an optimum pressure of 30 MPa, but it can also grow at atmospheric pressure. The latter has an optimum NaCl concentration of approximately 0.5 M, an optimum growth temperature of 8–12 °C, and an optimum pressure of 10 MPa, and it can also grow at atmospheric pressure. *S. benthica* strain DB21MT-2 and *Moritella yayanosii* strain DB21MT-5 are obligate piezophiles isolated from marine sediments at a depth of 10,898 meters in the Challenger Deep of the Mariana Trench (Kato et al., 1998; Nogi & Kato, 1999). The former had an optimum pressure of 70 MPa at 10 °C, whereas the latter had an optimum pressure of 80 MPa at 10 °C. Neither strain could grow at pressures below 50 MPa. *Profundibacter amoris* in the class Alphaproteobacteria is a piezophilic bacterium isolated from a microbial mat on a barite chimney in the low-temperature vent field "Loki's Castle" at a depth of approximately 2, 400 m on the Central Arctic Ridge. The optimum NaCl concentration was 0.34 M, the optimum growth temperature was 27–35 °C, and the optimum pressure was 30–35 MPa (Le Moine Bauer et al., 2019).

2.5 Extreme Environments Based on Salinity: Salt Lakes, Salt Pans, Rock Salt, Deep High-Salinity Anaerobic Basins

Halophilic microorganisms inhabit extreme environments based on their salinity. Salt lakes are enclosed bodies of water characterized by high concentrations of salt and dissolved minerals. The salinity of these lakes often exceeds that of seawater, and water levels and salinity can fluctuate drastically owing to evaporation and varying inflow. Representative salt lakes include the Dead Sea on the border of Israel and Jordan, Great Salt Lake in the United States, salt lakes in the Sahara Desert, and numerous salt lakes that are widely distributed across the world.

In 1936, Volcani (born Wilkansky) provided the first description of microbial communities in the Dead Sea (Wilkansky, 1936). The first halophilic archaea isolated from the Dead Sea were *Halobacterium salinarum* (formerly *Halobacterium halobium*) and *Haloarcula marismortui*, both of which belong to the class Halobacteria in the phylum Methanobacteriota (Volcani, 1944). These organisms are extremely halophilic archaea that can grow over 2.5 M NaCl. *Hb. salinarum* NRC34002 has been shown to be resistant to ultraviolet light (D37: 212 J/m^2), gamma rays (D37: 393 Gy), and mitomycin C (D37: 2.36 mg/mL/h) (Shahmohammadi et al., 1977). Furthermore, the *Hb. salinarum* NRC-1 has been shown to be resistant to gamma rays (D37: approximately 3.9 kGy) and desiccation (Kottemann et al., 2005). *Haloferax volcanii* (formerly *Halobacterium volcanii*), also in the class Halobacteria, was isolated from the coastal mud of the Dead Sea (Mullakhanbhai & Larsen, 1975). Its optimal NaCl concentration for growth at 30–40 °C is 1.7–2.5 M. Both *Hb. salinarum* and *Hf. volcanii* have established genetic manipulation systems and have been the subject of advanced molecular biological studies (Cline et al., 1989; Hackett & DasSarma, 1989).

Halorubrum sodomense (formerly *Halobacterium sodomense*), classified under the class Halobacteria, has been isolated from the surface waters of the Dead Sea.

Its optimal NaCl concentration at 35 °C is 1.7–2.5 M, and its optimal MgCl2 concentration under 2.0 M NaCl is 0.6–1.2 M (Oren, 1983a). *Halobaculum gomorrense*, also in the class Halobacteria, was isolated from a site 4 meters deep in the Dead Sea, with optimal NaCl concentration at 35–40 °C being 1.5–2.5 M and optimal $MgCl_2$ concentration under 2.1 M NaCl at 0.6–1 M (Oren et al., 1995). *Halorhabdus utahensis*, belonging to the class *Halobacteria*, is an extremely halophilic and thermophilic microorganism isolated from the sediments of the Great Salt Lake, with optimal growth at 50 °C with a NaCl concentration of 4.6 M and $MgCl_2$ concentration of 2–800 mM (Wainø et al., 2000). Kemp et al. (2018) examined microbial diversity in the Great Salt Lake, detecting *Halorubrum* spp. such as *Halorubrum arcis* and *Halorubrum californiense*, as well as *Haloarcula* spp. such as *Haloarcula argentinensis*, *Haloarcula hispanica*, and *Haloarcula japonica*. Boutaiba et al. (2011) studied the microbial diversity in the hypersaline Sidi Ameur and Himalatt Salt Lakes of the Algerian Sahara, and identified six genera of halophilic archaea within the *Halobacteria* classes Haloferax, Halobacterium, Haladaptatus, Halalkalicoccus, *Halorubrum*, and *Halosarcina*. Khallef et al. (2018) investigated the microbial diversity of hypersaline Chotts and Sebkhas aquatic ecosystems in Algeria, detecting seven genera of halophilic archaea within the class *Halobacteria*: *Halorubrum*, *Natribaculum*, *Haloarcula*, *Natrinema*, *Halostagnicola*, *Haloferax*, and *Haloterrigena*. *Methanohalobium evestigatum*, belonging to the class Methanomicrobia in the phylum Methanobacteriota, is an anaerobic methane-producing archaeon isolated from salt lagoon mud in Crimea. It grows optimally at 50 °C and in 4.3 M NaCl (Zhilina & Zavarzin, 1987).

Halomonas elongata, belonging to the class Gammaproteobacteria in the phylum Pseudomonadota, is a halophilic bacterium isolated from a solar salt production facility in Bonaire, Netherlands Antilles. The optimal growth temperature was 30 °C, with a range of 4–45 °C. It can survive at pH 5–9 and in the presence of 0.6–3.4 M NaCl (Vreeland et al., 1980). *Halobacteroides halobius*, belonging to the class Clostridia in the phylum Bacillota, is an anaerobic halophilic bacterium isolated from the deep sediments of the Dead Sea. Its optimal growth temperature is 37–42 °C, and its optimal NaCl concentration is 1.5–2.5 M (Oren et al., 1984). *Sporohalobacter lortetii* (formerly *Clostridium lortetii*), also in the class Clostridia, is an anaerobic, spore-forming halophilic bacterium isolated from the surface sediments of the Dead Sea. It shows optimal growth at 1.4–1.5 M NaCl at 30 °C and at 1.7 M NaCl at 42 °C (Oren, 1983b). *Orenia marismortui* (formerly *Sporohalobacter marismortui*), also in the class *Clostridia*, is an anaerobic halophilic bacterium isolated from the coastal sediments of the Dead Sea. It grows optimally at 36–45 °C and 0.5–2.0 M NaCl (Oren et al., 1987). *Salinibacter ruber*, belonging to the class Rhodothermia in the phylum Rhodothermota, is an extremely halophilic bacterium isolated from the crystallization ponds of saltwork in Mallorca, Balearic Islands, Spain. *The optimal growth* temperature was 37–47 °C, and the optimal salinity was 4.0–6.3 M NaCl. It accounts for 5–25% of the prokaryotic community in salt crystallization ponds (Antón et al., 2002). *Amycolatopsis halophila*, belonging to the class Actinomycetes in the phylum Actinomycetota, is a halophilic bacterium isolated from the sediments of Qijiaojing Lake, China. It grows at 25–45 °C in the presence of 0.2–2.6 M NaCl (Tang et al., 2010). Similarly, *Brevibacterium salitolerans*, also belonging to the class Actinomycetes, is a halophilic bacterium isolated from the sediments of the Lop Nur Salt Lake in China. It grows at 5–50 °C in the presence of 0–3.5 M NaCl (Guan et al., 2010). *Paracoccus halotolerans*, belonging to the class Alphaproteobacteria in the phylum Pseudomonadota, is a halophilic bacterium isolated from sediments of Yuncheng Salt Lake, China. It grows at 10–37 °C, up to a maximum concentration of 1 M NaCl (Meng et al., 2019).

Rock salt is formed when seawater trapped inland by ancient crustal movements crystallizes under layers of sediment over a long period of time. It is often mined in uplifted regions that were once seafloors, such as the Alps, Andes, and Himalayas, as well as in desert areas where the salt lakes have dried. From ancient rock salt layers, halophilic archaea and bacteria such as *Haloarcula*, *Halobacterium*, *Halococcus*, *Haloterrigena*, *Natronobacterium*, *Natronomonas*, and *Virgibacillus* have been isolated, with *Halobacterium* spp. being especially prevalent. *Halobacterium* spp. possess polyploid genomes and are resistant to DNA damage and cellular stress. Polyploidy is thought to be a key factor enabling these organisms to survive in salt for tens of millions of years (Jaakkola et al., 2014). Halophiles have also been isolated from commercially available edible rock salt. *Halarchaeum acidiphilum*, belonging to the class Halobacteria in the phylum Methanobacteriota, is a halophilic archaeon isolated from edible salt in Australia. It grows optimally at 37 °C, 3.6–4.1 M NaCl, 1 mM $MgCl_2$, and at pH 4.4–4.5, making it moderately acidophilic (Minegishi et al., 2010). Henriet et al. (2014) detected halophilic archaea in 14 out of 26 samples of edible salt. All isolates were related to 17 different genera of the family Halobacteriaceae, with *Haloarcula*, *Halobacterium*, and the genus *Halorubrum* being the most frequently detected. Gibtan et al. (2017) investigated the microbial communities in four types of commercial solar salt using amplicon sequencing targeting the 16S rRNA gene. *Halobacteria* accounted for 91.58% of all archaea. Among bacteria, *Pseudomonadota*, *Bacteroidota*, *Actinomycetota*, and *Bacillota* were the dominant phyla.

Deep hypersaline anoxic basins (DHABs) are extreme environments found on the seafloor in locations such as the Gulf of Mexico, Red Sea, and Mediterranean Sea. These basins are characterized by salt concentrations that are five to ten times higher than those of seawater, anoxic conditions, high pressure, and low light levels. In the sharply stratified region between seawater and the basins (the halocline), processes such as manganese cycling, sulfate reduction and sulfide oxidation, methanogenesis, aerobic (anaerobic) methane oxidation, and anammox and denitrification cycles occur and are sustained by diverse microbial communities (Merlino et al., 2018; Varrella et al., 2020). *Halanaeroarchaeum sulfurireducens* M27-SA2, a halophilic archaeon in the class Halobacteria with sulfur-reducing and acetate-oxidizing capabilities, has been isolated from the Medee Basin of the Mediterranean Sea (Messina et al., 2016). From the Shaban Deep in the Red Sea, *Halorhabdus tiamatea*, a halophilic archaeon in the class *Halobacteria*, *Haloplasma contractile*, a halophilic bacterium belonging to the class *Mollicutes* in the phylum *Mycoplasmatota* that reduces nitrate and nitrite, and *Marinobacter salsuginis*, a moderately halophilic bacterium belonging to the class Alphaproteobacteria in the phylum Pseudomonadota, have been isolated (Antunes et al., 2007, 2008a, b).

2.6 Extreme Environments Based on pH: Mine Drainage, Acidic Hot Springs, Alkaline Lakes

Acid mine drainage, generated by human activities, is a representative example of artificially created acidic environments. Naturally occurring acidic environments, such as volcanic hot springs and acidic wetlands, also exist. *Acidithiobacillus ferrooxidans* (formerly *Thiobacillus ferrooxidans*), which belongs to the class *Acidithiobacillia* in the phylum *Pseudomonadota*, is a prominent acidophilic iron-sulfur-oxidizing bacterium that thrives under acidic conditions (pH 2.0–3.5). It was first isolated from coal mine drainage in West Virginia, USA (Colmer & Hinkle, 1947). *Acidiphilium cryptum*, classified in the class Alphaproteobacteria of the Pseudomonadota phylum, is a heterotrophic bacterium that grows at 35–41 °C and pH 1.9–5.9 (Harrison, 1981). This bacterium was isolated either as a contaminant in cultures of the chemolithoautotrophic *Acidithiobacillus ferrooxidans* or from copper and coal mines. *Leptospirillum ferrooxidans,* belonging to the class Nitrospirota in the phylum Pseudomonadota, is a mesophilic acidophilic iron-oxidizing bacterium isolated from the Alaverda copper deposit in Armenia and grows at pH 1.5–4.0 (Markosyan, 1972; Hippe, 2000). This species plays a critical role in metal bioleaching. *Sulfobacillus thermosulfidooxidans*, classified as the class Clostridia in the phylum Bacillota, is an acidophilic, mixotrophic, moderately thermophilic, spore-forming, sulfur-oxidizing bacterium isolated from the Nikolaev copper-zinc-pyrite deposit in eastern Kazakhstan. Its optimum growth occurs at 50 °C and pH of 1.9–2.4 (Golovacheva & Karavaĭko, 1978). *Metallosphaera sedula*, belonging to the order Sulfolobales in the class Thermoprotei, the phylum Thermoproteota, is a thermoacidophilic, metal sulfide–oxidizing archaeon isolated from the Pisciarelli Solfatara in Italy. Optimal growth was observed at 65 °C and pH 3.0–3.5. It is considered for use in the removal of toxic metals from coal (Huber et al., 1989). *Thermoplasma acidophilum*, belonging to the order Thermoplasmatales in the class Thermoplasmata, the phylum Thermoproteota, is a thermoacidophilic archaeon isolated from a coal refuse pile in the United States. Its optimum growth temperature is 59 °C, and its optimum pH is 1–2 (Darland et al., 1970). In addition, in the order Thermoplasmatales, *Picrophilus oshimae* and *Picrophilus torridus* are thermoacidophilic archaea isolated from the solfataric field of the Kawayu hot springs in northern Japan. The optimum growth temperature is 60 °C, and the optimum pH is 0.7 (Schleper et al., 1985, 1996). *Ferroplasma acidiphilum*, also classified as Thermoplasmatales, was isolated from a gold-containing arsenopyrite/pyrite ore concentrate at a pilot plant in Russia. The optimum growth temperature is 35 °C with an optimum pH of 1.7 (Golyshina et al., 2000).

Alkaline environments include soda lakes with pH above 9, limestone caves, and hot springs. There are numerous alkaliphilic bacteria in the order Bacillales, class Bacilli, phylum Bacillota, such as *B. halmapalus*, *Alkalihalophilus pseudofirmus* (formerly *B. pseudofirmus*), *Halalkalibacterium halodurans* (formerly *B. halodurans*), *Salipaludibacillus agaradhaerens* (formerly *B. agaradhaerens*), *Sutcliffiella horikoshii* (formerly *Bacillus horikoshii*), *Alkalihalobacillus alcalophilus* (formerly *B. alcalophilus*), and *Evansella clarkii* (formerly *B. clarkii*) (Nielsen et al., 1995). *H. halodurans* C-125 and *A. pseudofirmus* OF4 are the most extensively studied facultative alkaliphilic bacteria (Takami & Horikoshi, 1999; Tamaki & Krulwich, 2000) and genetic manipulation systems have been established for both strains (Ito & Nagane, 2001; Wallace & Breaker, 2011). *Halorhodospira halochloris* (formerly *Ectothiorhodospira halochloris*), a halophilic, thermophilic, and alkaliphilic photosynthetic bacterium belonging to the class Gammaproteobacteria, phylum Pseudomonadota, was isolated from a hypersaline alkaline lake in Wadi El Natrun, Egypt. It grows optimally in the presence of 2.4–4.6 M NaCl, at 47–50 °C, and at pH 8.1–9.1 (Imhoff & Trüper, 1977). The obligate chemolithoautotrophic, halophilic, sulfur-oxidizing bacterium *Thiohalospira alkaliphile*, which belongs to the class Gammaproteobacteria, was isolated from the same hypersaline alkaline lake at Wadi El Natrun. It grows optimally at 2.0 M NaCl, 35 °C, and pH 8.5, with a growth pH range of 7.0–10.2 (Sorokin et al., 2008). *Cellulosibacter alkalithermophilus*, belonging to the class Clostridia, phylum Bacillota, is an anaerobic, thermophilic, and alkaliphilic bacterium isolated from the soil of a coconut plantation in Thailand. It grows optimally at 55 °C and pH 9.5, with a growth pH range of 8.0–10.0 (Watthanalamloet et al., 2012). Similarly, *Alkaliphilus transvaalensis*, also in the class Clostridia, was isolated from a highly alkaline water pool located 3.2 km deep in a gold mine in South Africa. It grows optimally at 40 °C and pH 10.0, with a growth pH range of 8.5–12.5 (Takai et al., 2001).

Many halophilic archaea in the family Natrialbaceae, order Natrialbales, class Halobacteria, phylum Methanobacteriota, display alkaliphilic properties. *Natronococcus occultus*, *Natronobacterium gregoryi*, and *Natrialba magadii* (formerly *Natronobacterium magadii*) are alkaliphilic, halophilic archaea isolated from the hypersaline alkaline Magadi Lake in Kenya. Their optimal growth temperature is 37 °C, optimal NaCl concentration is 3.5–4.8 M, and optimal pH is 9.0–10.0. *Nb. gregoryi* can grow at a pH of up to 12 (Tindall et al., 1984). *Natronobiforma cellulositropha*, an alkaliphilic and halophilic archaeon, was isolated from surface sediments in the hypersaline alkaline Tanatar-1 Lake in the Siberian region of Russia. Its optimal growth temperature is 43 °C, optimal NaCl concentration is 4 M, optimal pH is 8.5–9, and it can grow up to pH 9.9 (Sorokin et al., 2018a). *Natrarchaeobius halalkaliphilus*, an alkaliphilic and halophilic archaeon, was isolated from the hypersaline alkaline Searles Lake in the United States. It grows optimally at 45 °C, at 3.5 M NaCl, with an optimal pH of 8.5, and can grow up to pH 9.5 (Sorokin et al., 2019). *Natronomonas pharaonis* (formerly *Halobacterium pharaonis*), belonging to the order Halobacteriales, class Halobacteria, phylum Methanobacteriota, is an alkaliphilic and halophilic archaeon isolated from a hypersaline alkaline lake in Wadi El Natrun, Egypt. The optimal growth temperature is 37 °C, the optimal NaCl concentration is 3.5 M, and the optimal pH is 8.5. It can grow within the pH range of 7.7–9.3 (Soliman & Trüper, 1982). *Methanonatronarchaeum thermophilum*, belonging to the class Methanonatronarchaeia, phylum Methanobacteriota, is a thermophilic, alkaliphilic, and halophilic archaeon isolated from the sediments of a hypersaline soda lake in Russia. The optimal growth temperature was 50 °C, optimal NaCl concentration was 4 M, and optimal pH was 9.5–9.7. It can grow in the pH range of 8.2–10.2 (Sorokin et al., 2018b).

2.7 Extreme Environments Based on Aridity: Deserts

Xerophiles inhabit extreme environments that are characterized by dryness. These environments are distributed in extremely arid desert regions, such as the Atacama Desert in Chile, Namib Desert in southern Africa, Sahara Desert in northern Africa, Arabian Desert in Western Asia, and Gobi Desert in Central Asia. Warren-Rhodes et al. (2006) demonstrated that hypolithic cyanobacteria belonging to the class Cyanophyceae in the phylum Cyanobacteriota, specifically *Chroococcidiopsis* spp., exist in small, spatially isolated island-like patches in the soils of the Atacama Desert. Neilson et al. (2012) investigated the microbial communities in the unvegetated arid soils of the Atacama Desert and detected abundant bacteria such as the classes Rubrobacteria, Nitriliruptoria, Coriobacteriia, Actinomycetes, and Acidimicrobiia in the phylum Actinomycetota. Furthermore, Bartholomäus et al. (2024) reported that numerous viable and potentially active microorganisms such as *Acidimicrobiia*, *Geodermatophilaceae*, *Frankiales*, and *Burkholderiaceae* inhabit the hyperarid regions of the Atacama Desert. Ronca et al. (2015) examined the microbial communities of sand dunes in the Namib Desert and found that *Geodermatophilus* sp., *Blastococcus* sp., and *Arthrobacter* sp. in the phylum Actinomycetota, as well as *Microvirga* sp., *Massilia* sp., and *Novosphingobium* sp. in the phylum Pseudomonadota, were prevalent in dune habitats. For details on the diversity of extremophilic microorganisms isolated from Asian deserts, refer to Salam et al. (2018). *Chroococcidiopsis* spp. and *Anabaena* spp. are resistant to desiccation, ultraviolet radiation, and ionizing radiation (Billi et al., 2000; Singh et al., 2010). Additionally, bacteria of the genus *Deinococcus*, which are resistant to ionizing radiation, also exhibit resistance to ultraviolet radiation and desiccation. *D. deserti*, *D. peraridilitoris*, *D.apachensis*, *D. saudiensis*, *D. sahariensis*, and *D. gobiensis* have been isolated from deserts (de Groot et al., 2005; Rainey et al., 2005, 2007; Yuan et al., 2009; Bouraoui et al., 2012; Hossain et al., 2016). This suggests that desiccation, ultraviolet radiation, and ionizing radiation resistance are closely interconnected.

2.8 Concluding Remarks

Extreme environments on Earth highlight the remarkable adaptability of life owing to their diversity and wide distribution. Extremophilic microorganisms that thrive under such harsh conditions have evolved unique survival strategies from high-temperature hydrothermal vents to frigid permafrost, hypersaline lakes, environments with extreme pH, and high-pressure deep sea. Notably, many hyperthermophiles are considered to be extremely close to the last universal common ancestor of all life on Earth, making them critically important models for studying the evolution of life. Extreme environments such as hot springs and volcanic regions are thought to resemble the primitive conditions of the early Earth, offering insights into ancient life forms and the origins of life itself. The existence of extremophiles profoundly challenges traditional paradigms regarding fundamental biological questions such as "What is life?" and "Where are the limits of life?" This suggests that life is far more flexible and adaptable than was previously assumed. Research on these organisms has expanded our understanding of the fundamental limits and origins of life by demonstrating the ability of life to utilize energy from within the Earth and construct unique ecosystems independent of sunlight. Moreover, extremophilic microorganisms play a valuable role as model organisms in the field of astrobiology, exploring their potential for extraterrestrial life. Extreme environments such as the Atacama Desert, Antarctica's Dry Valleys, hot springs, deep-sea hydrothermal vents, and deep high-salinity anaerobic basins are considered analogous to the environments of Mars, Europa, and other celestial bodies that may harbor potential biospheres. Adaptations to complex stresses, such as radiation and desiccation, form the foundation for investigating the potential of extraterrestrial life. Furthermore, the unique enzymes (extremozymes) and molecules (extremolytes) produced by extremophiles offer immense potential for biotechnology. With wide-ranging applications in pharmaceuticals, food, energy, and environmental remediation, these organisms represent important resources contributing to the realization of a sustainable society. However, several extreme environments on Earth remain unexplored. Investigating these unknown realms will deepen our understanding of life's limits. Recent technological advances, particularly metagenomic analysis and culture-independent methods, have dramatically improved our understanding of the diversity and functions of extremophiles. In the future, as research on microbial communities in more complex extreme environments progresses, we expect new insights into the adaptive strategies of life.

References

Aanniz, T., Ouadghiri, M., Melloul, M., Swings, J., Elfahime, E., Ibijbijen, J., Ismaili, M., & Amar, M. (2015). Thermophilic bacteria in Moroccan hot spring, salt marches and desert soils. *Brazilian Journal of Microbiology, 46*, 443–453.

Albuquerque, L., Simões, C., Nobre, M. F., Pino, N. M., Battista, J. R., Silva, M. T., Rainey, F. A., & da Costa, M. S. (2005). *Truepera radiovictrix* gen. nov., sp. nov., a new radiation resistant species and the proposal of *Trueperaceae* fam. nov. *FEMS Microbiology Letters, 247*, 161–169.

Albuquerque, L., Johnson, M. M., Schumann, P., Rainey, F. A., & da Costa, M. S. (2014). Description of two new thermophilic species of the genus *Rubrobacter*, *Rubrobacter calidifluminis* sp. nov. and *Rubrobacter naiadicus* sp. nov., and emended description of the genus *Rubrobacter* and the species *Rubrobacter bracarensis*. *Systematic & Applied Microbiology, 37*, 235–243.

Antón, J., Oren, A., Benlloch, S., Rodríguez-Valera, F., Amann, R., & Rosselló-Mora, R. (2002). *Salinibacter ruber* gen. nov., sp. nov., a novel, extremely halophilic member of the *Bacteria* from saltern crystallizer ponds. *International Journal of Systematic & Environmental Microbiology, 52*, 485–491.

Antunes, A., França, L., Rainey, F. A., Huber, R., Nobre, M. F., Edwards, K. J., & da Costa, M. S. (2007). *Marinobacter salsuginis* sp. nov., isolated from the brine–seawater interface of the Shaban Deep, Red Sea. *International Journal of Systematic & Evolutionary Microbiology, 57*, 1035–1040.

Antunes, A., Rainey, F. A., Wanner, G., Taborda, M., Pätzold, J., Nobre, M. F., da Costa, M. S., & Huber, R. (2008a). A new lineage of halophilic, wall-less, contractile bacteria from a brine-filled deep of the Red Sea. *Journal of Bacteriology, 190*, 3580–3587.

Antunes, A., Taborda, M., Huber, R., Moissl, C., Nobre, M. F., & da Costa, M. S. (2008b). *Halorhabdus tiamatea* sp. nov., a non-pigmented, extremely halophilic archaeon from a deep-sea, hypersaline anoxic basin of the Red Sea, and emended description of the genus *Halorhabdus*. *International Journal of Systematic & Evolutionary Microbiology, 58*, 215–220.

Asker, D., Awad, T. S., Beppu, T., & Ueda, K. (2008). *Deinococcus misasensis* and *Deinococcus roseus*, novel members of the genus *Deinococcus*, isolated from a radioactive site in Japan. *Systematic & Applied Microbiology, 31*, 43–49.

Bakermans, C., Ayala-del-Río, H. L., Ponder, M. A., Vishnivetskaya, T., Gilichinsky, D., Thomashow, M. F., & Tiedje, J. M. (2006). *Pyschrobacter cryhalolentis* sp. nov. and *Psychrobacter arcticus* sp. nov., isolated from Siberian permafrost. *International Journal of Systematic & Environmental Microbiology, 56*, 1285–1291.

Bartholomäus, A., Genderjahn, S., Mangelsdorf, K., Schneider, B., Zamorano, P., Kounaves, S. P., Schulze-Makuch, D., & Wagner, D. (2024). Inside the Atacama Desert: Uncovering the living microbiome of an extreme environment. *Applied & Environmental Microbiology, 90*, 01443–01424.

Beblo, K., Rabbow, E., Rachel, R., Huber, H., & Rettberg, P. (2009). Tolerance of thermophilic and hyperthermophilic microorganisms to desiccation. *Extremophiles, 13*, 521–531.

Beblo, K., Douki, T., Schmalz, G., Rachel, E., Wirth, R., Huber, H., Reitz, G., & Rettberg, P. (2011). Survival of thermophilic and hyperthermophilic microorganisms after exposure to UV-C, ionizing radiation and desiccation. *Archives in Microbiology, 193*, 797–809.

Beblo-Vranesevic, K., Galinski, E. A., Rachel, R., Huber, H., & Rettberg, P. (2017a). Influence of osmotic stress on desiccation and irradiation tolerance of (hyper)-thermophilic microorganisms. *Archives in Microbiology, 199*, 17–28.

Beblo-Vranesevic, K., Huber, H., & Rettberg, P. (2017b). High tolerance of Hygrogenothermus marinus to sodium perchlorate. *Fronters in Microbiology, 8*, 1369.

Benammar, L., Bektaş, K. İ., Menasria, T., Beldüz, A. O., Güler, H. I., Bedaida, I. K., Gonzalez, J. M., & Ayachi, A. (2020). Diversity and enzymatic potential of thermophilic bacteria associated with terrestrial hot springs in Algeria. *Brazilian Journal of Microbiology, 51*, 1987–2007.

Berkner, S., Grogan, D., Albers, S. V., & Lipps, G. (2007). Small multicopy, non-integrative shuttle vectors based on the plasmid pRN1 for *Sulfolobus acidocaldarius* and *Sulfolobus solfataricus*, model organisms of the (cren-)archaea. *Nucleic Acids Research, 35*, e88.

Billi, D., Friedmann, E. I., Hofer, K. G., Caiola, M. G., & Ocampo-Friedmann, R. (2000). Ionizing-radiation resistance in the desiccation-tolerant cyanobacterium *Chroococcidiopsis*. *Applied & Environmental Microbiology, 66*, 1489–1492.

Birrien, J. L., Zeng, X., Jebbar, M., Cambon-Bonavita, M. A., Quérellou, J., Oger, P., Bienvenu, N., Xiao, X., & Prieur, D. (2011). *Pyrococcus yayanosii* sp. nov., an obligate piezophilic hyperthermophilic archaeon isolated from a deep-sea hydrothermal vent. *International Journal of Systematic & Environmental Microbiology, 61*, 2827–2831.

Blanc, M., Marilley, L., Beffa, T., & Aragno, M. (1999). Thermophilic bacterial communities in hot composts as revealed by most probable number counts and molecular (16S rDNA) methods. *FEMS Microbiology Ecology, 28*, 141–149.

Blöchl, E., Rachel, R., Burggraf, S., Hafenbradl, D., Jannasch, H. W., & Stetter, K. O. (1997). *Pyrolobus fumarii*, gen. and sp. nov., represents a novel group of archaea, extending the upper temperature limit for life to 113 degrees C. *Extremophiles, 1*, 14–21.

Bouraoui, H., Aissa, M. B., Abbassi, F., Touzel, J. P., O'donohue, M., & Manai, M. (2012). Characterization of *Deinococcus sahariens* sp. nov., a radiation-resistant bacterium isolated from a Saharan hot spring. *Archives in Microbiology, 194*, 315–322.

Boutaiba, S., Hacene, H., Bidle, K. A., & Maupin-Furlow, J. A. (2011). Microbial diversity of the hypersaline Sidi Ameur and Himalatt Salt Lakes of the Algerian Sahara. *Journal of Arid Environments, 75*, 909–916.

Bowman, J. P., Nichiles, D. S., & McMeekin, T. (1997). *Psychrobacter glacincola* sp. nov., a halotolerant, psychrophilic bacterium isolated from Antarctic sea ice. *Systematic & Applied Microbiology, 20*, 209–215.

Bowman, J. P., Gosink, J. J., McCammon, S. A., Lewis, T. E., Nichols, D. S., Nichols, P. D., Skerratt, J. H., Staley, J. T., & McMeekin, T. A. (1998a). *Colwellia demingiae* sp. nov., *Colwellia hornerae* sp. nov., *Colwellia rossensis* sp. nov. and *Colwellia psychrotropica* sp. nov.: psychrophilic Antarctic species with the ability to synthesize docosahexaenoic acid (22:6ω3). *International Journal of Systematic Bacteriology, 48*, 1171–1180.

Bowman, J. P., McCammon, S. A., Brownm, J. L., & McMeekin, T. A. (1998b). *Glaciecola punicea* gen. nov., sp. nov. and *Glaciecola pallidula* gen. nov., sp. nov.: psychrophilic bacteria from Antarctic sea-ice habitats. *International Journal of Systematic Bacteriology, 48*, 1213–1222.

Brock, T. D., & Freeze, H. (1969). *Thermus aquaticus*, a nonsporulating extreme thermophile. *Journal of Bacteriology, 98*, 289–297.

Brock, T. D., Brock, K. M., Belly, R. T., & Weiss, R. L. (1972). *Sulfolobus*: A new genus of sulfur-oxidizing bacteria living at low pH and high temperature. *Archives in Microbiology, 84*, 54–68.

Cannio, R., Contursi, P., Rossi, M., & Bartolucci, S. (1998). An autonomously replicating transforming vector for *Sulfolobus solfataricus*. *Journal of Bacteriology, 180*, 3237–3240.

Cava, F., Hidalgo, A., & Berenguer, J. (2009). *Thermus thermophilus* as biological model. *Extremophiles, 13*, 213–231.

Cha, M., Kim, J. K., Lee, W. H., Song, H., Lee, T. G., Kim, S. K., & Kim, S. J. (2024). Metabolic engineering of *Caldicellulosiruptor bescii* for hydrogen production. *Applied Microbiology & Biotechnology, 108*, 65.

Charbonneau, D. M., Meddeb-Mouelhi, F., Boissinot, M., Sirois, M., & Beauregard, M. (2012). Identification of thermophilic bacterial strains producing thermotolerant hydrolytic enzymes from manure compost. *Indian Journal of Microbiology, 52*, 41–47.

Chen, M. Y., Wu, S. H., Lin, G. H., Lu, C. P., Lin, Y. T., Chang, W. C., & Tsay, S. S. (2004). *Rubrobacter taiwanensis* sp. nov., a novel thermophilic, radiation-resistant species isolated from hot springs. *International Journal of Systematic & Evolutionary Microbiology, 54*, 1849–1855.

Cho, K. H., Hong, S. G., Cho, H. H., Lee, Y. K., Chun, J., & Lee, H. K. (2008). *Maribacter articus* sp. nov.. isolated from Arctic marine sediment. *International Journal of Systematic & Evolutionary Microbiology, 58*, 1300–1303.

Cline, S. W., Schalkwyk, L. C., & Doolittle, W. F. (1989). Transformation of the archaebacterium *Halobacterium volcanii* with genomic DNA. *Journal of Bacteriology, 171*, 4987–4991.

Colmer, A. R., & Hinkle, M. E. (1947). The role of microorganisms in acid mine drainage: A preliminary report. *Science, 106*, 253–256.

Darland, G., Brock, T. D., Samsonoff, W., & Conti, S. F. (1970). A thermophilic, acidophilic mycoplasma isolated from a coal refuse pile. *Science, 170*, 1416–1418.

de Groot, A., Chapon, V., Servant, P., Christen, R., Fischer-Le Saux, M., Sommer, S., & Heulin, T. (2005). *Deinococcus deserti* sp. nov., a gamma-radiation-tolerant bacterium isolated from the Sahara Desert. *International Journal of Systematic & Evolutionary Microbiology, 55*, 2441–2446.

de Rosa, M., Gambacorta, A., & Bu'lock, J. D. (1975). Extremely thermophilic acidophilic bacteria convergent with Sulfolobus acidocaldarius. *Journal of General Microbiology, 86*, 156–164.

Deming, J. W., Somers, L. K., Straube, W. L., Swarts, D. G., & Macdonell, M. T. (1988). Isolation of an obligately barophilic bacterium and description of a new genus, *Colwellia* gen. nov. *Systematic & Applied Microbiology, 10*, 152–160.

DiRuggiero, J., Santangelo, N., Nackerdien, Z., Ravel, J., & Robb, F. T. (1997). Repair of extensive ionizing-radiation DNA damage at 95 degrees C in the hyperthermophilic archaeon *Pyrococcus furiosus*. *Journal of Bacteriology, 179*, 4643–4645.

Ferreira, A. C., Nobre, M. F., Rainey, F. A., Silva, M. T., Wait, R., Burghardt, J., Chung, A. P., & da Costa, M. S. (1997). *Deinococcus geothermalis* sp. nov. and *Deinococcus murrayi* sp. nov., two extremely radiation-resistant and slightly thermophilic species from hot springs. *International Journal of Systematic Bacteriology, 47*, 939–947.

Fiala, G., & Stetter, K. O. (1986). *Pyrococcus furiosus* sp. nov. represents a novel genus of marine heterotrophic archaebacteria growing optimally at 100°C. *Archives in Microbiology, 145*, 56–61.

Ford, K. C., Bing, R. G., Manesh, M. J. H., Adams, M. W. W., & Kelly, R. M. (2025). *Anaericullus* (f. *Caldicellulosiruptor*) *bescii*. *Trends in Microbiology, 33*, 569–570.

Franzmann, P., Stackebrandt, E., Sanderson, K., Volkman, J., Cameron, D., Stevenson, P., Mcmeekin, T. A., & Burton, H. R. (1988). *Halobacterium lacusprofundi* sp. nov., a halophilic bacterium isolated from Deep Lake, Antarctica. *Systematic & Applied Microbiology, 11*, 20–27.

Gibtan, A., Park, K., Woo, M., Shin, J. K., Lee, D. W., Sohn, J. H., Song, M., Roh, S. W., Lee, S. J., & Lee, H. S. (2017). Diversity of extremely halophilic archaeal and bacterial communities from commercial salts. *Frontiers in Microbiology, 8*, 799.

Golovacheva, R. S., & Karavaĭko, G. I. (1978). Sulfobacillus, a new genus of thermophilic sporulating bacteria. *Mikrobiologiia, 47*, 815–822.

Golyshina, O. V., Pivovarova, T. A., Karavaiko, G. I., Kondratéva, T. F., Moore, E. R., Abraham, W. R., Lünsdorf, H., Timmis, K. N., Yakimov, M. M., & Golyshin, P. N. (2000). *Ferroplasma acidiphilum* gen. nov., sp. nov., an acidophilic, autotrophic, ferrous-iron-oxidizing, cell-wall-lacking, mesophilic member of the *Ferroplasmaceae* fam. nov., comprising a distinct lineage of the *Archaea*. *International Journal of Systematic and Evolutionary Microbiology, 50*, 997–1006.

Gosink, J. J., Woese, C. R., & Staley, J. T. (1998). *Polaribacter* gen. nov., with three new species, *P. irgensii* sp. nov., *P. franzmannii* sp. nov. and *P. filamentus* sp. nov., gas vacuolate polar marine bacteria of the *Cytophaga-Flavobacterium-Bacteroides* group and reclassification of '*Flectobacillus glomeratus*' as *Polaribacter glomeratus* comb. nov. *International Journal of Systematic Bacteriology, 48*, 223–235.

Guan, T. W., Zhao, K., Xiao, J., Liu, Y., Xia, Z. F., Zhang, X. P., & Zhang, L. L. (2010). *Brevibacterium salitolerans* sp. nov., an actinobacterium isolated from salt-lake sediment. *International Journal of Systematic & Environmental Microbiology, 60*, 2991–2995.

Hackett, N. R., & DasSarma, S. (1989). Characterization of the small endogenous plasmid of *Halobacterium* strain SB3 and its use in transformation of *H. halobium*. *Canadian Journal of Microbiology, 35*, 86–91.

Harrison, A. P. (1981). *Acidiphilium cryptum* gen. nov., sp. nov., heterotrophic bacterium from acidic mineral environments. *International Journal of Systematic Bacteriology, 31*, 327–332.

Heinen, U. J., & Heinen, W. (1972). Characteristics and properties of a caldo-active bacterium producing extracellular enzymes and two related strains. *Archives in Microbiology, 82*, 1–23.

Henriet, O., Fourmentin, J., Delincé, B., & Mahillon, J. (2014). Exploring the diversity of extremely halophilic archaea in food-grade salts. *International Journal of Food Microbiology, 191*, 36–44.

Hippe, H. (2000). *Leptospirillum* gen. nov. (ex Markosyan 1972), nom. rev., including *Leptospirillum ferrooxidans* sp. nov. (ex Markosyan 1972), nom. rev. and *Leptospirillum thermoferrooxidans* sp. nov. (Golovacheva et al. 1992). *International Journal of Systematic & Evolutionary Microbiology, 50*, 501–503.

Hossain, F., Khan, I. U., Habib, N., Xian, W. D., Hozzein, W. N., Zhang, Z. D., Zhi, Z. Y., & Li, W. J. (2016). *Deinococcus saudiensis* sp. nov., isolated from desert. *International Journal of Systematic and Evolutionary Microbiology, 66*, 5106–5111.

Huber, R., Langworthy, T. A., König, H., Thomm, M., Woese, C. R., Sleytr, U. B., & Stetter, K. O. (1986). *Thermotoga maritima* sp. nov. represents a new genus of unique extremely thermophilic eubacteria growing up to 90°C. *Archives in Microbiology, 144*, 324–333.

Huber, G., Spinnler, C., Gambacorta, A., & Stetter, K. O. (1989). *Metallosphaera sedula* gen. and sp. nov. represents a new genus of aerobic, metal-mobilizing, thermoacidophilic archaebacteria. *Systematic & Applied Microbiology, 12*, 38–47.

Huber, R., Wilharm, T., Huber, D., Trincone, A., Burggraf, S., Köning, H., Reinhard, R., Rockinger, I., Fricke, H., & Stetter, K. O. (1992). *Aquifex pyrophilus* gen. nov. sp. nov., represents a novel group of marine hyperthermophilic hydrogen-oxidizing bacteria. *Systematic & Applied Microbiology, 15*, 340–351.

Huber, H., Hohn, M. J., Rachel, R., Fuchs, T., Wimmer, V. C., & Stetter, K. O. (2002). A new phylum of Archaea represented by a nanosized hyperthermophilic symbiont. *Nature, 417*, 63–67.

Imanaka, T., Fujii, M., Aramori, I., & Aiba, S. (1982). Transformation of *Bacillus stearothermophilus* with plasmid DNA and characterization of shuttle vector plasmids between *Bacillus stearothermophilus* and *Bacillus subtilis*. *Journal of Bacteriology, 149*, 824–830.

Imhoff, J. F., & Trüper, H. G. (1977). *Ectothiorhodospira halochloris* sp. nov., a new extremely halophilic phototrophic bacterium containing bacteriochlorophyll b. *Archives in Microbiology, 114*, 115–121.

Ito, M., & Nagane, M. (2001). Improvement of the electro-transformation efficiency of facultatively alkaliphilic *Bacillus pseudofirmus* OF4 by high osmolarity and glycine treatment. *Bioscience, Biotechnology, & Biochemistry, 65*, 2773–2775.

Itoh, T., Onishi, M., Kato, S., Iino, T., Sakamoto, M., Kudo, T., Takashina, T., & Ohkuma, M. (2016). *Athalassotoga saccharophila* gen. nov., sp. nov., isolated from an acidic terrestrial hot spring, and proposal of *Mesoaciditogales* ord. nov. and *Mesoaciditogaceae* fam. nov. in the phylum *Thermotogae*. *International Journal of Systematic & Evolutionary Microbiology, 66*, 1045–1051.

Jaakkola, S. T., Zerulla, K., Guo, Q., Liu, Y., Ma, H., Yang, C., Bamford, D. H., Chen, X., Soppa, J., & Oksanen, H. M. (2014). Halophilic archaea cultivated from surface sterilized middle-late Eocene rock salt are polyploid. *PLoS One, 9*, e110533.

Jolivet, E., L'Haridon, S., Corre, E., Forterre, P., & Prieur, D. (2003). *Thermococcus gammatolerans* sp. nov., a hyperthermophilic archaeon from a deep-sea hydrothermal vent that resists ionizing radiation. *International Journal of Systematic & Evolutionary Microbiology, 53*, 847–851.

Jolivet, E., Corre, E., L'Haridon, S., Forterre, P., & Prieur, D. (2004). *Thermococcus marinus* sp. nov. and *Thermococcus radiotolerans* sp. nov., two hyperthermophilic archaea from deep-sea hydrothermal vents that resist ionizing radiation. *Extremophiles, 8*, 219–227.

Kashefi, K., & Lovley, D. R. (2003). Extending the upper temperature limit for life. *Science, 301*, 934.

Kato, C., Li, L., Nogi, Y., Nakamura, Y., Tamaoka, J., & Horikoshi, K. (1998). Extremely barophilic bacteria isolated from the Mariana Trench, Challenger Deep, at a depth of 11,000 meters. *Applied & Environmental Microbiology, 64*, 1510–1513.

Keller, M., Braun, F. J., Dirmeier, R., Hafenbradl, D., Burggraf, S., Rachel, R., & Stetter, K. O. (1995). *Thermococcus alcaliphilus* sp. nov., a new hyperthermophilic archaeum growing on polysulfide at alkaline pH. *Archives in Microbiology, 164*, 390–395.

Kemp, B. L., Tabish, E. M., Wolford, A. J., Jones, D. L., Butler, J. K., & Baxter, B. K. (2018). The biogeography of Great Salt Lake halophilic archaea: Testing the hypothesis of avian mechanical carriers. *Diversity, 10*, 124.

Khallef, S., Lestini, R., Myllykallio, H., & Houali, K. (2018). Isolation and identification of two extremely halophilic archaea from sebkhas in the Algerian Sahara. *Cellular and Molecular Biology, 64*, 83–91.

Korneev, K. V., Kondakova, A. N., Arbatsky, N. P., Novototskaya-Vlasova, K. A., Rivkina, E. M., Anisimov, A. P., Kruglov, A. A., Kuprash, D. V., Nedospasov, S. A., Knirel, Y. A., & Drutskaya, M. S. (2014). Distinct biological activity of lipopolysaccharides with different lipid A acylation status from mutant strains of *Yersinia pestis* and some members of genus *Psychrobacter*. *Biochemistry (Moscow), 79*, 1333–1338.

Kottemann, M., Kish, A., Iloanusi, C., Bjork, S., & DiRuggiero, J. (2005). Physiological responses of the halophilic archaeon *Halobacterium* sp. strain NRC1 to desiccation and gamma irradiation. *Extremophiles, 9*, 219–227.

Kurr, M., Huber, R., König, H., Jannasch, H. W., Fricke, H., Trineone, A., Kristjansson, J. K., & Stetter, K. O. (1991). *Methanopyrus kandleri*, gen. and sp. nov. represents a novel group of hyperthermophilic methanogens, growing at 110°C. *Archives in Microbiology, 156*, 239–247.

L'Haridon, S., Miroshnichenko, M. L., Hippe, H., Fardeau, M. L., Bonch-Osmolovskaya, E., Stackebrandt, E., & Jeanthon, C. (2001). *Thermosopho geolei* sp. nov., a thermophilic bacterium isolated from a continental petroleum reservoir in Western Siberia. *International Journal of Systematic & Evolutionary Microbiology, 51*, 1327–1334.

Le Moine Bauer, S., Sjøberg, A. G., L'Haridon, S., Stokke, R., Roalkvam, I., Steen, I. H., & eDahle, H. (2019). *Profundibacter amoris* gen. nov., sp. nov., a new member of the Roseobacter clade isolated from Loki's Castle Vent Field on the Arctic Mid-Ocean Ridge. *International Journal of Systematic & Environmental Microbiology, 69*, 975–981.

L'Haridon, S., Cilia, V., Messner, P., Raguénès, G., Gambacorta, A., Sleytr, U. B., Prieur, D., & Jeanthon, C. (1998). *Desulfurobacterium thermolithotrophum* gen. nov., sp. nov., a novel autotrophic, sulphur-reducing bacterium isolated from a deep-sea hydrothermal vent. *International Journal of Systematic Bacteriology, 48*, 701–711.

Liao, Y., Williams, T. J., Walsh, J. C., Ji, M., Poljak, A., Curmi, P. M. G., Duggin, I. G., & Cavicchioli, R. (2016). Developing a genetic manipulation system for the Antarctic archaeon, *Halorubrum lacusprofundi*: Investigating acetamidase gene function. *Scientific Reports, 6*, 34639.

Lipscomb, G. L., Conway, J. M., Blumer-Schuette, S. E., Kelly, R. M., & Adams, M. W. W. (2016). A highly thermostable kanamycin resistance marker expands the tool kit for genetic manipulation of *Caldicellulosiruptor bescii*. *Applied and Environmental Microbiology, 82*, 4421–4428.

Markosyan, G. E. (1972). A new iron-oxidizing bacterium *Leptospirillum ferrooxidans* nov. gen. nov. sp. *Biological Journal of Armenia, 25*, 26–29.

Meng, X. L., Ming, H., Huang, J. R., Zhang, L. Y., Cheng, L. J., Zhai, Z. L., Ji, W. L., Li, W. J., & Nie, G. X. (2019). *Paracoccus halotolerans* sp. nov., isolated from a salt lake. *International Journal of Systematic & Evolutionary Microbiology, 69*, 523–528.

Merlino, G., Barozzi, A., Michoud, G., Ngugi, D. K., & Daffoncho, D. (2018). Microbial ecology of deep-sea hypersaline anoxic basins. *FEMS Microbiology Ecology, 94*, fiy085.

Messina, E., Sorokin, D. Y., Kublanov, I. V., Toshchakov, S., Lopatina, A., Arcadi, E., Smedile, F., La Spada, G., La Cono, V., & Yakimov, M. M. (2016). Complete genome sequence of '*Halanaeroarchaeum sulfurireducens*' M27-SA2, a sulfur-reducing and acetate-oxidizing haloarchaeon from the deep-sea hypersaline anoxic lake Medee. *Standards in Genomic Sciences, 11*, 35.

Minegishi, H., Echigo, A., Nagaoka, S., Kamekura, M., & Usami, R. (2010). *Halarchaeum acidiphilum* gen. nov., sp. nov., a moderately acidophilic haloarchaeon isolated from commercial solar salt. *International Journal of Systematic & Evolutionary Microbiology, 60*, 2513–2516.

Mohagheghi, A., Grohmann, K., Himmel, M., Leighton, L., & Updegraff, D. M. (1986). Isolation and characterization of *Acidothermus cellulolyticus* gen. nov., sp. nov., a new genus of thermophilic, acidophilic, cellulolytic bacteria. *International Journal of Systematic Bacteriology, 36*, 435–443.

Mori, K., Yamazoe, A., Hosoyama, A., Ohji, S., Fujita, N., Ishibashi, J., Kimura, H., & Suzuki, K. (2014). *Thermotoga profunda* sp. nov. and *Thermotoga caldifontis* sp. nov., anaerobic thermophilic bacteria isolated from terrestrial hot springs. *International Journal of Systematic & Evolutionary Microbiology, 64*, 2128–2136.

Morikawa, M., Izawa, Y., Rashid, N., Hoaki, T., & Imanaka, T. (1994). Purification and characterization of a thermostable thiol protease from a newly isolated hyperthermophilic *Pyrococcus* sp. *Applied & Environmental Microbiology, 60*, 4559–4566.

Mountfort, D. O., Rainey, F. A., Burghardt, J., Kaspar, H. F., & Stackebrandt, E. (1998). *Psychromonas antarcticus* gen. nov., sp. nov., a new aerotolerant anaerobic, halophilic psychrophile isolated from pond sediment of the McMurdo Ice Shelf, Antarctica. *Archives in Microbiology, 169*, 231–238.

Mullakhanbhai, M. F., & Larsen, H. (1975). *Halobacterium volcanii* spec. nov., a Dead Sea halobacterium with a moderate salt requirement. *Archives in Microbiology, 104*, 207–214.

Narumi, I., Sawakami, K., Nakamoto, S., Nakayama, N., Yanagisawa, T., Takahashi, N., & Kihara, H. (1992). A newly isolated *Bacillus stearothermophilus* K1041 and its transformation by electroporation. *Biotechnology Techniques, 6*, 83–86.

Neilson, J. W., Quade, J., Ortiz, M., Nelson, W. M., Legatzki, A., Tian, F., LaComb, M., Betancourt, J. L., Wing, R. A., Soderlund, C. A., & Maier, R. M. (2012). Life at the hyperarid margin: Novel bacterial diversity in arid soils of the Atacama Desert, Chile. *Extremophiles, 16*, 553–566.

Nielsen, P., Fritze, D., & Priest, F. G. (1995). Phenetic diversity of alkaliphilic *Bacillus* strains: Proposal for nine new species. *Microbiology, 141*, 1745–1761.

Nogi, Y., & Kato, C. (1999). Taxonomic studies of extremely barophilic bacteria isolated from the Mariana trench and description of *Moritella yayanosii* sp. nov., a new barophilic bacterial isolate. *Extremophiles, 3*, 71–77.

Nogi, Y., Kato, C., & Horikoshi, K. (1998a). Taxonomic studies of deep-sea barophilic *Shewanella* strains and description of Shewanella violacea sp. nov. *Archives in Microbiology, 170*, 331–338.

Nogi, Y., Masui, N., & Kato, C. (1998b). *Photobacterium profundum* sp. nov., a new, moderately barophilic bacterial species isolated from a deep-sea sediment. *Extremophiles, 2*, 1–7.

Oren, A. (1983a). *Halobacterium sodomense* sp. nov., a Dead Sea halobacterium with an extremely high magnesium requirement. *International Journal of Systematic Bacteriology, 33*, 381–386.

Oren, A. (1983b). *Clostridium lortetii* sp. nov., a halophilic obligatory anaerobic bacterium producing endospores with attached gas vacuoles. *Archives in Microbiology, 136*, 42–48.

Oren, A., Weisburg, W. G., Kessel, M., & Woese, C. R. (1984). *Halobacteroides halobius* gen. Nov., sp. nov., a moderately halophic anaerobic bacterium from the bottom sediments of the Dead Sea. *Systematic & Applied Microbiology, 5*, 58–70.

Oren, A., Pohla, H., & Stackebrandt, E. (1987). Transfer of *Clostridium lortetii* to a new genus *Sporohalobacter* gen. nov. as *Sporohalobacter lortetii* comb. nov., and description of *Sporohalobacter marismortui* sp. nov. *Systematic & Applied Microbiology, 9*, 239–246.

Oren, A., Gurevich, P., Gemmell, R. T., & Teske, A. (1995). *Halobaculum gomorrense* gen. nov., sp. nov., a novel extremely halophilic archaeon from the Dead Sea. *International Journal of Systematic Bacteriology, 45*, 747–754.

Ortega-Villar, R., Escalante, A., Astudillo-Melgar, F., Lizárraga-Mendiola, L., Vázquez-Rodríguez, G. A., Hidalgo-Lara, M. E., & Coronel-Olivares, C. (2024). Isolation and characterization of thermophilic bacteria from a hot spring in the state of Hidalgo, Mexico, and geochemical analysis of the thermal water. *Microorganisms, 12*, 1066.

Oshima, T., & Imahori, K. (1971). Isolation of an extreme thermophile and thermostability of its transfer ribonucleic acid and ribosomes. *Journal of General & Applied Microbiology, 17*, 513–517.

Oshima, T., & Imahori, K. (1975). Description of *Thermus thermophilus* (Yoshida and Oshima) comb. nov., a nonsporulating thermophilic bacterium from a Japanese thermal spa. *International Journal of Systematic Bacteriology, 24*, 102–112.

Pal, M., Kumari, M., Kiran, S., Salwan, R., Mayilraj, S., Chhibber, S., & Gulati, A. (2018). *Chryseobacterium glaciei* sp. nov., isolated from the surface of a glacier in the Indian trans-Himalayas. *International Journal of Systematic & Environmental Microbiology, 68*, 865–870.

Papale, M., Romani, I., Finore, H., Lo Giudice, A., Piccolo, A., Cangemi, S., Di Meo, V., Nicolaus, B., & Poli, A. (2021). Prokaryotic diversity of the composting thermophilic phase: The case of ground coffee compost. *Microorganisms, 9*, 218.

Purwasena, I. A., Sugai, Y., & Sasaki, K. (2014). *Petrotoga japonica* sp. nov., a thermophilic, fermentative bacterium isolated from Yabase Oilfield in Japan. *Archives in Microbiology, 196*, 313–321.

Rainey, F. A., Ray, K., Ferreira, M., Gatz, B. Z., Nobre, M. F., Bagaley, D., Rash, B. A., Park, M. J., Earl, A. M., Shank, N. C., Small, A. M., Henk, M. C., Battista, J. R., Kämpfer, P., & da Costa, M. S. (2005). Extensive diversity of ionizing-radiation-resistant bacteria recovered from Sonoran Desert soil and description of nine new species of the genus *Deinococcus* obtained from a single soil sample. *Applied & Environmental Microbiology, 71*, 5225–5235.

Rainey, F. A., Ferreira, M., Nobre, M. F., Ray, K., Bagaley, D., Earl, A. M., Battista, J. R., Gómez-Silva, B., McKay, C. P., & da Costa, M. S. (2007). *Deinococcus peraridilitoris* sp. nov., isolated from a coastal desert. *International Journal of Systematic and Evolutionary Microbiology, 57*, 1408–1412.

Ronca, S., Ramond, J. B., Jones, B. E., Seely, M., & Cowan, D. A. (2015). Namib desert dune/interdune transects exhibit habitat-specific edaphic bacterial communities. *Frontiers in Microbiology, 6*, 845.

Salam, N., Yang, Z. W., Asem, M. D., Hozzein, W. N., & Li, W. J. (2018). Microbial diversity in Asian deserts: Distribution, biotechnological importance, and environmental impacts. In D. Egamberdieva, N. K. Birkeland, H. Panosyan, & W. J. Li (Eds.), *Extremophiles in Eurasian ecosystems: Ecology, diversity, and applications. Microorganisms for sustainability* (Vol. 8, pp. 365–387). Springer.

Santangelo, T. J., Cubonová, L., & Reeve, J. N. (2008). Shuttle vector expression in *Thermococcus kodakaraensis*: Contributions of cis elements to protein synthesis in a hyperthermophilic archaeon. *Applied & Environmental Microbiology, 74*, 3099–3104.

Sato, T., Fukui, T., Atomi, H., & Imanaka, T. (2003). Targeted gene disruption by homologous recombination in the hyperthermophilic archaeon *Thermococcus kodakaraensis* KOD1. *Journal of Bacteriology, 185*, 210–220.

Schleper, C., Puehler, G., Holz, I., Gambacarta, A., Janekovic, D., Santarius, U., Klenk, H. P., & Zilling, W. (1985). *Picrophilus* gen. nov., fam. nov.: A novel aerobic, heterotrophic, thermoacidophilic genus and family comprising archaea capable of growth around pH 0. *Journal of Bacteriology, 177*, 7050–7059.

Schleper, C., Pühler, G., & Klenk, H. P. (1996). *Picrophilus oshimae* and *Picrophilus torridus* fam. nov., gen. nov., sp. nov., two species of hyperacidophilic, thermophilic, heterotrophic, aerobic archaea. *International Journal of Systematic Bacteriology, 46*, 814–816.

Shahmohammadi, H. R., Asgarani, E., Terato, H., Ide, H., & Yamamoto, O. (1977). Effect of ^{60}Co gamma-rays, ultraviolet light, and mitomycin C on *Halobacterium salinarium* and *Thiobacillus intermedius*. *Journal of Radiation Research, 38*, 37–43.

Singh, H., Fernandes, T., & Apte, S. K. (2010). Unusual radioresistance of nitrogen-fixing cultures of anabaena strains. *Journal of Biosciences, 35*, 427–434.

Soliman, G. S. H., & Trüper, H. G. (1982). *Halobacterium pharaonis* sp. nov., a new, extremely haloalkaliphilic archaebacterium with low magnesium requirement. *Zentralblatt für Bakteriologie Mikrobiologie und Hygiene: I. Abt. Originale C: Allgemeine, angewandte und ökologische Mikrobiologie, 3*, 318–329.

Sorokin, D. Y., Tourova, T. P., Muyzer, G., & Kuenen, G. J. (2008). *Thiohalospira halophila* gen. nov., sp. nov. and *Thiohalospira alkaliphila* sp. nov., novel obligately chemolithoautotrophic, halophilic, sulfur-oxidizing gammaproteobacteria from hypersaline habitats. *International Journal of Systematic and Evolutionary Microbiology, 58*, 1685–1692.

Sorokin, D. Y., Khijiak, T. V., Kostrikina, N. A., Elcheninov, A. G., Toshchakov, S. V., Bale, N. J., Sinninghe Damsté, J. S., Kublanov, I., & V. (2018a). *Natronobiforma cellulositropha* gen. nov., sp. nov., a novel haloalkaliphilic member of the family *Natrialbaceae* (class *Halobacteria*) from hypersaline alkaline lakes. *Systematic & Applied Microbiology, 41*, 355–362.

Sorokin, D. Y., Merkel, A. Y., Abbas, B., Makarova, K., Rijpstra, I. C., Koenen, M., Sinninghe Damsté, J. S., Galinski, E. A., Koonin, E. V., & van Loosdrecht, M. C. M. (2018b). *Methanonatronarchaeum thermophilum* gen. nov., sp. nov. and "*Candidatus* Methanohalarchaeum thermophilum", extremely halo(natrono)philic methyl-reducing methanogens from hypersaline lakes comprising a new euryarchaeal class Methanonatronarchaeia classis nov. *International Journal of Systematic & Evolutionary Microbiology, 68*, 2199–2208.

Sorokin, D. Y., Elcheninov, A. G., Toshchakov, S. V., Bale, N. J., Sinninghe Damsté, J. S., Khijniak, T. V., & Kublanov, I. V. (2019). *Natrarchaeobius chitinivorans* gen. nov., sp. nov., and *Natrarchaeobius halalkaliphilus* sp. nov., alkaliphilic, chitin-utilizing haloarchaea from hypersaline alkaline lakes. *Systematic & Applied Microbiology, 42*, 309–318.
Stetter, K. O. (1982). Ultrathin mycelia-forming organisms from submarine volcanic areas having an optimum growth temperature of 105 °C. *Nature, 300*, 258–260.
Stetter, K. O. (1988). *Archaeoglobus fulgidus* gen. nov., sp. nov. a new taxon of extremely thermophilic archaebacteria. *Systematic & Applied Microbiology, 10*, 172–173.
Stöhr, R., Wabertski, A., Völker, H., Tindall, B. J., & Thomm, M. (2001). *Hydrogenothermus marinus* gen. nov., sp. nov., a novel thermophilic hydrogen-oxidizing bacterium, recognition of *Calderobacterium hydrogenophilum* as a member of the genus *Hydrogenobacter* and proposal of the reclassification of *Hydrogenobacter acidophilus* as *Hydrogenobaculum acidophilum* gen. nov., comb. nov., in the phylum '*Hydrogenobacter/Aquifex*'. *International Journal of Systematic & Evolutionary Microbiology, 51*, 1853–1862.
Strom, P. F. (1985). Identification of thermophilic bacteria in solid-waste composting. *Applied & Environmental Microbiology, 50*, 906–913.
Studholme, D. J., Jackson, R. A., & Leak, D. J. (1999). Phylogenetic analysis of transformable strains of thermophilic *Bacillus* species. *FEMS Microbiology Letters, 172*, 85–90.
Suzuki, H., & Yoshida, K. (2012). Genetic transformation of *Geobacillus kaustophilus* HTA426 by conjugative transfer of host-mimicking plasmids. *Journal of Microbiology and Biotechnology, 22*, 1279–1287.
Takai, K., Moser, D. P., Onstott, T. C., Spoelstra, N., Pfiffner, S. M., Dohnalkova, A., & Fredrickson, J. K. (2001). *Alkaliphilus transvaalensis* gen. nov., sp. nov., an extremely alkaliphilic bacterium isolated from a deep south African gold mine. *International Journal of Systematic and Evolutionary Microbiology, 51*, 1245–1256.
Takai, K., Nakamura, K., Toki, T., Tsunogai, U., Miyazaki, M., Miyazaki, J., Hirayama, H., Nakagawa, S., Nunoura, T., & Horikoshi, K. (2008). Cell proliferation at 122°C and isotopically heavy CH_4 production by a hyperthermophilic methanogen under high-pressure cultivation. *Proceedings of the National Academy of Sciences of the United States of America, 105*, 10949–10954.
Takami, H., & Horikoshi, K. (1999). Reidentification of facultatively alkaliphilic *Bacillus* sp. C-125 to *Bacillus halodurans*. *Bioscience, Biotechnology, & Biochemistry, 65*, 943–945.
Tamaki, H., & Krulwich, T. A. (2000). Reidentification of facultatively alkaliphilic *Bacillus firmus* OF4 as *Bacillus pseudofirmus* OF4. *Extremophiles, 4*, 19–22.
Tang, S. K., Wang, Y., Guan, T. W., Lee, J. C., Kim, C. J., & Li, W. J. (2010). *Amycolatopsis halophila* sp. nov., a halophilic actinomycete isolated from a salt lake. *International Journal of Systematic & Environmental Microbiology, 60*, 1073–1078.
Taylor, M. P., Esteban, C., & D., Leak, D. J. (2008). Development of a versatile shuttle vector for gene expression in *Geobacillus* spp. *Plasmid, 60*, 45–52.
Tindall, B. J., Ross, H. N. M., & Grant, W. D. (1984). *Natronobacterium* gen. nov. and *Natronococcus* gen. nov., two new genera of haloalkaliphilic archaebacteria. *Systematic & Applied Microbiology, 5*, 41–57.
Valenzuela, B., Solís-Cornejo, F., Araya, R., & Zamorano, P. (2024). Isolation of thermophilic bacteria from extreme environments in Northern Chili. *Microorganisms, 12*, 473.
Varrella, S., Tangherlini, M., & Corinaldesi, C. (2020). Deep hypersaline anoxic basins as untapped reservoir of polyextremophilic prokaryotes of biotechnological interest. *Marine Drug, 18*, 91.
Volcani, B. E. (1944). The microorganisms of the Dead Sea. In *Papers collected to commemorate the 70th anniversary of Dr Chaim Weizmann. Collective volume* (pp. 71–85). Daniel Sieff Research Institute.
Völkl, P., Huber, R., Drobner, E., Rachel, R., Burggraf, S., Trincone, A., & Stetter, K. O. (1993). *Pyrobaculum aerophilum* sp. nov., a novel nitrate-reducing hyperthermophilic archaeum. *Applied & Environmental Microbiology, 59*, 2918–2926.

Vreeland, R. H., Litchfield, C. D., Martin, E. L., & Elliot, E. (1980). *Halomonas elongata*, a new genus and species of extremely salt-tolerant bacteria. *International Journal of Systematic Bacteriology, 30*, 485–495.

Wainø, M., Tindall, B. J., & Ingvorsen, K. (2000). *Halorhabdus utahensis* gen. nov., sp. nov., an aerobic, extremely halophilic member of the Archaea from Great Salt Lake, Utah. *International Journal of Systematic & Environmental Microbiology, 50*, 183–190.

Wallace, J. G., & Breaker, R. R. (2011). Improved genetic transformation methods for the model alkaliphilic *Bacillus halodurans* C-125. *Letters in Applied Microbiology, 52*, 430–432.

Warren-Rhodes, K. A., Rhodes, K. L., Pointing, S. B., Ewing, S. A., Lacap, D. C., Gómez-Silva, B., Amundson, R., Friedmann, E. I., & MaKay, C. P. (2006). Hypolithic cyanobacteria, dry limit of photosynthesis, and microbial ecology in the hyperarid Atacama Desert. *Microbial Ecology, 52*, 389–398.

Watthanalamloet, A., Tachaapaikoon, C., Lee, Y. S., Kosugi, A., Mori, Y., Tanasupawat, S., Kyu, K. L., & Ratanakhanokchai, K. (2012). *Cellulosibacter alkalithermophilus* gen. nov., sp. nov., an anaerobic alkalithermophilic, cellulolytic-xylanolytic bacterium isolated from soil of a coconut garden. *International Journal of Systematic and Evolutionary Microbiology, 62*, 2330–2335.

Wilkansky, B. (1936). Life in the Dead Sea. *Nature, 138*, 467.

Windberger, E., Huber, R., Trincone, A., Fricke, H., & Stetter, K. O. (1989). *Thermotoga thermarum* sp. nov, and *Thermotoga neapolitana* occurring in African continental solfataric springs. *Archives of Microbiology, 151*, 506–512.

Wu, L. J., & Welker, N. E. (1989). Protoplast transformation of *Bacillus stearothermophilus* NUB36 by plasmid DNA. *Journal of General Microbiology, 135*, 1315–1324.

Yang, S. J., Kataeva, I., Wiegel, J., Yin, Y., Dam, P., Xu, Y., Westpheling, J., & Adams, M. W. W. (2010). Classification of '*Anaerocellum thermophilum*' strain DSM 6725 as *Caldicellulosiruptor bescii* sp. nov. *International Journal of Systematic & Evolutionary Microbiology, 60*, 2011–2015.

Yoshinaka, T., Yano, K., & Yamaguchi, H. (1973). Isolation of highly radioresistant bacterium, *Arthrobacter radiotolerans* nov. sp. *Agricultural & Biological Chemistry, 37*, 2269–2275.

Yu, J. S., Vargas, M., Mityas, C., & Noll, K. M. (2001). Liposome-mediated DNA uptake and transient expression in *Thermotoga*. *Extremophiles, 5*, 53–60.

Yuan, M., Zhang, W., Dai, S., Wu, J., Wang, Y., Tao, T., Chen, M., & Lin, M. (2009). *Deinococcus gobiensis* sp. nov., an extremely radiation-resistant bacterium. *International Journal of Systematic and Evolutionary Microbiology, 59*, 1513–1517.

Zhang, D. C., Wang, H. X., Cui, H. L., Yang, Y., Liu, H. C., Dong, X. Z., & Zhou, P. J. (2007). *Cryobacterium psychrotolerans* sp. nov., a novel psychrotolerant bacterium isolated from the China no. 1 glacier. *International Journal of Systematic & Environmental Microbiology, 57*, 866–869.

Zhang, D. C., Li, H. R., Xin, Y. H., Chi, Z. M., Zhou, P. J., & Yu, Y. (2008a). *Marinobacter psychrophilus* sp. nov., a psychrophilic bacterium isolated from the Arctic. *International Journal of Systematic Bacteriology, 58*, 1463–1466.

Zhang, D. C., Li, H. R., Xin, Y. H., Liu, H. C., Chi, Z. M., Zhou, P. J., & Yu, Y. (2008b). *Phaeobacter arcticus* sp. nov., a psychrophilic bacterium isolated from the Arctic. *International Journal of Systematic Bacteriology, 58*, 1384–1387.

Zhang, D. C., Liu, H. C., Xin, Y. H., Yu, Y., Zhou, P. J., & Zhou, Y. G. (2008c). *Salinibacterium xinjiangense* sp. nov., a psychrophilic bacterium isolated from the China no. 1 glacier. *International Journal of Systematic & Environmental Microbiology, 58*, 2739–2742.

Zhang, D. C., Liu, H. C., Xin, Y. H., Yu, Y., Zhou, P. J., & Zhou, Y. G. (2009). *Planomicrobium glaciei* sp. nov., a psychrotolerant bacterium isolated from a glacier. *International Journal of Systematic & Environmental Microbiology, 59*, 1387–1390.

Zhang, D. C., Busse, H. J., Liu, H. C., Zhou, Y. G., Schinner, F., & Margesin, R. (2011). *Sphingomonas glacialis* sp. nov., a psychrophilic bacterium isolated from alpine glacier cryoconite. *International Journal of Systematic Bacteriology, 61*, 587–591.

Zhilina, T. N., & Zavarzin, G. A. (1987). *Methanohalobium evestigatus*, n. gen., n. sp., the extremely halophilic methanogenic archaebacterium. *Doklady Akademii Nauk SSSR, 293*, 464–468.

Zillig, W., Stetter, K. O., Wunderl, S., Schulz, W., Priess, H., & Scholz, I. (1980). The *Sulfolobus*-"*Caldariella*" group: Taxonomy on the basis on the structure of DNA-dependent RNA polymerase. *Archives in Microbiology, 125*, 259–269.

Zillig, W., Stetter, K. O., Schäfer, W., Janekovic, D., Wunderl, S., Holz, I., & Palm, P. (1981). Thermoproteales: A novel type of extremely thermoacidophilic anaerobic archaebacteria isolated from Icelandic solfataras. *Zentralblatt für Bakteriologie Mikrobiologie und Hygine: I. Abt. Originale C: Allgemeine, angewandte und ökologiche Mikrobiologie, 2*, 205–227.

Zillig, W., Holz, I., Janekovic, D., Schäfer, W., & Reiter, W. D. (1983). The archaebacterium *Thermococcus celer* represents, a novel genus within the branch of the archaebacteria. *Systematic & Applied Microbiology, 4*, 88–94.

Part II
Characteristics of Enzymes and Catalysts Produced by Extremophiles and Their Industrial Applications

Chapter 3
Harnessing the Keratinolytic Potential of the Extremophilic Bacterium *Fervidobacterium islandicum* AW-1 for Sustainable Upcycling of Poultry Feather Waste

Nicole Enjeh Kim, Hyeon-Su Jin, and Dong-Woo Lee

Abstract Poultry feather waste is an abundant yet underutilized agro-industrial byproduct composed primarily of recalcitrant β-keratin, which exhibits significant resistance to proteolysis due to its high cysteine content and extensive hydrogen bonding. Conventional keratin processing methods, including chemical hydrolysis and thermal incineration, are energy-intensive, environmentally damaging, and often degrade essential amino acids. In this chapter, we highlight the keratinolytic potential of the extremely thermophilic anaerobe *Fervidobacterium islandicum* AW-1, which enables efficient bioconversion of feather keratin under high-temperature, anaerobic conditions. We present a comprehensive overview of the keratin degradation system in *F. islandicum* AW-1, supported by integrated genomic, proteomic, biochemical, and structural analyses. In addition, we introduce a novel soluble β-keratin substrate-based assay that enables high-resolution profiling of proteolytic specificity and enzyme kinetics. Downstream valorization pathways are also discussed, including the identification and function validation of low-molecular-weight bioactive peptides with antioxidant and skin-protective activities. This case study illustrates the broader potential of extremophile-derived enzymes in sustainable bioprocessing for high-value industrial and cosmeceutical applications. Finally, we propose how this microbial platform can be integrated into a closed-loop, zero-waste circular bioeconomy model to produce biostimulants, biodegradable plastics, and bioenergy from feather waste.

Keywords *Fervidobacterium islandicum* AW-1 · Keratin degradation · Extremophilic biocatalysts · Anaerobic fermentation · Circular bioeconomy · Feather waste · Bioactive peptides · Hydrogen production

N. E. Kim · H.-S. Jin · D.-W. Lee (✉)
Department of Biotechnology, Yonsei University, Seoul, South Korea
e-mail: leehicam@yonsei.ac.kr

M. Ito et al. (eds.), *Extremophiles for a Sustainable Future*,
https://doi.org/10.1007/978-981-95-6970-0_3

3.1 Introduction

The growing global population and increasing demand for industrial production have intensified the exploitation of finite natural resources, raising serious concerns about environmental sustainability and economic resilience (Elhacham et al., 2020; Williams et al., 2024; Ripple et al., 2024). One striking indicator of this imbalance is the recent estimate that anthropogenic mass, defined as the total weight of human-made materials, now surpasses Earth's living biomass (Elhacham et al., 2020). In response to these challenges, the transition from a linear "take-make-dispose" economy to a circular bioeconomy has gained momentum. This sustainable model emphasizes the regeneration and valorization of biological waste streams into high-value products, while minimizing environmental impact (Hadley Kershaw et al., 2021) (Table 3.1).

Among the most abundant yet underutilized biological residues is poultry feather waste, a major by-product of the agro-industrial sector (Lange et al., 2016; Rajabi et al., 2020). Feathers are composed of approximately 90% β-keratin by dry weight (Tesfaye et al., 2017). This highly resistant fibrous protein is stabilized by extensive disulfide cross-links due to its high cysteine content (7–13%) and dense hydrogen bonding (McKittrick et al., 2012). These structural features, while evolutionarily advantageous to birds, render keratin insoluble and resistant to enzymatic hydrolysis and microbial degradation. Common proteases are largely ineffective against keratin, and although several keratinases have been identified and characterized, their structural and functional features remain poorly understood (Bhagya Jyothi & Dhanasingh, 2025; Wu et al., 2017).

Thermophilic microorganisms have emerged as promising agents for keratin bioconversion due to their thermostable enzymes and compatibility with industrial processes (Zeldes et al., 2015). Of particular interest is *F. islandicum* AW-1, an extremely thermophilic anaerobe isolated from volcanic hot springs in Indonesia (Nam et al., 2002). This organism exhibits exceptional keratin-degrading ability under high-temperature, anaerobic conditions, without requiring any prior chemical or mechanical pretreatment (Yeo et al., 2018; Jin et al., 2019; Lee et al., 2015a). Its metabolic profile and enzymatic systems present a compelling model for the sustainable bioconversion of feather keratin.

In this chapter, we explore the keratinolytic system of *F. islandicum* AW-1 using an integrated multi-omics approach, including genomics, transcriptomics, proteomics, and biochemical analyses. We also introduce a custom soluble β-keratin-based assay developed to evaluate enzymatic activity with high resolution. Finally, we investigate downstream applications, such as the generation of bioactive peptides with cosmeceutical potential. This case study illustrates how extremophile-derived biocatalysts can underpin the development of circular, zero-waste biorefineries that transform recalcitrant keratinous waste into valuable biomaterials.

Table 3.1 Current status of valorized protein-rich wastes from various manufacturing industries

Industry	Product	Protein	Valorized products	Limitations of existing processing technologies
Meat, leather	Feather, hair, hides, horn, wool	Keratin	Bioactive peptides, biostimulants, animal feed, biofuel, bioplastic (Trojanowska et al., 2025), biosorbent (Zubair et al., 2022)	Costly processing methods, toxic effluent from chemical hydrolysis, kinetic constraints, and scale-up challenges
Meat	Blood, intestines, carcasses	Collagen, globin	Blood proteins for surgical use, bovine serum albumin for medical use, food processing agents, nutrition supplements (Hicks & Verbeek, 2016)	Limited use due to potential for prion contamination (Hicks & Verbeek, 2016)
Seafood	Bones, scales	Collagen, gelatin	Nutraceuticals, food processing agents, fish meal, hydrogels for tissue engineering, active food packaging, textiles (Rana et al., 2023)	Certain applications require proteins from endangered species (Rana et al., 2023), toxic effluents from pretreatments, and scale-up challenges (Gürdal & Çetinkaya, 2024)
Dairy	Whey, dairy wastewater	Globular proteins, albumin, casein, lactose	Biofuel, biopolymer (Chaudhary et al., 2023)	Casein is unstable at high temperatures (Chaudhary et al., 2023), composition variability of whey affects product quality and quantity (Liu et al., 2021)
Soy	Soy whey, soybean curd (okara), soy hull, soy meal	Soy protein, albumin, various storage proteins	Dietary supplements, food processing agents, biofuel, bioplastic, medical diagnostics, animal feed, biobased building materials (Karim et al., 2025)	Costly preprocessing steps, multiple (drying) pretreatments required, high waste output, toxic solvent requirements, allergen risk, many applications remain as proof-of-concept, scale-up challenges (Karim et al., 2025)
Food service, food retail, and solid municipal waste	Food scraps, kitchen waste	Mixed (highly variable depending on locale, season)	Biofuel (Peydayesh et al., 2023), volatile fatty acids, bioplastic (Karim et al., 2025)	Requires intensive upstream processing (sorting, pretreatments, component analysis), composition variability of feedstock affects product quality and quantity, processing design optimization challenges (Karim et al., 2025), suboptimal conversion (Zhou et al., 2023)

3.1.1 Current Poultry Feather Waste Treatment

The global poultry industry, which raises over 57 billion chickens annually, generates an estimated 13 million metric tons of feather waste (Cooper, 1970; Service UFA, 2025). This by-product, composed primarily of recalcitrant keratin, poses significant challenges for waste management due to its resistance to degradation. Conventional disposal methods, including incineration, landfilling, and chemical hydrolysis, are widely used but are energy-intensive, environmentally detrimental, and degrade essential amino acids while producing toxic by-products that hinder downstream processing (Peydayesh et al., 2023; Tursunova & Maksumova, 2024). Compounding the issue, feather waste is often intermixed with slaughterhouse residues, increasing biosafety risks through the potential spread of zoonotic pathogens and environmental contamination (Franke-Whittle & Insam, 2013). This convergence of environmental, economic, and health concerns highlights the urgent need for safer, more sustainable treatment alternatives.

Beyond waste management, feather disposal reflects broader sustainability concerns. Poultry production depends heavily on feed crops such as cereals and oilseeds, contributing to land use expansion, deforestation, biodiversity loss, and greenhouse gas emissions. In OECD countries alone, poultry farming utilizes approximately 93 million hectares of arable land, accounting for 44% of total cereal use for livestock, and these impacts are especially severe in ecologically sensitive regions such as Brazil and Argentina (Mottet & Tempio, 2017). Enhancing the resource efficiency and by-product recovery of poultry systems is thus essential for reducing their environmental footprint and advancing circular bioeconomy goals.

Despite these challenges, feather keratin has substantial biotechnological potential (Table 3.1). It serves as a raw material for producing a wide range of high-value products, including bioactive peptides (Yeo et al., 2018; Jin et al., 2018), cosmeceuticals (Lange et al., 2016), organic fertilizers (Gurav & Jadhav, 2013), animal feed additives (Chen et al., 2022), and biodegradable polymers (Wu et al., 2017). However, the lack of scalable and eco-friendly processing technologies remains a bottleneck. In this context, microbial keratinases present a promising solution. These enzymes are capable of cleaving disulfide bonds and hydrolyzing native keratin under mild conditions, enabling the recovery of amino acids and bioactive compounds without the drawbacks of conventional treatments (Bhagya Jyothi & Dhanasingh, 2025; Wu et al., 2017; Qiu et al., 2020; Kim et al., 2004). Their relevance is exemplified by keratin-degrading pathogens like dermatophytes, which utilize keratinases to colonize skin and nails (Keshwania et al., 2023).

Keratin degradation typically requires a multistep enzymatic cascade involving: (1) sulfitolysis to cleave disulfide bonds; (2) proteolysis to break peptide chains; and (3) deamination to release free amino acids. Because of this structural complexity, no single enzyme can achieve complete and efficient keratin hydrolysis (Wang et al., 2023). Effective degradation thus relies on the synergistic consortia comprising enzymes such as cysteine dioxygenases, thioredoxins, and glutathione reductases (Sharma et al., 2022; Kang et al., 2020a).

3.1.2 Thermophilic Bioconversion Strategies for a Circular Bioeconomy

The growing environmental impact of linear industrial systems has intensified interest in circular bioeconomy frameworks. This approach aims to minimize waste and maximize resource efficiency through the conversion of residual biomass into value-added products through renewable biotechnologies, thereby enabling sustainable growth within planetary boundaries (Kang et al., 2020a; Zeng & Li, 2021; Igbokwe et al., 2022). However, despite its conceptual appeal, the circular economy often faces criticism for lacking operational clarity and realistic scalability (Kirchherr et al., 2017). Idealized visions of zero-waste production frequently overlook thermodynamic constraints and the systemic inertia of existing infrastructure and consumer behavior (Igbokwe et al., 2022). As such, achieving full material circularity will require coordinated, large-scale transformation across policy, industry, and society. In the interim, deploying scalable, carbon-neutral technologies for sustainable waste valorization remains an urgent and achievable objective.

Among available strategies, microbial valorization of keratin-rich wastes represents a compelling solution. Thermophilic anaerobes such as *F. islandicum* AW-1 offer several practical and ecological advantages for integration into next-generation biorefinery systems. This extremophile can efficiently degrade poultry feathers without requiring chemical or mechanical pretreatment, operating under high-temperature, oxygen-free conditions (Yeo et al., 2018; Lee et al., 2015a). These features make it ideally suited for closed-loop systems where waste conversion must be rapid, safe, and environmentally benign. As shown in Fig. 3.1, poultry feathers rich in keratin and protein represent an abundant and underutilized feedstock for circular biomanufacturing.

Unlike conventional keratin extraction methods such as alkaline hydrolysis, steam explosion, microwave irradiation, and reduction chemistry, which often generate toxic byproducts and consume significant energy (Feroz et al., 2020), microbial degradation provides a sustainable, low-impact alternative. Our proposed framework adopts a production-centered model of circularity focused on thermophilic microbial bioprocessing. Specifically, we envision a closed-loop system in which feather waste is enzymatically transformed into bioactive peptides, agricultural biostimulants, and bioenergy carriers (Fig. 3.1). This targeted, scalable approach offers tangible progress toward circular bioeconomy goals without requiring large-scale sociopolitical overhauls (Vakulchuk et al., 2020).

Keratinolytic capacity is distributed across a wide range of bacteria, including *Bacillus*, *Streptomyces*, *Enterococcus*, and *Pseudomonas*, which harbor strains capable of attacking native keratin under diverse environmental conditions (Bhagya Jyothi & Dhanasingh, 2025; Bockle et al., 1995; Wang & Shih, 1999). While mesophilic species have historically dominated research and industrial applications, thermophilic microbes confer distinct benefits, particularly for anaerobic, high-temperature fermentation conditions (Zeldes et al., 2015; Jin et al., 2019). These benefits include: (1) reduced contamination risk—elevated process

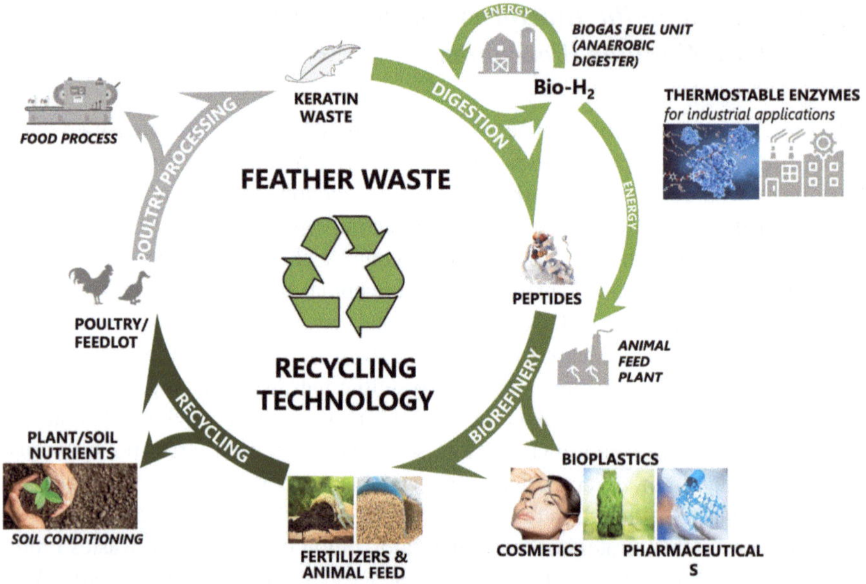

Fig. 3.1 Closed-loop biorefinery model enabled by *Fervidobacterium islandicum* AW-1. A closed-loop thermophilic bioprocess using *F. islandicum* AW-1 enables the valorization of poultry feather waste—a recalcitrant keratinous biomass—into multiple high-value outputs, including bioactive peptides, agricultural biostimulants, hydrogen gas, and precursors for biodegradable plastics. This extremophile-driven platform exemplifies a sustainable circular bioeconomy, linking waste-to-resource conversion with industrial biotechnology, cosmeceutical development, and regenerative agriculture

temperatures (≥60 °C) naturally suppress the growth of mesophilic pathogens and spoilage organisms; (2) faster reaction kinetics—thermostable enzymes exhibit catalytic activity, and elevated temperatures improve solubility of both substrates and hydrolysis products; (3) enhanced biosafety—high-temperature fermentation meets or exceeds thermal treatment guidelines for pathogen inactivation. For example, Sweden mandates pasteurization at 70 °C for 60 min for anaerobic digestion of slaughterhouse waste (Sahlström et al., 2008). Thermophilic systems inherently comply with these requirements while also offering better viral inactivation, and are shown to be effective even against heat-resistant viruses such as bovine enterovirus, parvovirus, rotavirus, and coxsackievirus B5 (Monteith et al., 1986; Spillmann et al., 1987).

Most importantly, *F. islandicum* AW-1 achieves near-complete degradation of feather waste within 48–72 h, significantly faster than the 5–6 days often required by mesophilic systems (Yeo et al., 2018; Lee et al., 2015a; Li et al., 2020). This reduction in process time improves throughput and feasibility for industrial deployment. Taken together, the keratinolytic efficiency, thermotolerance, and anaerobic metabolism of *F. islandicum* AW-1 make it a uniquely powerful chassis for thermophile-enabled circular bioeconomy platforms. Although industrial applications of this strain remain in early stages, its enzymatic capabilities and ecological robustness present promising opportunities for scale-up in sustainable biomanufacturing.

3.2 *Fervidobacterium islandicum* AW-1 and Keratinolytic Activity

3.2.1 Phenotypic and Genomic Features of F. islandicum AW-1

The extremophilic anaerobe *F. islandicum* AW-1 exhibits potent keratinolytic activity, capable of completely degrading native chicken feathers within 48 h at 70 °C in media supplemented with 0.8% (w/v) feather substrate (Yeo et al., 2018; Lee et al., 2015a). This fast-growing, non-sporulating, Gram-negative bacterium was originally isolated from a geothermal hot spring in Sileri Crater, Java, Indonesia (Nam et al., 2002). Phylogenetic analyses based on 16S rDNA sequences, cellular fatty acid profiles, and whole-genome sequencing place AW-1 within the order Thermotogales, genus *Fervidobacterium* (Lee et al., 2015a). As a representative of one of the most ancient bacterial lineages, Thermotogales offers insight into early microbial evolution (Bonch-Osmolovskaya, 2020). Consistent with this lineage, AW-1 exhibits the characteristic spheroid protrusion ("toga") structure, resulting in a distended periplasmic space (Petrus et al., 2012).

F. islandicum AW-1 is metabolically versatile, capable of utilizing diverse carbon sources such as arabinose, lactose, xylose, starch, and cellulose (Nam et al., 2002). Its growth follows a biphasic pattern: initially consuming soluble nutrients (e.g., yeast extract), then shifting to feather keratin as the primary substrate (Kang et al., 2020a). This transition is marked by increased release of free amino acids and sulfhydryl groups, likely reflecting disulfide bond reduction during keratin degradation. The proteases secreted by AW-1 exhibit remarkable thermostability and detergent resistance (Jin et al., 2017). A purified keratinase showed optimal activity at 100 °C and pH 9, retaining enzymatic function across 70–90 °C (Nam et al., 2002). Whole-cell extracts (WCEs) maintained substantial proteolytic activity for 48 h at 80 °C, outperforming commercial enzymes such as papain, trypsin, and proteinase K (Jin et al., 2017).

The complete genome of *F. islandicum* AW-1 revealed 2.36 Mbp, with ~91% coding DNA and 97% of annotated open reading frames (ORFs) encoding proteins (Lee et al., 2015a). 16S rRNA analysis shows closest similarity with *F. changbaicum* (99.3%), *F. pennivorans* (98.1%), *F. riparium* (96.1%), *F. nodosum* (95.4%), and *F. gondwanese* (94.7%) (Lee et al., 2015a; Javier-Lopez et al., 2024). Among these, *F. pennivorans*, *F. thailandense* FC2004CT, and *F. changbaicum* CBS-1 also exhibit keratinolytic properties (Friedrich & Antranikian, 1996; Kanoksilapatham et al., 2016; Cai et al., 2007). Comparative genomics with non-keratinolytic *F. nodosum* Rt17-B1 revealed that AW-1 harbors unique genes encoding carboxyl-terminal proteases, cysteine desulfurases, thioredoxin-disulfide reductases, and lipoprotein signal peptidases, many of which are associated with redox metabolism and keratin degradation (Lee et al., 2020).

Furthermore, *F. islandicum* AW-1 possesses exclusive genes involved in amino-sugar (chitin and N-acetylglucosamine) utilization, sugar alcohol (glycerol and glycerol-3-phosphate) metabolism, and alpha-linolenic acid degradation. Several key enzymes, including disulfide reductase, thioredoxin, and carboxypeptidases, exhibited less than 50% sequence identity compared to homologs in *F. nodosum*, suggesting a unique evolutionary trajectory underpinning its keratinolytic capability (Lee et al., 2015a).

3.2.2 Structural and Biochemical Characterization of Putative Keratinases

Annotation of the *F. islandicum* AW-1 genome using MEROPS and Pfam databases identified 57 protease-encoding genes, including 30 metalloproteases, 19 serine proteases, 2 cysteine proteases, and 2 threonine proteases. Over 50 of these were differentially expressed in response to feather keratin (Kang et al., 2020a). Reverse transcription PCR (RT-PCR) profiling under varying nutrient conditions revealed 21 proteases expressed under nitrogen supplementation. Among these, 16 genes were upregulated ($\geq$1.5 FC) after yeast extract depletion in feather-grown cultures. Quantitative real-time PCR (qRT-PCR) highlighted six keratinase candidates, including four metalloproteases and two peptidases, whose expression peaked under nutrient-limited conditions (NA23_RS01070, RS01075, RS08100, RS08750, RS09910, RS10455). Four of these (RS08100, RS08750, RS09910, and RS10455) were cloned, expressed in *Escherichia coli*, and shown to act synergistically with AW-1 WCE, supporting a multienzyme mechanism involving membrane-associated metalloproteases. Other keratinase-related genes, including RS02110, RS0455, RS05565, RS08155, and RS10350, were highly expressed before nutrient depletion, while five others (RS03580, RS05775, RS06425, RS09535, and RS09765) were constitutively expressed, suggesting a role in general nitrogen metabolism.

Among the keratinase-related proteases of *F. islandicum* AW-1, several have been structurally and functionally characterized, revealing distinct yet complementary features for efficient keratin degradation. A unique M32 family carboxypeptidase (FisCP; NA23_RS04280) was identified as a key component of the AW-1 keratinolytic system (Lee et al., 2015b). This enzyme, 107-kDa metalloprotease, is absent in closely related *Fervidobacterium* or *Thermotoga* species, suggesting a lineage-specific adaptation. FisCP exhibited optimal activity at 80 °C and pH 7 in the presence of Mn^{2+} or Co^{2+}, with complete inactivation by EDTA, confirming its dependence on divalent metal ions. Although it shares less than 38% sequence identity with known M32 homologs, FisCP retains the conserved HEXXH zinc-binding motif and features unique catalytic residues (His253, Glu254, His257, and Glu283).

It preferentially cleaves C-terminal basic residues and lacks a typical proline-rich loop, a structural feature that enhances substrate accessibility and broadens its catalytic repertoire (Lee et al., 2015b).

Another key enzyme, a β-aspartyl peptidase (FiBAP, NA23_RS08100), an M38 metalloprotease, was strongly upregulated under feather-based culture following yeast extract depletion (La et al., 2020). This enzyme plays a dual role in keratin hydrolysis and stress-responsive protein quality control. Recombinant FiBAP expressed in *E. coli* effectively reduced protein aggregation during heat shock, implicating its function in protein repair. Phylogenetic analysis classifies FiBAP within the IadA subfamily (Type I isoaspartyl dipeptidase), targeting isoaspartyl (isoAsp) residues, which are damage-associated motifs linked to protein aging and misfolding. FiBAP exhibits exceptional thermostability (T_m = 96.8 °C), optimal activity at 80 °C and pH 7, and requires Zn^{2+} as a cofactor. Its crystal structure revealed conserved inter-subunit salt bridges and a binuclear zinc catalytic center. LC-MS/MS of keratin hydrolysates confirmed FiBAP's preference for cleaving at isoAsp C-termini, though it also showed moderate activity toward α-peptide bonds. In contrast to the canonical protein isoAsp methyltransferase (PIMT)-mediated isoAsp repair pathway, FiBAP directly hydrolyzes isoAsp peptides, contributing to proteostasis and nutrient recycling under thermophilic, nutrient-limited conditions (La et al., 2020; Chen et al., 2011).

In addition, FiPcp, encoded by NA23_RS05565, is a pyrrolidone carboxypeptidase that targets N-terminal pyroglutamate (pGlu)-modified peptides, which are typically resistant to conventional proteases (Dhanasingh et al., 2021). Its catalytic triad (Cys156, His180, Glu93) facilitates acid-base hydrolysis of pGlu linkages. Notably, catalytic cysteine is susceptible to irreversible oxidation to cysteine sulfonic acid ($C156\text{-}SO_3H$), which inactivates the enzyme under oxidative stress. Structural analysis revealed a monomeric α/β fold with eight β-sheets (β1–β8) and flanking α-helices, accompanied by three functional loops: two for pGlu binding and inter-monomer contact, and one contributing to hydrophobic core formation. A distinctive lysine motif (^{139}KKK^{142}) mediates strong electrostatic inter-subunit interactions, stabilizing the homo-tetrameric structure. Mutagenesis and differential scanning calorimetry assays confirmed that this motif enhances thermostability. This was the first structurally characterized Pcp to demonstrate redox-sensitive inactivation, highlighting the role of intracellular redox balance during keratin degradation in stress-prone environments.

3.2.3 Development of a Recombinant β-Keratin Activity Assay

Conventional keratinase assays typically rely on heterogeneous and insoluble substrates such as azo-keratin (Lin et al., 1992), keratin azure (Letourneau et al., 1998), or ball-milled feather powder (Williams & Shih, 1989). These substrates suffer from critical limitations, including batch-to-batch variability, insolubility, and undefined amino acid sequences. Such properties hinder reproducibility, obscure cleavage site specificity, and prevent accurate quantification of keratinolytic activity. To overcome these limitations, we developed a synthetic assay system utilizing soluble, recombinant β-keratin as a defined substrate. This platform enables sensitive and quantitative evaluation of keratinolytic activity, substrate specificity, and cleavage patterns at high resolution. The recombinant substrate was constructed through a synthetic biology approach informed by genomic data and proteomics (Jin et al., 2017). Specifically, β-keratin genes from chromosomes 2 and 27 of *Gallus gallus* (red junglefowl), representing canonical feather keratin isoforms (Ng et al., 2014), were codon-optimized for *E. coli* expression, chemically synthesized, and cloned into suitable expression vectors.

Recombinant β-keratins were expressed in *E. coli* as inclusion bodies, solubilized under denaturing conditions, purified by Ni^{2+}-affinity chromatography, and refolded through stepwise dialysis. The resulting soluble β-keratin substrate was homogeneous, soluble, and reproducible, in stark contrast to traditional keratin substrates. Functional validation showed that crude whole-cell extracts (WCE) of *F. islandicum* AW-1 degraded the recombinant β-keratin in a concentration-dependent manner. In contrast, non-keratinolytic proteases such as trypsin (serine protease) and papain (cysteine protease) exhibited negligible activity, confirming the specificity of the assay system for keratinases. Due to the known and defined amino acid sequence of the substrate, this system enabled precise mapping of cleavage sites, including identification of P1 and P1' substrate recognition motifs. To enhance assay sensitivity, the recombinant β-keratin was fluorescently labeled with fluorescein-5-maleimide (FM), which covalently binds to accessible thiol groups. Proteolytic digestion of the FM-labeled substrate generated fluorescent peptide fragments, which allowed for real-time tracking of the keratinolytic process.

This recombinant keratin assay represents a significant advancement in keratinase characterization, facilitating high-throughput screening, accurate quantification, and detailed specificity profiling of candidate enzymes. Furthermore, the system is adaptable for engineering applications, inhibitor screening, and optimization of keratin bioconversion processes. By addressing the long-standing limitations of insoluble keratin substrates, this platform provides a robust foundation for future studies in microbial keratinolysis and enzyme discovery.

3.3 Keratin Degradation Mechanisms

3.3.1 Keratinolytic Enzyme Complex in F. islandicum AW-1

The keratinolytic machinery of *F. islandicum* AW-1 appears to function as a dynamic multienzyme system, potentially forming a transient or stable supercomplex that includes proteases, peptidases, redox regulators, and sulfur metabolism-related enzymes (Fig. 3.2). Integrated comparative genomics and biochemical assays indicate that this system relies heavily on a thioredoxin system and a sulfur mobilization

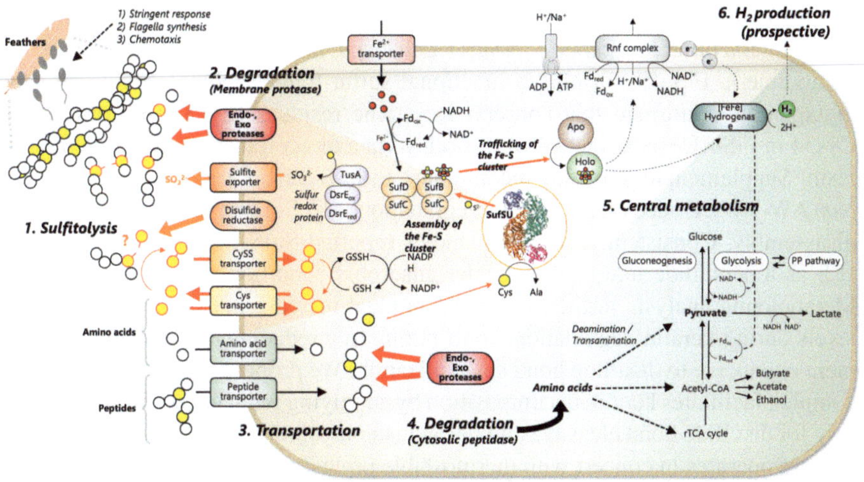

Fig. 3.2 Proposed keratin degradation mechanism and metabolic integration in *F. islandicum* AW-1. Under nutrient-limited conditions, *F. islandicum* AW-1 attaches to feather surfaces, initiating keratin degradation through membrane-associated and extracellular proteases that release peptides and amino acids. Disulfide bond cleavage is facilitated by disulfide reductases, sulfite-mediated sulfitolysis, and cysteine–cystine recycling. Redox buffering via glutathione and cysteine export maintains intracellular homeostasis. The liberated amino acids are catabolized into pyruvate and acetyl-CoA, fueling fermentative metabolism and partial rTCA pathways. Concurrently, sulfur and iron metabolism support [FeFe] hydrogenase maturation via Suf-dependent Fe–S cluster biogenesis. Together with the Rnf complex and ferredoxin, this machinery couples keratin degradation to hydrogen production and energy conservation under thermophilic anaerobic conditions

(Suf) complex, both of which likely initiate the reduction of disulfide bonds in keratin, facilitate cysteine-cystine recycling, and maintain intracellular redox homeostasis (Dhanasingh & Lee, 2019; Kang et al., 2020a, b).

Although Suf systems are traditionally known for their roles in Fe-S cluster biogenesis and stress responses, their involvement in keratin degradation, particularly in sulfur transfer and redox regulation, had not been previously established. *F. islandicum* AW-1 harbors a complete *suf* operon (*sufCBDSU*), in contrast to closely related *Thermotogaceae* species such as *F. nodosum* Rt17-B1 (Patel et al., 1985), which lack key components such as cysteine desulfurase (SufS) and Fe-S cluster assembly scaffold protein (SufU) (Jin et al., 2021). When grown on feathers in the presence of elemental sulfur (S^0), *F. islandicum* AW-1 exhibited significant upregulation of the suf operon and associated sulfur assimilation genes—including peptide/cysteine ABC transporters, sulfur formation proteins, the redox-regulated chaperone Hsp33, 2-oxoacid: ferredoxin oxidoreductase (OFOR), TauE/SafE-family sulfite exporter, and the stress-response chaperones DnaK and DnaJ. Interestingly, cells grown on feathers without S^0 supplementation demonstrated faster keratin degradation and enhanced growth, suggesting that environmental cues regulate the mobilization of the Suf system as part of a stress-adaptive metabolic response.

Functional characterization of recombinant FiSufS and FiSufU revealed a synergistic interaction. Monomeric FiSufU enhanced the desulfurase activity of the homodimeric FiSufS, forming a functional sulfur-transfer complex. Site-directed mutagenesis confirmed that conserved cysteine residues (Cys372 in FiSufS and Cys34 in FiSufU) participate in sulfur relay via a direct transpersulfuration mechanism. Supplementation of this recombinant FiSufS-SufU complex into *F. islandicum* AW-1 WCE accelerated feather degradation, highlighting its key role in keratin disassembly. Consistently, previous studies reported SufS-mediated generation of H_2S as a persulfide intermediate under anaerobic conditions (Wang et al., 2019). Metabolomic analysis further revealed a gradual increase in extracellular cysteine levels during keratin degradation, contributing to a reducing extracellular environment conducive to disulfide bond destabilization. We propose that the FiSufS-SufU complex facilitates keratin decomposition by supplying reducing equivalents necessary for disulfide bond cleavage and Fe-S cluster assembly. This sulfur mobilization process operates in concert with thermostable proteases to achieve efficient keratin disassembly.

In parallel, *F. islandicum* AW-1 harbors a functional thioredoxin (Trx) system, which may act complementarily in thiol-based redox regulation (Anjou et al., 2024) (Fig. 3.2). The canonical bacterial Trx system comprises thioredoxin (FiTrx) and thioredoxin reductase (FiTrxR), both of which were cloned, heterologously expressed, and structurally characterized from *F. islandicum* AW-1 (Dhanasingh & Lee, 2019). Structural attempts to solve FiTrx via molecular replacement using known bacterial Trx homologs (e.g., Trx-like protein from *Populus tremula* and Trx from *E. coli*) were unsuccessful, suggesting that FiTrx may possess a unique tertiary structure distinct from classical Trx folds. Nonetheless, functional assays indicated that the FiTrx system catalyzes thiol-disulfide exchange reactions, contributing to the reduction of oxidized thiol groups and initial destabilization of keratin matrices by facilitating disulfide bond cleavage (Katti et al., 1990).

Taken together, keratin degradation by *F. islandicum* AW-1 is mediated by a highly coordinated process network of redox-active and proteolytic enzymes. Redox-initiated sulfitolysis, driven by Suf and Trx systems, synergizes with thermostable keratinases to overcome the recalcitrance of β-keratin. This unique enzymatic strategy not only reflects evolutionary adaptation to nutrient-limited and high-temperature environments but also serves as a model for engineering thermophilic biocatalysts for keratin valorization (Jin et al., 2021).

3.3.2 Multi-omics Insights into Mechanistic Reprogramming

Multi-omics analyses have revealed that keratin degradation in *F. islandicum* AW-1 is a starvation-induced survival strategy governed by nutrient-dependent activation of keratinolytic machinery. Scanning electron microscopy (SEM) images captured *F. islandicum* AW-1 in direct physical contact with feather surfaces during degradation. Supporting the redox-dependence of the process, addition of dithiothreitol

(DTT), a reducing agent targeting disulfide bonds, enhanced feather degradation by *F. islandicum* AW-1 proteolytic extracts (Nam et al., 2002). Early subcellular fractionation localized keratinolytic activity to the solubilized membrane fraction, not the extracellular medium, indicating that cell-associated proteases are responsible. In two-chamber diffusion experiments, where the cell culture was partitioned by a semipermeable membrane, only feathers in direct contact with *F. islandicum* AW-1 cells underwent degradation, further supporting the requirement for physical proximity and membrane-localized enzymatic activity (Sung et al., 2025).

Integration of transcriptomic, proteomic, and metabolomic data revealed that *F. islandicum* AW-1 undergoes extensive physiological reprogramming to initiate keratin degradation under nutrient-limiting conditions (Sung et al., 2025). A hallmark of this reprogramming is the robust activation of redox-regulatory systems. Genes encoding thioredoxins, disulfide isomerases, and glutathione-dependent reductases were upregulated, indicating that sulfitolysis-disulfide bond cleavage via thiol exchange is a critical initial step in keratin destabilization (Fig. 3.2). Concurrently, genes involved in chemotaxis and motility, including *cheR* and *fliA*, were also induced, along with those involved in membrane remodeling, supporting a model wherein substrate sensing and surface colonization precede proteolytic activity.

This response is further characterized by the activation of a stringent stress adaptation program (Sung et al., 2025). Transcriptional regulators such as TetR and MarR, known to mediate starvation responses, were upregulated and likely coordinate the expression of keratinase gene clusters and redox-modulating enzymes. Metabolomic profiling indicated a shift toward metabolic pathways compatible with oligotrophic adaptation, including increased flux through the Entner-Doudoroff (ED) pathway and the reverse tricarboxylic acid (rTCA) cycle. These pathways likely maximize ATP yield and minimize carbon loss under nutrient-scarce conditions.

Moreover, dynamic changes in intracellular cyclic-di-GMP (c-di-GMP) levels suggest an additional regulatory layer. Oscillations in c-di-GMP, a secondary messenger that governs biofilm formation and sessility, imply coordinated regulation of stress responses, cell adherence, and proteolytic activity. This was supported by co-expression of genes encoding peptide transporters, carboxypeptidases, and methyl-accepting chemotaxis proteins (MCPs), suggesting a tightly integrated network coupling environmental sensing, redox regulation, and keratinolytic activation.

Collectively, keratin degradation in *F. islandicum* AW-1 is orchestrated by a membrane-associated enzyme system that integrates starvation-induced transcriptional reprogramming, chemotactic activation, redox-mediated sulfitolysis, and metabolic remodeling. This system exemplifies a thermophilic biodegradation strategy optimized for recalcitrant proteinaceous substrates and offers a robust model for microbial valorization of keratinous waste under extreme environmental conditions.

3.4 Anaerobic Fermentation Process Development

Bioprocessing plays a central role in a sustainable bioeconomy by replacing fossil-based production systems with renewable, biologically driven processes (Campbell et al., 2017). It involves the use of biological materials either as substrates or catalysts to produce value-added compounds for industries including bioenergy, biomaterials, pharmaceuticals, and agriculture (Lee et al., 2022). Unlike traditional chemical processes, bioprocessing adheres to green chemistry principles by minimizing solvent usage, reducing toxic effluents, and relying on renewable feedstocks (Liu, 2017). Currently, the valorization of lignocellulose biomass is a major focus, where plant-derived carbon polymers are hydrolyzed and funneled into microbial fermentation pathways to generate target products (Sulis et al., 2025).

Despite these advances, several bottlenecks limit bioprocessing scalability and economic competitiveness (Campbell et al., 2017). The high cost of developing microbial cell factories, limited cell chassis availability, inefficient metabolic rerouting, and susceptibility to contamination are key barriers. Rerouting metabolic fluxes can reduce conversion efficiency, while contamination during long fermentation runs often causes costly shutdowns. Moreover, raw substrate pretreatment and sterility requirements further inflate costs. These challenges have prompted interest in Next-Generation Industrial Biotechnology (NGIB) systems, which operate under unsterile (open) conditions with minimal energy and water input (Chen & Jiang, 2018). Contamination-resistant extremophiles are attractive for NGIB based on the fact that thermophiles exhibit rapid growth on a variety of unusual substrates at elevated temperatures and resistance to high concentrations of detergents and toxic compounds (Jin et al., 2017). Their elevated growth temperatures inherently suppress most mesophilic contaminants and accelerate reaction kinetics, reducing cooling costs and enabling continuous fermentation (Zeldes et al., 2015). Thermophiles also possess unique metabolisms capable of processing unconventional substrates, favoring native pathway augmentation over de novo pathway construction. Moreover, many industrially relevant products, such as hydrogen gas, are favored at higher fermentation temperatures (Verhaart et al., 2010).

Our proposed platform uses *F. islandicum* AW-1 in a keratinolytic anaerobic fermentation system to enable zero-waste valorization of feather biomass. This approach promotes material circularity in poultry farming by transforming keratin-rich waste into multiple high-value products, including bioactive low-molecular-weight peptides suitable for cosmeceutical, pharmaceutical, or biomaterial applications. Keratin hydrolysates derived from this process also show promise as biostimulants for enhanced soil fertility, as well as for high-nutrient feed supplements, potentially reducing reliance on soybean meal and enhancing food security by minimizing feed-food competition (Salehizadeh et al., 2025; Kaur et al., 2021). Functional genomic annotation revealed the presence of [FeFe] hydrogenases, indicating that the production of hydrogen gas during anaerobic fermentation can also be reclaimed during these processes (Chandrasekhar et al., 2015).

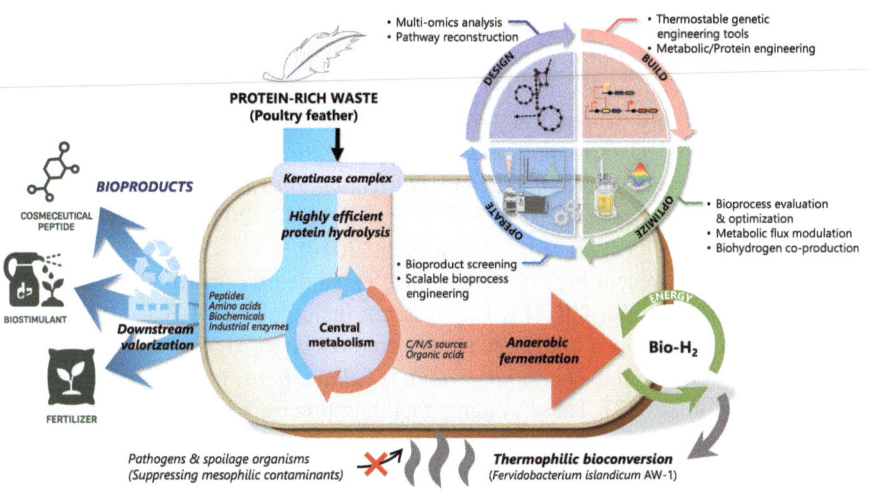

Fig. 3.3 Thermophilic bioprocess framework for a zero-waste feather valorization. Thermophilic enzymatic hydrolysis of feather keratin produces soluble peptides and amino acids, which enter central metabolic pathways to support energy generation and biosynthesis. These intermediates are funneled into anaerobic fermentation to yield organic acids, C/N/S sources, and molecular hydrogen (bio-H_2), alongside high-value coproducts including bioactive peptides, fertilizers, biochemicals, and industrial enzymes. Thermophilic conditions inherently suppress mesophilic contaminants, enhancing process efficiency and biosafety. Bio-H_2 production may offset process energy demands, contributing to self-sustaining operation. Coupling this system with genetic engineering and iterative design–build–optimize–operate (DBOO) cycles facilitates targeted bioproduct discovery and scalable bioreactor design, establishing *F. islandicum* AW-1 as a robust microbial platform for integrated thermophilic biorefinery systems

3.4.1 Prospects for the Genetic Engineering of Thermophiles

Despite the natural efficacy of *F. islandicum* AW-1's keratinolytic system, several barriers limit its direct industrial deployment. Even highly active thermophilic keratinases exhibit rate-limiting steps in large-scale feather degradation processes, and the requirement for nutrient-rich, expensive media restricts economic scalability (Bhagya Jyothi & Dhanasingh, 2025). To overcome these limitations, genetic engineering strategies are essential to improve enzyme kinetics, redirect metabolic fluxes, and enable cost-effective bioconversion of agro-industrial waste (Fig. 3.3). However, genetic engineering of thermophiles presents significant technical challenges as molecular tools compatible with high-temperature conditions remain limited. Conventional antibiotic resistance markers (e.g., chloramphenicol, ampicillin, tetracycline) are thermolabile, losing efficacy at the elevated temperatures required for thermophile cultivation (Brouns et al., 2005). Transformation protocols routinely used in mesophilic systems, such as conjugation, chemical transformation, or heat-shock induction, are less effective in thermophiles due to their unique membrane compositions and extreme growth conditions (Le et al., 2020; Grybchuk-Ieremenko et al., 2025). Furthermore, limited knowledge of thermophile-specific gene regulation, promoter architecture, and interspecies compatibility hampers the molecular design of synthetic biology tools. As a result, transformation efficiencies and heterologous expression in thermophiles remain low. To date, only a handful of thermophilic genera, including *Thermus*, *Geobacillus*, *Thermoanaerobacter*, *Thermococcus*, *Pyrococcus*, *Caldicellulosiruptor*, *Thermotoga*, and *Parageobacillus*, have been successfully engineered (Zeldes et al., 2015; Adalsteinsson et al., 2021; Kananavičiūtė & Čitavičius, 2015; Cha et al., 2013; Shaw et al., 2010).

Encouragingly, recent breakthroughs have expanded the thermophile engineering toolkit. Thermostable CRISPR-Cas systems, including type I-B and type II variants, have been developed to operate at temperatures up to 70 °C in *Parageobacillus thermoglucosidasius* and *Thermoanaerobacter ethanolicus*, respectively (Le et al., 2020; Yang et al., 2023). These systems facilitate precise and multiplexed genome editing, bypassing the limitations of traditional homologous recombination. Nutrient-based selection strategies have also proven useful in extreme thermophiles. One widely adopted method leverages uracil auxotrophy, where deletion of the *pyrF* gene renders cells dependent on exogenous uracil: deletion of the *pyrF* gene renders cells dependent on exogenous uracil, allowing selection using uracil or the toxic analog 5-fluoroorotic acid (5-FOA) for positive and counter-selection, respectively (Tripathi et al., 2010).

Recent advancements in thermostable selection systems, improved transformation efficiency strategies, and inducible promoter design will be critical to unlocking the full potential of thermophilic keratinases for industrial biotechnology. Realizing this potential will require continued development of thermostable selection markers, improved transformation protocols tailored to extremophilic physiology, and inducible promoter systems responsive at high temperatures. Such innovations will be key to unlocking the full industrial potential of thermophilic keratin degradationKeratin degradation systems.

Complementing these tools, advances in genome annotation pipelines, metabolic modeling, and systems biology are accelerating rational strain design in thermophiles (Schroeder & Saha, 2020a, b). Applying such tools to *F. islandicum* AW-1 could enable the creation of engineered chassis strains with enhanced keratinolytic performance. These could serve robust platforms for converting feather biomass into high-value bioproducts, including biogas, organic acids, ethanol, and bioactive peptides within a circular bioeconomy framework (Fig. 3.3).

3.4.2 *Keratin-Derived Bioactive Peptides as Novel Cosmeceuticals*

Bioactive peptides (BPs) are short protein fragments released through proteolytic cleavage that exert regulatory effects on a multitude of physiological systems, such as the immune, endocrine, and nervous systems (Kang et al., 2020b; Sánchez &

Vázquez, 2017). In the context of keratin waste valorization, anaerobic digestion of chicken feathers by *F. islandicum* AW-1 has led to the discovery of low molecular weight (LMW, <1 kDa) keratin-derived peptides (KPs) with promising antiaging properties for cosmeceutical applications (Yeo et al., 2018). Soluble keratin hydrolysates generated from feather fermentation were fractionated by ultrafiltration and chromatographic methods, and then analyzed via SDS-PAGE, amino acid profiling, and mass spectrometry (LC-MS/MS), resulting in the identification of 12 unique peptides in the hydrolysate fractions (Jin et al., 2018). These peptides were subsequently synthesized for functional and mechanistic characterization.

As a major extrinsic factor in skin aging, UVB radiation promotes the degradation of dermal matrix proteins by inducing matrix metalloproteinases (MMPs), such as MMP-1 and MMP-13, in human dermal fibroblasts (HDFs) (Fineschi et al., 2007). While existing antiaging compounds target MMPs directly, most are synthetic, costly, and suffer from issues such as low yield, immunogenicity, and limited efficacy. In contrast, several LMW peptides derived from *F. islandicum* AW-1 fermentation (each 5–6 amino acids long) exhibited significant, dose-dependent suppression of UVB-induced MMP-1 and MMP-13 expression in HDF assays. Mechanistic analysis of SP12 (sequence: GGFDL), one of the most potent peptides, revealed broad modulation of UVB-responsive signaling pathways. Phosphorylation profiling showed that SP12 significantly suppressed phosphorylation of MKK4/SEK1, P90RSK, and p38 MAPK, which are all key regulators upstream of MMP-1 expression in the MAPK cascade. Additionally, SP12 reduced phosphorylation of histone H3 (Ser10) and IKK-β, suggesting epigenetic suppression of MMP gene transcription through inhibition of both MAPK and NF-κB signaling. Notably, SP12 did not directly inhibit MMP catalytic activity, elastase, or collagenase, nor did it exhibit strong radical scavenging activity. Instead, its antiaging effect appears to arise from transcriptional repression of MMP genes via modulation of intracellular kinase signaling and chromatin modification (Jin et al., 2018). This represents a novel mechanism of action distinct from most traditional cosmeceuticals. The production of functional KPs from poultry waste offers a sustainable, zero-waste approach to generating high-value cosmetic ingredients. As a biodegradable and bioactive alternative to synthetic peptides, *F. islandicum* AW-1-derived KPs demonstrate commercial potential for the development of safe, effective, and eco-friendly cosmeceutical products.

3.4.3 Optimization of Anaerobic Digestion of Feathers

Challenges in scaling anaerobic fermentation and accurately quantifying active peptides in complex keratin hydrolysates have limited the development and commercialization of the use of feather-derived LMW KPs in antiaging cosmetics. We successfully scaled the anaerobic digestion of solid agro-waste to produce bioactive LMW KPs in a 5 L batch bioreactor, achieving a concentration of 0.8 g/L purified KPs, corresponding to a yield of 1.2% within 48 h (Jin et al., 2019). A fluorescence-based assay was developed, targeting N-terminal proline-containing sequences (PIXS motifs), using 3,4-dihydroxybenzoic acid derivatization to form fluorescent adducts derived from LMW KPs for selective quantification. The assay had high sensitivity, correlated well with LC-MS/MS data, and was compatible with common cosmetic formulations, thus verifying its potential for product quality control. These findings demonstrate the viability of leveraging extremophilic anaerobes at an industrial scale for converting feather waste into high-value compounds. The integration of a sensitive fluorescence-based assay with scalable bioprocessing and purification methods establishes a viable platform for sustainable peptide production with clear industrial relevance for pharmaceutical and cosmetic applications, while also serving as a blueprint for bioprocessing design frameworks for the valorization of other keratin-derived products across other industries.

3.4.4 Hydrogen Co-production

While debate continues over the future of renewable energy–particularly the viability of implementing a full transition to hydrogen gas–it remains essential to explore biotechnology-based solutions to emerging environmental challenges that will inevitably carry significant social consequences (Chandrasekhar et al., 2015). In considering the shift toward a circular bioeconomy-based future, it is crucial to recognize that no silver bullet or bioprocess within a sector can serve as a panacea for today's environmental challenges. For example, while shifting to an entirely hydrogen energy-based economic model would solve many of the issues encountered by the burning of fossil fuels in theory, recent studies have suggested that current technologies, infrastructure, and social participation are still insufficient to usher in this reality (Ehrhardt & Rothenberg, 2025). Despite this, it is integral in the upcoming bioeconomy to develop a range of viable bioprocessing strategies that can be used in tandem to tackle the environmental issues that exist today in the manufacturing sphere.

A key feature of *F. islandicum* AW-1-based bioprocess is the concurrent generation of hydrogen gas, offering a renewable energy output alongside keratin degradation. Members of the order *Thermotogales* are known for their robust hydrogen-producing capacity across diverse substrates (Van Ooteghem et al., 2002; Boileau et al., 2016). Genomic analysis of *F. islandicum* AW-1 reveals the presence of genes encoding group A [FeFe] hydrogenases, which are the most catalytically active and extensively studied among the four [FeFe] hydrogenase clades (A–D) (Land et al., 2020). In addition, the genome encodes radical S-adenosylmethionine (SAM) maturases HydE, HydF, and HydG, which are essential for assembling the enzyme's catalytic H-cluster (Britt et al., 2020).

We are currently investigating hydrogen yields during feather fermentation by *F. islandicum* AW-1. Traditionally, dark fermentation on glucose yields a theoretical maximum of 4 moles H_2 per mole of glucose, referred to as the Thauer limit (Thauer et al., 1977), based on EMP pathway stoichiometry. However, recent advances in intracellular redox management, metabolic engineering, and synthetic consortia design have demonstrated that this limit can be exceeded under optimized conditions (Kumaraswamy et al., 2019; Ergal et al., 2020). Related thermophilic strains have already exhibited high hydrogen productivity, reinforcing the promise of thermophiles as chassis for biohydrogen production (Basile et al., 2012; Sivaramakrishnan et al., 2021). Harnessing *F. islandicum* AW-1's innate hydrogenogenic potential could thus enable integrated waste valorization and energy recovery in circular bioprocessing systems.

3.5 Challenges and Future Directions

3.5.1 Technical and Economic Barriers in Scaling the Thermophilic Bioprocess

Despite the promising potential for extremophile-based biorefinery systems, several formidable challenges must be addressed before their full industrial realization. Three primary bottlenecks hinder progress: (i) scaling from laboratory to industrial-scale fermenters, (ii) rational strain engineering for optimized bioconversion, and (iii) overcoming regulatory and economic hurdles for commercialization (Hollinshead et al., 2014; Mahendrasinh Kosamia et al., 2022).

Scaling up biochemical processes from bench to industrial-scale bioreactors is intrinsically challenging. Microbial metabolisms are highly sensitive to heterogeneous environments that often arise in large fermenters. Issues such as mass transfer limitations, heterogeneity in feedstock composition, and spatial gradients in nutrients or oxygen result in microenvironmental fluctuations within the reactor (Lara et al., 2006). These nonuniform conditions can lead to physiological heterogeneity, metabolic instability, and ultimately reduced product yields. Additionally, microorganisms inherently allocate resources toward growth and survival rather than product biosynthesis. Target metabolites are frequently growth-inhibitory or energetically burdensome, leading to low substrate-to-product conversion efficiencies. To address these issues, various metabolic engineering strategies have been deployed, including overexpression of heterologous modules and modulation of redox or cofactor balances (Stephanopoulos, 2012). Nonetheless, synthetic rewiring of complex metabolic networks without compromising cell viability remains a central challenge in strain development.

Commercial viability represents another critical barrier. For industrial adoption, a bioprocess must be not only technically effective but also economically competitive and compliant with local regulatory framework (Soni & Soni, 2025). The high cost of upstream inputs, including high-purity substrates, enzymatic pretreatments, and strict sterility requirements, undermines the economic feasibility of many microbial systems. In response, the field is increasingly moving toward NGIB, which emphasizes the use of robust microbial hosts capable of operating under non-sterile, open conditions with inexpensive feedstocks and simplified downstream processing (Chen & Jiang, 2018).

To objectively evaluate commercialization potential, Techno-Economic Analysis (TEA) and Life Cycle Assessment (LCA) provide powerful frameworks. TEA integrates technical, economic, and risk-related parameters to estimate capital investments, operating costs, and potential revenue streams, aiding rational process design. LCA, on the other hand, quantifies the environmental footprint of the entire production pipeline, including greenhouse gas emissions, water use, and energy demand (Mahendrasinh Kosamia et al., 2022; Ganguly et al., 2025). However, limited policy incentives and incomplete datasets often hinder robust TEA/LCA studies for emerging bioprocesses. Furthermore, region-specific economic policies, such as renewable energy subsidies or waste valorization credits, play a decisive role in determining feasibility and return on investment.

3.5.2 Regulatory and Environmental Considerations

Transitioning away from fossil-fuel-based infrastructure is not only an environmental imperative but also a strategic geopolitical and economic necessity (Lee et al., 2022). Bioenergy technologies, such as hydrogen co-production via anaerobic thermophilic fermentation, present a decentralized and renewable alternative to consolidated energy systems dominated by petro-states (Chandrasekhar et al., 2015). For nations lacking natural fossil reserves, this decentralization could promote energy sovereignty, reduce vulnerability to global market fluctuations, and increase resilience in the face of supply disruptions (Vakulchuk et al., 2020).

By minimizing reliance on finite carbon-intensive resources, thermophilic bioprocessing contributes significantly to reducing the environmental burdens of both energy production and organic waste accumulation (Cucina, 2023). This includes lowering greenhouse gas emissions, reducing toxic effluent generation, and diverting agricultural residues such as feathers from landfilling or incineration. Nevertheless, large-scale deployment of such technologies will require substantial initial capital investment in infrastructure, including thermophilic fermenters, gas separation and capture systems (e.g., for biohydrogen), and regulatory-compliant biosafety containment units.

While the upfront costs are nontrivial, they may be mitigated over time through long-term gains in domestic energy security, reduced import dependence, and decreased environmental remediation expenditures. Moreover, adoption of these technologies may generate economic opportunities through job creation in local bio-manufacturing sectors and the development of value-added product pipelines (e.g., bioactive peptides, organic acids, and biodegradable polymers).

In a broader geopolitical context, the rise of decentralized, microbe-driven energy and materials production could contribute to democratizing access to energy and reducing the concentration of power historically associated with fossil fuel control (Vakulchuk et al., 2020). Although these implications remain speculative, the envisioned societal benefits in environmental, economic, and diplomatic contexts stress the urgency of accelerating the development and deployment of extremophile-based circular bioeconomy platforms.

3.6 Conclusion

The transition toward bio-based manufacturing and waste valorization represents a critical strategy for mitigating the environmental, economic, and societal burdens imposed by fossil fuel dependence. By embracing the principles of material circularity and next-generation bioprocessing, extremophilic microbial platforms, exemplified by *F. islandicum* AW-1, offer transformative potential for building sustainable, closed-loop industrial systems (Fig. 3.1).

F. islandicum AW-1, a hyperthermophilic anaerobe, exhibits exceptional keratinolytic activity, positioning it as a promising chassis organism for thermophilic biorefineries targeting recalcitrant proteinaceous wastes such as poultry feathers. Its application to feather recycling exemplifies a circular bioeconomy model, wherein agricultural by-products are converted into high-value outputs, including bioactive peptides, biohydrogen, and platform chemicals such as lactic acid. This multi-output valorization strategy not only addresses pressing waste management challenges but also delivers secondary benefits in the domains of energy generation, functional feed additives, and green chemistry.

However, to fully realize this potential, several technical and systemic barriers must be overcome. Continued progress is required in strain engineering, metabolic optimization, and thermophilic reactor design, particularly for enhancing process efficiency and scalability. Moreover, comprehensive TEA, LCA, and compliance with regulatory frameworks will be essential to ensure environmental safety, economic feasibility, and societal acceptance of these biotechnologies.

Ultimately, the development of extremophile-driven bioprocesses marks a compelling frontier in sustainable industrial biotechnology, aligning with global goals in climate change mitigation, energy transition, and resource-efficient agriculture. As tools for synthetic biology, omics integration, and bioprocess engineering continue to evolve, *F. islandicum* AW-1 and related extremophiles may pave the way for robust and decentralized solutions to the challenges of a post-carbon economy.

Author Contributions N.E.K., H.S.J., and D.W.L. formulated the research plan, analyzed and interpreted the data, and drafted the manuscript. N.E.K., H.S.J., and D.W.L. participated in the design of the study and analyzed and interpreted the data. D.W.L. conceived, planned, and supervised the study.

Competing Interests The authors declare no competing financial interests.

Funding Information This work was supported by the Bio & Medical Technology Development Program of the National Research Foundation (NRF) of Korea grant (RS-2023-NR076895 to DWL), funded by the Ministry of Science and ICT (MSIT), Republic of Korea.

References

Adalsteinsson, B. T., Kristjansdottir, T., Merre, W., Helleux, A., Dusaucy, J., Tourigny, M., et al. (2021). Efficient genome editing of an extreme thermophile, *Thermus thermophilus*, using a thermostable Cas9 variant. *Scientific Reports, 11*(1), 9586. https://doi.org/10.1038/s41598-021-89029-2

Anjou, C., Lotoux, A., Morvan, C., & Martin-Verstraete, I. (2024). From ubiquity to specificity: The diverse functions of bacterial thioredoxin systems. *Environmental Microbiology, 26*(6), e16668. https://doi.org/10.1111/1462-2920.16668

Basile, M. A., Carfagna, C., Cerruti, P., Gomez d'Ayala, G., Fontana, A., Gambacorta, A., et al. (2012). Continuous hydrogen production by immobilized cultures of *Thermotoga neapolitana* on an acrylic hydrogel with pH-buffering properties. *RSC Advances, 2*(9), 3611–3614. https://doi.org/10.1039/C2RA01025A

Bhagya Jyothi, J. L., & Dhanasingh, I. (2025). An update on thermostable keratinases for protein engineering against feather pollutants. *Applied Microbiology and Biotechnology, 109*(1), 75. https://doi.org/10.1007/s00253-025-13459-5

Bockle, B., Galunsky, B., & Muller, R. (1995). Characterization of a keratinolytic serine proteinase from *Streptomyces pactum* DSM 40530. *Applied and Environmental Microbiology, 61*(10), 3705–3710. https://doi.org/10.1128/aem.61.10.3705-3710.1995

Boileau, C., Auria, R., Davidson, S., Casalot, L., Christen, P., Liebgott, P.-P., et al. (2016). Hydrogen production by the hyperthermophilic bacterium *Thermotoga maritima* part I: Effects of sulfured nutriments, with thiosulfate as model, on hydrogen production and growth. *Biotechnology for Biofuels, 9*(1), 269. https://doi.org/10.1186/s13068-016-0678-8

Bonch-Osmolovskaya, E. (2020). Thermotogales. In *eLS* (pp. 802–808). Wiley.

Britt, R. D., Rao, G., & Tao, L. (2020). Biosynthesis of the catalytic H-cluster of [FeFe] hydrogenase: The roles of the Fe–S maturase proteins HydE, HydF, and HydG. *Chemical Science, 11*(38), 10313–10323. https://doi.org/10.1039/D0SC04216A

Brouns, S. J. J., Wu, H., Akerboom, J., Turnbull, A. P., de Vos, W. M., & van der Oost, J. (2005). Engineering a selectable marker for hyperthermophiles. *The Journal of Biological Chemistry, 280*(12), 11422–11431. https://doi.org/10.1074/jbc.M413623200

Cai, J., Wang, Y., Liu, D., Zeng, Y., Xue, Y., Ma, Y., et al. (2007). *Fervidobacterium changbaicum* sp. nov., a novel thermophilic anaerobic bacterium isolated from a hot spring of the Changbai Mountains, China. *International Journal of Systematic and Evolutionary Microbiology, 57*(Pt 10), 2333–2336. https://doi.org/10.1099/ijs.0.64758-0

Campbell, K., Xia, J., & Nielsen, J. (2017). The impact of systems biology on bioprocessing. *Trends in Biotechnology, 35*(12), 1156–1168. https://doi.org/10.1016/j.tibtech.2017.08.011

Cha, M., Chung, D., Elkins, J. G., Guss, A. M., & Westpheling, J. (2013). Metabolic engineering of *Caldicellulosiruptor bescii* yields increased hydrogen production from lignocellulosic biomass. *Biotechnology for Biofuels, 6*(1), 85. https://doi.org/10.1186/1754-6834-6-85

Chandrasekhar, K., Lee, Y. J., & Lee, D. W. (2015). Biohydrogen production: Strategies to improve process efficiency through microbial routes. *International Journal of Molecular Sciences, 16*(4), 8266–8293. https://doi.org/10.3390/ijms16048266

Chaudhary, V., Kajla, P., Verma, D., Singh, T. P., Kothakota, A., Prasath, V. A., et al. (2023). Valorization of dairy wastes into wonder products by the novel use of microbial cell factories. *Trends in Food Science and Technology, 142*, 104221. https://doi.org/10.1016/j.tifs.2023.104221

Chen, G.-Q., & Jiang, X.-R. (2018). Next generation industrial biotechnology based on extremophilic bacteria. *Current Opinion in Biotechnology, 50*, 94–100. https://doi.org/10.1016/j.copbio.2017.11.016

Chen, B., Retzlaff, M., Roos, T., & Frydman, J. (2011). Cellular strategies of protein quality control. *Cold Spring Harbor Perspectives in Biology, 3*(8), a004374. https://doi.org/10.1101/cshperspect.a004374

Chen, H., Gao, S., Li, Y., Xu, H.-J., Li, W., Wang, J., et al. (2022). Valorization of livestock keratin waste: Application in agricultural fields. *International Journal of Environmental Research and Public Health, 19*(11). https://doi.org/10.3390/ijerph19116681

Cooper, J. B. (1970). Proportion of feathers at different body locations of white leghorn pullets. *Poultry Science, 49*(3), 809–810. https://doi.org/10.3382/ps.0490809

Cucina, M. (2023). Integrating anaerobic digestion and composting to boost energy and material recovery from organic wastes in the circular economy framework in Europe: A review. *Bioresource Technology Reports, 24*, 101642. https://doi.org/10.1016/j.biteb.2023.101642

Dhanasingh, I., & Lee, S. H. (2019). Crystallization and preliminary X-ray diffraction analysis of thioredoxin from the feather-degrading thermophile *Fervidobacterium islandicum* AW-1. *Korean Society for Structural Biology, 7*(2), 47–51. https://doi.org/10.34184/kssb.2019.7.2.47

Dhanasingh, I., Sung, J. Y., La, J. W., Kang, E., Lee, D. W., & Lee, S. H. (2021). Structure of oxidized pyrrolidone carboxypeptidase from Fervidobacterium islandicum AW-1 reveals unique structural features for thermostability and keratinolysis. *Biochemical and Biophysical Research Communications, 540*, 101–107. https://doi.org/10.1016/j.bbrc.2020.12.056

Ehrhardt, T., & Rothenberg, G. (2025). The hydrogen economy fairytale. *Green Chemistry, 27*(23), 6690–6698. https://doi.org/10.1039/D5GC00946D

Elhacham, E., Ben-Uri, L., Grozovski, J., Bar-On, Y. M., & Milo, R. (2020). Global human-made mass exceeds all living biomass. *Nature, 588*(7838), 442–444. https://doi.org/10.1038/s41586-020-3010-5

Ergal, İ., Gräf, O., Hasibar, B., Steiner, M., Vukotić, S., Bochmann, G., et al. (2020). Biohydrogen production beyond the Thauer limit by precision design of artificial microbial consortia. *Communications Biology, 3*(1), 443. https://doi.org/10.1038/s42003-020-01159-x

Feroz, S., Muhammad, N., Ratnayake, J., & Dias, G. (2020). Keratin-based materials for biomedical applications. *Bioactive Materials, 5*(3), 496–509. https://doi.org/10.1016/j.bioactmat.2020.04.007

Fineschi, S., Cozzi, F., Burger, D., Dayer, J.-M., Meroni, P. L., & Chizzolini, C. (2007). Anti-fibroblast antibodies detected by cell-based ELISA in systemic sclerosis enhance the collagenolytic activity and matrix metalloproteinase-1 production in dermal fibroblasts. *Rheumatology, 46*(12), 1779–1785. https://doi.org/10.1093/rheumatology/kem241

Franke-Whittle, I. H., & Insam, H. (2013). Treatment alternatives of slaughterhouse wastes, and their effect on the inactivation of different pathogens: A review. *Critical Reviews in Microbiology, 39*(2), 139–151. https://doi.org/10.3109/1040841X.2012.694410

Friedrich, A. B., & Antranikian, G. (1996). Keratin degradation by *Fervidobacterium pennavorans*, a novel thermophilic anaerobic species of the order *Thermotogales*. *Applied and Environmental Microbiology, 62*(8), 2875–2882. https://doi.org/10.1128/aem.62.8.2875-2882.1996

Ganguly, A., Sun, P., Liu, X., Delgado, H. E., Sun, L., & Elgowainy, A. (2025). Techno-economic and life cycle analysis of bio-hydrogen production using bio-based waste streams through the integration of dark fermentation and microbial electrolysis. *Green Chemistry, 27*(21), 6213–6231. https://doi.org/10.1039/D4GC05020G

Grybchuk-Ieremenko, A., Lipovská, K., Kouřilová, X., Obruča, S., & Dvořák, P. (2025). An initial genome editing toolset for *Caldimonas thermodepolymerans*, the first model of thermophilic polyhydroxyalkanoates producer. *Microbial Biotechnology, 18*(2), e70103. https://doi.org/10.1111/1751-7915.70103

Gurav, R. G., & Jadhav, J. P. (2013). A novel source of biofertilizer from feather biomass for banana cultivation. *Environmental Science and Pollution Research, 20*(7), 4532–4539. https://doi.org/10.1007/s11356-012-1405-z

Gürdal, A. A., & Çetinkaya, T. (2024). Advancements in edible films for aquatic product preservation and packaging. *Reviews in Aquaculture, 16*(3), 997–1020. https://doi.org/10.1111/raq.12880

Hadley Kershaw, E., Hartley, S., McLeod, C., & Polson, P. (2021). The sustainable path to a circular bioeconomy. *Trends in Biotechnology, 39*(6), 542–545. https://doi.org/10.1016/j.tibtech.2020.10.015

Hicks, T. M., & Verbeek, C. J. R. (2016). Meat industry protein by-products: sources and characteristics. In *Protein byproducts: Transformation from environmental burden into value-added products*. Academic Press.

Hollinshead, W., He, L., & Tang, Y. J. (2014). Biofuel production: An odyssey from metabolic engineering to fermentation scale-up. *Frontiers in Microbiology, 5*, 344. https://doi.org/10.3389/fmicb.2014.00344

Igbokwe, V. C., Ezugworie, F. N., Onwosi, C. O., Aliyu, G. O., & Obi, C. J. (2022). Biochemical biorefinery: A low-cost and non-waste concept for promoting sustainable circular bioeconomy. *Journal of Environmental Management, 305*, 114333. https://doi.org/10.1016/j.jenvman.2021.114333

Javier-Lopez, R., Geliashvili, N., & Birkeland, N. K. (2024). Comparative genomics of *Fervidobacterium*: A new phylogenomic landscape of these wide-spread thermophilic anaerobes. *BMC Genomics, 25*(1), 1248. https://doi.org/10.1186/s12864-024-11128-x

Jin, H.-S., Park, S. Y., Kim, K., Lee, Y.-J., Nam, G.-W., Kang, N. J., et al. (2017). Development of a keratinase activity assay using recombinant chicken feather keratin substrates. *PLoS One, 12*(2), e0172712. https://doi.org/10.1371/journal.pone.0172712

Jin, H.-S., Song, K., Baek, J.-H., Lee, J.-E., Kim, D. J., Nam, G.-W., et al. (2018). Identification of matrix metalloproteinase-1-suppressive peptides in feather keratin hydrolysate. *Journal of Agricultural and Food Chemistry, 66*(48), 12719–12729. https://doi.org/10.1021/acs.jafc.8b05213

Jin, H.-S., Park, S. Y., Kim, J.-Y., Lee, J.-E., Lee, H.-S., Kang, N. J., et al. (2019). Fluorescence-based quantification of bioactive keratin peptides from feathers for optimizing large-scale anaerobic fermentation and purification. *Biotechnology and Bioprocess Engineering, 24*(1), 240–249. https://doi.org/10.1007/s12257-018-0400-8

Jin, H. S., Dhanasingh, I., Sung, J. Y., La, J. W., Lee, Y., Lee, E. M., et al. (2021). The sulfur formation system mediating extracellular cysteine-cystine recycling in *Fervidobacterium islandicum* AW-1 is associated with keratin degradation. *Microbial Biotechnology, 14*(3), 938–952. https://doi.org/10.1111/1751-7915.13717

Kananavičiūtė, R., & Čitavičius, D. (2015). Genetic engineering of *Geobacillus* spp. *Journal of Microbiological Methods, 111*, 31–39. https://doi.org/10.1016/j.mimet.2015.02.002

Kang, E., Jin, H. S., La, J. W., Sung, J. Y., Park, S. Y., Kim, W. C., et al. (2020a). Identification of keratinases from *Fervidobacterium islandicum* AW-1 using dynamic gene expression profiling. *Microbial Biotechnology, 13*(2), 442–457. https://doi.org/10.1111/1751-7915.13493

Kang, N. J., Jin, H.-S., Lee, S.-E., Kim, H. J., Koh, H., & Lee, D.-W. (2020b). New approaches towards the discovery and evaluation of bioactive peptides from natural resources. *Critical Reviews in Environmental Science and Technology, 50*(1), 72–103. https://doi.org/10.1080/10643389.2019.1619376

Kanoksilapatham, W., Pasomsup, P., Keawram, P., Cuecas, A., Portillo, M. C., & Gonzalez, J. M. (2016). *Fervidobacterium thailandense* sp. nov., an extremely thermophilic bacterium isolated from a hot spring. *International Journal of Systematic and Evolutionary Microbiology, 66*(12), 5023–5027. https://doi.org/10.1099/ijsem.0.001463

Karim, A., Osse, E. F., & Khalloufi, S. (2025). Innovative strategies for valorization of byproducts from soybean industry: A review on status, challenges, and sustainable approaches towards zero-waste processing systems. *Heliyon, 11*(3). https://doi.org/10.1016/j.heliyon.2025.e42118

Katti, S. K., LeMaster, D. M., & Eklund, H. (1990). Crystal structure of thioredoxin from *Escherichia coli* at 1.68 Å resolution. *Journal of Molecular Biology, 212*(1), 167–184. https://doi.org/10.1016/0022-2836(90)90313-B

Kaur, M., Bhari, R., & Singh, R. S. (2021). Chicken feather waste-derived protein hydrolysate as a potential biostimulant for cultivation of mung beans. *Biologia, 76*(6), 1807–1815. https://doi.org/10.1007/s11756-021-00724-x

Keshwania, P., Kaur, N., Chauhan, J., Sharma, G., Afzal, O., Alfawaz Altamimi, A. S., et al. (2023). Superficial dermatophytosis across the world's populations: Potential benefits from nanocarrier-based therapies and rising challenges. *ACS Omega, 8*(35), 31575–31599. https://doi.org/10.1021/acsomega.3c01988

Kim, J. S., Kluskens, L. D., de Vos, W. M., Huber, R., & van der Oost, J. (2004). Crystal structure of fervidolysin from Fervidobacterium pennivorans, a keratinolytic enzyme related to subtilisin. *Journal of Molecular Biology, 335*(3), 787–797. https://doi.org/10.1016/j.jmb.2003.11.006

Kirchherr, J., Reike, D., & Hekkert, M. (2017). Conceptualizing the circular economy: An analysis of 114 definitions. *Resources, Conservation and Recycling, 127*, 221–232. https://doi.org/10.1016/j.resconrec.2017.09.005

Kumaraswamy, K. G., Krishnan, A., Ananyev, G., Zhang, S., Bryant, D. A., & Dismukes, G. C. (2019). Crossing the Thauer limit: Rewiring cyanobacterial metabolism to maximize fermentative H2 production. *Energy & Environmental Science, 12*(3), 1035–1045. https://doi.org/10.1039/C8EE03606C

La, J. W., Dhanasingh, I., Jang, H., Lee, S. H., & Lee, D.-W. (2020). Functional characterization of primordial protein repair enzyme M38 metallo-peptidase from *Fervidobacterium islandicum* AW-1. *Frontiers in Molecular Biosciences, 7*, 600634. https://doi.org/10.3389/fmolb.2020.600634

Land, H., Sekretareva, A., Huang, P., Redman, H. J., Németh, B., Polidori, N., et al. (2020). Characterization of a putative sensory [FeFe]-hydrogenase provides new insight into the role of the active site architecture. *Chemical Science, 11*(47), 12789–12801. https://doi.org/10.1039/D0SC03319G

Lange, L., Huang, Y., & Busk, P. K. (2016). Microbial decomposition of keratin in nature—A new hypothesis of industrial relevance. *Applied Microbiology and Biotechnology, 100*(5), 2083–2096. https://doi.org/10.1007/s00253-015-7262-1

Lara, A. R., Galindo, E., Ramírez, O. T., & Palomares, L. A. (2006). Living with heterogeneities in bioreactors: Understanding the effects of environmental gradients on cells. *Molecular Biotechnology, 34*(3), 355–382. https://doi.org/10.1385/MB:34:3:355

Le, Y., Fu, Y., & Sun, J. (2020). Genome editing of the anaerobic thermophile *Thermoanaerobacter ethanolicus* using thermostable Cas9. *Applied and Environmental Microbiology, 87*(1), e01773–e01720. https://doi.org/10.1128/AEM.01773-20

Lee, Y.-J., Jeong, H., Park, G.-S., Kwak, Y., Lee, S.-J., Lee, S. J., et al. (2015a). Genome sequence of a native-feather degrading extremely thermophilic eubacterium, *Fervidobacterium islandicum* AW-1. *Standards in Genomic Sciences, 10*(1), 71. https://doi.org/10.1186/s40793-015-0063-4

Lee, Y.-J., Dhanasingh, I., Ahn, J.-S., Jin, H.-S., Choi, J. M., Lee, S. H., et al. (2015b). Biochemical and structural characterization of a keratin-degrading M32 carboxypeptidase from *Fervidobacterium islandicum* AW-1. *Biochemical and Biophysical Research Communications, 468*(4), 927–933. https://doi.org/10.1016/j.bbrc.2015.11.058

Lee, D. W., Lee, Y. J., Jin, H. S., Kim, J. Y., Nam, G. W., & Lee, S. J. (2020). Thermophile-derived keratinase and use thereof. In Patent U (Ed.), *US10,612,010B2*.

Lee, J.-Y., Lee, S.-E., & Lee, D.-W. (2022). Current status and future prospects of biological routes to bio-based products using raw materials, wastes, and residues as renewable resources. *Critical Reviews in Environmental Science and Technology, 52*(14), 2453–2509. https://doi.org/10.1080/10643389.2021.1880259

Letourneau, F., Soussotte, V., Bressollier, P., Branland, P., & Verneuil, B. (1998). Keratinolytic activity of *Streptomyces* sp. S.K1–02: A new isolated strain. *Letters in Applied Microbiology, 26*(1), 77–80. https://doi.org/10.1046/j.1472-765X.1998.00281.x

Li, Z.-W., Liang, S., Ke, Y., Deng, J.-J., Zhang, M.-S., Lu, D.-L., et al. (2020). The feather degradation mechanisms of a new *Streptomyces* sp. isolate SCUT-3. *Communications Biology, 3*(1), 191. https://doi.org/10.1038/s42003-020-0918-0

Lin, X., Lee, C.-G., Casale Ellen, S., & Shih Jason, C. H. (1992). Purification and characterization of a keratinase from a feather-degrading *Bacillus licheniformis* strain. *Applied Environmental Microbiology, 58*(10), 3271–3275. https://doi.org/10.1128/aem.58.10.3271-3275.1992

Liu, S. (2017). Chapter 1 – Introduction. In S. Liu (Ed.), *Bioprocess engineering* (Second ed., pp. 1–20). Elsevier.

Liu, W., Chen, X. D., Jeantet, R., André, C., Bellayer, S., & Delaplace, G. (2021). Effect of casein/whey ratio on the thermal denaturation of whey proteins and subsequent fouling in a plate heat exchanger. *Journal of Food Engineering, 289*, 110175. https://doi.org/10.1016/j.jfoodeng.2020.110175

Mahendrasinh Kosamia, N., Samavi, M., Piok, K., & Kumar Rakshit, S. (2022). Perspectives for scale up of biorefineries using biochemical conversion pathways: Technology status, techno-economic, and sustainable approaches. *Fuel, 324*, 124532. https://doi.org/10.1016/j.fuel.2022.124532

McKittrick, J., Chen, P. Y., Bodde, S. G., Yang, W., Novitskaya, E. E., & Meyers, M. A. (2012). The structure, functions, and mechanical properties of keratin. *Journal of Metals, 64*(4), 449–468. https://doi.org/10.1007/s11837-012-0302-8

Monteith, H. D., Shannon, E. E., & Derbyshire, J. B. (1986). The inactivation of a bovine enterovirus and a bovine parvovirus in cattle manure by anaerobic digestion, heat treatment, gamma irradiation, ensilage and composting. *The Journal of Hygiene, 97*(1), 175–184. https://doi.org/10.1017/S0022172400064457

Mottet, A., & Tempio, G. (2017). Global poultry production: Current state and future outlook and challenges. *World's Poultry Science Journal, 73*(2), 245–256. https://doi.org/10.1017/S0043933917000071

Nam, G.-W., Lee, D.-W., Lee, H.-S., Lee, N.-J., Kim, B.-C., Choe, E.-A., et al. (2002). Native-feather degradation by *Fervidobacterium islandicum* AW-1, a newly isolated keratinase-producing thermophilic anaerobe. *Archives of Microbiology, 178*(6), 538–547. https://doi.org/10.1007/s00203-002-0489-0

Ng, C. S., Wu, P., Fan, W. L., Yan, J., Chen, C. K., Lai, Y. T., et al. (2014). Genomic organization, transcriptomic analysis, and functional characterization of avian alpha- and beta-keratins in diverse feather forms. *Genome Biology and Evolution, 6*(9), 2258–2273. https://doi.org/10.1093/gbe/evu181

Patel, B. K. C., Morgan, H. W., & Daniel, R. M. (1985). *Fervidobacterium nodosum* gen. nov. and spec. nov., a new chemoorganotrophic, caldoactive, anaerobic bacterium. *Archives of Microbiology, 141*(1), 63–69. https://doi.org/10.1007/BF00446741

Petrus, A. K., Swithers, K. S., Ranjit, C., Wu, S., Brewer, H. M., Gogarten, J. P., et al. (2012). Genes for the major structural components of *Thermotogales* species' togas revealed by proteomic and evolutionary analyses of OmpA and OmpB homologs. *PLoS One, 7*(6), e40236. https://doi.org/10.1371/journal.pone.0040236

Peydayesh, M., Bagnani, M., Soon, W. L., & Mezzenga, R. (2023). Turning food protein waste into sustainable technologies. *Chemical Reviews, 123*(5), 2112–2154. https://doi.org/10.1021/acs.chemrev.2c00236

Qiu, J., Wilkens, C., Barrett, K., & Meyer, A. S. (2020). Microbial enzymes catalyzing keratin degradation: Classification, structure, function. *Biotechnology Advances, 44*, 107607. https://doi.org/10.1016/j.biotechadv.2020.107607

Rajabi, M., Ali, A., McConnell, M., & Cabral, J. (2020). Keratinous materials: Structures and functions in biomedical applications. *Materials Science and Engineering: C, 110*, 110612. https://doi.org/10.1016/j.msec.2019.110612

Rana, S., Singh, A., Surasani, V. K. R., Kapoor, S., Desai, A., & Kumar, S. (2023). Fish processing waste: A novel source of non-conventional functional proteins. *International Journal of Food Science and Technology, 58*(5), 2637–2644. https://doi.org/10.1111/ijfs.16104

Ripple, W. J., Wolf, C., Gregg, J. W., Rockström, J., Mann, M. E., Oreskes, N., et al. (2024). The 2024 state of the climate report: Perilous times on planet earth. *BioSci, 74*(12), 812–824. https://doi.org/10.1093/biosci/biae087

Sahlström, L., Bagge, E., Emmoth, E., Holmqvist, A., Danielsson-Tham, M.-L., & Albihn, A. (2008). A laboratory study of survival of selected microorganisms after heat treatment of biowaste used in biogas plants. *Bioresource Technology, 99*(16), 7859–7865. https://doi.org/10.1016/j.biortech.2007.09.071

Salehizadeh, M., Tajabadi Ebrahimi, M., Mousavi, S. N., Sepahi, A. A., & Orooji, R. (2025). Transforming feather meal into a high-performance feed for broilers. *Veterinary Medicine and Science, 11*(1), e70199. https://doi.org/10.1002/vms3.70199

Sánchez, A., & Vázquez, A. (2017). Bioactive peptides: A review. *Food Quality and Safety, 1*(1), 29–46. https://doi.org/10.1093/fqsafe/fyx006

Schroeder, W. L., & Saha, R. (2020a). Protocol for genome-scale reconstruction and melanogenesis analysis of *Exophiala dermatitidis*. *STAR Protocols, 1*(2), 100105. https://doi.org/10.1016/j.xpro.2020.100105

Schroeder, W. L., & Saha, R. (2020b). OptFill: A tool for infeasible cycle-free gapfilling of stoichiometric metabolic models. *iScience, 23*(1), 100783. https://doi.org/10.1016/j.isci.2019.100783

Service UFA (Ed.). (2025). *Livestock and poultry: World markets and trade*. United States Department of Agriculture FAS.

Sharma, I., Pranaw, K., Soni, H., Rawat, H. K., & Kango, N. (2022). Parametrically optimized feather degradation by *Bacillus velezensis* NCIM 5802 and delineation of keratin hydrolysis by multi-scale analysis for poultry waste management. *Scientific Reports, 12*(1), 17118. https://doi.org/10.1038/s41598-022-21351-9

Shaw, A. J., Hogsett, D. A., & Lynd, L. R. (2010). Natural competence in *Thermoanaerobacter* and *Thermoanaerobacterium* species. *Applied and Environmental Microbiology, 76*(14), 4713–4719. https://doi.org/10.1128/AEM.00402-10

Sivaramakrishnan, R., Shanmugam, S., Sekar, M., Mathimani, T., Incharoensakdi, A., Kim, S.-H., et al. (2021). Insights on biological hydrogen production routes and potential microorganisms for high hydrogen yield. *Fuel, 291*, 120136. https://doi.org/10.1016/j.fuel.2021.120136

Soni, S. K., & Soni, R. (2025). Policy and regulatory frameworks for green biorefinery: Driving sustainable waste valorization. In S. K. Soni & R. Soni (Eds.), *Green biorefinery solutions: Transforming biodegradable waste into resources* (pp. 315–349). Springer Nature Singapore.

Spillmann, S. K., Traub, F., Schwyzer, M., & Wyler, R. (1987). Inactivation of animal viruses during sewage sludge treatment. *Applied and Environmental Microbiology, 53*(9), 2077–2081. https://doi.org/10.1128/aem.53.9.2077-2081.1987

Stephanopoulos, G. (2012). Synthetic biology and metabolic engineering. *ACS Synthetic Biology, 1*(11), 514–525. https://doi.org/10.1021/sb300094q

Sulis, D. B., Lavoine, N., Sederoff, H., Jiang, X., Marques, B. M., Lan, K., et al. (2025). Advances in lignocellulosic feedstocks for bioenergy and bioproducts. *Nature Communications, 16*(1), 1244. https://doi.org/10.1038/s41467-025-56472-y

Sung, J.-Y., Kim, J.-Y., Jin, H.-S., Baek, J.-H., Kim, N. E., Song, H. H., et al. (2025). Keratin degradation reflects a starvation survival strategy in *Fervidobacterium islandicum* AW-1. *bioRxiv*, 2025.07.04.663136. https://doi.org/10.1101/2025.07.04.663136

Tesfaye, T., Sithole, B., & Ramjugernath, D. (2017). Valorisation of chicken feathers: A review on recycling and recovery route—Current status and future prospects. *Clean Technologies and Environmental Policy, 19*(10), 2363–2378. https://doi.org/10.1007/s10098-017-1443-9

Thauer, R. K., Jungermann, K., & Decker, K. (1977). Energy conservation in chemotrophic anaerobic bacteria. *Bacteriological Reviews, 41*(1), 100–180. https://doi.org/10.1128/br.41.1.100-180.1977

Tripathi, S. A., Olson, D. G., Argyros, D. A., Miller, B. B., Barrett, T. F., Murphy, D. M., et al. (2010). Development of pyrF-based genetic system for targeted gene deletion in clostridium thermocellum and creation of a pta mutant. *Applied and Environmental Microbiology, 76*(19), 6591–6599. https://doi.org/10.1128/AEM.01484-10

Trojanowska, D. J., Zych, A., Sganga, S., Tirelli, N., Boventi, M., Rinaldi, C., et al. (2025). Upgrading keratin into a moldable bioplastic. *Matter, 8*(4), 102039. https://doi.org/10.1016/j.matt.2025.102039

Tursunova, D., & Maksumova, O. (2024). Analysis of the amino acid and protein composition of feather keratin hydrolyzed using alkali and sodium hydrosulfite. *Results in Chemistry, 9*, 101621. https://doi.org/10.1016/j.rechem.2024.101621

Vakulchuk, R., Overland, I., & Scholten, D. (2020). Renewable energy and geopolitics: A review. *Renewable and Sustainable Energy Reviews, 122*, 109547. https://doi.org/10.1016/j.rser.2019.109547

Van Ooteghem, S. A., Beer, S. K., & Yue, P. C. (2002). Hydrogen production by the thermophilic bacterium *Thermotoga neapolitana. Applied Biochemistry and Biotechnology, 98*, 177–189.

Verhaart, M. R. A., Bielen, A. A. M., Jvd, O., Stams, A. J. M., & Kengen, S. W. M. (2010). Hydrogen production by hyperthermophilic and extremely thermophilic bacteria and archaea: Mechanisms for reductant disposal. *Environmental Technology, 31*(8–9), 993–1003. https://doi.org/10.1080/09593331003710244

Wang, J. J., & Shih, J. (1999). Fermentation production of keratinase from *Bacillus licheniformis* PWD-1 and a recombinant *B. subtilis* FDB-29. *Journal of Industrial Microbiology & Biotechnology, 22*(6), 608–616. https://doi.org/10.1038/sj.jim.2900667

Wang, J., Guo, X., Li, H., Qi, H., Qian, J., Yan, S., et al. (2019). Hydrogen sulfide from cysteine desulfurase, not 3-mercaptopyruvate sulfurtransferase, contributes to sustaining cell growth and bioenergetics in *E. coli* under anaerobic conditions. *Frontiers in Microbiology, 10*, 2357. https://doi.org/10.3389/fmicb.2019.02357

Wang, Z., Chen, Y., Yan, M., Li, K., Okoye, C. O., Fang, Z., et al. (2023). Research progress on the degradation mechanism and modification of keratinase. *Applied Microbiology and Biotechnology, 107*(4), 1003–1017. https://doi.org/10.1007/s00253-023-12360-3

Williams, C. M., & Shih, J. C. H. (1989). Enumeration of some microbial groups in thermophilic poultry waste digesters and enrichment of a feather-degrading culture. *The Journal of Applied Bacteriology, 67*(1), 25–35. https://doi.org/10.1111/j.1365-2672.1989.tb04951.x

Williams, S. L., Toda, M., Chiller, T., Brunkard, J. M., & Litvintseva, A. P. (2024). Effects of climate change on fungal infections. *PLoS Pathogens, 20*(5), e1012219. https://doi.org/10.1371/journal.ppat.1012219

Wu, W.-L., Chen, M.-Y., Tu, I. F., Lin, Y.-C., EswarKumar, N., Chen, M.-Y., et al. (2017). The discovery of novel heat-stable keratinases from *Meiothermus taiwanensis* WR-220 and other extremophiles. *Scientific Reports, 7*(1), 4658. https://doi.org/10.1038/s41598-017-04723-4

Yang, Z., Li, Z., Li, B., Bu, R., Tan, G.-Y., Wang, Z., et al. (2023). A thermostable type I-B CRISPR-Cas system for orthogonal and multiplexed genetic engineering. *Nature Communications, 14*(1), 6193. https://doi.org/10.1038/s41467-023-41973-5

Yeo, I., Lee, Y.-J., Song, K., Jin, H.-S., Lee, J.-E., Kim, D., et al. (2018). Low-molecular weight keratins with anti-skin aging activity produced by anaerobic digestion of poultry feathers with *Fervidobacterium islandicum* AW-1. *Journal of Biotechnology, 271*, 17–25. https://doi.org/10.1016/j.jbiotec.2018.02.003

Zeldes, B. M., Keller, M. W., Loder, A. J., Straub, C. T., Adams, M. W. W., & Kelly, R. M. (2015). Extremely thermophilic microorganisms as metabolic engineering platforms for production of fuels and industrial chemicals. *Frontiers in Microbiology, 6*. https://doi.org/10.3389/fmicb.2015.01209

Zeng, X., & Li, J. (2021). Emerging anthropogenic circularity science: Principles, practices, and challenges. *iScience, 24*(3), 102237. https://doi.org/10.1016/j.isci.2021.102237

Zhou, J., Li, D., Zhang, X., Liu, C., & Chen, Y. (2023). Valorization of protein-rich waste and its application. *Science of The Total Environment, 901*, 166141. https://doi.org/10.1016/j.scitotenv.2023.166141

Zubair, M., Roopesh, M. S., & Ullah, A. (2022). Nano-modified feather keratin derived green and sustainable biosorbents for the remediation of heavy metals from synthetic wastewater. *Chemosphere, 308*, 136339. https://doi.org/10.1016/j.chemosphere.2022.136339

Chapter 4
Enhancing Thermostability of Extremozymes: Strategies and Industrial Applications

Aya Kubo

Abstract Enzymes are renowned for their high specificity and efficiency, making them valuable in numerous industries. Nonetheless, a significant challenge is thermal instability, which leads to denaturation of the protein structure and subsequent loss of activity at elevated temperatures. This issue imposes considerable constraints on the efficiency and cost-effectiveness of industrial processes. Consequently, enhancing the thermostability of enzymes can yield substantial benefits, such as increased reaction rates, cost savings, improved product quality, enhanced long-term storage stability, and improved hygiene. Several strategies have been developed to enhance enzyme thermostability. These include protein engineering techniques aimed at optimizing the function of existing enzymes, such as rational design, directed evolution, semirational design, and ancestral protein reconstruction (APRC). Additionally, enzyme immobilization techniques, which confine enzymes to specific locations and facilitate continuous reactions and reuse, have been employed. These techniques encompass carrier-binding, crosslinking, and entrapment methods. Other approaches include PEGylation, which involves the attachment of polyethylene glycol (PEG) molecules, and solvent engineering, which modifies the solvent environment of enzymes. Thermostable enzymes are already being applied in various industries, including biofuel and bioethanol production, detergent, food, pharmaceutical, chemical, textile, and environmental biotechnology. By enabling reactions under high-temperature conditions and offering excellent storage stability, these enzymes contribute to the advancement of more efficient and environmentally sustainable "clean and green" technologies. Research and development on thermostable enzymes extend beyond scientific interest to encompass significant economic incentives.

Keywords Thermostability · Extremozymes · Protein engineering · Industrial applications

A. Kubo (✉)
Bio-Resilience Research Project, Toyo University Top Priority Research Program, Toyo University, Saitama, Japan
e-mail: kubo071@toyo.jp

M. Ito et al. (eds.), *Extremophiles for a Sustainable Future*,
https://doi.org/10.1007/978-981-95-6970-0_4

4.1 Introduction

Enzymes serve as catalysts for specific chemical reactions within living organisms. Owing to their remarkable specificity and efficiency, they are anticipated to have extensive applications in various industries. Notably, enzymes sourced from extremophiles are termed "extremozymes"and their exceptional environmental adaptability makes them extensively utilized in industry (Bhatt et al., 2024). These enzymes retain high activity and stability even under severe conditions, such as extreme temperatures, high salinity, elevated pH levels, and the presence of organic solvents, thereby enabling processes that are challenging to achieve with conventional enzymes. However, many natural enzymes are optimized to function at temperatures conducive to the survival of the organisms from which they originate, resulting in denaturation of their three-dimensional structures and loss of activity under high-temperature conditions. This thermal instability significantly constrains the efficiency and cost-effectiveness of industrial processes in which they are intended to be used. Solutions to this challenge include employing enzymes derived from thermophilic bacteria, which are inherently stable at elevated temperatures, or modifying the enzymes of interest to enhance their thermostability. A prominent example of an industrial application of enzymes from thermophilic bacteria is the thermostable polymerase used in polymerase chain reaction (PCR) (Mullis et al., 1986). Thirty-three years have elapsed since Mullis and Smith were awarded the Nobel Prize in 1993, and the advancement of contemporary molecular biology research would be inconceivable without this technique. Although enzymes from thermophilic bacteria and archaea represent a groundbreaking discovery, they do not necessarily fulfill all commercial requirements. Numerous industrial processes necessitate not only high temperatures but also the simultaneous presence of multiple harsh conditions, such as specific pH levels and high concentrations of salts, surfactants, or organic solvents (Mesbah, 2022). Although natural thermophilic enzymes are thermostable, their other properties may not be optimal for industrial applications. Furthermore, the cultivation of thermophilic bacteria is challenging, rendering large-scale commercial production of their enzymes inefficient and costly, which poses a significant obstacle to their practical applications.

An artificial engineering approach is crucial for developing enzymes with specific desired properties, such as enhanced thermostability. This process involves designing enzymes to meet specific industrial requirements, which is of paramount importance in industrial applications. According to the Arrhenius law, an increase in the reaction temperature by 10 °C approximately doubles the reaction rate; therefore, improving thermostability directly enhances enzyme activity (Daintith, 2008). Furthermore, thermostable enzymes offer long-term storage stability and durability at ambient temperatures, thereby reducing the need for frequent enzyme addition as required with conventional enzymes and significantly lowering the costs. The ability to easily remove undesired activities during purification enhances the product quality. Additionally, there are significant hygienic benefits, as the conditions in which these enzymes are used inhibit the contamination and proliferation of undesirable microorganisms, eliminating the need for special anti-spoilage measures. Thermostable enzymes also demonstrate high tolerance to organic solvents and maintain their activity at the interfaces between immiscible substances, such as water and oil, rendering them adaptable to a diverse range of reaction systems.

This chapter examines strategies for enhancing enzyme thermostability and explores the industrial applications of thermostable enzymes. The global market for industrial enzymes has been experiencing consistent growth, with a valuation of USD 7.5 billion in 2024 and projections indicating an increase to USD 12.3 billion by 2033 (Markets & Markets, 2025). This expansion is driven by the rising demand for more efficient and environmentally sustainable processes, the advent of novel enzyme-based application technologies across various industries, and the growing demand for enzymes in detergent and cleaning sectors. Thermostable enzymes are identified as significant catalysts in these market transformations because of their efficiency and advantages to minimize environmental impact. This underscores that research and development in the field of thermostable enzymes is not only a scientific endeavor but also an area with substantial economic potential.

4.2 Strategies to Improve the Thermostability of Enzymes

4.2.1 Protein Engineering

Protein engineering represents a significant advancement in biotechnology, involving the manipulation of the natural macromolecular structure of existing enzymes to enhance their functionality or generate novel functions. The primary objective of this process is to address the inherent limitations of enzymes under extreme conditions, thereby improving their stability and activity. A variety of methods have been developed to enhance enzyme thermostability, with the most commonly employed techniques in protein engineering being rational design, directed evolution, semirational design, and ancestral sequence reconstruction (Fig. 4.1).

4.2.1.1 Rational Design

Rational design is an approach aimed at enhancing enzyme stability through the introduction of site-specific mutations, which facilitate targeted modifications based on comprehensive structural and functional insights, including crystal structure data, biophysical characteristics, functional attributes, and sequence information of the enzyme (Pongsupasa et al., 2022). In numerous instances, thermostability is augmented by amino acid substitutions that stabilize secondary structures, increase ion pair formation on the molecular surface, strengthen subunit interactions, introduce salt bridges, form disulfide bonds, and improve packing of the hydrophobic core within the molecule (Rigoldi et al., 2018). Additionally, strategies such as loop

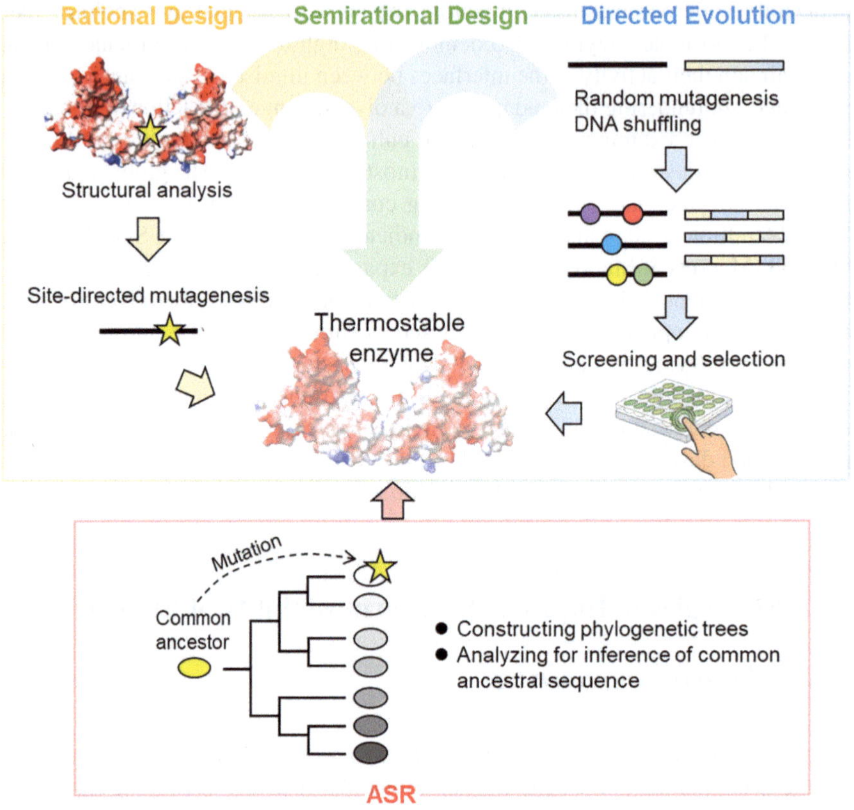

Fig. 4.1 Schematic illustration of protein engineering methods for constructing thermostability-enhanced enzymes

shortening, glycine substitutions, and the incorporation of proline residues have been employed to restrict flexible regions, thereby enhancing enzyme rigidity and preventing thermal denaturation (Watanabe & Suzuki, 1998; Suzuki, 1999). To facilitate these objectives, various in silico tools, including FireProt and PoPMuSiC, have been developed (Capriotti et al., 2005; Folkman et al., 2016; Musil et al., 2017; Schymkowitz et al., 2005; Pires et al., 2014; Thomas Ó Conchúir et al., 2015; Yin et al., 2007; Craig & Dombkowski, 2013; Goldenzweig et al., 2018; Kwasigroch et al., 2002; Wijma et al., 2014). The primary advantages of this approach include a high likelihood of yielding beneficial modifications and a significant reduction in the size of the mutation library, thereby conserving time and effort required for screening. However, given that the impact of each amino acid substitution is contingent upon the surrounding structural environment, a substitution that enhances the thermostability of one enzyme may not necessarily do so for another, resulting in a variable success rate (Sanchez-Ruiz & Makhatadze, 2001). Furthermore, the need for prior knowledge of the enzyme structure and mechanism poses a challenge for applying this method to enzymes lacking such information.

One study reported the application of a computational method to identify amino acid substitutions in the alcohol dehydrogenase from *Candida magnoliae*, which has a Tm of approximately 43 °C; through the introduction of mutations, the thermostability of the enzyme was enhanced to approximately 94 °C (Aalbers et al., 2020). Furthermore, Watanabe and colleagues successfully increased the thermostability of xylanase from *Talaromyces cellulolyticus* by over 20 °C. This was achieved by conducting sequence comparisons with xylanases from other organisms to identify substitution sites and generate mutants (Watanabe et al., 2016). Additionally, a report utilizing molecular dynamics (MD) simulations revealed that the N-terminus of phenylalanine hydroxylase from *Chromobacterium violaceum* exhibits flexibility, and substituting this region with proline improved thermostability (Liu et al., 2019).

4.2.1.2 Directed Evolution

Directed evolution is an iterative process that emulates Darwinian evolution in vitro by introducing random mutations into genes encoding enzymes to construct diverse gene libraries, followed by the selection or screening of thermostable mutant enzymes (Thomson et al., 2022). The mechanism of this technique involves constructing gene libraries using random mutagenesis methods, such as error-prone PCR or DNA shuffling, or by inducing random mutations in organisms harboring the genes encoding the target enzymes in the presence of mutagens. Subsequently, high-throughput screening, which combines heat treatment and activity assays, is employed to select enzymes with enhanced thermostability. Furthermore, by gradually increasing the selection pressure, it is possible to develop more effective and resistant enzymes (Dalby, 2011). A significant advantage of directed evolution is that prior knowledge of the enzyme structure or function is not required. However, this method requires extensive experimentation and large-scale screening. Not all enzyme activities are amenable to straightforward high-throughput screening methods, and not all screening techniques can be readily employed at the required scale. Additionally, enzymes obtained through random mutagenesis may occasionally exhibit undesirable effects that differ from their intended properties.

As a concrete example of success, it has been documented that the thermostability of *Bacillus subtilis p*-nitrobenzyl esterase was enhanced by over 14 °C in Tm through six generations of random mutagenesis and recombination with DNA shuffling (Giver et al., 1998). Additionally, the thermostable kanamycin resistance gene currently employed as a selection marker for *Thermus thermophilus* was developed by introducing mutations into the kanamycin resistance gene from *Staphylococcus aureus* via DNA shuffling, followed by screening *T. thermophilus* in the presence of kanamycin. Consequently, a kanamycin resistance gene with thermostability exceeding that of the wild type by over 20 °C was obtained (Hoseki et al., 1999). Furthermore, mutants exhibiting improved half-lives, optimal temperatures, and melting temperatures have been developed for various enzymes, including polyethylene terephthalate (PET)-degrading enzymes, cholesterol oxidase, and maltogenic amylase (Brott et al., 2022; Ebrahimi Fana et al., 2023; Ruan et al., 2022; Shi et al., 2023; Tamakoshi et al., 2001).

4.2.1.3 Semirational Design

Semirational design is a protein engineering strategy that integrates rational design and directed evolution. This approach leverages the existing knowledge of protein structure and function to strategically identify amino acid positions for mutation, thereby creating a more targeted and potentially more effective protein library than random mutagenesis methods. By identifying residues that directly interact with the substrate and the active site cavity through structural information and introducing random mutations into these amino acid residues, it is possible to achieve highly effective outcomes that could not have been anticipated through rational selection (Mildvan, 2004). In instances where structural information is limited, semirational design can be employed by utilizing molecular modeling to identify the residues that are most likely to interact with the substrate molecule (Schmitzer et al., 2004). Furthermore, random mutagenesis can be used to identify mutational hot spots within the target gene, and by introducing rational mutations into multiple hot spots, it is feasible to achieve smaller-scale but broader enzyme modification (Reetz et al., 2001).

For instance, in the case of pyranose 2-oxidase derived from *Trametes multicolor*, random mutagenesis has been employed to identify amino acids whose substitutions enhance the thermostability of the enzyme. Subsequently, saturation mutagenesis, in which a target amino acid is systematically substituted with all other naturally occurring amino acids, was used to determine the thermostability-enhancing amino acids post-substitution. The combination of multiple site-directed mutations results in alterations in substrate specificity and improvements in thermostability (Masuda-Nishimura et al., 1999; Spadiut et al., 2010). Furthermore, in the context of arginine deiminase derived from *Enterococcus faecalis*, which is utilized in the production of L-citrulline, it has been documented that substituting amino acids predicted to be significant based on molecular dynamics (MD) simulations into a highly active enzyme mutant obtained through random mutagenesis further enhanced the enzyme stability.

4.2.1.4 Ancestral Sequence Reconstruction (ASR)

Ancestral sequence reconstruction (ASR) is an emerging technique in molecular evolution research that aims to estimate the gene sequences of common ancestors by utilizing homologous gene sequences from multiple organisms (Thornton, 2004). For instance, by analyzing 160 phylogenetically diverse elongation factor (EF) homologues, the ancestral EF protein sequence has been inferred to have existed in life forms dating back to the Precambrian era. It has been discovered that this protein exhibited a progressive increase in thermostability during track back through time (Gaucher et al., 2003, 2008). Furthermore, the ancestral nucleoside diphosphate kinase (NDK), estimated from the NDK sequences of Archaea and Bacteria, also exhibits high thermostability (Akanuma et al., 2013). These findings support the hypothesis that primordial organisms were thermophilic and that ancestral proteins were more thermostable than their modern counterparts (Thomson et al., 2022). In ASR, the ancestral sequence is inferred by comparing the target enzyme sequences of multiple organisms, followed by synthesis, expression, and evaluation of the enzyme properties. The primary advantage of this method is that it does not require structural data for the enzyme or screening of extensive mutant libraries; instead, it relies solely on abundant sequence information. However, adapting this technique to species-specific proteins remains a challenge.

For instance, in the case of 3-isopropylmalate dehydrogenase (IPMDH) from *T. thermophilus*, the introduction of a sequence from a putative ancestral enzyme, inferred from homologous sequences of 18 bacteria and archaea, resulted in enhanced thermostability (Watanabe et al., 2006). It was demonstrated that the ancestral residues contributing to thermostability are independent of conservation degree. Furthermore, for β-amylase from *B. circulans*, an ancestral type was deduced from plant and bacterial amylase sequences, and through the introduction of mutations, mutant variants exhibiting improved thermostability and enzymatic activity were successfully generated (Yamashiro et al., 2010).

In addition to ASR, another method, called Reconstructing Evolutionary Adaptive Paths (REAP), has been proposed for enzyme engineering by identifying ancestral enzyme sequences from homologous sequences (Cole & Gaucher, 2011). Although there have been reports of modifying DNA polymerase and altering its adaptive substrate using this method (Chen et al., 2010), to the best of our knowledge, there have been no reports of enhanced thermostability resulting from enzyme engineering using REAP.

The successful enhancement of thermostability through protein engineering offers significant insights into the fundamental molecular determinants of protein stability. By systematically altering amino acid sequences and observing the consequent changes in thermal properties, it is feasible to evaluate the theories concerning protein rigidity. The interplay between these engineering approaches and fundamental understanding is crucial for establishing the principles of future enzyme designs. Moreover, when comparing the aforementioned protein engineering methods, rational design exhibits targeted efficiency when structural data are accessible, whereas directed evolution is suitable for exploring extensive sequence spaces without prior knowledge. Semirational design and ASR are hybrid approaches that employ evolutionary insights to guide targeted mutagenesis and reconstruct intrinsically stable ancestral types. Among these methods, bioinformatics and artificial intelligence-based approaches, such as AlphaFold (Jumper et al., 2021), have demonstrated the ability to overcome individual limitations, thereby enhancing the robustness of predictions and designs (Thomson et al., 2022). The fact that these methods are complementary rather than mutually exclusive suggests a future where more efficient and higher success rates of enzyme thermostability can be achieved.

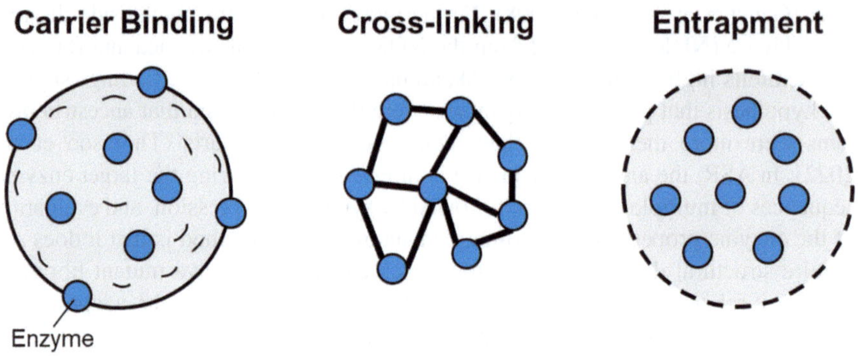

Fig. 4.2 Schematic illustration of enzyme immobilization methods

4.2.2 Enzyme Immobilization

Enzyme immobilization is a technique that confines enzymes to a specific location, facilitating continuous reactions and allowing for the recovery and reuse of enzymes after the reaction (Robescu & Bavaro, 2025). This technology is pivotal in enhancing enzyme stability, prolonging storage life, and simplifying the separation of enzymes from products, thereby playing a significant role in industrial applications such as bioreactors and biosensors (Yamaguchi & Miyazaki, 2024). Numerous methods have been investigated, which can be broadly categorized into three distinct groups: carrier binding, cross-linking, and entrapment (Fig. 4.2).

4.2.2.1 Carrier Binding

Carrier binding involves the immobilization of water-soluble enzymes onto insoluble carriers, including cellulose, resin beads, activated carbon, silica gel, and magnetic particles. This method can be categorized into covalent, ionic, physical adsorption, and affinity binding, depending on the nature of the binding interaction. For a comprehensive analysis of each method, refer to the review by Yamaguchi and Miyazaki (2024). Covalent binding, while forming the most robust bonds and preventing enzyme detachment, poses a risk of enzyme inactivation and denaturation due to chemical modification. This approach limits the conformational flexibility of the enzyme, thereby enhancing its resistance to denaturation induced by heat, pH fluctuations, and organic solvents (Singh et al., 2013). Notably, multipoint covalent binding, in which the enzyme is attached to the carrier at multiple sites, substantially increases the rigidity of the enzyme structure and augments its stability. In contrast, ionic binding and physical adsorption methods immobilize enzymes under milder conditions than covalent binding, which may lead to enzyme detachment from the carrier. However, these methods offer advantages such as reduced enzyme inactivation and carrier reusability.

As an example of enhanced enzyme thermostability through this approach, the immobilization of β-glucosidase from *Trichoderma reesei* on zeolite resulted in an expanded pH range and increased thermostability compared to the free enzyme (Ramírez-Ramírez et al., 2022). Furthermore, cellulase from *Aspergillus niger*, when immobilized on glutaraldehyde-egg protein-carrageenan beads, demonstrated superior thermostability compared to free cellulase and exhibited notable efficacy in clarifying products such as mango and orange juice (Wahba et al., 2025).

4.2.2.2 Cross-Linking

Cross-linking represents an immobilization strategy that avoids the use of a carrier, instead establishing direct covalent bonds between enzyme molecules to form insoluble macroparticles known as Cross-Linked Enzyme Aggregates (CLEAs). This approach imparts substantial rigidity to the enzyme structure, thereby mitigating thermal denaturation and aggregation (Singh et al., 2013). Similar to the aforementioned covalent bonding technique, multipoint covalent bonds confer rigidity and significantly enhance the thermostability of enzymes and their resistance to denaturants (Datta et al., 2013). Despite the benefits of improved thermostability and the prevention of enzyme leaching due to robust bonding, there is a potential risk of enzyme inactivation or diminished activity resulting from chemical modification. Consequently, the judicious selection of cross-linking agents and reaction conditions is imperative (Robescu & Bavaro, 2025). Moreover, the absence of a carrier facilitates higher enzyme loading, allowing concurrent purification and immobilization (Sheldon & van Pelt, 2013).

As an example of enhanced enzyme thermostability through this approach, a study reported the development of CLEAs of cyclodextrin glucosyltransferase derived from *Alkalihalobacillus lehensis* (formerly *B. lehensis*) for the production of cyclodextrin, which forms inclusion complexes and is utilized in a wide range of fields, such as pharmaceuticals, cosmetics, and supplements (Jailani et al., 2022). Utilizing chitosan as a cross-linking agent resulted in CLEAs with superior enzyme activity, demonstrating an extended half-life and enhanced thermostability compared to the free enzyme.

4.2.2.3 Entrapment

Entrapment is a technique in which enzymes are confined within materials, such as polymer gels or semipermeable membranes, without direct chemical modification (Robescu & Bavaro, 2025; Datta et al., 2013). This physical entrapment protects the enzymes from adverse environmental conditions, including heat, extreme pH, and protease degradation. By carefully adjusting the pore size within the polymer network, enzyme leakage can be minimized while allowing substrates and products to diffuse freely. The absence of chemical interactions helps preserve the three-dimensional structure of the enzyme, thereby reducing the risk of denaturation.

However, an increase in the thickness of the polymer matrix can result in mass transfer resistance, impeding substrate access to the active site of the enzyme. Additionally, if the pore size is excessively large, there is a potential risk of enzyme leakage from the carriers.

Studies have indicated that the immobilization of protease from fish viscera in calcium alginate gel and the entrapment of laccase from *Trametes versicolor* in a sponge-like hydrogel significantly enhance their thermostability and reproducibility compared to their free enzyme counterparts (Geethanjali & Subash, 2013; Sun et al., 2015).

4.2.3 Others

4.2.3.1 PEGylation

PEGylation is a technique that involves the covalent attachment of polyethylene glycol (PEG) molecules to enzymes and is predominantly utilized in pharmaceutical development (Makharadze et al., 2025). Mechanistically, the substantial hydrophilic PEG chains establish a "steric barrier" around the enzyme, thereby preventing enzyme aggregation, reducing proteolytic degradation, and minimizing nonspecific protein adsorption. This steric bulk may also impede immune recognition (Harris et al., 2001). PEG modification "rigidifies" the native three-dimensional structure of the enzyme, rendering it less susceptible to heat-induced unfolding and irreversible structural alterations (Yang et al., 2024). Consequently, the thermal vibrations of the internal enzyme groups were diminished, resulting in enhanced thermostability. The hydrophilicity of PEG influences the hydration layer of the enzyme, which is crucial for its stability (Ogino & Ishikawa, 2001). Unlike immobilization, which renders enzymes insoluble and facilitates their reuse, PEGylation aims to enhance stability and in vivo half-life while preserving enzyme solubility.

PEGylation is expected to enhance enzyme activity, thermostability, resistance to inhibitors, and increase stability and half-life of the enzyme. It is also expected to improve solubility in water or organic solvents and reduce its immunogenicity. Conversely, it has been noted that the three-dimensional structure of enzymes may undergo significant alterations, potentially leading to diminished biological activity, and that an immune response to PEG, specifically the development of "anti-PEG antibodies," may occur (Yang et al., 2024; Makharadze et al., 2025).

For instance, it has been demonstrated that Bst DNA polymerase exhibits enhanced activity, thermostability, and resistance to inhibitors through PEGylation, thereby expediting the detection of *Salmonella* in crude samples using the loop-mediated isothermal amplification (LAMP) method (Yang et al., 2024). The PEGylated enzyme retained over 80% of its activity even at 70 °C, whereas the unmodified enzyme rapidly lost activity above 65 °C. Furthermore, a study employing α-chymotrypsin indicated that PEGylation increased the thermal denaturation temperature by up to 6 °C; this effect was independent of the molecular weight of

PEG and was associated with a reduction in the structural dynamics of the protein (Rodríguez-Martínez et al., 2008).

4.2.3.2 Solvent Engineering

Solvent engineering is a technique employed to regulate the functionality and stability of enzymes by modifying the solvent, which acts as a reaction milieu for enzymes (Ogino & Ishikawa, 2001). This approach has been extensively investigated as a principal strategy for enhancing the thermostability of enzymes and plays an important role in significantly broadening the scope of enzyme applications in the bioindustry. The addition of polyols or ionic liquids to aqueous enzyme solutions results in strong interactions with water molecules, thereby reducing the number of water molecules surrounding the enzyme. Consequently, the hydration layer essential for maintaining the tertiary structure of the enzyme is fortified, thereby diminishing the likelihood of thermal denaturation (Arakawa, 2002; Khachatrian et al., 2025). Moreover, numerous organic solvents and ionic liquids (alkyl chains on the cation) interact with the hydrophobic regions within the enzyme, aiding in preserving its internal structure. This interaction enables the enzyme to maintain a stable tertiary structure, even under elevated temperatures (Mohtashami et al., 2019; Piccoli & Martínez, 2025). The presence of bulky molecules, such as polyols, around the enzyme physically impedes excessive molecular movement and aggregation at high temperatures (Miklos et al., 2010). Additionally, enzyme stabilization can be modulated by altering the combination of cations and anions in ionic liquids (Zhao, 2016). However, while solvent adjustment can enhance enzyme stability, it may also reduce enzyme flexibility, potentially diminishing enzyme activity, particularly in the presence of organic solvents, or decreasing substrate solubility, thereby resulting in a slower reaction rate (Ogino & Ishikawa, 2001).

4.3 Industrial Applications of Thermostable Enzymes

While each chapter provides a detailed examination of the industrial applications of various extremozymes, this chapter specifically focuses on thermostable enzymes (Table 4.1).

4.3.1 Biofuel and Bioethanol Production

Current research endeavors are focused on the development of biofuels derived from plant-based biomass that does not compete with food crops to establish sustainable energy sources (Chaudhari et al., 2023). A primary challenge in this endeavor is the prohibitively high cost of enzymes required for biomass

Table 4.1 Thermostable enzymes and their industrial applications mentioned in this chapter

Enzyme	Microorganism	Industrial application	Reference
Amylase	*Bacillus* spp., and *B. licheniformis*	Detergent, Food and Textile	Cui et al. (2022), Nguyen et al. (2025), Rigoldi et al. (2018)
Cellulase	*Pyrococcus horikoshii, Talaromyces emersonii, Thermotoga naphthophila* and others	Biofuel and Textile	Ajeje et al. (2021), Ando et al. (2002), Voutilainen et al. (2010), Rigoldi et al. (2018), Yu et al. (2015)
Laccase	*Brevibacillus agri,* and *Agaricus brasiliensis*	Biotechnology	Hamuro et al. (2017), Panwar et al. (2023)
Lactase	*Pyrococcus furiosus,* and *Sulfolobus solfataricus*	Food	Sun et al. (2014), Synowiecki et al. (2006)
Lactonase	*Clonostachys rosea, S. acidocaldarius, and S. solfataricus*	Biotechnology	Restaino et al. (2018), Qiu et al. (2023)
Lignocellulase	*Thermotoga maritima*	Biofuel and Biotechnology	Yeoman et al. (2010)
Lipase	*Anoxybacillus flavithermus, B. sonorensis, Bacillus* spp., *Geobacillus stearothermophilus, G. zalihae,* and *Thermomyces lanuginosus*	Biofuel, Detergent, Food, Pharmaceutical and Biotechnology	Ben Bacha et al. (2015), Hamdan et al. (2023), Nguyen et al. (2025), Rigoldi et al. (2018)
MHETase	*Ideonella sakaiensis*	Biotechnology	Gulati et al. (2025)
Nitrilase	*Acidovorax facilis*	Pharmaceutical	Xu et al. (2018)
Oxidase	*Bacillus* spp.	Detergent	Nguyen et al. (2025), Rigoldi et al. (2018)
Peroxidase	*Bacillus* spp.	Detergent	Nguyen et al. (2025), Rigoldi et al. (2018)
PETase	*Ideonella sakaiensis,* and *Thermobifida fusca*	Biotechnology	Joho et al. (2024), Roth et al. (2014)
Protease	*Bacillus* spp., and *B. licheniformis*	Detergent and Pharmaceutical	Nguyen et al. (2025), Rigoldi et al. (2018), Yuan et al. (2025)
Xylanase	*Bacillus* spp., *Dictyoglomus turgidum* and others	Biofuel and Textile	Ajeje et al. (2021), Nguyen et al. (2025), Rigoldi et al. (2018)

saccharification. Nonetheless, it has been proposed that this economic obstacle can be mitigated through the utilization of thermostable cellulases and xylanases (Ajeje et al., 2021). Conducting saccharification processes at elevated temperatures markedly enhances the reaction efficiency by reducing the viscosity of the biomass substrate and increasing the diffusion rate between the enzyme and substrate. Consequently, enzyme activity is augmented, and the quantity of enzyme required is reduced, thereby significantly reducing the production costs (Yeoman et al., 2010; Rigoldi et al., 2018). For instance, a mutant cellobiohydrolase, isolated from *T. emersonii*, engineered with additional disulfide bridges exhibited improved thermostability, facilitating efficient cellulose hydrolysis at 80 °C (Voutilainen et al., 2010). Furthermore, given that this enzyme has been successfully expressed in *Saccharomyces cerevisiae*, it exemplifies a successful case of artificial thermostability, combining the exceptional property of heat resistance with the commercial viability of large-scale production of the enzyme.

4.3.2 Detergent Industry

In the detergent industry, thermostable enzymes significantly enhance cleaning efficiency while mitigating environmental impact (Rigoldi et al., 2018). Enzymes such as proteases, lipases, amylases, oxidases, peroxidases, and cellulases are incorporated into detergents to facilitate the breakdown of various chemical bonds in an aqueous environment. These enzymes are engineered through protein engineering to maintain stability under elevated temperature, high alkalinity, and the presence of surfactants, thereby ensuring robust cleaning performance (Vojcic et al., 2015). Many enzymes used in detergents are derived from *Bacillus* species, and ongoing research continues to explore novel enzymes and modify existing ones (Nguyen et al., 2025). For instance, thermostable lipase from *Geobacillus stearothermophilus* (formerly *B. stearothermophilus*) strains has been shown to retain over 70% of its activity after incubation at 70 °C for 30 min and demonstrates high compatibility with various commercial detergents (Ben Bacha et al., 2015). Furthermore, alkaline protease AprE2709 from *B. licheniformis* strain 2709, which is extensively used in the Chinese detergent industry, was introduced to enhance electrostatic interactions and hydrophobic bonds, resulting in improved termostability and catalytic activity (Yuan et al., 2025).

4.3.3 Food Industry

Thermostable enzymes play an important role in enhancing product quality and optimizing manufacturing processes in food production. Among these, thermostable amylases are particularly significant in starch processing and bread production, representing approximately 30% of the global enzyme market (Kholikov et al., 2025). Ongoing research aims to further enhance the performance of thermostable enzymes. A notable study reported the modification of α-amylase from *B. licheniformis* through rational design, which led to mutant strains exhibiting improved hydrophobic interactions and reduced disruption of the conformational catalytic residues. This modification resulted in a marked increase in thermostability at pH 8.5 compared to that of the wild-type enzyme (Cui et al., 2022).

Furthermore, thermostable enzymes have been used to produce lactose-free products (Li et al., 2013). Approximately 70% of the global population experiences lactose intolerance, which is characterized by a reduced capacity to digest lactose (Lomer et al., 2008). Lactase, an enzyme responsible for hydrolyzing lactose into glucose and galactose, is crucial for producing lactose-free products. However, traditional lactase is susceptible to high temperatures, necessitating its aseptic addition after pasteurization. Recently, thermostable lactases have been discovered from various sources, including the hyperthermophilic bacterium *Pyrococcus furiosus* (Synowiecki et al., 2006). The utilization of these enzymes facilitates lactose hydrolysis during pasteurization or concurrently with yogurt fermentation, potentially streamlining the process and reducing manufacturing time. The primary enzyme used for lactose hydrolysis is β-galactosidase, and it has been observed that β-galactosidase derived from *Sulfolobus solfataricus* exhibits enhanced enzymatic activity at elevated temperatures due to a mutation from Phe441 to tyrosine (Sun et al., 2014). Additionally, β-galactosidase is employed in the biosynthesis of galacto-oligosaccharides, a category of prebiotics.

4.3.4 Pharmaceuticals and Chemical Industry

Chirality is a critical factor in the pharmaceutical industry because the safety and efficacy of many drugs are contingent upon a specific stereoisomer or enantiomer (Patel, 2006). Traditional chemical synthesis methods often necessitate the use of costly chiral catalysts and harsh reaction conditions, which can lead to the formation of undesirable by-products. In contrast, enzymes serve as biocatalysts in the production of pharmaceutical intermediates, offering remarkable advantages in terms of chemoselectivity, regioselectivity, and stereoselectivity. Enzymes, such as lipases and nitrilases, are employed in the chiral synthesis of these intermediates, and ongoing efforts are being made to enhance their thermostability. For instance, the thermostability of lipase from *G. zalihae* was improved by introducing disulfide bonds, whereas the nitrilase from *Acidovorax facilis* was stabilized by introducing positive charges into the α-helix to fortify the substrate pocket (Xu et al., 2018; Hamdan et al., 2023).

4.3.5 Textile Industry

α-Amylase is a crucial enzyme in the textile industry, primarily utilized for the removal of sizing agents (desizing) applied during the weaving process. Thermostable enzymes facilitate their application at high temperatures (Madhu & Chakraborty, 2017). Furthermore, cellulases are employed in scouring, bleaching, dyeing, and modification of fabric textures (biopolishing). Thermostable cellulases sourced from thermophilic bacteria and those with enhanced thermostability through immobilization have also been developed. It has been demonstrated that the use of these enzymes in the biopolishing of cotton fabrics results in softer textiles with minimal fiber damage (Yu et al., 2015; Ando et al., 2002).

4.3.6 Environmental Biotechnology

Enzymes such as thermostable lipases, which can degrade oil-based pollutants in the environment, hold significant potential for application in environmental bioremediation. For instance, ultra-thermostable alkaline lipase derived from *B. sonorensis* 4R, along with thermostable lipases from *Anoxybacillus flavithermus* HBB 134 and *Thermomyces lanuginosus*, are anticipated to be utilized not only as additives in detergents and the food industry but also for environmental bioremediation. Furthermore, laccases employed for the degradation of phenols and dyes are predominantly of fungal origin for industrial applications; however, a thermostable laccase from *Brevibacillus agri* has been documented (Panwar et al., 2023). Additionally, a highly thermostable mutant has been developed using ASR from laccase derived from the medicinal mushroom *Agaricus brasiliensis* (Hamuro et al., 2017). Similarly, in the context of lactonases used for the degradation of organophosphorus compounds, efforts have been made to isolate thermostable enzymes and to engineer artificially thermostable enzymes (Restaino et al., 2018; Qiu et al., 2023). Moreover, thermostable lignocellulases employed in biofuel production, as previously mentioned, are also used in bioremediation (Yeoman et al., 2010).

In recent years, plastic degradation has emerged as a critical challenge in mitigating ecological pollution caused by microplastics, prompting active research on plastic-degrading enzymes (Amobonye et al., 2021). Current efforts utilizing directed evolution are focused on enhancing the catalytic efficiency and thermostability of polyethylene terephthalate-degrading enzymes (PETase) (Joho et al., 2024). Furthermore, structural analyses of thermostable cutinase, an enzyme capable of effectively degrading PET, have elucidated the structural basis for its remarkable thermostability. These findings suggest potential avenues for further enhancing PETase thermostability through rational design strategies (Roth et al., 2014). In an alternative engineering approach, a whole-cell biocatalyst based on *S. cerevisiae* was developed to display PETase and MHETase, which degrades the PET degradation intermediate compound mono(2-hydroxyethyl) terephthalate (Gulati & Sun, 2025). This innovation enabled the complete depolymerization of PET into its monomeric components while enhancing the stability of the enzymes involved.

Furthermore, a shared characteristic across the previously mentioned industrial sectors is the adoption of enzyme-based technologies, which are often regarded as environmentally friendly alternatives to conventional chemical methods. These technologies facilitate the more efficient utilization of raw materials, minimize waste production, and eliminate the need for hazardous chemicals (Sharma et al., 2019). Notably, thermostable enzymes are distinguished by their ability to operate effectively at elevated temperatures, resulting in higher conversion rates, improved substrate solubility, and diminished likelihood of microbial contamination.

4.4 Concluding Remarks

Enzymes facilitate specific chemical reactions with remarkable specificity and efficiency, making them indispensable across various industrial sectors. Extremozymes are highly sought after for industrial applications because of their ability to retain high activity and stability under extreme conditions, such as high or low temperatures, high salinity, extreme pH levels, and the presence of organic solvents. These properties enable the execution of processes that were previously challenging. However, a significant challenge remains: many natural enzymes, except those from thermophilic bacteria, experience structural alterations and lose activity at high temperatures, posing a problem for their thermostability. To overcome this limitation, enzymes are commonly modified to enhance their thermostability for industrial applications. This chapter introduced two primary strategies—protein engineering and enzyme immobilization—along with the industrial applications of artificially thermostable enzymes. By employing these methods to enhance the thermostability of extremozymes that have not yet been stabilized, their industrial utility as highly efficient enzymes is anticipated to increase in the future. For instance, the radiation-resistant bacterium *Deinococcus radiodurans*, which is the focus of our study, possesses a unique and extraordinary DNA repair mechanism. Among its critical enzymes, PprA is pivotal and is marketed in Japan as a ligation enhancer because of its ability to facilitate the reactions of commercially available ligases. It also has the potential to be used as a medical reagent to identify cells with DNA damage. Enhancing the thermostability of this enzyme could lead to the development of more stable and user-friendly reagents. Furthermore, new applications are anticipated, such as serving as an accelerator for molecular biology experiments conducted under high-temperature conditions, such as LAMP. Globally, as the demand for more efficient and environmentally sustainable processes increases, the significance of thermostable enzymes is expected to increase. Research and development in the field of thermostable enzymes holds substantial promise, both scientifically and economically.

References

Aalbers, F. S., Fürst, M. J., Rovida, S., Trajkovic, M., Rubén Gómez Castellanos, J., Bartsch, S., Vogel, A., Mattevi, A., & Fraaije, M. W. (2020). Approaching boiling point stability of an alcohol dehydrogenase through computationally-guided enzyme engineering. *eLife, 9*, e54639.
Ajeje, S. B., Hu, Y., Song, G., Peter, S. B., Afful, R. G., Sun, F., Asadollahi, M. A., Amiri, H., Abdulkhani, A., & Sun, H. (2021). Thermostable cellulases/xylanases from thermophilic

and hyperthermophilic microorganisms: Current perspective. *Frontiers in Bioengineering & Biotechnology, 9*, 794304.

Akanuma, S., Nakajima, Y., Yokobori, S., Kimura, M., Nemoto, N., Mase, T., Miyazono, K., Tanokura, M., & Yamagishi, A. (2013). Experimental evidence for the thermophilicity of ancestral life. *Proceedings of the National Academy of Sciences of the United States of America, 110*, 11067–11072.

Amobonye, A., Bhagwat, P., Singh, S., & Pillai, S. (2021). Plastic biodegradation: Frontline microbes and their enzymes. *Science of the Total Environment, 759*, 143536.

Ando, S., Ishida, H., Kosugi, Y., & Ishikawa, K. (2002). Hyperthermostable endoglucanase from *Pyrococcus horikoshii. Applied & Environmental Microbiology, 68*, 430–433.

Arakawa, T. (2002). Hydration as a major factor in preferential solvent–protein interactions. *Crystal Growth & Design, 2*, 549–551.

Ben Bacha, A., Moubayed, N. M. S., & Abid, I. (2015). Thermostable, alkaline and detergent-tolerant lipase from a newly isolated thermophilic *Bacillus stearothermophilus. Indian Journal of Biochemistry & Biophysics, 52*, 179–188.

Bhatt, H. B., Sani, R. K., Amoozegar, M. A., & Singh, S. P. (2024). Editorial: Extremozymes: Characteristics, structure, protein engineering and applications. *Frontiers in Microbiology, 15*, 1423463.

Brott, S., Pfaff, L., Schuricht, J., Schwarz, J. N., Böttcher, D., Badenhorst, C. P. S., Wei, R., & Bornscheuer, U. T. (2022). Engineering and evaluation of thermostable IsPETase variants for PET degradation. *Engineering in Life Sciences, 22*, 192–203.

Capriotti, E., Fariselli, P., & Casadio, R. (2005). I-Mutant2.0: Predicting stability changes upon mutation from the protein sequence or structure. *Nucleic Acids Research, 33*, W306–W310.

Chaudhari, Y. B., Várnai, A., Sørlie, M., Horn, S. J., & Eijsink, V. G. H. (2023). Engineering cellulases for conversion of lignocellulosic biomass. *Protein Engineering, Design & Selection, 36*, gzad002.

Chen, F., Gaucher, E. A., Leal, N. A., Hutter, D., Havemann, S. A., Govindarajan, S., Ortlund, E. A., & Benner, S. A. (2010). Reconstructed evolutionary adaptive paths give polymerases accepting reversible terminators for sequencing and SNP detection. *Proceedings of the National Academy of Sciences of the United States of America, 107*, 1948–1953.

Chica, R. A., Doucet, N., & Pelletier, J. N. (2005). Semi-rational approaches to engineering enzyme activity: Combining the benefits of directed evolution and rational design. *Current Opinion in Biotechnology, 16*, 378–384.

Cole, M. F., & Gaucher, E. A. (2011). Exploiting models of molecular evolution to efficiently direct protein engineering. *Journal of Molecular Evolution, 72*, 193–203.

Craig, D. B., & Dombkowski, A. A. (2013). Disulfide by design 2.0: A web-based tool for disulfide engineering in proteins. *BMC Bioinformatics, 14*, 346.

Cui, X., Yuan, X., Li, S., Hu, X., Zhao, J., & Zhang, G. (2022). Simultaneously improving the specific activity and thermostability of α-amylase BLA by rational design. *Bioprocess and Biosystems Engineering, 45*, 1839–1848.

Daintith, J. (2008). *A dictionary of chemistry*. Oxford University Press.

Dalby, P. A. (2011). Strategy and success for the directed evolution of enzymes. *Current Opinion in Structural Biology, 21*, 473–480.

Datta, S., Christena, L. R., & Rajaram, Y. R. (2013). Enzyme immobilization: an overview on techniques and support materials. *3 Biotech, 3*, 1–9.

Ebrahimi Fana, S., Fazaeli, A., & Aminian, M. (2023). Directed evolution of cholesterol oxidase with improved thermostability using error-prone PCR. *Biotechnology Letters, 45*, 1159–1167.

Folkman, L., Stantic, B., Sattar, A., & Zhou, Y. (2016). EASE-MM: Sequence-based prediction of mutation-induced stability changes with feature-based multiple models. *Journal of Molecular Biology, 428*, 1394–1405.

Gaucher, E. A., Thomson, J. M., Burgan, M. F., & Benner, S. A. (2003). Inferring the palaeoenvironment of ancient bacteria on the basis of resurrected proteins. *Nature, 425*, 285–288.

Gaucher, E. A., Govindarajan, S., & Ganesh, O. K. (2008). Palaeotemperature trend for Precambrian life inferred from resurrected proteins. *Nature, 451*, 704–707.

Geethanjali, S., & Subash, A. (2013). Optimization and immobilization of purified *Labeo rohita* visceral protease by entrapment method. *Enzyme Research, 2013*, 874050.

Giver, L., Gershenson, A., Freskgard, P. O., & Arnold, F. H. (1998). Directed evolution of a thermostable esterase. *Proceedings of the National Academy of Sciences of the United States of America, 95*, 12809–12813.

Goldenzweig, A., Goldsmith, M., Hill, S. E., Gertman, O., Laurino, P., Ashani, Y., Dym, O., Unger, T., Albeck, S., Prilusky, J., Lieberman, R. L., Aharoni, A., Silman, I., Sussman, J. L., Tawfik, D. S., & Fleishman, S. J. (2018). Automated structure- and sequence-based design of proteins for high bacterial expression and stability. *Molecular Cell, 70*, 380.

Gulati, S., & Sun, Q. (2025). Complete Enzymatic Depolymerization of Polyethylene Terephthalate (PET) Plastic Using a Saccharomyces cerevisiae-Based Whole-Cell Biocatalyst. *Environmental Science & Technology Letters, 12*(4), 419-424.

Hamdan, S. H., Maiangwa, J., Nezhad, N. G., Ali, M. S. M., Normi, Y. M., Shariff, F. M., Rahman, R., & Leow, T. C. (2023). Knotting terminal ends of mutant T1 lipase with disulfide bond improved structure rigidity and stability. *Applied Microbiology and Biotechnology, 107*, 1673–1686.

Hamuro, Y., Tajima, K., Matsumoto-Akanuma, A., Sakamoto, S., Furukawa, R., Yamagishi, A., Ohno, N., & Akanuma, S. (2017). Characterization of a thermostable mutant of *agaricus brasiliensis* laccase created by phylogeny-based design. *Journal of Bioscience and Bioengineering, 124*, 623–629.

Harris, J. M., Martin, N. E., & Modi, M. (2001). Pegylation: A novel process for modifying pharmacokinetics. *Clinical Pharmacokinetics, 40*, 539–551.

Hoseki, J., Yano, T., Koyama, Y., Kuramitsu, S., & Kagamiyama, H. (1999). Directed evolution of thermostable kanamycin-resistance gene: A convenient selection marker for *Thermus thermophilus*. *Journal of Biochemistry, 126*, 951–956.

Jailani, N., Jaafar, N. R., Suhaimi, S., Mackeen, M. M., Bakar, F. D. A., & Illias, R. M. (2022). Cross-linked cyclodextrin glucanotransferase aggregates from *Bacillus lehensis* G1 for cyclodextrin production: Molecular modeling, developmental, physicochemical, kinetic and thermodynamic properties. *International Journal of Biological Macromolecules, 213*, 516–533.

Joho, Y., Vongsouthi, V., Gomez, C., Larsen, J. S., Ardevol, A., & Jackson, C. J. (2024). Improving plastic degrading enzymes *via* directed evolution. *Protein Engineering, Design & Selection, 37*, gzae009.

Jumper, J., Evans, R., Pritzel, A., Green, T., Figurnov, M., Ronneberger, O., Tunyasuvunakool, K., Bates, R., Žídek, A., Potapenko, A., Bridgland, A., Meyer, C., Kohl, S. A. A., Ballard, A. J., Cowie, A., Romera-Paredes, B., Nikolov, S., Jain, R., Adler, J., Back, T., Petersen, S., Reiman, D., Clancy, E., Zielinski, M., Steinegger, M., Pacholska, M., Berghammer, T., Bodenstein, S., Silver, D., Vinyals, O., Senior, A. W., Kavukcuoglu, K., Kohli, P., & Hassabis, D. (2021). Highly accurate protein structure prediction with AlphaFold. *Nature, 596*, 583–589.

Khachatrian, A. A., Mukhametzyanov, T. A., Salikhov, R. Z., Klimova, A. E., Yakhvarov, D. G., Garifullin, B. F., Terenteva, O. S., Padnya, P. L., Stoikov, I. I., & Solomonov, B. N. (2025). Cholinium-based ionic liquids modulate protein stability: A comparative study of enzymes and albumins. *Molecules, 30*(7), 1574.

Kholikov, A., Vokhidov, K., Murtozoyev, A., Tóth, Z. S., Nagy, G. N., Vértessy, B. G., & Makhsumkhanov, A. (2025). Characterization of a thermostable α-amylase from *Bacillus licheniformis* 104.K for industrial applications. *Microorganisms, 13*, 1757.

Kwasigroch, J. M., Gilis, D., Dehouck, Y., & Rooman, M. (2002). PoPMuSiC, rationally designing point mutations in protein structures. *Bioinformatics, 18*, 1701–1702.

Li, B., Wang, Z., Li, S., Donelan, W., Wang, X., Cui, T., & Tang, D. (2013). Preparation of lactose-free pasteurized milk with a recombinant thermostable β-glucosidase from *Pyrococcus furiosus*. *BMC Biotechnology, 13*, 73.

Liu, Z., Cheng, Z., Ye, S., Zhou, L., & Zhou, Z. (2019). Catalytic ability improvement of phenylalanine hydroxylase from *Chromobacterium violaceum* by N-terminal truncation and proline introduction. *Journal of Microbiology and Biotechnology, 29*, 1375–1382.

Lomer, M. C., Parkes, G. C., & Sanderson, J. D. (2008). Review article: lactose intolerance in clinical practice—myths and realities. *Alimentary Pharmacology & Therapeutics, 27*, 93–103.

Madhu, A., & Chakraborty, J. N. (2017). Developments in application of enzymes for textile processing. *Journal of Cleaner Production, 145*, 114–133.

Makharadze, D., Del Valle, L. J., Katsarava, R., & Puiggalí, J. (2025). The art of PEGylation: From simple polymer to sophisticated drug delivery system. *International Journal of Molecular Sciences, 26*(7), 3102.

Markets & Markets. (2025). *Industrial enzymes market*. https://www.marketsandmarkets.com/Market-Reports/industrial-enzymes-market-237327836.html. Accessed 1 July.

Masuda-Nishimura, I., Minamihara, T., & Koyama, Y. (1999). Improvement in thermal stability and reactivity of pyranose oxidase from *Coriolus versicolor* by random mutagenesis. *Biotechnology Letters, 21*, 203–207.

Mesbah, N. M. (2022). Industrial biotechnology based on enzymes from extreme environments. *Frontiers in Bioengineering and Biotechnology, 10*, 870083.

Miklos, A. C., Li, C., Sharaf, N. G., & Pielak, G. J. (2010). Volume exclusion and soft interaction effects on protein stability under crowded conditions. *Biochemistry, 49*, 6984–6991.

Mildvan, A. S. (2004). Inverse thinking about double mutants of enzymes. *Biochemistry, 43*, 14517–14520.

Mohtashami, M., Fooladi, J., Haddad-Mashadrizeh, A., Housaindokht, M. R., & Monhemi, H. (2019). Molecular mechanism of enzyme tolerance against organic solvents: Insights from molecular dynamics simulation. *International Journal of Biological Macromolecules, 122*, 914–923.

Mullis, K., Faloona, F., Scharf, S., Saiki, R., Horn, G., & Erlich, H. (1986). Specific enzymatic amplification of DNA *in vitro*: The polymerase chain reaction. *Cold Spring Harbor Symposia on Quantitative Biology, 51*(Pt 1), 263–273.

Musil, M., Stourac, J., Bendl, J., Brezovsky, J., Prokop, Z., Zendulka, J., Martinek, T., Bednar, D., & Damborsky, J. (2017). FireProt: Web server for automated design of thermostable proteins. *Nucleic Acids Research, 45*, W393–w399.

Nguyen, V. M. L., Ndao, A., Peterson, E. C., Blais, J. F., & Adjallé, K. (2025). *Bacillus* species: Evolving roles in bio-based detergents. *PRO, 13*, 1885.

Ogino, H., & Ishikawa, H. (2001). Enzymes which are stable in the presence of organic solvents. *Journal of Bioscience and Bioengineering, 91*, 109–116.

Panwar, V., Lzaod, S., & Dutta, T. (2023). Thermostable bacterial laccase: Catalytic properties and its application in biotransformation of emerging pollutants. *ACS Omega, 8*, 34710–34719.

Patel, R. N. (2006). Biocatalysis: Synthesis of chiral intermediates for drugs. *Current Opinion in Drug Discovery & Development, 9*, 741–764.

Piccoli, V., & Martínez, L. (2025). Cation hydrophobicity effects on protein solvation in aqueous ionic liquids. *The Journal of Physical Chemistry. B, 129*, 6765–6776.

Pires, D. E., Ascher, D. B., & Blundell, T. L. (2014). mCSM: Predicting the effects of mutations in proteins using graph-based signatures. *Bioinformatics, 30*, 335–342.

Pongsupasa, V., Anuwan, P., Maenpuen, S., & Wongnate, T. (2022). Rational-design engineering to improve enzyme thermostability. In F. Magnani, C. Marabelli, & F. Paradisi (Eds.), *Enzyme engineering: Methods and protocols* (pp. 159–178). Springer US.

Qiu, Y., Xu, H., Ji, Q., Xu, R., Zhu, M., Dang, Y., Shi, X., Zhang, L., & Xia, Y. (2023). Mutation, food-grade expression, and characterization of a lactonase for zearalenone degradation. *Applied Microbiology and Biotechnology, 107*, 5107–5118.

Ramírez-Ramírez, L. G., Zazueta-Álvarez, D. E., Fileto-Pérez, H. A., Reyes-Jáquez, D., Núñez-Núñez, C. M., Galindo-De la Rosa, J. D., López-Miranda, J., & Vázquez-Ortega, P. G. (2022). Improvement in the thermostability of a recombinant β-glucosidase immobilized in zeolite under different conditions. *Molecules, 27*(13), 4105.

Reetz, M. T., Wilensek, S., Zha, D., & Jaeger, K. E. (2001). Directed evolution of an enantioselective enzyme through combinatorial multiple-cassette mutagenesis. *Angewandte Chemie (International Ed. in English), 40*, 3589–3591.

Restaino, O. F., Borzacchiello, M. G., Scognamiglio, I., Fedele, L., Alfano, A., Porzio, E., Manco, G., De Rosa, M., & Schiraldi, C. (2018). High yield production and purification of two recombinant thermostable phosphotriesterase-like lactonases from *Sulfolobus acidocaldarius* and *Sulfolobus solfataricus* useful as bioremediation tools and bioscavengers. *BMC Biotechnology, 18*, 18.

Rigoldi, F., Donini, S., Redaelli, A., Parisini, E., & Gautieri, A. (2018). Review: Engineering of thermostable enzymes for industrial applications. *APL Bioengineering, 2*, 011501.

Robescu, M. S., & Bavaro, T. (2025). A comprehensive guide to enzyme immobilization: All you need to know. *Molecules, 30*(4), 939.

Rodríguez-Martínez, J. A., Solá, R. J., Castillo, B., Cintrón-Colón, H. R., Rivera-Rivera, I., Barletta, G., & Griebenow, K. (2008). Stabilization of alpha-chymotrypsin upon PEGylation correlates with reduced structural dynamics. *Biotechnology and Bioengineering, 101*, 1142–1149.

Roth, C., Wei, R., Oeser, T., Then, J., Föllner, C., Zimmermann, W., & Sträter, N. (2014). Structural and functional studies on a thermostable polyethylene terephthalate degrading hydrolase from *Thermobifida fusca. Applied Microbiology and Biotechnology, 98*, 7815–7823.

Ruan, Y., Zhang, R., & Xu, Y. (2022). Directed evolution of maltogenic amylase from B*acillus licheniformis* R-53: Enhancing activity and thermostability improves bread quality and extends shelf life. *Food Chemistry, 381*, 132222.

Sanchez-Ruiz, J. M., & Makhatadze, G. I. (2001). To charge or not to charge? *Trends in Biotechnology, 19*, 132–135.

Schmitzer, A. R., Lépine, F., & Pelletier, J. N. (2004). Combinatorial exploration of the catalytic site of a drug-resistant dihydrofolate reductase: Creating alternative functional configurations. *Protein Engineering, Design & Selection, 17*, 809–819.

Schymkowitz, J., Borg, J., Stricher, F., Nys, R., Rousseau, F., & Serrano, L. (2005). The FoldX web server: An online force field. *Nucleic Acids Research, 33*, W382–W388.

Sharma, S., Vaid, S., Bhat, B., Singh, S., & Bajaj, B. K. (2019). Thermostable enzymes for industrial biotechnology. In R. S. Singh, R. R. Singhania, A. Pandey, & C. Larroche (Eds.), *Advances in enzyme technology* (pp. 469–495). Elsevier.

Sheldon, R. A., & van Pelt, S. (2013). Enzyme immobilisation in biocatalysis: Why, what and how. *Chemical Society Reviews, 42*, 6223–6235.

Shi, L., Liu, P., Tan, Z., Zhao, W., Gao, J., Gu, Q., Ma, H., Liu, H., & Zhu, L. (2023). Complete depolymerization of PET wastes by an evolved PET hydrolase from directed evolution. *Angewandte Chemie (International Ed. in English), 62*, e202218390.

Singh, R. K., Tiwari, M. K., Singh, R., & Lee, J. K. (2013). From protein engineering to immobilization: Promising strategies for the upgrade of industrial enzymes. *International Journal of Molecular Sciences, 14*, 1232–1277.

Spadiut, O., Nguyen, T. T., & Haltrich, D. (2010). Thermostable variants of pyranose 2-oxidase showing altered substrate selectivity for glucose and galactose. *Journal of Agricultural and Food Chemistry, 58*, 3465–3471.

Sun, X., Duan, X., Wu, D., Chen, J., & Wu, J. (2014). Characterization of *Sulfolobus solfataricus* β-galactosidase mutant F441Y expressed in *Pichia pastoris. Journal of the Science of Food and Agriculture, 94*, 1359–1365.

Sun, H., Yang, H., Huang, W., & Zhang, S. (2015). Immobilization of laccase in a sponge-like hydrogel for enhanced durability in enzymatic degradation of dye pollutants. *Journal of Colloid and Interface Science, 450*, 353–360.

Suzuki, Y. (1999). The proline rule – A strategy for protein thermal stabilization. *Proceedings of the Japan Academy, Series B, 75*, 133–137.

Synowiecki, J., Grzybowska, B., & Zdziebło, A. (2006). Sources, properties and suitability of new thermostable enzymes in food processing. *Critical Reviews in Food Science and Nutrition, 46*, 197–205.

Tamakoshi, M., Nakano, Y., Kakizawa, S., Yamagishi, A., & Oshima, T. (2001). Selection of stabilized 3-isopropylmalate dehydrogenase of *Saccharomyces cerevisiae* using the host-vector system of an extreme thermophile, *Thermus thermophilus*. *Extremophiles, 5*, 17–22.

Thomas Ó Conchúir, S., Barlow, K. A., Pache, R. A., Ollikainen, N., Kundert, K., O'Meara, M. J., Smith, C. A., & Kortemme, T. (2015). A web resource for standardized benchmark datasets, metrics, and rosetta protocols for macromolecular modeling and design. *PLoS One, 10*, e0130433.

Thomson, R. E. S., Carrera-Pacheco, S. E., & Gillam, E. M. J. (2022). Engineering functional thermostable proteins using ancestral sequence reconstruction. *The Journal of Biological Chemistry, 298*, 102435.

Thornton, J. W. (2004). Resurrecting ancient genes: Experimental analysis of extinct molecules. *Nature Reviews. Genetics, 5*, 366–375.

Vojcic, L., Pitzler, C., Körfer, G., Jakob, F., Ronny, M., Maurer, K. H., & Schwaneberg, U. (2015). Advances in protease engineering for laundry detergents. *New Biotechnology, 32*, 629–634.

Voutilainen, S. P., Murray, P. G., Tuohy, M. G., & Koivula, A. (2010). Expression of *Talaromyces emersonii* cellobiohydrolase Cel7A in *Saccharomyces cerevisiae* and rational mutagenesis to improve its thermostability and activity. *Protein Engineering, Design & Selection, 23*, 69–79.

Wahba, M. I., Saleh, S. A. A., Wahab, W. A. A., & Mostafa, F. A. (2025). Studies on the preparation of a sufficient carrier from egg protein and carrageenan for cellulase with optimization and application. *Scientific Reports, 15*, 3868.

Watanabe, K., & Suzuki, Y. (1998). Protein thermostabilization by proline substitutions. *Journal of Molecular Catalysis B: Enzymatic, 4*, 167–180.

Watanabe, K., Ohkuri, T., Yokobori, S., & Yamagishi, A. (2006). Designing thermostable proteins: Ancestral mutants of 3-isopropylmalate dehydrogenase designed by using a phylogenetic tree. *Journal of Molecular Biology, 355*, 664–674.

Watanabe, M., Fukada, H., & Ishikawa, K. (2016). Construction of thermophilic xylanase and its structural analysis. *Biochemistry, 55*, 4399–4409.

Wijma, H. J., Floor, R. J., Jekel, P. A., Baker, D., Marrink, S. J., & Janssen, D. B. (2014). Computationally designed libraries for rapid enzyme stabilization. *Protein Engineering, Design & Selection, 27*, 49–58.

Xu, Z., Cai, T., Xiong, N., Zou, S. P., Xue, Y. P., & Zheng, Y. G. (2018). Engineering the residues on "A" surface and C-terminal region to improve thermostability of nitrilase. *Enzyme and Microbial Technology, 113*, 52–58.

Yamaguchi, H., & Miyazaki, M. (2024). Bioremediation of hazardous pollutants using enzyme-immobilized reactors. *Molecules, 29*(9), 2021.

Yamashiro, K., Yokobori, S., Koikeda, S., & Yamagishi, A. (2010). Improvement of *Bacillus circulans* beta-amylase activity attained using the ancestral mutation method. *Protein Engineering, Design & Selection, 23*, 519–528.

Yang, M., Li, Z., Ren, H., Lu, C., Gao, X., & Xu, H. (2024). PEG modification increases thermostability and inhibitor resistance of Bst DNA polymerase. *Bioscience, Biotechnology, and Biochemistry, 88*, 768–775.

Yeoman, C. J., Han, Y., Dodd, D., Schroeder, C. M., Mackie, R. I., & Cann, I. K. (2010). Thermostable enzymes as biocatalysts in the biofuel industry. *Advances in Applied Microbiology, 70*, 1–55.

Yin, S., Ding, F., & Dokholyan, N. V. (2007). Eris: An automated estimator of protein stability. *Nature Methods, 4*, 466–467.

Yu, Y., Yuan, J., Wang, Q., Fan, X., Wang, P., & Cui, L. (2015). Noncovalent immobilization of cellulases using the reversibly soluble polymers for biopolishing of cotton fabric. *Biotechnology and Applied Biochemistry, 62*, 494–501.

Yuan, Y., Zhao, G., Lu, J., Wang, L., Shi, Y., & Zhang, J. (2025). Enhancing the thermostability of *Bacillus licheniformis* alkaline protease 2709 by computation-based rational design. *Molecules, 30*, 1160.

Zhao, H. (2016). Protein stabilization and enzyme activation in ionic liquids: Specific ion effects. *Journal of Chemical Technology and Biotechnology, 91*, 25–50.

Chapter 5
Lignin and Its Fragment Degradation for the Valorization of Lignin by Microorganisms: Emphasis on Marine-Derived Isolates

Eri Kumagawa and Yukari Ohta

Abstract Lignin, a structurally complex aromatic polymer comprising 15–30% of woody plant biomass, plays a pivotal role in plant integrity and global carbon cycling. This section investigates the degradation of lignin and its oligomeric fragments by microorganisms, emphasizing the unique enzymatic systems adapted to the physicochemical challenges of marine environments. The structural heterogeneity of lignin complicates its valorization; however, marine bacteria have evolved specialized metabolic pathways that enable efficient processing under both aerobic and anaerobic conditions. Particular attention is given to key ligninolytic enzymes, including intracellular glutathione-S-transferases, which facilitate the conversion of lignin-derived compounds. The ecological significance of marine microbial communities in mediating lignin degradation and contributing to coastal and oceanic carbon dynamics is also addressed. Advances in molecular biology have revealed novel metabolic pathways and genetic architectures, offering promising avenues for biotechnological applications including the production of bio-based materials and bioplastics. As global interest in sustainable resource utilization intensifies, elucidating the metabolic potential of marine bacteria for lignin valorization is crucial for promoting environmental resilience and advancing a circular bioeconomy.

Keywords Lignin microbial metabolism · Enzymatic conversion · Marine environment · Bio-based material

---

E. Kumagawa
Amano Enzyme, Inc., Business Development Division, Nagoya, Aichi, Japan
e-mail: eri_kumagawa@amano-enzyme.com

Y. Ohta (✉)
School of Life and Environmental Sciences, Azabu University, Sagamihara, Kanagawa, Japan
e-mail: y-ohta@azabu-u.ac.jp

M. Ito et al. (eds.), *Extremophiles for a Sustainable Future*,
https://doi.org/10.1007/978-981-95-6970-0_5

5.1 Introduction

5.1.1 *Lignin in Environments*

Lignin represents one of the abundant biopolymers on Earth, comprising 15–30% of woody plant biomass and serving as a crucial structural component of vascular plants (Ralph et al., 2004). This complex aromatic polymer arises through radical-mediated oxidative coupling of three principal 4-hydroxyphenylpropanoid monomers: *p*-coumaryl alcohol, coniferyl alcohol, and sinapyl alcohol. These monomers give rise to p-hydroxyphenyl (H), guaiacyl (G), and syringyl (S) lignin units, respectively, which are connected through various interunit linkages including β-O-4 aryl ether bonds (50–60% of total linkages), β-5 phenylcoumarin bonds (9–12%), β-β resinol bonds (3–12%), and 5–5 biphenyl bonds (4–8%). Among them, β-O-4 aryl ether is the most abundant linkage. This heterogeneity complicates spectroscopic characterization and chemical valorization (Happs et al., 2021), as bond types and monomer ratios vary by species, tissue type, and developmental stage. The ratios of H, G, and S lignin units significantly influence the mechanical strength of lignocellulosic biomass, as well as its thermal, chemical, and biological stability and reactivity.

The three-dimensional, highly branched, and chemically heterogeneous architecture of lignin confers mechanical strength, water impermeability, and resistance to microbial attack to plant cell walls. This structural complexity makes lignin one of the most recalcitrant organic compounds in nature, significantly impacting its valorization potential and material performance—even though it remains the most abundant renewable aromatic compound in nature (Umezawa, 2018; Yoo et al., 2018).

5.1.2 *Marine Environment Characteristics and Lignin Input*

Lignin in plant residues forms a recalcitrant carbon pool, moderating nutrient mineralization and enhancing long-term carbon sequestration. Marine environments receive substantial inputs of terrestrial organic matter (Gough et al., 1993; Li et al., 2012), with an estimated approximately 420 Tg C/yr of organic carbon of terrestrial carbon entering ocean systems through riverine discharge, coastal erosion, and atmospheric deposition (Chen et al., 2025). Lignin comprises approximately 15–25% of the terrestrial organic carbon input, plausibly making it a significant component of marine dissolved organic matter (DOM). The role of heterotrophic bacteria and archaea in lignin transformation in the open ocean was assessed (McDonald et al., 2019). This study investigated the role of marine heterotrophic bacteria and archaea in the transformation of lignin in the open ocean, specifically within the Sargasso Sea. The research highlights that these microbes can alter the optical properties of lignin and may serve as a potential sink for a portion of this macromolecule, thereby influencing the oceanic carbon cycle. The other study (Lu et al., 2016) investigated the sources and transformations of terrigenous dissolved organic matter (tDOM), specifically dissolved lignin phenols and chromophoric DOM, in Otsuchi Bay, Japan. Otsuchi Bay is a semi-closed coastal area in northern Japan, significantly impacted by the 2011 Tohoku earthquake and tsunami. The research aimed to understand how physical mixing, photochemical processes, and biodegradation influence DOM dynamics in this coastal environment. The tDOM in Otsuchi Bay primarily originates from nonwoody angiosperms, and its distribution is largely controlled by physical mixing. While both photochemical and biological processes contribute to tDOM degradation, biodegradation emerges as the dominant removal pathway for lignin phenols in this coastal system.

The marine environments present unique challenges for organic matter degradation, including the following factors: high salinity (typically 35 ppt), variable pH levels (7.5–8.5 in surface waters), fluctuating pressure with depth, regions of limited oxygen availability (oxygen minimum zones), often nutrient-limited conditions, and temperature variations with depth and latitude. These environmental conditions create distinct selective pressures that have influenced the evolution of specialized microbial communities adapted to marine lignin degradation. Fate of lignin in ecosystems has profound implications for global carbon cycling and biogeochemical processes (Gough et al., 1993; Li et al., 2012).

5.1.3 Marine Bacteria as Potential Lignin Degraders

Marine bacteria have evolved diverse enzymatic systems and metabolic pathways to process lignin under the challenging conditions of marine environments (Ohta et al., 2012; Wang et al., 2021; Peng & Lin, 2025). Additionally, the ocean serves as a crucial reservoir for a diverse array of microorganisms, many of which can enter a dormant state. This dormancy enables them to withstand the fluctuating environmental conditions of their habitat, effectively pausing their metabolic processes until more favorable circumstances arise. The ocean provides a unique refuge for these microorganisms, allowing them to remain shielded from external threats, thereby ensuring their longevity and potential for reactivation. This intricate relationship between the marine environments and its microbial inhabitants plays a vital role in maintaining ecological balance and supporting various biogeochemical cycles (McDonald et al., 2019). Understanding the metabolism and genetic potential of these microorganisms is essential not only for ecological research but also for exploring biotechnological applications that could arise from their unique capabilities. Recent advances in molecular biology techniques, including metagenomics, proteomics, and synthetic biology, have revolutionized our understanding of marine bacterial lignin degradation. These tools have revealed previously unknown metabolic pathways, enzyme families, and genetic regulatory mechanisms that govern lignin metabolism in marine environments (Leadbeater et al., 2021; Peng et al., 2023; Peng & Lin, 2025).

5.1.4 *Ecological Significance*

Although terrestrial lignocellulose degradation has been extensively characterized, the contributions of marine and estuarine microbial communities—particularly under fluctuating physicochemical conditions—remain comparatively underexplored (Lam et al., 2022; Peng et al., 2023; Peng & Lin, 2025). Coastal wetlands and mangrove ecosystems function as significant reservoirs of organic carbon, with mangroves alone accounting for approximately 30% of global coastal carbon sequestration. Wherein microbial consortia facilitate the transformation of complex lignocellulosic substrates (Peng et al., 2023; Ghose et al., 2024; Ding et al., 2025). Central to these inquiries are the characterization of microbial assemblages and the examination of how coastal environmental gradients influence microbial community structure and functional capacity. These ecosystems play a pivotal role in blue carbon dynamics (Leadbeater et al., 2021). Over the past several decades, research has progressed from descriptive surveys of microbial diversity within these habitats to mechanistic investigations into the functional potential and metabolic pathways governing lignocellulose degradation.

5.2 Diversity of Lignin Degradation Pathways

5.2.1 *Aerobic Degradation Pathways*

Under aerobic conditions, the majority of bacteria in both marine and terrestrial environments primarily employ oxidative enzymes to break down lignin fragments (Xu et al., 2019). These systems typically consist of activation of the aromatic ring and cleavage of tri- or diol-intermediates by oxidative enzymes. The β-ketoadipate pathway represents one of the most well-characterized routes for energy intake from aromatic compounds including lignin-derived compounds, which is analogous to fatty acid degradation (Wells Jr & Ragauskas, 2012).

Ferulic acid serves as a model substrate for studying bacterial aromatic catabolism. Its microbial breakdown proceeds via four mechanistically distinct routes (Xu et al., 2019). First, the non-oxidative decarboxylation pathway, in which decarboxylases remove the carboxyl group without prior oxidation, begins with activation of ferulic acid by coenzyme A (CoA) ligation and proceeds through hydration/aldol cleavage steps rather than classical β-oxidation. Second, the CoA-dependent β-oxidation pathway shortens the side chain through iterative β-oxidation steps. Third, the CoA-dependent non-β-oxidation pathway involves feruloyl-CoA formation followed by hydration and aldol cleavage. Fourth, the side-chain reduction pathway reduces the α,β-unsaturated side chain of ferulic acid prior to subsequent decarboxylation and oxidation steps. All four pathways converge on vanillic acid before further funneling into central aromatic metabolism and the tricarboxylic acid cycle.

In the recent study (Ley et al., 2023), two bacterial consortia were enriched from nearshore sediments of the East and South China Seas. These consortia demonstrated the ability to degrade alkali lignin, a model compound for lignin. Consortiums were primarily composed of *Nitratireductor* sp. and *Halomonas* sp., both belonging to the Proteobacteria phylum. The study revealed that marine microorganisms in nearshore regions predominantly utilize pathways for lignin degradation, employing protocatechuate and catechol as ring-cleavage substrates, which is similar to that of terrestrial microbial systems. These processes are considered to form a crucial part of the oceanic carbon cycle, regardless of whether these microorganisms are found in sediments or the water column (Ley et al., 2023).

The phenylacetyl-CoA pathway provides an alternative route for lignin degradation, especially for marine bacteria, particularly effective for processing side-chain modified lignin fragments, which differs from the beta-ketoadipate pathway that is typical for terrestrial lignin-degrading fungi (Woo & Hazen, 2018). 16S rRNA amplicon sequencing showed an increased abundance of Gammaproteobacteria and Alphaproteobacteria (e.g., families Idiomarinaceae, Alcanivoracaceae, and Alteromonadaceae) in the lignin-adapted bacterial communities in the Mediterranean Sea. Metagenomic analysis revealed a higher abundance of genes involved in aromatic compound degradation, specifically the phenylacetyl-CoA pathway, after lignin enrichment (Woo & Hazen, 2018).

5.2.2 Anaerobic Degradation Pathways

In oxygen-limited environments, such as sediments and deep-sea habitats, marine bacteria have developed certain lignin degradation mechanisms. These pathways often involve reductive processes and the production of specialized enzymes capable of functioning without molecular oxygen (Wang et al., 2021). Recent study (Peng et al., 2023) conducting in situ lignocellulose enrichment experiments coupled with 16S rRNA amplicon sequencing and shotgun metagenomics sequencing aimed to identify and characterize bacterial consortia responsible for degrading different lignin/lignocellulosic substrates in the southern-east intertidal zone of the East China Sea. The incubation experiments using lignin-amended oligotrophic seawater from the Sargasso Sea showed that the populations of specific bacterial and archaeal phyla, particularly the archaeal groups Euryarchaeota and Thaumarchaeota, increased markedly compared with the unamended control treatments, with the strongest response observed under simultaneous nitrogen and phosphorus supplementation (McDonald et al., 2019). The research highlights the substrate-dependent taxonomic groups and the comprehensive set of genes associated with lignin aerobic/anaerobic degradation, emphasizing the critical yet understudied ecological roles of these bacteria in marine ecosystems (Hou et al., 2023).

5.2.3 *Key Ligninolytic Enzymes*

5.2.3.1 DyP-Type, Laccases and Manganese Peroxidases

Dye-decolorizing peroxidases (DyPs) represent a unique class of heme-containing enzymes that play crucial roles in lignin degradation by terrestrial bacteria (Catucci et al., 2020; Jiang et al., 2022). A recent study established models for identifying marine lignin-utilizing bacteria by linking extracellular enzyme activity (specifically DyP), reactive oxygen species production, and cell growth to lignin degradation, by linking these phenotypes to actual organisms in *Bacillus* and *Vibrio* strains (Wang et al., 2021).

Laccases, copper-containing oxidases, represent another important class of ligninolytic enzymes in marine bacteria. These enzymes catalyze the oxidation of various phenolic and non-phenolic compounds, contributing to lignin breakdown through radical-mediated mechanisms.

Manganese peroxidases, though less common in marine bacteria compared to fungi, have been identified in several marine bacterial species. These enzymes utilize manganese as a mediator to oxidize lignin fragments, particularly those containing non-phenolic structures.

5.2.3.2 Ferulic Acid Esterases

Ferulic acid (FA), *p*-coumaric acid, and 4-hydroxybenzoic acid are hydroxycinnamic acids (HCAs) abundant in grasses (Chandrakanth et al., 2023). HCAs integrated into polysaccharides and lignin contribute to the recalcitrance of plants linking the cell wall polymers.

Microbial enzymes can enhance the efficient decomposition of plant biomass by processing HCAs. Numerous HCA esterases have been identified and characterized at the protein level. The depolymerization of lignocellulosic plant biomass is facilitated by the synergistic actions of auxiliary enzymes in concert with those targeting the primary structural components. Among HCA esterases, FA esterases (FAEs) are the most frequently reported (Guo et al., 2025). FAEs are primarily derived from fungi (e.g., *Anaeromyces*, *Aspergillus*, *Neurospora*, *Thermothelomyces*) and bacteria (e.g., *Cellulosilyticum*, *Cellvibrio*, *Prevotella*, *Ruminiclostridium*) (Leontakianakou et al., 2025). Hundreds of FAE sequences are registered in databases like Carbohydrate-Active enZYmes Database (CAZy) (Drula et al., 2022) and α/β hydrolase fold superfamily ESTHER database (Lenfant et al., 2013). In the ESTHER database, FAEs are classified into A85-FAE, FaeC, and tannase families.

5.2.3.3 Case Study of a FAE from the Marine-Derived Sphingomonadales Bacterium

In our study, a *Sphingobium* strain degrading wood components from submerged wood collected from the seabed of Otsuchi Bay, Tohoku region, Japan was investigated (Ohta et al., 2021). Microscopic observation of the wood revealed numerous shipworms and polychaetes. Feces collected from shipworm burrows showed various bacteria (rod-shaped, short rod-shaped, coccoid) attached to wood fragments. Submerged wood serves as both a nutrient source and a habitat for diverse organisms (Ohta et al., 2021).

Approximately 60 bacterial strains were isolated from solid cultures using rice straw and wood flour as carbon sources. While FAE activity in aquatic microorganisms is rarely reported, about two-thirds of the isolates from submerged wood showed FAE activity (Ohta et al., 2021). A *Sphingobium* strain OW59 with high ferulic acid esterase activity was selected and named. OW59 could grow on various aromatic compounds and sugars as sole carbon sources. Genomic analysis of OW59 suggested its involvement in hemicellulose degradation.

The non-secretary tannase family α/β hydrolase gene encoding FAE activity was identified (Ohta et al., 2021). This enzyme shares a catalytic triad consensus (Ser-His-Asp) within the tannase family block X of the ESTHER database (Lenfant et al., 2013). Before this study, this gene was annotated as mono(2-hydroxyethyl) terephthalic acid hydrolase (MHETase), involved in the downstream pathway of polyethylene terephthalate (PET) biodegradation (Maity et al., 2021). This research showed that this enzyme cleaves FA esters, CA esters, and other HCA esters but not MHET. Verifying enzyme function at the protein level is important for evaluating the biodegradability of plastics like PET that may enter aquatic environments.

Liquid chromatography-mass spectrometry analysis of reactions with crude rice straw extract and either the enzyme or OW59 showed common peaks, indicating the FAE activity contributes to plant biomass degradation.

5.2.3.4 Methyltransferase System

The discovery of unique methyltransferase systems in marine bacteria has opened new avenues for understanding lignin degradation mechanisms. This enzymatic system operates intracellularly and is responsible for downstream metabolism in bacterial lignin degradation, following the depolymerization of lignin into fragments small enough to be internalized. The mtgB methyltransferase identified in Bathyarchaeia (Yu et al., 2023) represents a novel approach to lignin modification, enabling the conversion of recalcitrant lignin fragments into more easily degradable forms (Venkatesagowda & Dekker, 2021).

Bathyarchaeia, abundant microorganisms vital to the global carbon cycle, remain poorly understood in terms of their origins and ecological roles. This study presents the largest dataset of Bathyarchaeia metagenome assembled genomes and reorganizes them into eight order-level units (Yu et al., 2023). It reveals diverse carbon metabolic processes, particularly unique C1 pathways, indicating their importance as often overlooked methylotrophs. Molecular dating suggests Bathyarchaeia diverged from other groups around 3.3 billion years ago, with significant diversification events occurring at 3.0, 2.5, and between 1.8 and 1.7 billion years ago, likely driven by geological changes. The lignin-degrading Bathyarchaeia group emerged around 300 million years ago, potentially linked to decreased carbon sequestration during the Late Carboniferous period, highlighting the influence of geological factors on their evolution and Earth's environment. Understanding Bathyarchaeia's unique adaptations and their role in lignin degradation can provide insights into their ecological significance and potential applications in biotechnology and environmental management (Hou et al., 2023; Yu et al., 2023).

5.2.3.5 Glutathione-S-Transferase System

Since lignin constitutes approximately 15–30% of plant biomass and its β-aryl ether linkages represent 45–70% of interunit bonds, understanding bacterial enzymatic pathways is vital for advancing sustainable bioconversion technologies. Early studies characterized bacterial lignin degradation enzymes such as β-etherases (Fig. 5.1), i.e., Glutathione-S-transferases (GSTs), in *Sphingomonas* sp. SYK-6 (Masai et al., 2007; Kamimura et al., 2017) and the other diverse sources (Kontur et al., 2019) including deposited sequences in public databases (Voß et al., 2020). This enzymatic system functions after the depolymerization of lignin into sufficiently small fragments that can be transported into the cell.

The conceptual framework integrates the GST system as a glutathione-dependent enzymatic mechanism catalyzing the reductive cleavage of β-aryl ether bonds, a predominant linkage in lignin fragments. This enzymatic synergy underpins bacterial lignin degradation pathways. Understanding the stereochemical specificity and catalytic kinetics of GSTs is fundamental to elucidating their role (Husarcíková et al., 2018).

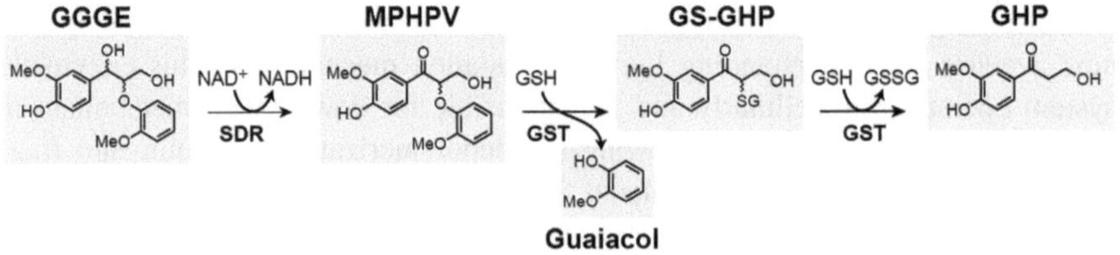

Fig. 5.1 Reaction scheme of the β-etherase pathway. Abbreviations: GGGE, guaiacylglycerol-β-guaiacyl ether; MPHPV, (2-methoxyphenoxy) hydroxypropiovanillone; GS-GHP, glutathione conjugate of GHP; GHP, guaiacylhydroxypropanone; NAD^+, oxidized form of nicotinamide adenine dinucleotide; NADH, reduced form of NAD; GSH, reduced glutathione; GSSG, oxidized glutathione; SDR, short-chain dehydrogenase/reductase; GST, glutathione S-transferase. (Reproduced from Kumagawa et al. (2023))

5.2.3.6 Case Study of an Etherase System from the Marine-Derived Sphingomonadale Bacterium

Sphingomonadales bacteria are distinguished by their extraordinary survival capabilities under extreme environmental conditions (Aylward et al., 2013; Asaf et al., 2020). These microorganisms demonstrate remarkable resilience in environments characterized by nutrient limitation, oxygen depletion, heavy metal contamination, the presence of polycyclic aromatic hydrocarbons, and petroleum exposure. Their distinctive physiological architecture is exemplified by the absence of lipopolysaccharides characteristic of Gram-negative bacteria, replaced instead by hydrophobic sphingoglycolipids in their cell walls. This unique structural configuration confers upon these organisms an exceptional capacity for the internalization of hydrophobic compounds (Kumar et al., 2017).

Our previous investigations have successfully identified six distinct enzymes from *Novosphingobium* sp. MBES04 that catalyze a cascade reaction involving the specific reductive cleavage of major bonds within lignin oligomers (Ohta et al., 2015). This enzymatic consortium, designated as the etherase metabolic system, represents a paradigmatic example of biological lignin processing. The revolutionary aspect of this system lies in its capacity to efficiently produce functional aromatic phenylpropanone monomers from lignin through a one-pot reaction utilizing five of these characterized enzymes (Ohta et al., 2017).

This technological advancement has enabled the selective production of specific aromatic monomers from lignin under mild reaction conditions. Specifically, our studies have demonstrated the targeted production of guaiacylhydroxypropanone (GHP) from lignin extracted from softwood (*Cryptomeria japonica*), and both GHP and syringylhydroxypropanone (SHP) from lignin derived from hardwood (*Eucalyptus globulus*) (Ohta et al., 2017). Furthermore, innovative methodologies have been developed to derive novel bio-based plastics and coating materials from these enzymatically produced monomers.

5.2.3.7 Novel GSTs in Marine-Derived Sphingomonadale Bacterium

Altererythrobacter sp. B11 was isolated from deep-sea sediments using spent mushroom compost as a carbon source (Maeda et al., 2018). Its 16S rRNA sequence showed 98.1% identity to the marine species *Altererythrobacter atlanticus* 26DY36^{T}. *B11* requires magnesium and calcium, major minerals in seawater, for growth and thrives at seawater salinity, indicating its adaptation to the marine environment. However, as it was isolated from deep-sea sediments in an area with rapid sedimentation, the possibility that it sustained life after flowing into the ocean along with terrestrial organic matter cannot be ruled out. Recent findings also suggest that *Altererythrobacter* species contribute to lignin degradation in terrestrial wetlands, implying a wider distribution beyond marine environments.

When B11 was cultured in a medium containing guaiacylglycerol-β-guaiacyl ether (GGGE), GGGE was degraded, and guaiacol accumulated without the accumulation of intermediate 2-methoxyphenoxyhydroxypropiovanillone (MPHPV) or GHP (Kumagawa et al., 2023). This demonstrated B11's superior lignin degradation capability compared to MBES04, with approximately three times higher degradation rate.

The enzyme responsible for MPHPV conversion in B11 was identified as a GST, which catalyzes the addition of glutathione to the 3-position carbon. Through BLASTP searches using the *ligF* gene from SYK-6 as a template, five candidate proteins were found. Detailed amino acid alignment analysis classified these into three clades: LigF1, BaeA (dimers with different stereoselectivity), and LigF2. The LigF2 clade, which includes the protein BBC71095, showed amino acid substitutions in the highly conserved γ-glutamyl recognition site of GSTs. The recombinant protein, named AbLigF2, exhibited GST activity and produced glutathione adducts when MPHPV was used as a substrate (Kumagawa et al., 2023, Fig. 5.2).

AbLigF2 showed optimal activity at 40 °C and maintained 30% activity after 2 h at 50 °C. Its optimal pH was 10–11, making it the most heat-tolerant and alkali-tolerant enzyme reported to date. AbLigF2 also maintained 50% and 40% activity after 30 min of exposure to 1% Tween 20 and 1% ethanol, respectively, indicating

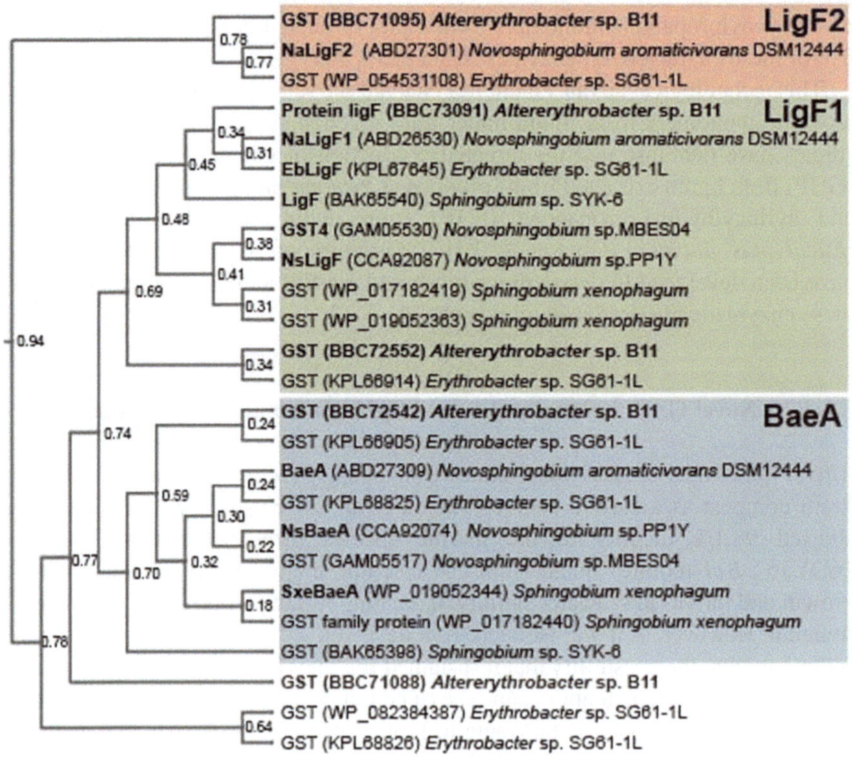

Fig. 5.2 Phylogenetic tree of reported and candidate GSTs from Sphingomonadale bacterium. (Reproduced from Kumagawa et al. (2023))

its structural stability. These findings highlight B11 as a source of highly efficient and stable enzymes for lignin model compound conversion (Kumagawa et al., 2023).

5.2.3.8 Physiological Significance of the Etherase System of Sphingomonadales

Sphingomonadales genera—including *Sphingomonas*, *Sphingobium*, *Novosphingobium*, and *Sphingopyxis*—are ubiquitously distributed across the Earth's biosphere. They exhibit remarkable metabolic versatility, as their genomes encode a wide array of oxidative enzymes and hydrolases involved in the degradation of aromatic compounds and the other recalcitrant and toxic compounds (Kumar et al., 2017; Cecil et al., 2018; Asaf et al., 2020).

Novosphingobium sp. MBES04 possesses an active etherase system that produces phenylpropanone monomers as end-products by specifically and reductively cleaving the β-O-4 ether bond (Ohta et al., 2015), a major bond in lignin molecules. Despite producing these monomers, the strain does not utilize them as an energy source and instead discharges them outside the cell. This observation is consistent with the absence of an ortholog of the *hpvZ* gene, which is essential for GHP catabolism in *Sphingobium* sp. SYK-6, a well-studied lignin-converting bacterium (Kamimura et al., 2017). The etherase system consumes reducing power, specifically 2 moles of reduced glutathione for every 1 mole of phenylpropanone monomer produced. Energy supply from other carbon sources, such as glucose or alanine, enhances the production of guaiacylhydroxypropanone (GHP) from GGGE, suggesting that energy from other sources activates the etherase system of MBES04.

Genomic analysis of MBES04 revealed three circular DNA molecules (one chromosome and two plasmids). While the Embden-Meyerhof pathway was found to lack 6-phosphofructokinase, and the pentose phosphate pathway lacked 6-phosphogluconate dehydrogenase, all enzymes necessary for the Entner-Doudoroff pathway were present, suggesting it plays a central role in sugar metabolism. The presence of genes encoding all five complexes required for oxidative phosphorylation indicated its importance for energy acquisition (Kumagawa et al., 2024).

Transcriptome analysis revealed that two gene clusters, each containing four tandemly linked genes, are specifically induced by a hardwood lignin preparation (*Eucalyptus globulus*) and a β-O-4-type lignin model biaryl compound (GGGE), but not by vanillin. The most highly induced gene in cluster G-II is estimated to be a 2,4′-dihydroxyacetophenone (DHA) dioxygenase, which produces 4-hydroxybenzoate from DHA and is involved in energy production through oxidative degradation. Another induced cluster (G-III) is related to multidrug resistance, encoding a transcriptional regulator (TetR) and transporters of the resistance-nodulation-division (RND) family and the EmrAB complex. These clusters are not only structurally distinct but also functionally responsive to specific signals derived from lignin. Cluster G-II is regulated by a shared promoter and is transcriptionally activated by phenylpropanone monomers such as GHP, which function as effector molecules. This suggests that the etherase system plays a dual role: it generates these monomers and senses them to trigger downstream gene expression. Cluster G-III, on the other hand, appears to serve as a protective function, enhancing the strain's resistance to potentially toxic lignin-derived compounds through efflux mechanisms. The selective induction of both clusters by GGGE and lignin preparations, but not by simpler aromatic compounds like vanillin, demonstrates MBES04's ability to discriminate between structurally complex lignin fragments and to respond with tailored transcriptional programs.

Importantly, GHP, a common product of GGGE and APA-lignin degradation by the etherase system, acts not only as a metabolic by-product but also as an effector molecule that induces the expression of genes in clusters G-II and G-III. This dual role of GHP highlights that the etherase system of MBES04 functions as a sensor for lignin fragments. Rather than serving solely as a catabolic pathway, the etherase system enables MBES04 to detect the presence of lignin-derived compounds and activate transcriptional responses that support environmental adaptation. These responses include oxidative degradation of aromatic compounds and enhanced tolerance to toxic substances via efflux systems. This sensory function is considered a survival strategy, allowing MBES04 to detect nutrient availability and respond to environmental stressors. The selective induction of gene clusters by GGGE and lignin preparations suggests that MBES04 can distinguish structurally complex lignin fragments and respond accordingly (Kumagawa et al., 2024). The strain preferentially utilizes easily degradable aromatic compounds with lower energy demands for catabolism. The degradation of phenylpropanone monomers is energetically demanding, requiring activation of the C3 side chain with CoA and ATP. Instead, MBES04 may prefer to assimilate enzymatically degradable substances like glucose, DHA, vanillin, cresol, and benzoate, which have low energy costs (Kumagawa et al., 2024, Fig. 5.3).

5.3 Biotechnological Applications

5.3.1 Technical Lignin from Industrial Processing of Lignocelluloses

Technical lignin preparations, obtained through the industrial processing of lignocellulosic substrates, are sustainable feedstocks that pose distinct challenges because of their altered structural characteristics and the presence of contaminants, which restrict its utilization across diverse industrial applications (Ekielski & Mishra, 2020).

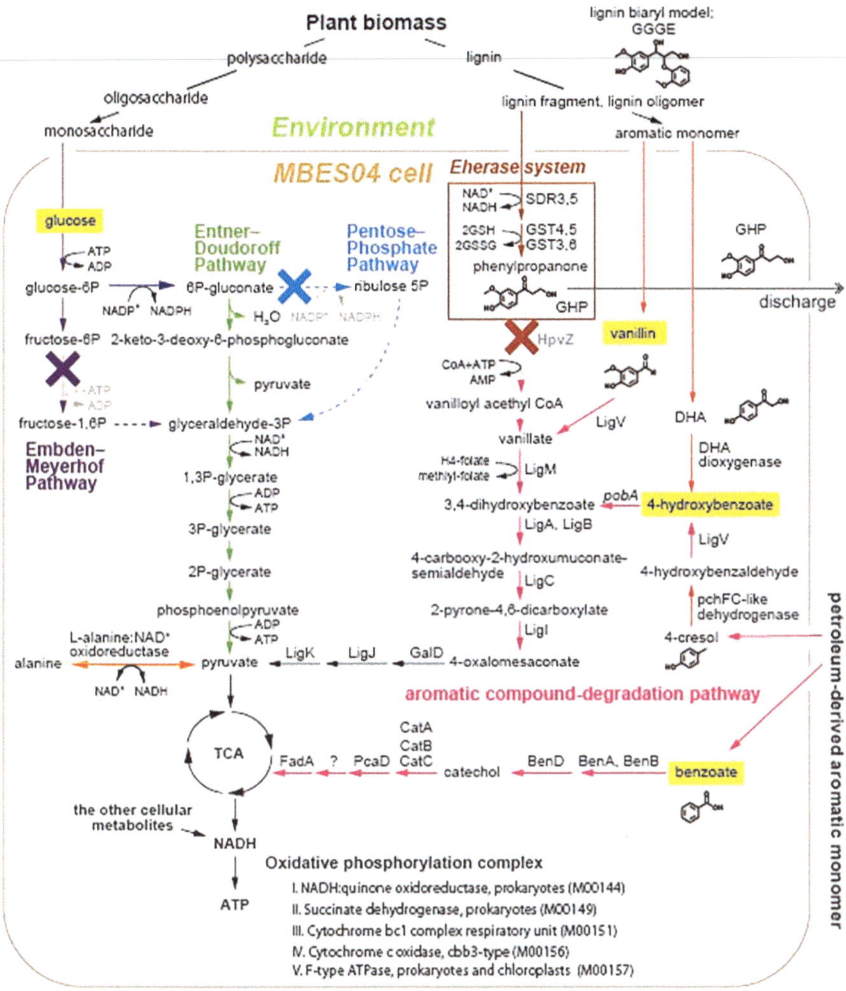

Fig. 5.3 Overview of metabolic pathway in MBES04 with glycolysis, aromatic compound degradation, and the etherase system. Abbreviations: DHA, 2,4′-dihydroxyacetophenone; GGGE, guaiacylglycerol-β-guaiacyl ether; GHP, guaiacylhydroxypropanone; GSH, reduced glutathione; GSSG, oxidized glutathione; GST, glutathione S-transferase; SDR, short-chain dehydrogenase/reductase (Cα-dehydrogenase). (Reproduced from Kumagawa et al. (2024))

5.3.2 *Challenge in Lignin Valorization*

Research on the applications and uses of different lignin types in industry has emerged as a critical area of inquiry due to lignin's abundance as the second-most prevalent biopolymer and its potential to serve as a renewable feedstock for biofuels, materials, and chemicals (Holladay et al., 2007; Ragauskas et al., 2014; Beckham et al., 2016; Liu et al., 2021). Historically, lignin was primarily considered a low-value by-product in the pulp and paper industries, mainly used for combustion to generate heat and power. However, advances over the past two decades have shifted the focus toward lignin valorization, emphasizing its conversion into high-value products such as bio-based polymers, carbon fibers, and specialty chemicals (Rinaldi et al., 2016; Wong et al., 2020). The global production of lignin exceeds 50 million tons annually, underscoring its significance for sustainable biorefineries and circular economy initiatives (Chen et al., 2024). This growing interest aligns with societal demands for renewable resources and reduced reliance on fossil fuels.

Despite its potential, lignin's complex and heterogeneous structure, varying by plant source and extraction method, presents challenges for its efficient industrial utilization (Ekielski & Mishra, 2020; Jiju et al., 2025). Current knowledge gaps include a limited understanding of how different lignin types influence downstream processing and product performance.

The lack of standardized characterization techniques further complicates comparisons across studies. These gaps hinder the development of scalable, economically viable lignin valorization pathways, restricting lignin's role beyond low-value applications. Addressing these issues is crucial to unlocking lignin's full potential in biofuel production, advanced materials, and chemical synthesis (Jiju et al., 2025).

Conceptually, lignin valorization integrates the understanding of lignin's chemical structure, extraction processes, and conversion technologies. The relationship between lignin type—such as Kraft, organosolv, or lignosulfonate—and its suitability for specific industrial applications forms the foundation for targeted valorization strategies. This framework guides the selection of depolymerization methods and subsequent upgrading to value-added products, linking biomass fractionation to end-use performance (Rinaldi et al., 2016). Theoretical models of lignin's polymeric architecture and reactivity underpin these approaches, facilitating the design of efficient catalytic and biochemical processes (Yu et al., 2025).

Translating laboratory findings to industrial-scale applications presents significant challenges, including process optimization, cost-effectiveness, and regulatory considerations. The development of economically viable bioprocesses requires careful consideration of substrate costs, product yields, and downstream processing requirements (Wong et al., 2020).

5.3.3 Synthetic Urushiols from Lignin-Derived Precursors: A Sustainable Approach to High-Performance Coatings

Isozaki et al. (2024) reported the first successful synthesis of urushiol analogs from lignin-derived phenylpropanoids, specifically GHP and SHP, obtained via the enzymatic degradation of woody biomass. Through a five-step synthetic route, polyunsaturated side chains derived from vegetable oils were installed onto these aromatic cores, yielding catechol-based compounds structurally analogous to natural urushiol.

The resulting artificial urushiols exhibited curing behavior and mechanical properties comparable to traditional Asian lacquer. Notably, the S-type analog demonstrated rapid enzymatic polymerization, while the G-type analog achieved a Martens hardness of 147.4 ± 0.9 N/mm^2 after 73 days—approaching the performance of commercial PVC coatings. FT-IR analysis confirmed that the molecular transformations during curing closely resemble those of natural urushi, including quinone formation and oxidative cross-linking of unsaturated side chains.

The work defines three key concepts: enzymatic lignin degradation, which involves selective cleavage of lignin's complex polymeric bonds; polyunsaturated side chain installation, referring to the chemical modification of lignin-derived monomers to introduce unsaturated alkenyl groups; and synthetic route optimization, encompassing the integration of biocatalytic and chemical steps to achieve efficient urushiol analogue synthesis (Isozaki et al., 2024). These concepts are interrelated, as enzymatic depolymerization provides the aromatic precursors that undergo side chain functionalization, culminating in synthetic urushiols with targeted substitution patterns. This work demonstrated the potential of lignin as a renewable aromatic feedstock for advanced materials and aligns with global sustainability goals by offering a bio-based alternative to petroleum-derived coatings.

5.3.4 Platform Chemical Production

In the concept of the "biological funnel" in microbial lignin valorization (Linger et al., 2014; Liu et al., 2021; Liu et al., 2022), diverse lignin-derived aromatic compounds are metabolized into central intermediates such as protocatechuate and catechol, which then serve as precursors for value-added products like polyhydroxyalkanoates (PHAs) (Kumar et al., 2018; Zhao et al., 2024) and muconic acid. This supports existing theories of microbial aromatic catabolism and highlights the metabolic versatility of strains like *Bacillus* (Gu et al., 2024) and *Ralstonia* (Zhao et al., 2024) isolates in lignin bioconversion. Advances in metabolic engineering have expanded the theoretical framework by demonstrating that non-lignin-degrading microbes can be reprogrammed to efficiently convert lignin-derived compounds into bioplastics, challenging the traditional view that only native ligninolytic microbes are suitable for lignin valorization. The integration of both ortho- and meta-cleavage pathways within single microbial strains, as observed in some engineered *Pseudomonas* species (Salvachúa et al., 2020; Borrero-de Acuña et al., 2021), provides new insights into optimizing carbon flux for enhanced PHA production, suggesting that pathway balancing is critical to overcoming bottlenecks such as toxic intermediate accumulation.

The heterogeneity and recalcitrance of lignin remain central theoretical challenges. Engineering approaches that reduces the molecular weight and complexity of lignin fragments can significantly improve microbial assimilation and conversion efficiency (Zhao et al., 2023).

The role of regulatory elements, such as the protocatechuate-responsive autoregulation system in engineered microbes, introduces a novel layer of metabolic control theory that can enhance lignin-derived bioproduct yields by dynamically adjusting gene expression in response to substrate availability. Theoretical models of microbial consortia and co-culture systems overcome limitations of single strains, providing a systems biology perspective on lignin valorization that aligns with ecological and synthetic biology principles (Saratale et al., 2022). The integration of synthetic biology tools opens new avenues for sustainable lignin valorization with reduced freshwater and organic carbon inputs, aligning with circular bioeconomy goals and resource conservation policies.

5.4 Conclusion

The exploration of lignin degradation by marine-derived microorganisms reveals a complex interplay between unique enzymatic systems and the challenging marine environments. Lignin, a significant component of woody plant biomass, presents substantial valorization potential, yet its structural heterogeneity complicates its breakdown. Marine bacteria have evolved specialized metabolic pathways for lignin degradation, utilizing both aerobic and anaerobic mechanisms, and employing key ligninolytic enzymes such as dye-decolorizing peroxidases, laccases, and glutathione-S-transferases. The ecological significance of these microbial communities is underscored by their role in blue carbon dynamics and global carbon cycling. Advances in molecular biology techniques have enhanced our understanding of these metabolic capabilities, paving the way for biotechnological applications, including the production of bio-based materials and bioplastics. As global demand for sustainable resources increases, further research into the intricate relationships between marine microorganisms and lignin degradation processes is essential for fostering environmental sustainability and promoting a circular economy. This compilation of research offers valuable insights into the potential of marine bacteria for lignin valorization and highlights the need for continued investigation in this promising field.

References

Asaf, S., Numan, M., Khan, A. L., & Al-Harrasi, A. (2020). *Sphingomonas*: From diversity and genomics to functional role in environmental remediation and plant growth. *Critical Reviews in Biotechnology, 40*(2), 138–152. https://doi.org/10.1080/07388551.2019.1709793

Aylward, F. O., McDonald, B. R., Adams, S. M., Valenzuela, A., Schmidt, R. A., Goodwin, L. A., Woyke, T., Currie, C. R., Suen, G., & Poulsen, M. (2013). Comparison of 26 sphingomonad genomes reveals diverse environmental adaptations and biodegradative capabilities. *Applied and Environmental Microbiology, 79*(12), 3724–3733. https://doi.org/10.1128/AEM.00518-13

Beckham, G. T., Johnson, C. W., Karp, E. M., Salvachúa, D., & Vardon, D. R. (2016). Opportunities and challenges in biological lignin valorization. *Current Opinion in Biotechnology, 42*, 40–53. https://doi.org/10.1016/j.copbio.2016.02.030

Borrero-de Acuña, J. M., Gutierrez-Urrutia, I., Hidalgo-Dumont, C., Aravena-Carrasco, C., Orellana-Saez, M., Palominos-Gonzalez, N., van Duuren, J. B. J. H., Wagner, V., Gläser, L., Becker, J., Kohlstedt, M., Zacconi, F. C., Wittmann, C., & Poblete-Castro, I. (2021). Channelling carbon flux through the meta-cleavage route for improved poly(3-hydroxyalkanoate) production from benzoate and lignin-based aromatics in *pseudomonas putida. Microbial Biotechnology, 14*(6), 2385–2402. https://doi.org/10.1111/1751-7915.13705

Catucci, G., Valetti, F., Sadeghi, S. J., & Gilardi, G. (2020). Biochemical features of dye-decolorizing peroxidases: Current impact on lignin degradation. *Biotechnology and Applied Biochemistry, 67*(5), 751–759. https://doi.org/10.1002/bab.2015

Cecil, J. H., Garcia, D. C., Giannone, R. J., & Michener, J. K. (2018). Rapid, parallel identification of catabolism pathways of lignin-derived aromatic compounds in *Novosphingobium aromaticivorans. Applied and Environmental Microbiology, 84*(22), e01185–e01118. https://doi.org/10.1128/AEM.01185-18

Chandrakanth, N. N., Zhang, C., Freeman, J., de Souza, W. R., Bartley, L. E., & Mitchell, R. A. C. (2023). Modification of plant cell walls with hydroxycinnamic acids by BAHD acyltransferases. *Frontiers in Plant Science, 13*, 1088879. https://doi.org/10.3389/fpls.2022.1088879

Chen, X., Mosier, N., & Ladisch, M. (2024). Valorization of lignin from aqueous-based lignocellulosic biorefineries. *Trends in Biotechnology, 42*(11), 1348–1362. https://doi.org/10.1016/j.tibtech.2024.07.004

Chen, F., Bai, X., Luo, G., Zhang, G., Ran, C., & Luo, X. (2025). Assessing the global flux of organic carbon transported from terrestrial surfaces to oceans by rivers. *Carbon Balance and Management, 20*(1), 29. https://doi.org/10.1186/s13021-025-00318-z

Ding, J., Liu, F., Zeng, J., Gu, H., Huang, J., Wu, B., Shu, L., Yan, Q., He, Z., & Wang, C. (2025). Depth heterogeneity of lignin-degrading microbiome and organic carbon processing in mangrove sediments. *npj Biofilms and Microbiomes, 11*(1), 5. https://doi.org/10.1038/s41522-024-00638-x

Drula, E., Garron, M. L., Dogan, S., Lombard, V., Henrissat, B., & Terrapon, N. (2022). The carbohydrate-active enzyme database: Functions and literature. *Nucleic Acids Research, 50*(D1), D571–D577. https://doi.org/10.1093/nar/gkab1045

Ekielski, A., & Mishra, P. K. (2020). Lignin for bioeconomy: The present and future role of technical lignin. *International Journal of Molecular Sciences, 22*(1), 63. https://doi.org/10.3390/ijms22010063

Ghose, M., Parab, A. S., Manohar, C. S., Mohanan, D., & Toraskar, A. D. (2024). Unraveling the role of bacterial communities in mangrove habitats under the urban influence, using a next-generation sequencing approach. *Journal of Sea Research, 198*, 102469. https://doi.org/10.1016/j.seares.2024.102469

Gough, M. A., Fauzi, R., Mantoura, C., & Preston, M. R. (1993). Terrestrial plant biopolymers in marine sediments. *Geochimica et Cosmochimica Acta, 57*, 945–964. https://doi.org/10.1016/0016-7037(93)90032-R

Gu, J., Qiu, Q., Yu, Y., Sun, X., Tian, K., Chang, M., Wang, Y., Zhang, F., & Huo, H. (2024). Bacterial transformation of lignin: Key enzymes and high-value products. *Biotechnology for Biofuels and Bioproducts, 17*(1), 2. https://doi.org/10.1186/s13068-023-02447-4

Guo, L., Dong, Y., Zhang, D., Pan, X., Jin, X., Yan, X., & Lu, Y. (2025). Mining versatile feruloyl esterases: Phylogenetic classification, structural features, and deep learning model. *Bioresources and Bioprocessing, 12*(1), 7. https://doi.org/10.1186/s40643-024-00835-8

Happs, R. M., Addison, B., Doeppke, C., Donohoe, B. S., Davis, M. F., & Harman-Ware, A. E. (2021). Comparison of methodologies used to determine aromatic lignin unit ratios in lignocellulosic biomass. *Biotechnology for Biofuels, 14*(1), 58. https://doi.org/10.1186/s13068-021-01897-y

Holladay, J. E., White, J. F., Bozell, J. J., & Johnson, D. (2007). *Top value-added chemicals from biomass – volume II—Results of screening for potential candidates from biorefinery lignin*. Pacific Northwest National Laboratory.

Hou, J., Wang, Y., Zhu, P., Yang, N., Liang, L., Yu, T., Niu, M., Konhauser, K., Woodcroft, B. J., & Wang, F. (2023). Taxonomic and carbon metabolic diversification of Bathyarchaeia during its coevolution history with early earth surface environment. *Science Advances, 9*(27), eadf5069. https://doi.org/10.1126/sciadv.adf5069

Husarcíková, J., Voß, H., Domínguez de María, P., & Schallmey, A. (2018). Microbial β-etherases and glutathione lyases for lignin valorisation in biorefineries: Current state and future perspectives. *Applied Microbiology and Biotechnology, 102*(13), 5391–5401. https://doi.org/10.1007/s00253-018-9040-3

Isozaki, K., Matsuda, H., Agata, R., Jeon, J., Wu, B., Pincella, F., Ikenaga, M., Tachibana, Y., Ohta, Y., & Nakamura, M. (2024). Synthetic Urushiols from biorenewable carbon resources: Chemical conversion of enzymatic degradation products of wood lignin to an ancient yet future coating material. *RSC Sustainability, 2*(5), 1358–1362. https://doi.org/10.1039/D3SU00446E

Jiang, C., Yan, H., Shen, X., Zhang, Y., Wang, Y., Sun, S., Jiang, H., Zang, H., Zhao, X., Hou, N., Li, Z., Wang, L., Wang, H., & Li, C. (2022). Genome functional analysis of the psychrotrophic lignin-degrading bacterium *Arthrobacter* sp. C2 and the role of DyP in catalyzing lignin degradation. *Frontiers in Microbiology, 13*, 921549. https://doi.org/10.3389/fmicb.2022.921549

Jiju, P. S., Patel, A. K., Shruthy, N. S., Shalu, S., Dong, C. D., & Singhania, R. R. (2025). Sustainability through lignin valorization: Recent innovations and applications driving industrial transformation. *Bioresources and Bioprocessing, 12*(1), 88. https://doi.org/10.1186/s40643-025-00929-x

Kamimura, N., Takahashi, K., Mori, K., Araki, T., Fujita, M., Higuchi, Y., & Masai, E. (2017). Bacterial catabolism of lignin-derived aromatics: New findings in a recent decade: Update on bacterial lignin catabolism. *Environmental Microbiology Reports, 9*(6), 679–705. https://doi.org/10.1111/1758-2229.12597

Kontur, W. S., Olmsted, C. N., Yusko, L. M., Niles, A. V., Walters, K. A., Beebe, E. T., Vander Meulen, K. A., Karlen, S. D., Gall, D. L., Noguera, D. R., & Donohue, T. J. (2019). A heterodimeric glutathione S-transferase that stereospecifically breaks lignin's β(R)-aryl ether bond reveals the diversity of bacterial β-etherases. *The Journal of Biological Chemistry, 294*(6), 1877–1890. https://doi.org/10.1074/jbc.RA118.006548

Kumagawa, E., Katsumata, M., & Ohta, Y. (2023). Catalytic and molecular properties of alkaliphilic and thermotolerant β-etherase from *Altererythrobacter* sp. B11. *Bioscience Biotechnology and Biochemistry, 87*(10), 1183–1192. https://doi.org/10.1093/bbb/zbad091

Kumagawa, E., Katsumata, M., Nishimura, H., Watanabe, T., Ishii, S., & Ohta, Y. (2024). The etherase system of *Novosphingobium* sp. MBES04 functions as a sensor of lignin fragments through phenylpropanone production to induce specific transcriptional responses. *Environmental Microbiology Reports, 16*(1), e13210. https://doi.org/10.1111/1758-2229.13210

Kumar, R., Verma, H., Haider, S., Bajaj, A., Sood, U., Ponnusamy, K., Nagar, S., Shakarad, M. N., Negi, R. K., Singh, Y., Khurana, J. P., Gilbert, J. A., & Lal, R. (2017). Comparative genomic analysis reveals habitat-specific genes and regulatory hubs within the genus *Novosphingobium*. *mSystems, 2*(3), e00020–e00017. https://doi.org/10.1128/mSystems.00020-17

Kumar, M., Verma, S., Gazara, R. K., Kumar, M., Pandey, A., Verma, P. K., & Thakur, I. S. (2018). Genomic and proteomic analysis of lignin degrading and polyhydroxyalkanoate accumulating β-proteobacterium *Pandoraea* sp. ISTKB. *Biotechnology for Biofuels, 11*, 154. https://doi.org/10.1186/s13068-018-1148-2

Lam, M. Q., Oates, N. C., Leadbeater, D. R., Goh, K. M., Yahya, A., Md Salleh, M., Ibrahim, Z., Bruce, N. C., & Chong, C. S. (2022). Genomic analysis to elucidate the lignocellulose degrading capability of a new halophile *Robertkochia solimangrovi*. *Genes, 13*(11), 2135. https://doi.org/10.3390/genes13112135

Leadbeater, D. R., Oates, N. C., Bennett, J. P., Li, Y., Dowle, A. A., Taylor, J. D., Alponti, J. S., Setchfield, A. T., Alessi, A. M., Helgason, T., McQueen-Mason, S. J., & Bruce, N. C. (2021).

Mechanistic strategies of microbial communities regulating lignocellulose deconstruction in a UK salt marsh. *Microbiome, 9*(1), 48. https://doi.org/10.1186/s40168-020-00964-0

Lenfant, N., Hotelier, T., Velluet, E., Bourne, Y., Marchot, P., & Chatonnet, A. (2013). ESTHER, the database of the α/β-hydrolase fold superfamily of proteins: Tools to explore diversity of functions. *Nucleic Acids Research, 41*(Database issue), D423–D429. https://doi.org/10.1093/nar/gks1154

Leontakianakou, S., Adlercreutz, P., & Nordberg Karlsson, E. (2025). Fantastic ferulic acid esterases and their functions. *International Journal of Molecular Sciences, 26*(15), 7474. https://doi.org/10.3390/ijms26157474

Ley, Y., Cheng, X. Y., Ying, Z. Y., Zhou, N. Y., & Xu, Y. (2023). Characterization of two marine lignin-degrading consortia and the potential microbial lignin degradation network in nearshore regions. *Microbiology Spectrum, 11*(3), e0442422. https://doi.org/10.1128/spectrum.04424-22

Li, X., Zhang, T., Sun, S., Lan, H., & Yu, T. (2012). Lignin in marine environment and its analysis-a review. *Journal of Ocean University of China, 11*, 501–506. https://doi.org/10.1007/s11802-012-1834-9

Linger, J. G., Vardon, D. R., Guarnieri, M. T., Karp, E. M., Hunsinger, G. B., Franden, M. A., Johnson, C. W., Chupka, G., Strathmann, T. J., Pienkos, P. T., & Beckham, G. T. (2014). Lignin valorization through integrated biological funneling and chemical catalysis. *Proceedings of the National Academy of Sciences of the United States of America, 111*(33), 12013–12018. https://doi.org/10.1073/pnas.1410657111

Liu, Z. H., Hao, N., Wang, Y. Y., Dou, C., Lin, F., Shen, R., Bura, R., Hodge, D. B., Dale, B. E., Ragauskas, A. J., Yang, B., & Yuan, J. S. (2021). Transforming biorefinery designs with 'plug-in processes of lignin' to enable economic waste valorization. *Nature Communications, 12*(1), 3912. https://doi.org/10.1038/s41467-021-23920-4

Liu, Z. H., Li, B. Z., Yuan, J. S., & Yuan, Y. J. (2022). Creative biological lignin conversion routes toward lignin valorization. *Trends in Biotechnology, 40*(12), 1550–1566. https://doi.org/10.1016/j.tibtech.2022.09.014

Lu, C. J., Benner, R., Fichot, C. G., Fukuda, H., Yamashita, Y., & Ogawa, H. (2016). Sources and transformations of dissolved lignin phenols and chromophoric dissolved organic matter in Otsuchi bay, Japan. *Frontiers in Marine Science, 3*, 85. https://doi.org/10.3389/fmars.2016.00085

Maeda, A. H., Nishi, S., Ishii, S., Shimane, Y., Kobayashi, H., Ichikawa, J., Kurosawa, K., Arai, W., Takami, H., & Ohta, Y. (2018). Complete genome sequence of *Altererythrobacter* sp. strain B11, an aromatic monomerdegrading bacterium, isolated from deep-sea sediment under the seabed off Kashima, Japan. *Genome Announcements, 6*(12), e00200–e00218. https://doi.org/10.1128/genomeA.00200-18

Maity, W., Maity, S., Bera, S., & Roy, A. (2021). Emerging roles of PETase and MHETase in the biodegradation of plastic wastes. *Applied Biochemistry and Biotechnology, 193*(8), 2699–2716. https://doi.org/10.1007/s12010-021-03562-4

Masai, E., Katayama, Y., & Fukuda, M. (2007). Genetic and biochemical investigations on bacterial catabolic pathways for lignin-derived aromatic compounds. *Bioscience Biotechnology and Biochemistry, 71*(1), 1–15. https://doi.org/10.1271/bbb.60437

McDonald, N., Achterberg, E. P., Carlson, C. A., Gledhill, M., Liu, S., Matheson-Barker, J. R., Nelson, N. B., & Parsons, R. (2019). The role of heterotrophic bacteria and archaea in the transformation of lignin in the open ocean. *Frontiers in Marine Science, 6*, 743. https://doi.org/10.3389/fmars.2019.00743

Ohta, Y., Nishi, S., Haga, T., Tsubouchi, T., Hasegawa, R., Konishi, M., Nagano, Y., Tsuruwaka, Y., Shimane, Y., Mori, K., Usui, K., Suda, E., Tsutsui, K., Nishimoto, A., Fujiwara, Y., Maruyama, T., & Hatada, Y. (2012). Screening and phylogenetic analysis of deep-sea bacteria capable of metabolizing lignin-derived aromatic compounds. *Open Journal of Marine Science, 2*, 177–187. https://doi.org/10.4236/ojms.2012.24021

Ohta, Y., Nishi, S., Hasegawa, R., & Hatada, Y. (2015). Combination of six enzymes of a marine *Novosphingobium* converts the stereoisomers of β-O-4 lignin model dimers into the respective monomers. *Scientific Reports, 5*, 15105. https://doi.org/10.1038/srep15105

Ohta, Y., Hasegawa, R., Kurosawa, K., Maeda, A. H., Koizumi, T., Nishimura, H., Okada, H., Qu, C., Saito, K., Watanabe, T., & Hatada, Y. (2017). Enzymatic specific production and chemical functionalization of phenylpropanone platform monomers from lignin. *ChemSusChem, 10*(2), 425–433. https://doi.org/10.1002/cssc.201601235

Ohta, Y., Katsumata, M., Kurosawa, K., Takaki, Y., Nishimura, H., Watanabe, T., & Kasuya, K. I. (2021). Degradation of ester linkages in rice straw components by *Sphingobium* species recovered from the sea bottom using a non-secretory tannase-family α/β hydrolase. *Environmental Microbiology, 23*(8), 4151–4167. https://doi.org/10.1111/1462-2920.15551

Peng, Q., & Lin, L. (2025). Comparative metagenomics reveals the metabolic flexibility of coastal prokaryotic microbiomes contributing to lignin degradation. *Biotechnology for Biofuels and Bioproducts, 18*(1), 9. https://doi.org/10.1186/s13068-025-02605-w

Peng, Q., Lin, L., Tu, Q., Wang, X., Zhou, Y., Chen, J., Jiao, N., & Zhou, J. (2023). Unraveling the roles of coastal bacterial consortia in degradation of various lignocellulosic substrates. *mSystems, 8*(4), e0128322. https://doi.org/10.1128/msystems.01283-22

Ragauskas, A. J., Beckham, G. T., Biddy, M. J., Chandra, R., Chen, F., Davis, M. F., Davison, B. H., Dixon, R. A., Gilna, P., Keller, M., Langan, P., Naskar, A. K., Saddler, J. N., Tschaplinski, T. J., Tuskan, G. A., & Wyman, C. E. (2014). Lignin valorization: Improving lignin processing in the biorefinery. *Science (New York, N.Y.), 344*(6185), 1246843. https://doi.org/10.1126/science.1246843

Ralph, J., Lundquist, K., Brunow, G., Lu, F., Kim, H., Schatz, P. F., Marita, J. M., Hatfield, R. D., Ralph, S. A., Christensen, J. H., & Boerjan, W. (2004). Lignins: Natural polymers from oxidative coupling of 4-hydroxyphenyl- propanoids. *Phytochemistry Reviews, 3*, 29–60. https://doi.org/10.1023/B:PHYT.0000047809.65444.a4

Rinaldi, R., Jastrzebski, R., Clough, M. T., Ralph, J., Kennema, M., Bruijnincx, P. C., & Weckhuysen, B. M. (2016). Paving the way for lignin valorisation: Recent advances in bioengineering, biorefining and catalysis. *Angewandte Chemie (International ed. in English), 55*(29), 8164–8215. https://doi.org/10.1002/anie.201510351

Salvachúa, D., Rydzak, T., Auwae, R., De Capite, A., Black, B. A., Bouvier, J. T., Cleveland, N. S., Elmore, J. R., Huenemann, J. D., Katahira, R., Michener, W. E., Peterson, D. J., Rohrer, H., Vardon, D. R., Beckham, G. T., & Guss, A. M. (2020). Metabolic engineering of pseudomonas putida for increased polyhydroxyalkanoate production from lignin. *Microbial Biotechnology, 13*(1), 290–298. https://doi.org/10.1111/1751-7915.13481

Saratale, R. G., Cho, S. K., Kadam, A. A., Ghodake, G. S., Kumar, M., Bharagava, R. N., Varjani, S., Nair, S., Kim, D. S., Shin, H. S., & Saratale, G. D. (2022). Developing microbial co-culture system for enhanced polyhydroxyalkanoates (PHA) production using acid pretreated lignocellulosic biomass. *Polymers, 14*(4), 726. https://doi.org/10.3390/polym14040726

Umezawa, T. (2018). Lignin modification in planta for valorization. *Phytochemistry Reviews, 17*, 1305–1327. https://doi.org/10.1007/s11101-017-9545-x

Venkatesagowda, B., & Dekker, R. F. H. (2021). Microbial demethylation of lignin: Evidence of enzymes participating in the removal of methyl/methoxyl groups. *Enzyme and Microbial Technology, 147*, 109780. https://doi.org/10.1016/j.enzmictec.2021.109780

Voß, H., Heck, C. A., Schallmey, M., & Schallmey, A. (2020). Database mining for novel bacterial β-Etherases, glutathione-dependent lignin-degrading enzymes. *Applied and Environmental Microbiology, 86*(2), e02026–e02019. https://doi.org/10.1128/AEM.02026-19

Wang, X., Lin, L., & Zhou, J. (2021). Links among extracellular enzymes, lignin degradation and cell growth establish the models to identify marine lignin-utilizing bacteria. *Environmental Microbiology, 23*(1), 160–173. https://doi.org/10.1111/1462-2920.15289

Wells, T., Jr., & Ragauskas, A. J. (2012). Biotechnological opportunities with the β-ketoadipate pathway. *Trends in Biotechnology, 30*(12), 627–637. https://doi.org/10.1016/j.tibtech.2012.09.008

Wong, S. S., Shu, R., Zhang, J., Liu, H., & Yan, N. (2020). Downstream processing of lignin derived feedstock into end products. *Chemical Society Reviews, 49*(15), 5510–5560. https://doi.org/10.1039/d0cs00134a

Woo, H. L., & Hazen, T. C. (2018). Enrichment of bacteria from Eastern Mediterranean Sea involved in lignin degradation via the phenylacetyl-CoA pathway. *Frontiers in Microbiology, 9*, 922. https://doi.org/10.3389/fmicb.2018.00922

Xu, Z., Lei, P., Zhai, R., Wen, Z., & Jin, M. (2019). Recent advances in lignin valorization with bacterial cultures: Microorganisms, metabolic pathways, and bio-products. *Biotechnology for Biofuels, 12*, 32. https://doi.org/10.1186/s13068-019-1376-0

Yoo, C. G., Dumitrache, A., Muchero, W., Natzke, J., Akinosho, H., Li, M., Sykes, R., Brown, S. D., Davison, B. H., Tuskan, G. A., Pu, Y., Ragauskas, A. J., & Ragauskas, A. J. (2018). Significance of lignin S/G ratio in biomass recalcitrance of *Populus trichocarpa* variants for bioethanol production. *ACS Sustainable Chemistry & Engineering, 6*, 2162–2168. https://doi.org/10.1021/acssuschemeng.7b03586

Yu, T., Hu, H., Zeng, X., Wang, Y., Pan, D., Deng, L., Liang, L., Hou, J., & Wang, F. (2023). Widespread Bathyarchaeia encode a novel methyltransferase utilizing lignin-derived aromatics. *mLife, 2*(3), 272–282. https://doi.org/10.1002/mlf2.12082

Yu, Z., Kong, W., Liang, W., Guo, Y., Cui, J., Hu, Y., Sun, Z., Elangovan, S., & Xu, F. (2025). Heterogeneously catalyzed reductive depolymerization of lignin to value-added chemicals. *ChemSusChem, 18*(3), e202401399. https://doi.org/10.1002/cssc.202401399

Zhao, Z. M., Meng, X., Pu, Y., Li, M., Li, Y., Zhang, Y., Chen, F., & Ragauskas, A. J. (2023). Bioconversion of homogeneous linear C-lignin to polyhydroxyalkanoates. *Biomacromolecules, 24*(9), 3996–4004. https://doi.org/10.1021/acs.biomac.3c00288

Zhao, Y., Xue, L., Huang, Z., Lei, Z., Xie, S., Cai, Z., Rao, X., Zheng, Z., Xiao, N., Zhang, X., Ma, F., Yu, H., & Xie, S. (2024). Lignin valorization to bioplastics with an aromatic hub metabolite-based autoregulation system. *Nature Communications, 15*(1), 9288. https://doi.org/10.1038/s41467-024-53609-3

Chapter 6
Recent Advances in Haloarchaea: Enzymes and Industrial Applications

Hiroaki Minegishi

Abstract Halophiles inhabit environments from moderate salinity to saturation, and the most hypersaline systems are chiefly populated by haloarchaea (class Halobacteria). This chapter synthesizes their physiology, diversity, and industrial potential. Most haloarchaea grow at mesophilic temperatures and near-neutral pH, but the lineage includes alkaliphiles, acidophiles, psychrotolerant strains, and magnesium-requiring specialists. Hallmarks include ether lipid membranes, S-layers or polysaccharide walls, and intracellular KCl that stabilizes proteins. Haloarchaea are isolated from salt lakes, salterns, halite deposits, commercial salts, and increasingly marine settings. Recent taxonomic revisions expanded orders and families and recognized nano-haloarchaea and halophilic methanogens. Functionally, haloarchaea furnish robust biocatalysts and cellular chassis for high-salt operations: proteases, lipases/esterases, and amylases retain activity at molar salinity and in organic solvents, enabling leather processing, cleaning, lipid conversions, and starch hydrolysis. Several genera accumulate PHAs; *Haloferax mediterranei* produces PHBV from diverse feedstocks, with seawater media and hypoosmotic lysis simplifying recovery. In bioremediation, haloarchaea degrade hydrocarbons, decolorize azo dyes at up to 5 M NaCl, and mediate metal detoxification via transport, transformation, and extracellular polymers. High salinity suppresses contamination and can obviate desalination pretreatment. We outline opportunities in enzyme engineering, strain domestication, and process intensification and challenges in continuous operation and long-term stability.

Keywords Haloarchaea · Halobacteria · Halophilic enzymes · Proteases · Lipases/Esterases · Amylases · Polyhydroxyalkanoates (PHA PHBV) · Bioremediation · Hypersaline environments · Soy sauce and miso fermentation

---

H. Minegishi (✉)
Department of Applied Chemistry, Faculty of Science and Engineering, Toyo University, Kawagoe, Saitama, Japan
e-mail: minehiro@toyo.jp

M. Ito et al. (eds.), *Extremophiles for a Sustainable Future*,
https://doi.org/10.1007/978-981-95-6970-0_6

6.1 Introduction

Microorganisms capable of survival and growth are found across the full range of salinities from 0 M NaCl to saturation, and they can be broadly classified into non-halophiles and halophiles based on their optimal NaCl concentration for growth. Non-halophiles (optimal NaCl < 0.2 M) include soil microorganisms and some marine bacteria; among these are halotolerant species that can grow even at salt concentrations exceeding 2 M. Halophiles, defined by an optimal NaCl concentration ≥ 0.2 M, are further subdivided according to their NaCl requirement. Slightly halophilic microorganisms (0.2–0.5 M) are largely marine; they include Gram-negative genera such as *Vibrio*, *Pseudomonas*, *Alteromonas*, and some *Alcaligenes* species, as well as Gram-positive taxa such as *Micrococcus*.

Slightly halophilic bacteria are distinguished from halotolerant organisms by an obligatory requirement for Na^+ for growth. Moderately halophilic microorganisms (0.5–2.5 M) are frequently isolated in large numbers from a variety of salted materials, such as cured meats and fish (Kushner, 1978; Oren, 2008). The genus *Halomonas* is well known in this group; other examples include organisms that specifically require NaCl, such as *Paracoccus halodenitrificans* (*Halomonas halodenitrificans*), and those that can grow in media containing salts other than NaCl, such as *Vibrio costicola* (*Salinivibrio costicola*). Extremely halophilic microorganisms (≥ 2.5 M) are isolated from hypersaline environments including the Dead Sea, Great Salt Lake, and Lake Magadi (Kenya's Rift Valley), as well as rock-salt deposits and solar salterns; most belong to the class Halobacteria within the phylum Euryarchaeota (Oren, 2008; Ventosa et al., 1998).

In recent years, numerous isolates have also been reported from commercial salts (Minegishi, 2008; Satari et al., 2021; Zhang et al., 2024), fermented seafoods (Aslam et al., 2007; Roh et al., 2007a, b; Whon et al., 2021), and marine isolates have been reported (Cho et al., 2021; Inoue et al., 2011). Outside the class Halobacteria, known extremely halophilic bacteria include *Actinopolyspora halophila* (Gochnauer, 1975), *Ectothiorhodospira halophila (Halorhodospira halophila), and Ectothiorhodospira halochloris (Halorhodospira halochloris)* (Ollivier et al., 1994), as well as *Salinibacter ruber* (Antón et al., 2002). Among archaea, *Methanohalobium evestigatum*—an obligately anaerobic methanogen isolated from sediments of a Crimean salt lagoon—grows at 2.6–5.1 M NaCl with an optimum at 4.3 M (Zhilina & Zavarzin, 1987).

Halophilic microorganisms are promising drivers of next-generation industrial biotechnology, with direct applications in fermentation, biodegradation, and soil remediation (Singh & Singh, 2017). Chemical, food, and environmental production sites inevitably encounter high salt concentrations (changes in ionic strength), organic solvents, and fluctuations in pH and temperature resulting from raw materials, processing, and cleaning operations. These conditions can alter reaction rates, solubility, protein stability, and microbial activity. Using halophilic organisms and the enzymes they produce, which can function and grow under these stressful conditions, can enable robust bioprocesses that help maintain product quality and process efficiency (Cabrera & Blamey, 2018; Moopantakath et al., 2023). In recent years, the molecular mechanisms of salt tolerance and halophilicity—such as the density of surface acidic residues, rearrangement of the hydrophobic core, retention of surface-bound water, and utilization of the Hofmeister effect—have been elucidated (DasSarma & DasSarma, 2015; Ortega et al., 2015; Lo Nostro & Ninham, 2012). Furthermore, the integration of genome-driven research, protein engineering, and process design has suggested the potential for halophilic microorganisms and enzymes to function in actual process industries (Hogg et al., 2024; Mesbah, 2022). This chapter reviews the characteristics of halophilic archaea and assesses their potential for industrial application.

6.2 Characteristics of Halophilic Archaea

Members of the class Halobacteria (haloarchaea) utilize oxygen for growth and, as aerobic chemoorganotrophs or photoheterotrophs, require organic substrates such as casamino acids, yeast extract, and peptone. In contrast, some haloarchaea are capable of proliferating under anoxic conditions by using dimethyl sulfoxide (DMSO), arginine, or fumarate. As noted above, many haloarchaea are extreme halophiles that require ≥2.5 M NaCl for growth; however, moderately halophilic taxa with optimal NaCl concentrations ≤1.7 M (9.9% w/v) have also been reported, including *Halobellus clavatus* strain TNN18 (Cui et al., 2011), *Haloferax volcanii* strain DS2 (Torreblanca et al., 1986), *Halohasta litchfieldiae* strain tADL (Mou et al., 2012), *Halorussus ruber* strain YC25 (Xu et al., 2015), and *Halostagnicola kamekurae* strain 194-10 (Nagaoka et al., 2010). Moreover, reports of species adapted to relatively low salinities are increasing, such as *Haladaptatus paucihalophilus* strain DX253, which can grow at 0.8 M NaCl (4.7% w/v; optimal 2.6–3.1 M) (Savage et al., 2007), and *Haloferax namakaokahaiae* strain Mke2.3^{T}, which can grow at 0.5 M NaCl (2.9% w/v; optimal 1.7 M) (McDuff et al., 2016). In 2016, an anaerobic haloarchaeon that grows exclusively by sulfur respiration, *Halanaeroarchaeum sulfurireducens* strain HSR2, was isolated from a salt lake in the Kulunda Steppe, Russia (Sorokin et al., 2016), further expanding the concept of haloarchaea beyond traditional definitions.

Many haloarchaea grow under otherwise moderate conditions—mesophilic temperatures and near-neutral pH—apart from their requirement for high salinity. Most species have growth ranges around 20–50 °C and exhibit optimal growth near 37 °C. The lowest growth temperature reported for a haloarchaeon is 4 °C for *Halorubrum lacusprofundi* strain ACAM 34, isolated from deep Antarctic sediments (optimum, 37 °C) (McGenity & Grant, 1995). At the opposite extreme, only nine species are known to grow at ≥60 °C; among these, the highest recorded maximum growth temperature is 62 °C for *Natribaculum breve* TRM20010 and *Natribaculum longum* TRM20345 (both with an optimum of ~37 °C), isolated from Lop Nur (the "wandering lake") in northwestern China (Liu et al., 2015). In addition, *Halopiger goleamassiliensis* strain IIH3, isolated from sediments of Lake El

Goléa, Algeria, grows at 40–60 °C with an optimum of 55 °C (Ikram et al., 2014), making it one of the most thermophilic haloarchaea currently known.

With respect to pH, most haloarchaea have near-neutral optima. Alkaliphilic taxa with optimal growth pH $\geq$ 9.0 are known but relatively few (about 30 species as of July 2023), and only about five tolerate $\geq$ pH 10.0. Many of these belong to the family Natrialbaceae; among them, *Natronobacterium gregoryi* can grow at pH 12 (the type strain SP2 grows up to pH 11.0) (Seckbach & Oren, 2004). Prior to the discovery in 2001 of the alkaliphilic bacterium *Alkaliphilus transvaalensis* strain SAGM1 (Takai et al., 2001), *Nbt. gregoryi* had been regarded as the most alkaliphilic organism known. By contrast, acidophilic haloarchaea capable of growing at pH $\leq$ 4.5 are limited to five species of the genus *Halarchaeum*, and of these only *Halarchaeum acidiphilum* strain MH1-52-1 has an optimal growth pH $\leq$ 5 (Minegishi et al., 2008). Notably, MH1-52-1 was isolated from an Okinawa-produced solar salt supplemented with coral powder; dissolution of this salt yields a slightly alkaline solution, making the survival of MH1-52-1 in such an environment particularly intriguing (Minegishi et al., 2008; 2010a, b).

Some haloarchaea additionally require high magnesium concentrations for growth. Whereas many species need ~0.1–0.2 M Mg^{2+}, *Halobaculum gomorrense* strain DS2807 from the Dead Sea requires at least 0.8 M $MgCl_2 \cdot 6H_2O$ together with 1.0 M NaCl (Oren et al., 1995). *Halobaculum magnesiiphilum* strain MGY-184, isolated in Japan from commercially available salts, grows on medium containing 1.48 M $MgCl_2 \cdot 6H_2O$ and 0.3 M NaCl (Shimoshige et al., 2013).

Halophilic archaea that can grow under hypersaline conditions are constantly influenced by external salt. Many lack a conventional cell wall and instead possess an S-layer composed of glycoprotein (Rodrigues-Oliveira et al., 2017). By contrast, coccoid taxa such as *Halococcus* have a heteropolysaccharide cell wall and do not lyse in hypotonic solutions (Legat et al., 2013; Steber & Schleifer, 1975).

Like other archaea, halophilic archaea possess ether lipids in which isoprenoid alcohols are ether-linked to glycerol (Koga, 2007). Among their membrane lipids, the principal phospholipids include archaetidylglycerol, archaetidylglycerophosphate, and archaetidylglycerosulfate; the glycolipids include diglycosyl archaeol, triglycosyl archaeol, and related species (Cui et al., 2024). To date, roughly ten glycolipid structures have been elucidated, distinguished by the type and number of sugars attached at the *sn*-1 position of glycerol, the glycosidic linkage patterns, and the positions of sulfate substitution (Cui et al., 2024). Membrane lipids in halophilic archaea are exclusively diether species; tetraether lipids have not been detected (Bale et al., 2019). Their membranes are stabilized by sodium ions, which are thought to neutralize excess negative charge and stabilize hydrophobic structures (e.g., via long-range/kosmotropic effects) (Tenchov et al., 2006). As an exception, in genera that require high magnesium concentrations for growth (e.g., *Halobaculum*), Mg^{2+} appears to play the principal membrane-stabilizing role (Shimoshige et al., 2013).

Intracellularly, high concentrations of salt (predominantly K^+) are accumulated to counterbalance the elevated external osmotic pressure (DasSarma & DasSarma, 2015; Oren, 1999). Consistent with this, many intracellular enzymes in halophilic archaea depend more strongly on potassium than on sodium, and the physiologically critical salt is considered to be KCl (Madern et al., 2000; Oren, 1999). Certain halophilic archaea also harbor bacteriorhodopsin, a purple retinal pigment resembling both the proteorhodopsin of aerobic anoxygenic phototrophic bacteria and the visual pigments of eukaryotes. Bacteriorhodopsin is a light-driven proton pump; the proton motive force it generates supports Na^+/H^+ antiport and active K^+ uptake, helping cells maintain ion homeostasis under high salinity (Béjà et al., 2000; Lanyi et al., 1979; Wagner et al., 1978).

6.3 Isolation of Halophilic Archaea

Isolation sources of haloarchaea include hypersaline environments such as halite, salt lakes, and salterns, with numerous isolation reports from salt lakes distributed worldwide. For example, many novel genera and species have been isolated from Utah's Great Salt Lake (USA), the Dead Sea on the Jordan–Israel border (Levant), Lake Magadi (an alkaline salt lake in Kenya), and multiple salt lakes in inland China and on the Tibetan Plateau. Haloarchaea have also been recovered from halite deposits spanning geological periods from the Silurian (~419 Ma) to the Gelasian Stage of the Pleistocene (~1.8 Ma) (Gramain et al., 2011). For instance, *Halosimplex carlsbadense* strain 2-9-1 was isolated from within halite crystals of the Permian (~250 Ma) Salado Formation in southeastern New Mexico (Vreeland et al., 2002). Although questions have been raised as to whether brines from surface salt lakes overlying halite-bearing regions may have percolated into fissures relatively recently, then crystallized, and trapped cells at that time, there is no doubt that halite is an effective source for isolating haloarchaea.

Haloarchaea are isolated not only from natural but also from artificial hypersaline environments. In addition to salterns worldwide—such as those at Guérande (France) and Guerrero Negro (Mexico)—many haloarchaea have been isolated from commercial solar salts, pan-evaporated ("boiled") salts, and salt-preserved foods (Dillon et al., 2013; Henriet et al., 2014; Robertson et al., 2009; Roh et al., 2010).

Even in Japan, where natural hypersaline environments are scarce, several isolations have been reported. In 1985, strain TR-1 was isolated from sand at the Noto salterns by the groups of Horikoshi and Grant, and in 1990, it was named *Haloarcula japonica* (Takashina et al., 1990), representing the first report of a novel haloarchaeal species from Japan. Notably, strain No. 172 (172P1), isolated in 1980 from salt-encrusted sand near a seashore in Kagoshima Prefecture, was later described as *Natrialba asiatica* (Kamekura & Dyall-Smith, 1995).

Jeotgal, a traditional Korean fermented food, is prepared by fermenting small shrimp with rock salt for several months. For example, *Halalkalicoccus jeotgali* (Roh et al., 2007a) and *Natronococcus jeotgali* (Roh et al., 2007b) have been isolated from this product. Use of rock salt as an ingredient likely introduced these haloarchaea. Accordingly, haloarchaea are ubiquitous in hypersaline environments worldwide, and because salt lakes, halite, and solar (sea) salts ultimately derive from seawater, their origin has been inferred to be marine. As a marine example, *Halomarina oriensis* strain KeC-11 was described in 2011 as a new genus and species. Although not a direct isolation from seawater, KeC-11 was obtained from a seawater aquarium at the Atmosphere and Ocean Research Institute (The University of Tokyo) that was maintained with northwestern Pacific seawater collected 280 km south of Tokyo; the seawater was passed through a 0.2 μm filter, and cells retained on the filter served as the source material for isolation. Nevertheless, reports remain scarce, and the origin of haloarchaea is still enigmatic (Inoue et al., 2011).

Haloarchaea have also been detected or isolated from settings outside hypersaline environments. For example, clones affiliated with *Haloarcula* were detected at hydrothermal vents at the Pacmanus site in the Manus Basin (Papua New Guinea), and *Haloarcula* has likewise been detected in volcanic plume aerosols from active volcanoes in Russia (Kamchatka) and the United States (Hawaii, New Mexico, California, and Wyoming) (Takai et al., 2001). From Yellowstone vents, strains capable of withstanding 75 °C for 5–30 min have been isolated (Ellis et al., 2008).

Isolations from plants and animals have also been reported. A strain closely related to *Halococcus* was recovered from the nasal salt gland (an excretory gland that removes excess salt taken up from seawater) of the Cory's shearwater (*Calonectris diomedea)*, a seabird distributed on Mediterranean and Atlantic islands that can dive to depths of ~15 m (Brito-Echeverría et al., 2009). In humans, haloarchaea have been detected in the intestinal mucosa of patients with inflammatory bowel disease (Oxley et al., 2010).

Moreover, a team from Kagoshima University and Kyoto University reported archaea-specific sequences in portions of brain tissue from patients with dementia (Sakiyama et al., 2015). The detection of organisms long regarded as nonpathogenic in the human gastrointestinal tract and brain—neither of which is a hypersaline environment—is noteworthy.

6.4 Taxonomy of Halophilic Archaea

The class Halobacteria has seen a marked rise in recognized taxa in recent years. As of December 2023, taxa with validly published names comprised nine families, 82 genera, and 357 species—approximately 50% more than the May 2017 census (6 families, 57 genera, and 233 species) (Oren, 2024). Approximately 20 years ago, the group comprised only one order, one family, and ten genera, reflecting limited diversity. As reports of new genera and species accumulated, the need to reexamine their phylogenetic relationships was noted. Many haloarchaea possess multiple 16S rRNA genes with relatively low mutual similarity, which obscured their phylogeny. In 2010, Minegishi et al. conducted phylogenetic analyses using the RNA polymerase subunit B′ gene (*rpoB′*) as a molecular marker and pointed out issues within the haloarchaeal lineage. Furthermore, in 2012, Minegishi et al. showed that haloarchaea separate into two groups according to the orientation of a 16-gene array extending from the MC1 gene (encoding a non-histone chromosomal protein) located upstream of the rRNA operon to the *lpdA* gene (encoding dihydrolipoamide dehydrogenase) (Minegishi et al., 2012). In 2015, based on genome sequences, Gupta et al. reclassified the class Halobacteria—previously treated as a single order and a single family—into three orders and three families: order Halobacteriales (family Halobacteriaceae), order Haloferacales (family Haloferacaceae), and order Natrialbales (family Natrialbaceae)(Gupta et al., 2015). In 2016, Halobacteriales was further subdivided into three families (Halobacteriaceae, Haloarculaceae, and Halococcaceae), and Haloferacales into two families (Haloferacaceae and Halorubraceae) (Gupta et al., 2016). Genus-level revisions were also undertaken, with especially extensive changes involving *Haloterrigena* and *Natrinema*. Because these genera were proposed at roughly the same time, the accumulation of new species had led to significant inconsistencies in molecular phylogenetic analyses. In 2021, de la Haba et al. reassigned many *Haloterrigena* species to Natrinema (de la Haba et al. (2021)). Additional reclassifications—such as within *Halorubrum* and *Natronolimnobius*—have likewise been made, and haloarchaeal systematics has undergone major changes in recent years (Infante-Domínguez et al., 2020; Sorokin et al., 2020). Historically, the class Halobacteria was treated as three orders and six families. Recent genome-based proposals merged Haloferacales and Natrialbales into Halobacteriales; together with the newly proposed Halorutilales, the class is currently treated as two orders (Halobacteriales and Halorutilales). *Halorutilus salinus* is, at present, the sole representative of Halorutilales (Cui et al., 2023; Durán-Viseras et al., 2023; Oren, 2024).

Traditionally, the term halophilic archaea referred to members of the class Halobacteria; however, in recent years, halophilic archaea from a phylogenetically distinct lineage have been reported. These so-called nano-haloarchaea belong to the DPANN superphylum proposed by Rinke et al. in 2013. The first reported taxa, "*Candidatus* Nanosalina" sp. J07AB43 (Narasingarao et al., 2012) and "*Candidatus* Nanosalinarum" sp. J07AB56 (Rinke et al., 2013), were subjected to whole-genome analysis, yielding estimated genome sizes of ~1.2 Mbp; their cells are likewise very small (~0.6 μm). Subsequently, in 2016, "*Candidatus* Nanopetramus" sp. SG9 was reported from rock salt in the Atacama Desert (Crits-Christoph et al., 2016). In 2019, Antarctic "*Candidatus* Nanohaloarchaeum antarcticus" strains CHI and R1 were reported to require *Halorubrum lacusprofundi* for growth (Hamm et al., 2019), and in 2020 "*Candidatus* Nanohalobium constans" LC1Nh, which lives in symbiosis with *Halomicrobium* sp. LC1Hm, was described (La Cono et al., 2020). To date, nano-haloarchaea have been recovered from the same hypersaline settings as haloarchaea of the class Halobacteria, including rock salt, salt lakes, and solar salterns.

In addition, in 2018 a halophilic methanogen positioned phylogenetically between the classes Methanomicrobia and Halobacteria was discovered: *Methanonatronarchaeum thermophilum*, for which the new class Methanonatronarchaeia, order Methanonatronarchaeales, and family Methanonatronarchaeaceae were proposed. The type strain, AMET1, requires ≥3 M NaCl for growth (optimum 4 M), is moderately thermophilic (optimum 50 °C; growth up to 60 °C), and grows at pH 8.2–10.2 (optimum 9.5–9.7) (Sorokin et al.,

2018). It is phylogenetically closer to the Halobacteria than *M. evestigatum* (order Methanosarcinales, family Methanosarcinaceae) and has been suggested as a potential ancestor of haloarchaea.

6.5 Industrial Enzymes from Halophilic Archaea

Halophilic enzymes produced by halophilic archaea are known to exhibit thermostability and tolerance to organic solvents. Because they often retain activity under conditions that inactivate conventional enzymes, they are regarded as promising new tools for industrial applications (Cabrera & Blamey, 2018; DasSarma & DasSarma, 2015; Ortega et al., 2011). For example, multiple proteases and lipases have already been reported to sustain catalysis at molar-level salt concentrations, in the presence of surfactants, and even in biphasic systems containing an organic phase (Akmoussi-Toumi et al., 2018; De Castro et al., 2006; Haque et al., 2019; Li & Yu, 2014). Recent reviews emphasize that members of the Haloferacaceae—particularly *Haloferax*—are compatible with high-salinity processes run under open (non-sterile) fermentation and may therefore be candidates for process substitution (Griffiths et al., 2025). They also highlight an engineering caveat: many of these enzymes exhibit a lower-limit salt concentration for activity (Herrero-Alfonso et al., 2024).

Proteases play key roles in many biological processes and have numerous applications in biotechnology and industry. Haloarchaea produce abundant extracellular proteases and are well suited to high-salinity, often high-pH contexts such as leather processing, cleaning, and food processing (De Castro et al., 2006; Song et al., 2023; Yao et al., 2023). As comprehensively compiled by De Castro et al. in 2006, secretory pathways (Sec/Tat) and protease gene repertoires have been documented across multiple genera—including *Haloferax*, *Halobacterium*, *Natrinema*, and *Haloarcula*—delineating a nutrient-acquisition strategy in which extracellular proteolysis proceeds to oligopeptides that feed into central metabolism. More recently, halolysin R4 from *Haloferax mediterranei* has been shown to function as an antagonistic/defensive factor (Chen et al., 2021), and halolysins have been implicated in activating halocins (anti-haloarchaeal peptides) (Chen et al., 2024), underscoring the multipurpose roles of proteases in hypersaline ecosystems. From an applications standpoint, deployment in intrinsically high-salinity unit operations—such as high-salinity cleaning and leather pre-treatments—is particularly attractive because it may obviate separate desalting steps (Griffiths et al., 2025).

In applications that presuppose organic-solvent tolerance—such as lipid transformations and ester synthesis—lipid-hydrolyzing enzymes from haloarchaea have attracted considerable attention (Cabrera & Blamey, 2018; Haque et al., 2019). For example, an organic-solvent-tolerant lipase from *Haloarcula* sp. G41 functions under biodiesel-synthesis conditions (Li & Yu, 2014), and an organic-solvent-tolerant extracellular lipase (HML) from *Haloferax mediterranei* has been purified and characterized (Akmoussi-Toumi et al., 2018). Progress has also been made with esterases: a carboxylesterase from *Halobacterium salinarum* was heterologously expressed in *Haloferax volcanii*, enabling catalyst designs that include immobilization (Ortega-de la Rosa et al., 2024). In addition, direct analyses from native archaeal strains are increasing, as exemplified by the report of a novel family IV esterase (HjEstP1) from *Haloarcula japonica* (Kato et al., 2024). Collectively, these systems support stereoselective synthesis and fatty acid ester conversions under multi-stress conditions combining high salinity and organic solvents with potential applications to the production of flavor esters and functional lipids in the food sector (Akmoussi-Toumi et al., 2018; Li & Yu, 2014).

The main applications of amylases include food and beverage manufacturing that harness starch hydrolysis, animal feed production, and uses in medicine and biotechnology. Starch degradation in haloarchaea is largely mediated by extracellular amylases reported in genera such as *Haloarcula* and *Haloterrigena*, with many studies describing optima at 1–5 M salt, 45–70 °C, and pH 6.5–8.7 (Gómez-Villegas et al., 2021). For example, an organic-solvent-tolerant α-amylase from *Haloarcula* sp. S-1 exhibits high activity at 50 °C, 4.3 M NaCl, and pH 7.0, and retains activity and stability in solvents including benzene, toluene, and chloroform (Fukushima et al., 2005). Furthermore, the secreted α-amylase (AmyH) of *Haloarcula hispanica* is extremely stable under high-salt conditions yet—unusually—maintains activity even at low salinity, highlighting the plasticity of the activity range (Hutcheon et al., 2005). More recently, detailed analyses of a starch-degrading enzyme from *Haloarcula* sp. HS have progressed, and amylase production under high-salt conditions has also been verified in *Haloterrigena turkmenica* (Gómez-Villegas et al., 2021; Santorelli et al., 2016).

Other enzymes produced by halophilic archaea include cellulases and chitinases. In chitin degradation, chitinases have been identified in multiple haloarchaea (Minegishi et al., 2017; Sorokin et al., 2019), and functional chitinases have been demonstrated in *Halobacterium salinarum* and *Hfx. mediterranei* (Sorokin et al., 2015). In *Hfx. mediterranei*, the identification of chitin-metabolism gene clusters and the corresponding catabolic pathway has advanced, opening prospects for the pretreatment and hydrolysis of shellfish wastes under multi-stress conditions such as high salinity with coexisting organic phases (Hou et al., 2014). Moreover, the potential for chitinase production at high salt has been reported in other genera, including *Natrinema*, further broadening the search space (Yavari-Bafghi et al., 2019). By contrast, for cellulose degradation, comparative genomics indicates that certain haloarchaea broadly encode cellulase gene sets associated with cellulotrophy, suggesting the presence of extracellular depolymerization repertoires in genera such as *Natronobiforma*, *Natronolimnobius*, *Natrarchaeobius*, *Halosimplex*, *Halomicrobium*, and *Halococcoides* (Elcheninov et al., 2023).

6.6 Production of Bioplastics by Haloarchaea

Several haloarchaeal genera are known to accumulate polyhydroxyalkanoates (PHAs)—biodegradable plastics—as intracellular storage polymers (Koller, 2019; Simó-Cabrera et al., 2021). The principal products are PHB and PHBV (the 3-hydroxybutyrate/3-hydroxyvalerate copolymer), and adjusting the 3 HV fraction of PHBV enables control over processing behavior and mechanical toughness (Abbasi et al., 2022). Recent reviews position haloarchaea as PHA production hosts that combine salt-enabled process advantages with practical engineering tractability (Koller, 2019; Simó-Cabrera et al., 2021).

For example, *Haloarcula marismortui* accumulates PHB up to 21% of cell dry weight (CDW), and identification of a class III PHA synthase (PhaC/PhaE) established a foundation for archaeal PHA biosynthesis (Han et al., 2007). *Halogeometricum borinquense* strain E3 synthesizes PHBV directly from sugarcane bagasse hydrolysate (~50% CDW, ~13 mol% 3 HV), demonstrating a direct route from unrefined waste sugars to copolymer (Salgaonkar & Bragança, 2017). In addition, *Halolamina* sp. strains NRS-35/38 have been reported to produce 27–33% PHB from starch, and PHB production on soluble starch has also been confirmed for *Halorubrum* and *Natrinema* (Hagagy et al., 2022; Karray et al., 2021).

These process advantages are distinctive to haloarchaea. High-salinity media strongly suppress adventitious contamination, making non-sterile operation a realistic option (Koller, 2019). Seawater can be used for media preparation, and harvested cells undergo lysis by hypoosmotic shock (water treatment), enabling water-based extraction that minimizes solvent use (Montemurro et al., 2022; Wang et al., 2020).

Haloferax mediterranei is a well-established model organism for PHBV biosynthesis. It synthesizes PHBV with 3-hydroxyvalerate incorporated from unrelated substrates without the addition of exogenous precursors (Han et al., 2013). Genomic and biochemical analyses indicate that four propionyl-CoA–supplying routes operate in parallel in this strain (citramalate, aspartate-derived, methylmalonyl-CoA, and 3-hydroxypropionate pathways) (Han et al., 2013). The genome encodes multiple *phaC*-like loci together with a class III PhaEC synthase (Han et al., 2010), and endogenous type I-B CRISPR interference has been used to redirect carbon flux, improving both titer and productivity (Lin et al., 2021). Process demonstrations have progressed stepwise. Conversion of ricotta cheese exhausted whey (RCEW) to PHBV has been validated from laboratory to pilot scale (Raho et al., 2020), and non-sterile fermentation using waste silkworm excrement has likewise yielded PHBV with increased 3 HV fractions (Cai et al., 2022).

6.7 Haloarchaeal Bioremediation

Halophilic archaea possess metabolic, membranous, and enzymatic systems optimized for salinities exceeding that of seawater, enabling them to degrade, transform, and immobilize pollutants in environments where conventional microorganisms perform poorly—such as hypersaline produced water from oil fields and saline effluents originating from dyeing, leather processing, salt production, and salt-preserved foods (Dutta, 2022; Martínez-Espinosa, 2024). Recent reviews synthesize their applicability to hydrocarbons and other aromatics, synthetic dyes, and metals and metalloids, and highlight their compatibility with circular, integrated process designs (Martínez-Espinosa, 2024).

With respect to hydrocarbons and PAHs (polycyclic aromatic hydrocarbons), a foundational study on *Haloferax* sp. D1227 first demonstrated the use of an aromatic compound as a sole carbon source in archaea, delineating an aerobic pathway in which 3-phenylpropionic acid undergoes β-oxidation-like side-chain shortening to benzoyl-CoA, followed by ring cleavage via the gentisate pathway (Fu & Oriel, 1999). This constitutes key evidence that the canonical architecture of aerobic aromatic degradation can operate under high salinity. In addition, *Haloarcula* sp. D1 metabolizes 4-hydroxybenzoate through the gentisate pathway, further underscoring the diversity of aerobic aromatic catabolism among archaea (Fairley et al., 2002). From an application standpoint, biodegradation of aromatics in hypersaline produced water and significant reductions in chemical oxygen demand (COD) have been achieved by haloarchaea, with removal efficiencies reported to surpass those of hydrogen peroxide treatment (Bonfá et al., 2011).

With respect to azo dyes, isolates of *Halogeometricum* and *Haloferax* have been shown to decolorize Remazol Black B (and related dyes) at salinities up to 5 M NaCl, operating effectively across a broad range with optima at 2.5–3.4 M NaCl, pH 7, and 30–45 °C. These traits suggest the potential to minimize pretreatment steps when tackling the coupled challenges of high salinity and high dye loads in textile effluents (Kiadehi et al., 2018).

In the context of metals and metalloids, functional analyses of the *Halobacterium* sp. NRC-1 *ars* gene cluster have established that transport, detoxification, and transformation—including the As(III) methyltransferase ArsM—contribute to arsenic resistance (Wang et al., 2004). These mechanisms provide a molecular basis for microbially mediated immobilization and speciation control in hypersaline wastewaters containing mixed metals. More broadly, metal resistance in halophiles encompasses polymer-based chelation and external immobilization, efflux via dedicated transporters, and enzymatic detoxification, offering a conceptual framework for manipulating metal fate under high-salinity conditions (Voica et al., 2016; Vera-Bernal & Martínez-Espinosa, 2021).

Natrialba sp. C21 produces a biosurfactant that remains stable at high salinity while concurrently degrading aromatics, potentially improving solubilization and mobility of hydrophobic pollutants and thereby facilitating biodegradation (Khemili-Talbi et al., 2015). Likewise, the exopolysaccharide produced by *Haloferax mucosum* exhibits high molecular weight and strong emulsifying capacity with stability across wide temperature and operational ranges, indicating potential utility as an immobilizing or auxiliary matrix for metals and hydrophobic compounds (López-Ortega et al., 2020).

From a process-engineering perspective, high salinity itself suppresses exogenous contaminants, enabling simplified equipment and non-sterile operation; moreover, treating streams at their native salinity can obviate desalination pretreatment (Bonfá et al., 2011; Martínez-Espinosa, 2024). Outstanding challenges include optimization of continuous operation, reactor design, and control strategies, together with accumulation of long-term stability and scale-up data (Martínez-Espinosa, 2024).

6.8 Applications in Soy Sauce and *Miso* Fermentation

Fermentation of soy sauce (shoyu) and *miso* is a long-term, high-salt process in which umami and aroma are progressively built through the concerted actions of hydrolytic enzymes from koji (mold-fermented grain substrate; steamed grains inoculated with *Aspergillus oryzae*/*A. sojae*), halotolerant yeasts such as *Zygosaccharomyces rouxii*, and halophilic lactic acid bacteria such as *Tetragenococcus halophilus* (Devanthi & Gkatzionis, 2019; Tan et al., 2022). In particular, the moromi (brined mash) stage of soy sauce typically proceeds at 18–22% (w/v) salt (Zheng et al., 2024), while *miso* matures within a medium-to-high salinity range depending on product type (Kusumoto et al., 2021). This "high-ionic-strength ecosystem" is intrinsically compatible with halophilic archaea (haloarchaea), which stabilize themselves by maintaining high intracellular K^+ and by presenting protein surfaces enriched in acidic amino acids (Gunde-Cimerman et al., 2018; Saini & Patel, 2023).

Many haloarchaea operate optimally at roughly 2.5–4 M NaCl and are known to secrete or display extracellular enzymes—proteases, lipases/esterases, and α-amylases—that resist inactivation under high salt, organic solvents, and temperature fluctuations (DasSarma & DasSarma, 2015). Accordingly, they could provide a "late-stage catalytic reinforcement" to moromi or miso ripening by supplying catalytic stability that is difficult to obtain from conventional microorganisms (DasSarma & DasSarma, 2015; Nwankwo et al., 2023). Focusing on deep proteolysis, high-salt-tolerant archaeal proteases could continue cleavage downstream of the primary hydrolysis executed by koji proteases, driving peptides toward lower-molecular-weight fractions without excessively increasing bitterness; this would support the accumulation of taste-active amino acids such as glutamate, alanine, and proline, and potentially synergize with nucleotide-derived umami (Díez-Simón et al., 2020; Higuchi et al., 1991). In parallel, lipases/esterases can act on the pools of fatty acids, alcohols, and organic acids, thereby indirectly shaping key aroma outcomes—e.g., ester formation by halotolerant yeasts during alcoholic fermentation and the balance of pyrazine- and phenolic-type notes that emerge during maturation (Tan et al.,

2022; Díez-Simón et al., 2020). In response to the contemporary challenge of flavor retention in reduced-salt products, one feasible approach is to formulate haloarchaeal enzymes "in-process": even if the overall salt level is lowered, high-salt micro-niches can be engineered so that robust, salt-activated enzymes are engaged only when and where needed.

From safety and regulatory perspectives, although pathogenicity reports on haloarchaea are scarce, public track records as food starters remain limited; therefore, systematic gating is required—whole-genome screening for toxin-related and antimicrobial resistance genes, evaluation of biofilm formation and impacts on other microorganisms, and verification of survival/residuals in the final product (Chokesajjawatee et al., 2020; Heo et al., 2022). Taken together, haloarchaea can be viewed as a source of stable biocatalysts that function reliably under high salt and high ionic strength, with the potential to streamline deep proteolysis and the precursor networks of aroma—i.e., the "core of flavor"—in shoyu and miso.

Moreover, their application could extend to the degradation and valorization of high-salinity by-products generated during soy sauce manufacture, such as soy sauce lees (fermentation mash of soybeans) (Dutta, 2022).

References

Abbasi, M., Pokhrel, D., Coats, E. R., Guho, N. M., & McDonald, A. G. (2022). Effect of 3-hydroxyvalerate content on thermal, mechanical, and rheological properties of PHBV. *Polymers, 14*, 4183.

Akmoussi-Toumi, S., Khemili-Talbi, S., Ferioune, I., & Kebbouche-Gana, S. (2018). Purification and characterization of an organic solvent-tolerant and detergent-stable lipase (HML) from *Haloferax mediterranei. International Journal of Biological Macromolecules, 116*, 817–830.

Antón, J., Oren, A., Benlloch, S., Rodríguez-Valera, F., Amann, R., & Rosselló-Mora, R. (2002). *Salinibacter ruber* gen. nov., sp. nov., a novel, extremely halophilic member of the Bacteria from saltern crystallizer ponds. *International Journal of Systematic and Evolutionary Microbiology, 52*, 485–491.

Aslam, Z., Lim, J. H., Im, W.-T., Yasir, M., Chung, Y. R., & Lee, S.-T. (2007). *Salinicoccus jeotgali* sp. nov., isolated from jeotgal, a traditional Korean fermented seafood. *International Journal of Systematic and Evolutionary Microbiology, 57*, 633–638.

Bale, N. J., Sorokin, D. Y., Hopmans, E. C., Koenen, M., Rijpstra, W. I. C., Villanueva, L., Wienk, H., & Sinninghe Damsté, J. S. (2019). New insights into the polar lipid composition of extremely halo(alkali)philic euryarchaea from hypersaline lakes. *Frontiers in Microbiology, 10*, 377.

Béjà, O., Aravind, L., Koonin, E. V., Suzuki, M. T., Hadd, A., Nguyen, L. P., Jovanovich, S. B., Gates, C. M., Feldman, R. A., Spudich, J. L., et al. (2000). Bacterial rhodopsin: Evidence for a new type of phototrophy in the sea. *Science, 289*, 1902–1906.

Bonfá, M. R. L., Grossman, M. J., Mellado, E., & Durrant, L. R. (2011). Biodegradation of aromatic hydrocarbons by Haloarchaea and their use for the reduction of the chemical oxygen demand of hypersaline petroleum produced water. *Chemosphere, 84*, 1671–1676.

Brito-Echeverría, J., López-López, A., Yarza, P., Antón, J., & Rosselló-Móra, R. (2009). Occurrence of *Halococcus* spp. in the nostrils salt glands of the seabird *Calonectris diomedea. Extremophiles, 13*, 557–565.

Cabrera, M. Á., & Blamey, J. M. (2018). Biotechnological applications of archaeal enzymes from extreme environments. *Biological Research, 51*, 37.

Cai, S., Wu, Y., Li, Y., Yang, S., Liu, Z., Ma, Y., Lv, J., Shao, Y., Jia, H., Zhao, Y., & Cai, L. (2022). Open (non-sterile) fermentation with waste silkworm excrement to obtain high-3HV PHBV by *Haloferax mediterranei*. *Frontiers in Microbiology, 13*, 981605.

Chen, S., Sun, S., Wang, R., Feng, H., & Xiang, H. (2021). *Haloferax mediterranei* halolysin R4 confers antagonistic and defensive capabilities. *Applied and Environmental Microbiology, 87*, e02889–e02820.

Chen, S., Sun, S., Wang, R., Feng, H., & Xiang, H. (2024). Halocin H4 is activated through cleavage by halolysin HlyR4. *Applied and Environmental Microbiology, 90*, e02284–e02223.

Cho, E.-S., Cha, I.-T., Roh, S. W., & Seo, M.-J. (2021). *Haloferax litoreum* sp. nov., *Haloferax marinisediminis* sp. nov., and *Haloferax marinum* sp. nov., low salt-tolerant haloarchaea isolated from seawater and sediment. *Antonie Van Leeuwenhoek, 114*, 2065–2082.

Chokesajjawatee, N., Santiyanont, P., Vajrodaya, S., Ruengvisesh, S., Rattanavichai, W., & Suanyuk, N. (2020). Safety assessment of a *lactobacillus plantarum* starter culture for fermented pork sausage (nham) production. *Scientific Reports, 10*, 13752.

Crits-Christoph, A., Gelsinger, D. R., Ma, B., Wierzchos, J., Ravel, J., Davila, A., Casero, M. C., & DiRuggiero, J. (2016). Functional interactions of archaea, bacteria and viruses in a hypersaline endolithic community. *Environmental Microbiology, 18*, 2064–2077.

Cui, H. L., Yang, X., Gao, X., & Xu, X.-W. (2011). *Halobellus clavatus* gen. nov., sp. nov., and *Halorientalis regularis* gen. nov., sp. nov., two new members of the family Halobacteriaceae. *International Journal of Systematic and Evolutionary Microbiology, 61*, 2682–2689.

Cui, C., Han, D., Hou, J., & Cui, H.-L. (2023). Genome-based classification of the class Halobacteria and description of Haladaptataceae fam. nov. and Halorubellaceae fam. nov. *International Journal of Systematic and Evolutionary Microbiology, 73*, 005984.

Cui, H.-L., Hou, J., Amoozegar, M. A., Dyall-Smith, M. L., de la Haba, R. R., Minegishi, H., Montalvo-Rodriguez, R., Oren, A., Sanchez-Porro, C., Ventosa, A., & Vreeland, R. H. (2024). Proposed minimal standards for description of new taxa of the class Halobacteria. *International Journal of Systematic and Evolutionary Microbiology, 74*, 006290.

DasSarma, S., & DasSarma, P. (2015). Halophiles and their enzymes: Negativity put to good use. *Current Opinion in Microbiology, 25*, 120–126.

De Castro, R. E., Maupin-Furlow, J. A., Giménez, M. I., Herrera Seitz, M. K., & Sánchez, J. J. (2006). Haloarchaeal proteases and proteolytic systems. *FEMS Microbiology Reviews, 30*, 17–35.

de la Haba, R. R., Minegishi, H., Kamekura, M., Shimane, Y., & Ventosa, A. (2021). Phylogenomics of Haloarchaea: The controversy of the genera *Natrinema–Haloterrigena*. *Frontiers in Microbiology, 12*, 740909.

Devanthi, P. V. P., & Gkatzionis, K. (2019). Soy sauce fermentation: Microorganisms, aroma formation, and process modification. *Food Research International, 120*, 364–374.

Díez-Simón, C., Eichelsheim, C., & van Ruth, S. (2020). Chemical and sensory characteristics of soy sauce: A review. *Journal of Agricultural and Food Chemistry, 68*, 11651–11673.

Dillon, J. G., Carlin, M., Gutierrez, A., Nguyen, V., & McLain, N. (2013). Patterns of microbial diversity along a salinity gradient in the Guerrero Negro solar saltern, Baja CA Sur, Mexico. *Frontiers in Microbiology, 4*, 399.

Durán-Viseras, A., Sánchez-Porro, C., Viver, T., Konstantinidis, K. T., & Ventosa, A. (2023). Discovery of the streamlined haloarchaeon *Halorutilus salinus*, comprising a new order widespread in hypersaline environments across the world. *MSystems, 8*, e0119822.

Dutta, B. (2022). Biotechnological potentials of halophilic microorganisms and their adaptive strategies. *Beni-Suef University Journal of Basic and Applied Sciences, 11*, 52.

Elcheninov, A. G., Sorokin, D. Y., Toshchakov, S. V., Khijniak, T. V., & Kublanov, I. V. (2023). Cellulose metabolism in halo(natrono)archaea: A comparative genomics study. *Frontiers in Microbiology, 14*, 1112247.

Ellis, D. G., Bizzoco, R. W., & Kelley, S. T. (2008). Halophilic archaea determined from geothermal steam vent aerosols. *Environmental Microbiology, 10*, 1582–1590.

Fairley, D. J., Boyd, D. R., Sharma, N. D., Allen, C. C. R., Morgan, P., Larkin, M. J., Woodward, J. R., & Colby, J. (2002). Aerobic metabolism of 4-hydroxybenzoic acid in archaea via the gentisate pathway by *Haloarcula* sp. strain D1. *Journal of Bacteriology, 184*, 370–373.

Fu, W., & Oriel, P. (1999). Degradation of 3-phenylpropionic acid by *Haloferax* sp. D1227. *Extremophiles, 3*, 45–53.

Fukushima, T., Mizuki, T., Echigo, A., Inoue, A., & Usami, R. (2005). Organic solvent tolerance of halophilic alpha-amylase from a haloarchaeon, *Haloarcula* sp. strain S-1. *Extremophiles, 9*, 85–89.

Gochnauer, M. B., Leppard, G. G., Komaratat, P., Kates, M., Novitsky, T., & Kushner, D. J. (1975). Isolation and characterization of *Actinopolyspora halophila*, gen. et sp. nov., an extremely halophilic actinomycete. *Canadian Journal of Microbiology, 21*, 1500–1511.

Gómez-Villegas, P., Vigara, J., Romero, L., Gotor, C., Raposo, S., Gonçalves, B., & Léon, R. (2021). Biochemical characterization of the amylase activity from the new haloarchaeal strain *Haloarcula* sp. HS isolated in the Odiel Marshlands. *Biology (Basel), 10*, 337.

Gramain, A., Díaz, G. C., Demergasso, C., Lowenstein, T. K., & McGenity, T. J. (2011). Archaeal diversity along a subterranean salt core from the Salar Grande (Chile). *Environmental Microbiology, 13*, 2105–2121.

Griffiths, D. B., Tiwari, R. P., Murphy, D. V., Scott, C., & Turner, N. J. (2025). *Haloferax* and the Haloferacaceae: Potential role in bioindustry. *Biotechnology Advances, 84*, 108666.

Gunde-Cimerman, N., Plemenitaš, A., & Oren, A. (2018). Strategies of adaptation of microorganisms of the three domains of life to high salt concentrations. *FEMS Microbiology Reviews, 42*, 353–375.

Gupta, R. S., Naushad, S., & Baker, S. (2015). Phylogenomic analyses and molecular signatures for the class Halobacteria and its two major clades: A proposal for division of the class Halobacteria into an emended order Halobacteriales and two new orders, Haloferacales ord. nov. and Natrialbales ord. nov., containing the novel families Haloferacaceae fam. nov. and Natrialbaceae fam. nov. *International Journal of Systematic and Evolutionary Microbiology, 65*, 1050–1069.

Gupta, R. S., Naushad, S., Fabros, R., & Adeolu, M. (2016). A phylogenomic reappraisal of family-level divisions within the class Halobacteria: Proposal to divide the order Halobacteriales into the families Halobacteriaceae, Haloarculaceae fam. nov., and Halococcaceae fam. nov., and the order Haloferacales into the families, Haloferacaceae and Halorubraceae fam. nov. *Antonie Van Leeuwenhoek, 109*, 565–587.

Hagagy, N., Saddiq, A. A., Tag, H. M., Selim, S., AbdElgawad, H., & Martínez-Espinosa, R. M. (2022). Characterization of polyhydroxybutyrate (PHB) synthesized by newly isolated haloarchaea *Halolamina* spp. *Molecules, 27*, 7366.

Hamm, J. N., Erdmann, S., Eloe-Fadrosh, E. A., Angeloni, A., Zhong, L., Brownlee, C., Williams, T. J., Barton, K., Carswell, S., Smith, M. A., et al. (2019). Unexpected host dependency of Antarctic Nanohaloarchaeota. *Proceedings of the National Academy of Sciences of the United States of America, 116*, 14661–14670.

Han, J., Lu, Q., Zhou, L., Zhou, J., & Xiang, H. (2007). Molecular characterization of the $phaEC_{Hm}$ genes required for PHB biosynthesis in *Haloarcula marismortui*. *Applied and Environmental Microbiology, 73*, 6058–6065.

Han, J., Li, M., Hou, J., Wu, L.-P., Zhou, D., & Xiang, H. (2010). Comparison of four *phaC* genes from *Haloferax* mediterranei and their function in different PHBV copolymer biosyntheses in *Haloarcula hispanica*. *Saline Systems, 6*, 9.

Han, J., Hou, J., Zhang, F., Ai, G., Li, M., Cai, S., Liu, H., Wang, L., Wang, Z., Zhang, S., et al. (2013). Multiple propionyl coenzyme A-supplying pathways for production of the bioplastic poly(3-hydroxybutyrate-co-3-hydroxyvalerate) in *Haloferax mediterranei*. *Applied and Environmental Microbiology, 79*, 2922–2931.

Haque, R. U., Paradisi, F., & Allers, T. (2019). *Haloferax volcanii* as immobilised whole-cell biocatalyst: New applications for halophilic systems. *Applied Microbiology and Biotechnology, 103*, 3807–3817.

Henriet, O., Fourmentin, J., Delincé, B., & Mahillon, J. (2014). Exploring the diversity of extremely halophilic archaea in food-grade salts. *International Journal of Food Microbiology, 191*, 36–44.

Heo, S., Kim, T., & Kim, S. (2022). Safety assessment systems for microbial starters derived from fermented foods. *Journal of Microbiology and Biotechnology, 32*, 1219–1229.

Herrero-Alfonso, P., Pejenaute, A., Millet, O., & Ortega-Quintanilla, G. (2024). Electrostatics introduce a trade-off between mesophilic stability and adaptation in halophilic proteins. *Protein Science, 33*, e5003.

Higuchi, T., Aoki, T., & Uchida, K. (1991). Esterase activity in soy sauce moromi as a factor hydrolyzing flavor esters. *Journal of Fermentation and Bioengineering, 71*, 163–167.

Hogg, B. N., Schnepel, C., Finnigan, J. D., & Turner, N. J. (2024). The impact of metagenomics on biocatalysis. *Angewandte Chemie International Edition, 63*, e202402316.

Hou, J., Han, J., Cai, L., Zhou, J., Lü, Y., Jin, C., Liu, J., & Xiang, H. (2014). Characterization of genes for chitin catabolism in *Haloferax mediterranei*. *Applied Microbiology and Biotechnology, 98*, 1185–1194.

Hutcheon, G. W., Vasisht, N., & Bolhuis, A. (2005). Characterisation of a highly stable alpha-amylase from the halophilic archaeon *Haloarcula hispanica*. *Extremophiles, 9*, 487–495.

Ikram, H. I., Robert, C., Michelle, C., Raoult, D., Hacène, H., & Desnues, C. (2014). Non-contiguous finished genome sequence and description of *Halopiger goleamassiliensis* sp. nov. *Standards in Genomic Sciences, 9*, 956–969.

Infante-Domínguez, C., de la Haba, R. R., Corral, P., Sánchez-Porro, C., Arahal, D. R., & Ventosa, A. (2020). Genome-based analyses reveal a synonymy among *Halorubrum distributum*, *Halorubrum terrestre*, *Halorubrum arcis* and *Halorubrum litoreum*; emended description of *Halorubrum distributum*. *International Journal of Systematic and Evolutionary Microbiology, 70*, 1698–1705.

Inoue, K., Itoh, T., Ohkuma, M., & Kogure, K. (2011). *Halomarina oriensis* gen. nov., sp. nov., a halophilic archaeon isolated from a seawater aquarium. *International Journal of Systematic and Evolutionary Microbiology, 61*, 942–946.

Kamekura, M., & Dyall-Smith, M. L. (1995). Taxonomy of the family Halobacteriaceae and the description of two new genera *Halorubrobacterium* and *Natrialba*. *Journal of General and Applied Microbiology, 41*, 333–350.

Karray, F., Ben Abdallah, M., Baccar, N., Zaghden, H., & Sayadi, S. (2021). Production of poly(3-hydroxybutyrate) by *Haloarcula*, *Halorubrum* and *Natrinema* haloarchaeal genera using starch as a carbon source. *Archaea, 2021*, 8888712.

Kato, H., Ambai, S., Ikeda, F., Abe, K., Nakamura, S., & Yatsunami, R. (2024). Characterization of a family IV esterase from extremely halophilic archaeon *Haloarcula japonica*. *Extremophiles, 29*, 7.

Khemili-Talbi, S., Kebbouche-Gana, S., Akmoussi-Toumi, S., Angar, Y., & Gana, M. L. (2015). Isolation of an extremely halophilic archaeon *Natrialba* sp. C21 able to degrade aromatic compounds and to produce stable biosurfactant at high salinity. *Extremophiles, 19*, 1109–1120.

Kiadehi, M. S. H., Amoozegar, M. A., Asad, S., & Siroosi, M. (2018). Exploring the potential of halophilic archaea for the decolorization of azo dyes. *Water Science and Technology, 77*, 1602–1611.

Koga, Y. (2007). Biosynthesis of ether-type polar lipids in archaea. *Journal of Biochemistry, 141*, 449–455.

Koller, M. (2019). Polyhydroxyalkanoate biosynthesis at the edge of water activity—Haloarchaea as biopolyester factories. *Bioengineering, 6*, 34.

Kushner, D. J. (1978). Life in high salt and solute concentrations. In D. J. Kushner (Ed.), *Microbial life in extreme environments* (pp. 317–368). Academic Press.

Kusumoto, K.-I., Yamagata, Y., Tazawa, R., Kitagawa, M., Kato, T., Isobe, K., & Kashiwagi, Y. (2021). Japanese traditional miso and koji making. *Journal of Fungi, 7*(7), 579.

La Cono, V., Messina, E., Rohde, M., Arcadi, E., Ciordia, S., Crisafi, F., Denaro, R., Ferrer, M., Giuliano, L., Golyshin, P. N., et al. (2020). Symbiosis between nanohaloarchaeon and haloarchaeon is based on utilization of different polysaccharides. *Proceedings of the National Academy of Sciences of the United States of America, 117*, 20223–20234.

Lanyi, J. K., Helgerson, S. L., & Silverman, M. P. (1979). Relationship between proton motive force and potassium ion transport in *Halobacterium halobium* envelope vesicles. *Archives of Biochemistry and Biophysics, 193*, 329–339.

Legat, A., Denner, E. B. M., Dornmayr-Pfaffenhuemer, M., Pfeiffer, P., Knopf, B., Claus, H., Gruber, C., König, H., Wanner, G., & Stan-Lotter, H. (2013). Properties of *Halococcus salifodinae*, an isolate from Permian rock salt deposits, compared with halococci from surface waters. *Life (Basel), 3*, 244–259.

Li, X., & Yu, H.-Y. (2014). Characterization of an organic solvent-tolerant lipase from *Haloarcula* sp. G41 and its application for biodiesel production. *Folia Microbiologica, 59*, 455–463.

Lin, L., Chen, J., Mitra, R., Gao, Q., Cheng, F., Xu, T., Zuo, Z., Xiang, H., & Han, J. (2021). Optimising PHBV biopolymer production in haloarchaea via CRISPRi-mediated redirection of carbon flux. *Communications Biology, 4*, 1007.

Liu, Q., Ren, M., & Zhang, L. L. (2015). *Natribaculum breve* gen. nov., sp. nov., and *Natribaculum longum* sp. nov., halophilic archaea isolated from saline soil. *International Journal of Systematic and Evolutionary Microbiology, 65*, 604–608.

Lo Nostro, P., & Ninham, B. W. (2012). Hofmeister phenomena: An update on ion specificity in biology. *Chemical Reviews, 112*, 2286–2322.

López-Ortega, M. A., Rodríguez-Hernández, A. I., Camacho-Ruíz, R. M., Córdova, J., López-Cuellar, M. d. R., Chavarría-Hernández, N., & González-García, Y. (2020). Physicochemical characterization and emulsifying properties of a novel exopolysaccharide produced by haloarchaeon *Haloferax mucosum. International Journal of Biological Macromolecules, 142*, 152–162.

Madern, D., Ebel, C., & Zaccai, G. (2000). Halophilic adaptation of enzymes. *Extremophiles, 4*, 91–98.

Martínez-Espinosa, R. M. (2024). Halophilic archaea as tools for bioremediation technologies. *Applied Microbiology and Biotechnology, 108*, 401.

McDuff, S., King, G. M., & Adams, M. W. W. (2016). Isolation and characterization of extremely halophilic CO-oxidizing Euryarchaeota from hypersaline cinders, sediments and soils, and description of a novel CO oxidizer, *Haloferax namakaokahaiae* Mke2.3T, sp. nov. *FEMS Microbiology Ecology, 92*, fiw028.

McGenity, T. J., & Grant, W. D. (1995). Transfer of *Halobacterium saccharovorum, Halobacterium sodomense, Halobacterium trapanicum* NRC 34021 and *Halobacterium lacusprofundi* to the genus *Halorubrum* gen. nov., as *Halorubrum saccharovorum* comb. nov., *Halorubrum sodomense* comb. nov., *Halorubrum trapanicum* comb. nov., and *Halorubrum lacusprofundi* comb. nov. *Systematic and Applied Microbiology, 18*, 237–243.

Mesbah, N. M. (2022). Industrial biotechnology based on enzymes from extreme environments. *Frontiers in Bioengineering and Biotechnology, 10*, 870083.

Minegishi, H., Mizuki, T., Echigo, A., Fukushima, T., Kamekura, M., & Usami, R. (2008). Acidophilic haloarchaeal strains are isolated from various solar salts. *Saline Systems, 4*, 16.

Minegishi, H., Echigo, A., Nagaoka, S., Kamekura, M., & Usami, R. (2010a). *Halarchaeum acidiphilum* gen. nov., sp. nov., a moderately acidophilic haloarchaeon isolated from commercial solar salt. *International Journal of Systematic and Evolutionary Microbiology, 60*, 2513–2516.

Minegishi, H., Kamekura, M., Itoh, T., Echigo, A., Usami, R., & Hashimoto, T. (2010b). Further refinement of the phylogeny of the Halobacteriaceae based on the full-length RNA polymerase subunit B′ (*rpoB*′) gene. *International Journal of Systematic and Evolutionary Microbiology, 60*, 2398–2408.

Minegishi, H., Kamekura, M., Kitajima-Ihara, T., Nakasone, K., Echigo, A., Shimane, Y., Usami, R., Itoh, T., & Ihara, K. (2012). Gene orders in the upstream of 16S rRNA genes divide genera of the family Halobacteriaceae into two groups. *International Journal of Systematic and Evolutionary Microbiology, 62*, 188–195.

Minegishi, H., Enomoto, S., Echigo, A., Shimane, Y., Kondo, Y., Inoma, A., Kamekura, M., Takai, K., Itoh, T., Ohkuma, M., et al. (2017). *Salinarchaeum chitinilyticum* sp. nov., a chitin-degrading haloarchaeon isolated from commercial salt. *International Journal of Systematic and Evolutionary Microbiology, 67*, 2274–2278.

Montemurro, M., Salvatori, G., Alfano, S., Martinelli, A., Verni, M., Pontonio, E., Villano, M., & Rizzello, C. G. (2022). Exploitation of wasted bread as substrate for polyhydroxyalkanoates production through the use of *Haloferax mediterranei* and seawater, and water-based extraction for PHBV recovery. *Frontiers in Microbiology, 13*, 1000962.

Moopantakath, J., Imchen, M., Anju, V. T., Busi, S., Dyavaiah, M., Martínez-Espinosa, R. M., & Kumavath, R. (2023). Bioactive molecules from haloarchaea: Scope and prospects for industrial and therapeutic applications. *Frontiers in Microbiology, 14*, 1113540.

Mou, Y.-Z., Qiu, X.-X., Zhao, M.-L., Cui, H.-L., Oh, D., & Dyall-Smith, M. L. (2012). *Halohasta litorea* gen. nov., sp. nov., and *Halohasta litchfieldiae* sp. nov., isolated from the Daliang aquaculture farm, China, and from Deep Lake, Antarctica, respectively. *Extremophiles, 16*, 895–901.

Nagaoka, S., Minegishi, H., Echigo, A., Shimane, Y., Kamekura, M., & Usami, R. (2010). *Halostagnicola kamekurae* sp. nov., an extremely halophilic archaeon from solar salt. *International Journal of Systematic and Evolutionary Microbiology, 60*, 2828–2831.

Narasingarao, P., Podell, S., Ugalde, J. A., Brochier-Armanet, C., Emerson, J. B., Brocks, J. J., Heidelberg, K. B., Banfield, J. F., & Allen, E. E. (2012). De novo metagenomic assembly reveals abundant novel major lineage of archaea in hypersaline microbial communities. *ISME Journal, 6*, 81–93.

Nwankwo, C. E. I., Hou, J., & Cui, H.-L. (2023). Extracellular proteases from halophiles: Diversity and application challenges. *Applied Microbiology and Biotechnology, 107*, 5923–5934.

Ollivier, B., Caumette, P., Garcia, J. L., & Mah, R. A. (1994). Anaerobic bacteria from hypersaline environments. *Microbiological Reviews, 58*, 27–38.

Oren, A. (1999). Bioenergetic aspects of halophilism. *Microbiology and Molecular Biology Reviews, 63*, 334–348.

Oren, A. (2008). Microbial life at high salt concentrations: Phylogenetic and metabolic diversity. *Saline Systems, 4*, 2.

Oren, A. (2024). Novel insights into the diversity of halophilic microorganisms and their functioning in hypersaline ecosystems. *npj Biodiversity, 3*, 18.

Oren, A., Gurevich, P., Gemmell, R. T., & Teske, A. (1995). *Halobaculum gomorrense* gen. nov., sp. nov., a novel extremely halophilic archaeon from the Dead Sea. *International Journal of Systematic Bacteriology, 45*, 747–754.

Ortega, G., Laín, A., Tadeo, X., López-Méndez, B., Castaño, D., & Millet, O. (2011). Halophilic enzyme activation induced by salts. *Scientific Reports, 1*, 6.

Ortega, G., Diercks, T., & Millet, O. (2015). Halophilic protein adaptation results from synergistic residue–ion interactions in the folded and unfolded states. *Chemistry & Biology, 22*, 1597–1607.

Ortega-de la Rosa, N. D., Romero-Borbón, E., Rodríguez, J. A., Camacho-Ruiz, A., & Córdova, J. (2024). Cloning, expression, characterization and immobilization of a recombinant carboxylesterase from the halophilic archaeon, *Halobacterium salinarum* NCR-1. *Biomolecules, 14*, 534.

Oxley, A. P. A., Lanfranconi, M. P., Würdemann, D., Ott, S., Schreiber, S., McGenity, T. J., Timmis, K. N., & Nogales, B. (2010). Halophilic archaea in the human intestinal mucosa. *Environmental Microbiology, 12*, 2398–2410.

Raho, S., Carofiglio, V. E., Montemurro, M., Miceli, V., Centrone, D., Stufano, P., Schioppa, M., Pontonio, E., & Rizzello, C. G. (2020). Production of the polyhydroxyalkanoate PHBV from ricotta cheese exhausted whey by *Haloferax mediterranei* fermentation. *Food, 9*, 1459.

Rinke, C., Schwientek, P., Sczyrba, A., Ivanova, N. N., Anderson, I. J., Cheng, J.-F., Darling, A., Malfatti, S., Swan, B. K., Gies, E. A., et al. (2013). Insights into the phylogeny and coding potential of microbial dark matter. *Nature, 499*, 431–437.

Robertson, C. E., Spear, J. R., Harris, J. K., & Pace, N. R. (2009). Diversity and stratification of archaea in a hypersaline microbial mat. *Applied and Environmental Microbiology, 75*, 1801–1810.

Rodrigues-Oliveira, T., Belmok, A., Vasconcellos, D., Schuster, B., & Kyaw, C. M. (2017). Archaeal S-layers: Overview and current state of the art. *Frontiers in Microbiology, 8*, 2597.

Roh, S. W., Nam, Y.-D., Chang, H.-W., Sung, Y., Kim, K.-H., Lee, H.-J., Oh, H.-M., & Bae, J.-W. (2007a). *Natronococcus jeotgali* sp. nov., a halophilic archaeon isolated from shrimp jeotgal, a traditional fermented seafood from Korea. *International Journal of Systematic and Evolutionary Microbiology, 57*, 2129–2131.

Roh, S. W., Nam, Y.-D., Chang, H.-W., Sung, Y., Kim, K.-H., Oh, H.-M., & Bae, J.-W. (2007b). *Halalkalicoccus jeotgali* sp. nov., a halophilic archaeon from shrimp jeotgal, a traditional Korean fermented seafood. *International Journal of Systematic and Evolutionary Microbiology, 57*, 2296–2298.

Roh, S. W., Kim, K.-H., Nam, Y.-D., Chang, H.-W., Park, E.-J., & Bae, J.-W. (2010). Investigation of archaeal and bacterial diversity in fermented seafood using barcoded pyrosequencing. *ISME Journal, 4*, 1–16.

Saini, P., & Patel, A. B. (2023). Survival strategies and stress adaptations in halophilic archaebacteria. In B. P. Tripathi (Ed.), *Microbial adaptive mechanisms: Survival strategies in extreme environments* (ACS Symposium Series). American Chemical Society.

Sakiyama, Y., Kanda, N., Higuchi, Y., et al. (2015). New type of encephalomyelitis responsive to trimethoprim/sulfamethoxazole treatment in Japan. *Neurology: Neuroimmunology & Neuroinflammation, 2*, e143.

Salgaonkar, B. B., & Bragança, J. M. (2017). Utilization of sugarcane bagasse by *Halogeometricum borinquense* strain E3 for biosynthesis of poly(3-hydroxybutyrate-co-3-hydroxyvalerate). *Bioengineering, 4*, 50.

Santorelli, M., Maurelli, L., Pocsfalvi, G., Fiume, I., Squillaci, G., La Cara, F., Del Monaco, G., & Morana, A. (2016). Isolation and characterisation of a novel alpha-amylase from the extreme haloarchaeon *Haloterrigena turkmenica*. *International Journal of Biological Macromolecules, 92*, 174–184.

Satari, L., Guillén, A., Latorre-Pérez, A., & Porcar, M. (2021). Beyond archaea: The table salt bacteriome. *Frontiers in Microbiology, 12*, 714110.

Savage, K. N., Krumholz, L. R., Oren, A., Elshahed, M. S., & Knight, V. K. (2007). *Haladaptatus paucihalophilus* gen. nov., sp. nov., a halophilic archaeon isolated from a low-salt, sulfide-rich spring. *International Journal of Systematic and Evolutionary Microbiology, 57*, 19–27.

Seckbach, J., & Oren, A. (2004). Introduction to the extremophiles. In J. Seckbach (Ed.), *Origins: Genesis, evolution and diversity of life* (pp. 373–393). Kluwer Academic Publishers.

Shimoshige, H., Yamada, T., Minegishi, H., Echigo, A., Shimane, Y., Kamekura, M., Itoh, T., & Usami, R. (2013). *Halobaculum magnesiiphilum* sp. nov., a magnesium-dependent haloarchaeon isolated from commercial salt. *International Journal of Systematic and Evolutionary Microbiology, 63*, 861–866.

Simó-Cabrera, L., García-Chumillas, S., Hagagy, N., Saddiq, A., Tag, H., Selim, S., AbdElgawad, H., Arribas Agüero, A., Monzó Sánchez, F., Cánovas, V., Pire, C., & Martínez-Espinosa, R. M. (2021). Haloarchaea as cell factories to produce bioplastics. *Marine Drugs, 19*, 159.

Singh, A., & Singh, A. K. (2017). Haloarchaea: Worth exploring for their biotechnological potential. *Biotechnology Letters, 39*, 1793–1800.

Song, P., Zhang, X., Wang, S., Xu, W., Wang, F., Fu, R., & Wei, F. (2023). Microbial proteases and their applications. *Frontiers in Microbiology, 14*, 1236368.

Sorokin, D. Y., Toshchakov, S. V., Kolganova, T. V., & Kublanov, I. V. (2015). Halo(natrono) archaea isolated from hypersaline lakes utilize cellulose and chitin as growth substrates. *Frontiers in Microbiology, 6*, 942.

Sorokin, D. Y., Kublanov, I. V., Yakimov, M. M., Rijpstra, W. I. C., & Sinninghe Damsté, J. S. (2016). *Halanaeroarchaeum sulfurireducens* gen. nov., sp. nov., the first obligately anaerobic sulfur-respiring haloarchaeon, isolated from a hypersaline lake. *International Journal of Systematic and Evolutionary Microbiology, 66*, 2377–2381.

Sorokin, D. Y., Merkel, A. Y., Abbas, B., Makarova, K. S., Rijpstra, W. I. C., Koenen, M., Sinninghe Damsté, J. S., Galinski, E. A., Koonin, E. V., & van Loosdrecht, M. C. M. (2018). *Methanonatronarchaeum thermophilum* gen. nov., sp. nov. and '*Candidatus* Methanohalarchaeum thermophilum', extremely halo(natrono)philic methyl-reducing methanogens from hypersaline lakes comprising a new euryarchaeal class *Methanonatronarchaeia* classis nov. *International Journal of Systematic and Evolutionary Microbiology, 68*, 2199–2208.

Sorokin, D. Y., Elcheninov, A. G., Toshchakov, S. V., Bale, N. J., Sinninghe Damsté, J. S., Khijniak, T. V., & Kublanov, I. V. (2019). *Natrarchaeobius chitinivorans* gen. nov., sp. nov., and *Natrarchaeobius halalkaliphilus* sp. nov., alkaliphilic, chitin-utilizing haloarchaea from hypersaline alkaline lakes. *Systematic and Applied Microbiology, 42*, 309–318.

Sorokin, D. Y., Merkel, A. Y., Messina, E., Yakimov, M. M., Itoh, T., Mesbah, N. M., Wiegel, J., & Oren, A. (2020). Reclassification of the genus *Natronolimnobius*: Proposal of two new genera, *Natronolimnohabitans* gen. nov. to accommodate *Natronolimnobius innermongolicus* and *Natrarchaeobaculum* gen. nov. to accommodate *Natronolimnobius aegyptiacus* and *Natronolimnobius sulfurireducens*. *International Journal of Systematic and Evolutionary Microbiology, 70*, 3399–3405.

Steber, J., & Schleifer, K. H. (1975). *Halococcus morrhuae*: A sulfated heteropolysaccharide as the structural component of the bacterial cell wall. *Archives of Microbiology, 105*, 173–177.

Takai, K., Moser, D. P., Onstott, T. C., Spoelstra, N., Pfiffner, S. M., Dohnalkova, A., & Fredrickson, J. K. (2001). *Alkaliphilus transvaalensis* gen. nov., sp. nov., an extremely alkaliphilic bacterium isolated from a deep South African gold mine. *International Journal of Systematic and Evolutionary Microbiology, 51*, 1245–1256.

Takashina, T., Hamamoto, T., Otozai, K., Grant, W. D., & Horikoshi, K. (1990). *Haloarcula japonica* sp. nov., a new triangular halophilic archaebacterium. *Systematic and Applied Microbiology, 13*, 177–181.

Tan, G., Wang, Y., Hu, M., Li, X., Li, X., Pan, Z., Li, M., Li, L., & Zheng, Z. (2022). Comparative evaluation of the microbial diversity and metabolite profiles of Japanese-style and Cantonese-style soy sauce fermentation. *Frontiers in Microbiology, 13*, 976206.

Tenchov, B., Vescio, E. M., Sprott, G. D., Zeidel, M. L., & Mathai, J. C. (2006). Salt tolerance of archaeal extremely halophilic lipid membranes. *Journal of Biological Chemistry, 281*, 10016–10023.

Torreblanca, M., Rodríguez-Valera, F., Juez, G., Ventosa, A., Kamekura, M., & Kates, M. (1986). Classification of non-alkaliphilic Halobacteria based on numerical taxonomy. *Systematic and Applied Microbiology, 8*, 89–99.

Ventosa, A., Nieto, J. J., & Oren, A. (1998). Biology of moderately halophilic aerobic bacteria. *Microbiology and Molecular Biology Reviews, 62*, 504–544.

Vera-Bernal, M., & Martínez-Espinosa, R. M. (2021). Insights on cadmium removal by bioremediation: The case of haloarchaea. *Microbiology Research, 12*, 354–375.

Voica, D. M., Bartha, L., Banciu, H. L., & Oren, A. (2016). Heavy metal resistance in halophilic bacteria and archaea. *FEMS Microbiology Letters, 363*, fnw146.

Vreeland, R. H., Straight, S., Krammes, J., Dougherty, K., Rosenzweig, W. D., & Kamekura, M. (2002). *Halosimplex carlsbadense* gen. nov., sp. nov., a unique halophilic archaeon, with three 16S rRNA genes, that grows only in defined medium with glycerol and acetate or pyruvate. *Extremophiles, 6*, 445–452.

Wagner, G., Hartmann, R., & Oesterhelt, D. (1978). Potassium uniport and ATP synthesis in *Halobacterium halobium*: Light-driven K^+ uptake mediated by bacteriorhodopsin. *European Journal of Biochemistry, 89*, 169–179.

Wang, G., Kennedy, S. P., Fasiludeen, S., Rensing, C., & DasSarma, S. (2004). Arsenic resistance in *Halobacterium* sp. strain NRC-1 examined by using an improved gene knockout system. *Journal of Bacteriology, 186*, 3187–3194.

Wang, K., Chen, C., & Zhang, R. (2020). Production of polyhydroxyalkanoates by *Haloferax mediterranei*: Osmotic-shock-facilitated extraction among advantages. *Bioengineering, 7*, 123.

Whon, T. W., Ahn, S. W., Yang, S., Kim, J. Y., Kim, Y. B., Kim, Y., Hong, J.-M., Jung, H., Choi, Y.-E., Lee, S. H., & Roh, S. W. (2021). ODFM, an omics data resource from microorganisms associated with fermented foods. *Scientific Data, 8*, 113.

Xu, W.-D., Zhang, W.-J., Han, D., Cui, H.-L., & Yang, K. (2015). *Halorussus ruber* sp. nov., isolated from an inland salt lake of China. *Archives of Microbiology, 197*, 91–95.

Yao, H., Liu, S., Liu, T., Ren, D., Zhou, Z., Yang, Q., & Mao, J. (2023). Microbial-derived salt-tolerant proteases and their applications in high-salt traditional soybean fermented foods: A review. *Bioresources and Bioprocessing, 10*, 82.

Yavari-Bafghi, M., Babavalian, H., & Amoozegar, M. A. (2019). Isolation, screening and identification of haloarchaea with chitinolytic activity from hypersaline lakes of Iran. *Archives of Biological Sciences, 71*, 71–81.

Zhang, Q.-K., Zhu, L.-R., Dong, X.-Y., Yang, X.-Y., Hou, J., & Cui, H.-L. (2024). *Salinirarus marinus* gen. nov., sp. nov., *Haloplanus salilacus* sp. nov., *Haloplanus pelagicus* sp. nov., *Haloplanus halophilus* sp. nov., and *Haloplanus halobius* sp. nov., halophilic archaea isolated from commercial coarse salts with potential as starter cultures for salt-fermented foods. *FEMS Microbiology Letters, 371*, fnae075.

Zheng, S., Liu, H., Chen, L., Guo, Z., & Lu, M. (2024). A method of reducing salt content in fermented soy sauce. *Food, 13*, 807.

Zhilina, T. N., & Zavarzin, G. A. (1987). Methanohalobium evestigatus [sic], n. gen., n. sp.: The extremely halophilic methanogenic archaebacterium. *Doklady Akademii Nauk SSSR, 293*, 464–468.

Chapter 7
Recent Developments in the Study of the Biology of Sulfolobales: Acidothermophilic Archaea as Microbial Models and Platforms for Biomass Degradation

Pengju Wu, Yunfeng Yang, Qi Gan, Shikuan Liang, Haodun Li, Shuai Li, Yan Xie, Qihong Huang, Jinfeng Ni, and Yulong Shen

Abstract The extremophilic microbes belonging to the Sulfolobales order of archaea are known to thrive in acidic and high temperature environments. Their unique ability to withstand such harsh conditions is largely attributed to the enzymes they produce, which are highly resistant to extreme temperatures and pH levels. Due to their distinct genetic makeup and evolutionary position, these prokaryotic microorganisms have emerged as valuable models for studying the evolution of eukaryotic traits. Sulfolobales archaea can be broadly categorized into two groups based on their metabolic capabilities: chemolithotrophic ion and sulfur oxidizers that can fix carbon dioxide, and non-ion and sulfur oxidizers that rely on heterotrophic growth. Examples of the former group include the genus *Metallosphaera*, while the latter group is represented by the genera *Saccharolobus* and *Sulfolobus*. *Metallosphaera* species are recognized for their potential applications in bioleaching, waste gas treatment, and production of bio-based chemicals from CO_2, whereas *Saccharolobus* and *Sulfolobus* species hold promise in biomass degradation and utilization due to their unique biocatalytic properties and ability to grow on various carbon sources including hexose and pentose sugars. Despite their potential industrial significance, the practical exploitation of Sulfolobales microorganisms necessitates a comprehensive understanding of their biology and the establishment of genetic manipulation systems. Several comprehensive reviews have been published on various aspects of Sulfolobales biology, including their potential biotechnological applications such as bioleaching, waste treatment, bio-based chemical production, and biomass degradation. In this section, we will provide an overview of recent research advancements in Sulfolobales biology and biotechnological applications,

P. Wu · Y. Yang · Q. Gan · S. Liang · H. Li · S. Li · Y. Xie · Q. Huang · J. Ni · Y. Shen (✉)
State Key Laboratory of Microbial Technology, Institute of Microbial Technology, Shandong University, Shandong, Jinan, China
e-mail: yulgshen@sdu.edu.cn

M. Ito et al. (eds.), *Extremophiles for a Sustainable Future*,
https://doi.org/10.1007/978-981-95-6970-0_7

with a specific emphasis on biomass degradation utilizing obligate heterotrophic Sulfolobales species.

Keywords Archaea · Sulfolobales · Acidothermophile · Biomass degradation and transformation · Bioleaching · CO_2 fixation

7.1 Introduction

The extremophilic microbes belonging to the Sulfolobales order of archaea are known to thrive in acidic and high temperature environments. Their unique ability to withstand such harsh conditions is largely attributed to the enzymes they produce, which are highly resistant to extreme temperatures and pH levels. Due to their distinct genetic makeup and evolutionary position, these prokaryotic microorganisms have emerged as valuable models for studying the evolution of eukaryotic traits. Sulfolobales archaea can be broadly categorized into two groups based on their metabolic capabilities: chemolithotrophic ion and sulfur oxidizers that can fix carbon dioxide, and non-ion and sulfur oxidizers that rely on heterotrophic growth. Examples of the former group include the genus *Metallosphera*, while the latter group is represented by the genera *Saccharolobus* and *Sulfolobus*. *Metallosphera* species are recognized for their potential applications in bioleaching, waste gas treatment, and production of bio-based chemicals from CO_2, whereas *Saccharolobus* and *Sulfolobus* species hold promise in biomass degradation and utilization due to their unique biocatalytic properties and ability to grow on various carbon sources including hexose and pentose sugars. Despite their potential industrial significance, the practical exploitation of Sulfolobales microorganisms necessitates a comprehensive understanding of their biology and the establishment of genetic manipulation systems. Several comprehensive reviews have been published on various aspects of Sulfolobales biology, including their potential biotechnological applications such as bioleaching, waste treatment, bio-based chemical production, and biomass degradation. In this section, we will provide an overview of recent research advancements in Sulfolobales biology and biotechnological applications, with a specific emphasis on biomass degradation utilizing obligate heterotrophic Sulfolobales species.

7.2 Diversity and Biological Features of Sulfolobales

The first species of Sulfolobales, *Sulfolobus acidocaldarius*, was discovered in 1972 by Thomas Brock in Yellowstone National Park (Brock et al., 1972). Since then, numerous species have been isolated from hot springs worldwide, characterized by their ability to thrive in extreme conditions with pH levels below 4 and temperatures exceeding 65 °C. Among the many species that have been identified are *Saccharolobus solfataricus* (original name *Sulfolobus solfataricus*) (Zillig et al., 1980), *Acidianus ambivalens* (Zillig et al., 1985) (original name *Desulfurolobus ambivalens*), *Sa. shibatae* (Grogan et al., 1990), *A. infernus* (Segerer et al., 1986), *Metallosphaera sedula* (Huber et al., 1989), *M. prunae* (Itoh et al., 2001), *M. hakonensis* (Takayanagi et al., 1996), *Sulfurisphaera ohwakuensis* (Kurosawa et al., 1998), *Sulfurisphaera tokodaii* (Suzuki et al., 2002), *Sa. tengchongensis* (original name *Sulfolobus tengchongensis*) (Xiang et al., 2003), *M. cuprina* (Liu et al., 2011), *A. copahuensis* in 2014 (Urbieta et al., 2014), *M. tengchongensis* (Peng et al., 2015), *Sa. caldissimus* (Sakai & Kurosawa, 2018), *Sulfurisphaera javensis* (Tsuboi et al., 2018), and *Stygiolobus caldivivus* (Sakai et al., 2022). It is worth noting that several species, particularly those originally classified under the genus *Sulfolobus*, have been reclassified based on physiological and phylogenetic studies (Sakai & Kurosawa, 2018). Through the analysis of 16S rDNA and whole-genome sequences, it has been determined that the Sulfolobales comprise eight validated genera (*Sulfolobus*, *Saccharolobus*, *Metallosphaera*, *Acidianus*, *Sulfurisphaera*, *Stygiolobus*, *Sulfodiicoccus*, *Sulfuracidifex*) and one metagenomics-assembled genome (MAG) designated as *Aramenus*, encompassing over 23 species. These organisms exhibit a wide range of metabolic capabilities, from lithotrophy based on inorganic iron- and sulfur biotransformations, to autotrophy, to chemoheterotrophy in less acidophilic species. This diversity in metabolic pathways highlights the adaptability of these organisms to varying environmental conditions (Counts et al., 2021; Liu et al., 2021a).

Similar to other archaea, Sulfolobales archaea exhibit a combination of characteristics found in both bacteria and eukaryotes. The mechanisms of DNA replication, transcription, and translation in Sulfolobales show more similarities to those in eukaryotes than bacteria, while genes involved in metabolic processes display greater resemblance to bacterial counterparts (Koonin et al., 1997). Furthermore, members of the Sulfolobales order demonstrate more eukaryotic-like traits, such as a eukaryotic cell cycle and proteins related to cell division like ESCRT-III and Vps4 homologs, than euryarchaea, a major culturable lineage of archaea (Lindas et al., 2008; Lundgren et al., 2008), aligning with their closer evolutionary relationship to Asgard archaea and eukaryotes (Liu et al., 2021b; Zhang et al., 2025a). Although these characteristics make Sulfolobales ideal models for investigating the origin and evolution of eukaryotic-like features, challenges persist in fully comprehending the biology of these intriguing prokaryotic organisms. Within the Sulfolobales archaea, strains utilizing lithotrophy and mixotrophy hold promise in biomining applications, whereas chemoheterotrophic isolates such as *Sulfolobus* and *Saccharolobus* provide potentials for biomass degradation and utilization (Lewis et al., 2021; Manesh et al., 2024; Schocke et al., 2019).

7.3 Nucleoid-Associated Proteins (NAPs) and Structural Organization of Sulfolobales Chromatin

The archaea belonging to the order Sulfolobales lack histones but possess a variety of small, basic DNA-binding proteins, including Sul7d, Cren7, and Alba, which have been the subject of extensive research (Cajili & Prieto, 2022; Kalichuk et al., 2016; Tang et al., 2024; Zhang et al., 2020a, b, 2019). These proteins play a crucial role in organizing the chromatin structure, with recent studies providing insights into the mechanisms involved. Structural studies of Cren7-DNA complexes have shown that Cren7 binds to DNA as monomers, predominantly occupying the minor groove, with adjacent molecules arranged in a head-to-tail configuration on opposite side of the DNA duplex. Each monomer induces a sharp bend in the DNA toward the major groove, resulting in an overall S-shaped conformation (Zhang et al., 2019). Both Cren7 and Sis7d are capable of compacting DNA into highly condensed structures, with Cren7 demonstrating greater efficiency due to its combined DNA bending and bridging abilities (Zhang et al., 2020b). Although Alba was initially identified as a chromatin protein, its function has been redefined as an RNA chaperone in *Sa. islandicus*, operating optimally at the organism's growth temperatures (Cao et al., 2018; Zhang et al., 2020b). In *Sa. solfataricus* P2, two distinct Cren7-binding regions have been identified upstream of the TATA-box and near the TSS, suggesting that Cren7 may provide a structural basis for the binding of transcriptional machinery (Blombach et al., 2024; Blombach & Werner, 2025). Furthermore, a recent study has demonstrated that Cren7 can be recruited by the CRISPR-specific transcription factor Cbp1 to form chromatin-like structures at CRISPR arrays, facilitating the expression of long CRISPR arrays (Blombach et al., 2024).

In *S. acidocaldarius*, the Lrs14 family proteins display nonspecific DNA-binding properties, with a preference for AT-rich regions, thereby suggesting their potential as chromatin-organizing factors (De Kock et al., 2025). As highly abundant DNA-binding proteins (Baes et al., 2023), the Lrs14 members share structural similarity with Sso10a, Sul12a, and TrmBL2 (Ahmad et al., 2015; Lemmens et al., 2022) and modulate the transcription of various genes (Abella et al., 2007; Kessler et al., 2006; Orell et al., 2013). These findings collectively indicate their role in maintaining chromatin structure, establishing them as novel chromatin-organizing proteins. A recent study in our group has demonstrated the essential nature of the TrmB1 gene in *Sa. islandicus*, with transcription occurring in a cell cycle-dependent manner. Transcriptomic and chromatin immunoprecipitation sequencing (ChIP-seq) analyses suggest that TrmB1 plays a key role in orchestrating chromatin dynamics and transcription regulation (Gan et al., Communication Biology, in revision).

The cells of the *Saccharolobus* and *Sulfolobus* archaea exhibit chromosome structure similar to the chromosome interaction domain (CID) to bacterial genomes (Takemata & Bell, 2021). Long-range loop structures have also been observed, along with evidence of a hub-like structure that colocalizes genes involved in ribosome biogenesis (Takemata & Bell, 2021). These structures also further give rise to distinct A and B compartments, in which genes display high and low expression levels, respectively (Takemata et al., 2019). The distribution of ClsN, a SMC-like chromatin protein, and the global gene expression patterns is casually linked to A/B compartmentation (Takemata et al., 2019). The A compartment contains the origins of replication. To evaluate the impact of three-dimensional organization on genome evolution, Bell's group characterized the effect of replication origins and compartmentalization on primary sequence evolution in 11 *Sulfolobus* species (Badel et al., 2022). Using single-nucleotide polymorphism analyses, they found that the A compartment is less likely to be mutated. The authors showed that the B compartment is less accessible by transposon than the A compartment. The results suggest that the A compartment serves as a haven for stable maintenance of gene sequences, while sequences in the B compartment can be diversified. However, the mechanisms underlying the formation and dynamic changes of these compartments, and their impact on gene transcription, remain to be fully understood. Our recent study has shown that chromatin compartmentalization patterns undergo global changes during a cell cycle, correlating with fluctuations in the transcription level of *trmB1* in *Sa. islandicus* (Gan et al., Communication Biology, in revision).

7.4 DNA Replication and Repair

Although Sulfolobales possess circular genome similar to bacteria, their DNA replication machinery and associated proteins show greater similarity to eukaryotic systems, as other archaea. For example, they have multiple origins (Lundgren et al., 2004), similar initiation and DNA replication proteins (Duggin et al., 2008; Samson et al., 2013), replisome components, and structures (Makarova et al., 2012, 2024; Samson et al., 2016). Research on DNA replication proteins and mechanisms in Sulfolobales has been extensive since the discovery of archaea (Lewis et al., 2021). A new replication initiation-related chromatin protein, UBP (*ucm*-binding protein), has been identified in Sulfolobales and Desulfurococcales (Dhanaraju et al., 2025). ChIP-Seq identified >300 genomic-binding sites, indicating widespread chromatin regulatory functions. UBP specifically targets the *ucm* elements—conserved sequences required for replication origin activation—and binds to the N-terminus of MCM (mini-chromosome maintaining helicase). UBP likely functions as a molecular bridge that stabilizes MCM at replication origins before timely dissociated to initiate DNA replication (Dhanaraju et al., 2025). Structural studies of the UBP-DNA complex revealed that, unlike known archaeal NAPs, UBP does not induce significant DNA bending. Instead, it exhibits distinct sequence specificity, resembling bacterial NAPs (such as IHF or Fis) in both binding preference and functional localization (Dhanaraju et al., 2025).

Single-stranded DNA-binding proteins (SSBs) have been regarded as indispensable replication factors. Recently, we reported that the gene encoding the canonical SSB (SisSSB) in *Sa. islandicus* REY15A is not essential for cell viability (Xiao et al., 2023). Interestingly, at a lower temperature (55 °C), the protein level of

SisSSB increases and the growth of *ssb* deletion strains is retarded. SisSSB exhibits melting activity on dsRNA and DNA/RNA hybrid in vitro and is able to melt RNA hairpin in *Escherichia coli*. Furthermore, the core SisSSB domain is able to complement the absence of cold-shock proteins in *E. coli*. Importantly, these activities are conserved in the canonical SSBs from Crenarchaeota species that lack bacterial Csp homologs. Overall, this study has clarified the function of the archaeal canonical SSBs, which do not function as a DNA-processing factor essential for replication but play a role in the processes requiring melting of dsRNA or DNA/RNA hybrid. The study also raises a question about which protein is involved in single-stranded DNA binding during the process of DNA replication in Sulfolobales archaea if there is.

As other archaea, the DNA repair systems of Sulfolobales archaea exhibit a mosaic of the eukaryotic and bacterial counterparts and have their own unique characteristics (Lewis et al., 2021; Quaiser et al., 2008; Rouillon & White, 2010; White, 2003). They utilize the Mre11/Rad50 complex to facilitate the repair of double-stranded DNA breaks via the recombinational repair pathway, a mechanism that exhibits similarities to that observed in eukaryotic organisms. The nucleotide excision repair (NER) pathway proteins XPD, XPB, and XPF are encoded in Sulfolobales genomes. It has been observed that these proteins are able to interact with archaeal-specific factors, such as Bax1, in order to effectively repair DNA damage (Constantinescu-Aruxandei et al., 2016; Wettasinghe et al., 2023). It is noteworthy that the in vivo functions of the NER system have not been fully confirmed. Additionally, the proteins involved in base excision repair (BER) exhibit similarity to those found in both bacteria and eukaryotes (Lewis et al., 2021).

In the realm of the BER pathway, Zhang et al. (2025b) unveiled a discovery regarding a novel bi-functional glycosylase belonging to the HhH-GPD (helix-hairpin-helix-glycine/proline/aspartate) family protein from *Sa. islandicus.* This glycosylase exhibited remarkable capability of effectively removing 1-methyladenine (1-meA) from DNA, as well as alkylated bases from DNA generated by methyl methanesulfonate (MMS), both in vitro and in vivo. The researchers were able to pinpoint the crucial residue responsible for this function. It was determined that Sis-HhH-GPD likely leverages the aromatic rings of Y154 and W57 to form a stacking interaction with the 1-meA base, facilitating its expulsion. Moreover, an interesting observation was made that three out of the four cysteine residues, which are implicated in coordinating with the Fe-S cluster loop, play a role in the catalytic process (Zhang et al., 2025b).

Homologous recombination repair (HRR), a vital pathway in Sulfolobales archaea, plays a crucial role in repairing DNA double-strand breaks (DSBs) and in response to DNA damages (Huang et al., 2015; Sun et al., 2018). Until recently, it was unclear whether Sulfolobales archaea possessed additional pathways for DSB repair. However, a recent study conducted by Li Huang's group has suggested that the replicative primase, composed of the catalytic subunit PriS and noncatalytic subunits PriL and PriX from *Sa. islandicus*, is implicated in DSB repair (Xiong et al., 2025). The overproduction or knockdown of PriL increases or decreases, respectively, the rate of survival and mutation frequency of *S. islandicus* cells following treatment with a DNA damaging agent. The increase in mutation is attributed primarily to an increase in small insertions or deletions. Further, overproduction of PriL enhances the repair of CRISPR-generated DSBs in vivo. These results are consistent with the extraordinary ability of PriSL to promote annealing between DNA strands sharing microhomology in addition to the activity of the heterodimer in terminal transfer and primer extension. The primase-mediated DSB repair is cell-cycle dependent since PriL is barely detectable during the S/G2 transition. The study demonstrated that replicative primase is involved in DSB repair through microhomology-mediated end joining in archaea (Xiong et al., 2025).

Translesion synthesis (TLS) is a vital and efficient pathway for repairing DNA damage in a timely manner. Research has shown that the Y-family DNA polymerase Dpo4 plays a crucial role in TLS in Sulfolobales (Ling et al., 2001). Additionally, one of the three B-family DNA polymerases in *Sa. islandicus*, DNA polymerase 2 (Dpo2), has been identified as specialized for DNA damage repair. Dpo2 demonstrates high efficiency in extending mismatches and mis-paired primer termini following nucleotide placement opposite a damaged site, functioning similarly to the eukaryotic DNA polymerase Pol zeta, which is dedicated to DNA damage repair (Feng et al., 2022). Through its actions, Dpo2 can enhance cell survival when the cellular DNA damage occurs, albeit at the cost of increased mutagenesis. Interestingly, Dpo2 acts as a damage-inducible TLS enzyme that is solely responsible for targeted mutagenesis, specifically facilitating GC to AT/TA conversions. These findings suggest that Dpo2 is the primary DNA polymerase involved in DNA damage tolerance and targeted mutagenesis. Furthermore, it has been suggested that the Dpo2-dependent DNA repair pathway may be conserved in crenarchaea with a low-GC composition genome, indicating its importance in this archaeal lineage (Feng et al., 2020).

The *E. coli* Pol V mutasome is a complex consisting of UmuC, an UmuD' dimer, and RecA, serving as a multimeric Y-family TLS polymerase (Timinskas et al., 2025). Through the utilization of AlphaFold structural prediction, Timinskas et al. were able to construct a hypothetical Sulfolobales mutasome comprised of PolB2, associated proteins PPB2 and iRadA. This heteropentameric PolB2 complex displayed high confidence modeling scores, with PolB2 demonstrating unexpected binding to iRadA via a structural motif reminiscent of the RadA/Rad51 oligomerization motif. The authors identified clamp binding motifs within some mutasomes, present in either iRadA or PolB2, yet rarely in both proteins concurrently. Through their analysis, they were able to provide clarification for previously perplexing experimental outcomes by elucidating the structural characteristics of Pol V. Although there are structural distinctions between PolB2 and Pol V mutasomes, striking similarities were found in their overall architectural organization and interactions (Timinskas et al., 2025). It would be interesting to verify the prediction experimentally.

7.5 Chromosome Segregation

The mechanism of chromosome segregation in Sulfolobales remains poorly understood; however, recent research has shed new light on this process. In a study conducted by Daniela Barillà's group in 2012, it was discovered that the genome of Sulfolobales archaea, such as *Sa. solfataricus* and *S. acidocaldarious*, contains genes encoding ParA-like proteins known as SegA. This SegA protein exhibits 34% sequence similarity to the bacterial ParA protein. Additionally, an archaeon-specific protein called SegB (encoded by an adjacent gene) was also identified, which does not show any homology to the bacterial ParB protein (Kalliomaa-Sanford et al., 2012). Recently, it has been demonstrated by Kabli et al. (2025) that SegB forms multiple foci within the nucleoid, displaying a preference for localizing toward the cell periphery. This protein plays a role in compacting DNA over short distances and facilitating long-range looping between distant genomic sites (Kabli et al., 2025). The findings indicate SegB may function in the process of chromosome condensation and facilitate efficient genome segregation. Furthermore, another research group reached similar conclusions demonstrating that SegB interacts with a centromeric DNA sequence located near the segAB operon in vitro. In addition, they observed the formation of distinct foci on individual segregating chromosomes in vivo (Charles-Orszag et al., 2025). SegA binds DNA nonspecifically in vitro and promotes chromosome compaction and segregation in vivo. During cell division, SegA shifts from chromosome associated puncta to form a single, elongated structure that runs between separating SegB foci. Late in division, SegA retreats to regions surrounding separated SegB foci. It was proposed that SegA and SegB interact to form a bipolar, DNA segregating structure radically different from bacterial ParABS systems (Charles-Orszag et al., 2025). Another study conducted by Parham et al. (2025) demonstrates that the compaction of the genome into two spatially segregated nucleoids occurs only after cells have assembled a division ring. This finding suggests that regulatory checkpoints, similar to those found in eukaryotes, play a role in ensuring the high-fidelity division process in Sulfolobale archaea (Parham et al., 2025).

7.6 Cell Division and Cell Cycle Regulation

7.6.1 Cell Division

The study of cell division in Sulfolobales archaea represents a cutting-edge frontier within the field of archaeal research. Cell division is a crucial process essential for the continuity of all cellular life forms. Throughout the course of evolution, two predominant mechanisms of cell division have emerged: one centered around the FtsZ-mediated contraction mechanism, while the other relies on the endosomal sorting complex required for transport (ESCRT) system (Ithurbide et al., 2022;

Jover & Dekker, 2023). Bacteria primarily utilize the FtsZ mechanism for cell division, while different archaeal lineages employ either of these two systems. Eukaryotes have inherited the ESCRT-based cell division system from their archaeal ancestors. Sulfolobales undergo division through a simplified version of the eukaryotic ESCRT mechanism. This process involves three ESCRT-III homologous proteins (ESCRT-III, ESCRT-III-1, and ESCRT-III-2), the AAA + ATPase Vps4, and an archaeon-specific component, CdvA (Jover & Dekker, 2023). CdvA interacts with the cellular membrane to facilitate the recruitment of ESCRT-III homologous proteins toward the midpoint of the cell, leading to the formation of a ring-shaped structure that initiates the process of cell division. Research conducted on *S. acidocaldarius* has demonstrated that ESCRT-III (or CdvB) assembles into a noncontractile ring-like formation at the cellular midpoint, acting as a scaffold for the recruitment of ESCRT-III-1 (CdvB1) and ESCRT-III-2 (CdvB2). These three proteins organize themselves into a ring-like structure. Upon depolymerization and degradation of the static ESCRT-III ring by Vps4 and proteasome, the tension stored within the ESCRT-III-1 and ESCRT-III-2 polymers is released, prompting the contraction of the cellular membrane and completing the cell division process (Risa et al., 2020). However, our recent study on the *Sa. islandicus* REY15A have not fully aligned with the phenomenon observed in *S. acidocaldarius* (Liu et al., 2025). Our study revealed that all the three ESCRT-III proteins in *Sa. islandicus*, exhibit the ability to form contracting rings, which contract in a sequential manner (the so-called “relay race” model). Specifically, CdvB is initially contracted, followed by contracting of ESCRT-III-1 and ESCRT-III-2 complexes during the “pre-late” and “late” phases of cytokinesis, respectively. These stages correspond to the point in the cell division process where the narrow membrane tube connecting the two daughter cells undergoes detachment (Liu et al., 2017, 2025). Recently, Kuo et al. reported that the C-terminal region of CdvB, known to interact with CdvA, plays a crucial role in destabilizing CdvB as cells transition from the division phase to the G1 phase (Kuo et al., 2026). The degradation rate of CdvB increases during division, partially attributed to the upregulation of the proteasome-activating nucleotidase (PAN), which is under the control of a newly identified cyclically expressed transcription factor, “CCTF1” (Saci_0800), capable of repressing PAN expression. These findings shed light on how archaea regulate proteasome-mediated degradation during cell division, despite the absence of CDK/cyclin or APC/C proteins. While our understanding of the cell division process has advanced, there remain unresolved questions, such as the accurate localization of CdvA at the division ring, its disassembly and degradation at the division ring, and the involvement of post-translational modifications in cell division. These gaps in knowledge necessitate further investigation in the near future.

Furthermore, a role of the surface layer or “S-layer” in cell division has been discovered (Foo et al., 2025). S-layer is a two-dimensional lattice of proteins that coats a wide range of archaea and bacteria in place of a cell wall or capsular polysaccharides. Foo et al. showed that the S-layer flattens the membrane at cytokinesis to accelerate ESCRT-III-dependent cell division and is important for robust, successful cell divisions under conditions of mechanical stress (Foo et al., 2025).

7.6.2 Cell Cycle Regulation

The cell cycle of Sulfolobales exhibits a pattern resembling that of the eukaryotes (Cezanne et al., 2024; Lindås & Bernander, 2013). It begins with a brief (approximately 5% of the cell cycle) pre-replication phase-the G1 phase, during which the cell increases in size to prepare for chromosome replication. This is followed by the S phase (approximately 30–35% of the cell cycle), during which chromosome replication occurs. The G2 phase (accounting for ~50–60% of the cell cycle) is a period of active growth, followed by the M phase, during which cells carrying two sets of neatly arranged chromosomes begin to separate, ultimately completing cell division in the D phase. The M and D phases of Sulfolobales cells are both very short, each accounting for only ~5% of the cell cycle (Cezanne et al., 2024; Lindås & Bernander, 2013). The molecular mechanisms driving its progression remain poorly understood. Since many TACK archaea exhibit an ordered cell cycle despite lacking CDK-cyclin homologs, the evolution of their cell cycle regulatory systems may predate the emergence of eukaryotic cell cycle mechanisms. Recent studies have revealed that the proteasome influences cell cycle progression by regulating the degradation of CdvB during cytoplasmic division in *S. acidocaldarius* (Risa et al., 2020) and *Sa. islandicus* (Liu et al., 2025). This proteasome-dependent cell division regulatory mechanism is similar to what was described in eukaryotic cells (Litsios et al., 2024).

We and others recently discovered a transcription factor named aCcr1 (archaeal cell cycle regulator 1), which contains a ribbon-helix-helix (RHH) domain, regulates the process of cell division by inhibiting the expression of *cdvA* (Li et al., 2023; Yang et al., 2023). Two more transcription factors, aCcr2 and aCcr3, that regulate the cell cycle process were identified and characterized (Yang et al., bioRxiv). The cell cycle regulatory network composed of aCcr1, aCcr2, and aCcr3 likely coordinates the processes such as chromosome dimer resolution, chromosome segregation, cell division, genome replication and repair, amino acid synthesis, defense systems, and motility (Yang et al., 2025, bioRxiv). In addition, a transcriptomic analysis of synchronized *Sa. islandicus* REY15A population has revealed new insights into the complexity of the archaeal cell cycle (Gomez-Raya-Vilanova et al., 2025). In addition to core physiological processes such as cell division, chromosome separation, and replication, numerous biological processes-including various metabolic pathways, protein synthesis, cell motility, and even antiviral defense systems are regulated by the cell cycle.

After chromosome segregation and cell division, the chromatin needs to be relaxed to facilitate DNA replication and transcription. The mechanism of this process is also currently unclear. Recently, we identified a novel OLD (overcoming lysogenization defect) family protein, named as Cran1 (Cell cycle-related ATPase and nickase 1) that probably participates in chromatin regulation during the transition from M-G1 to S phase (Yang et al., 2025, bioRxiv). OLD proteins are diverse ATPases-nucleases functioning in antiphage defense in bacteria. The study shows that this new ATPase and nuclease protein does not appear to be involved in antiviral defense but plays an essential role in cell cycle progression. The gene in *Sa. islandicus* cannot be deleted and exhibits cyclic expression patterns at both transcriptional and translational levels, displaying with a peak expression from M-G1 to S cell cycle phase. Cran1 overexpression leads to significant growth retardation, cell size enlargement, and increased cellular DNA content. Cran1 displays potent nickase and ATPase activities in vitro, with the nickase activity being dependent on the presence of the functional ATPase domain. Notably, Cran1 copurifies with chromatin-associated proteins, such as Cren7 and a histone deacetylase homolog, suggesting its involvement in chromatin related activities. Collectively our results suggest that Cran1 plays an important role in cell cycle progression (Yang et al., 2025, bioRxiv).

7.7 Cell Surface Appendages

The surface appendages of microorganisms play a crucial role in adaptation to environmental stress and serve various functions such as nutrient uptake, biofilm formation, and intercellular communication. In Sulfolobales, these appendages consist of Type IV pili (T4Ps), Type IV secretion system pili (T4SS pili), bindosome, and threads (Beltran et al., 2023; Frols et al., 2008; Gaines et al., 2022; Lassak et al., 2012; van Wolferen et al., 2016; Zolghadr et al., 2010). They participate in processes such as motility, adhesion, species-specific recognition, contact, sugar binding, DNA transport, and other pathways whose mechanisms remain unidentified (Beltran et al., 2023; Frols et al., 2008; Gaines et al., 2022; Lassak et al., 2012; van Wolferen et al., 2016; Zolghadr et al., 2010). In recent years, utilizing novel microscopy technologies, scientists have revealed more insights into the structures and mechanisms of these appendages (Table 7.1) and have raised new questions for future research.

7.7.1 Structure and Functions of T4Ps in Sulfolobales

Model *Sulfolobales*, including *S. acidocaldarius*, *Sa. solfataricus*, and *Sa. islandicus*, generally possess four kinds of T4Ps, archaella, Archaea adhesion pili (Aap), UV-inducible pili of Sulfolobales (Ups), and bindosome (Frols et al., 2008; Lassak et al., 2012; Zolghadr et al., 2010, 2011). The archaella are composed of motor (ArlH/ArlI/ArlJ/ArlX), periplasm stator (ArlF/ArlG), and archaellin (ArlB) (Lassak et al., 2012). Recent structural analyses of ArlB filaments from different crenarchaea including *Sa. islandicus* have revealed that the filaments are supercoiled and consist of ten protofilaments (Gaines et al., 2024a; Kreutzberger et al., 2023, 2022), a feature considered to represent convergent evolution with bacterial flagella (Kreutzberger et al., 2022). Some bacteria have acquired archaella for motility through horizontal gene transfer from archaea (Sivabalasarma et al., 2025). These findings provide a broad framework for understanding archaella assembly and

Table 7.1 Structure of cell appendages in Crenarchaeota

Appendages	Type	Subunit	PDB ID	Year	Organism	Method	References
Archaella	T4P	ArlX	N	2012	*S. acidocaldarius* DSM639	Cryo-EM	Banerjee et al. (2012)
		ArlI	4II7, 4IHQ	2013	*S. acidocaldarius* DSM639	X-Ray	Reindl et al. (2013)
		ArlH	4YDS	2015	*S. acidocaldarius* DSM639	X-Ray	Chaudhury et al. (2016)
		ArlH-ArlX	N	2015	*S. acidocaldarius* DSM639	Cryo-EM	Chaudhury et al. (2016)
		ArlF	4ZBH, 4P94	2015	*S. acidocaldarius* DSM639	X-Ray	Banerjee et al. (2015)
		ArlG	5TUH	2020	*S. acidocaldarius* DSM639	X-Ray	Tsai et al. (2020)
		ArlF-ArlG	6PBK, 5TUG	2020	*S. acidocaldarius* DSM639	X-Ray	Tsai et al. (2020)
		ArlB	8CWM	2022	*Sa. islandicus* REY15A	Cryo-EM	Kreutzberger et al. (2022)
		ArlB	7TXI	2023	*Aeropyrum pernix* K1	Cryo-EM	Kreutzberger et al. (2023)
		ArlB	9ETT, 8QX4	2024	*S. acidocaldarius* DSM639	Cryo-EM	Gaines et al. (2024a)
Archaella-like	T4P	Iho670	3J1R	2012	*Ignicoccus hospitalis* KIN4/I	Cryo-EM	Yu et al. (2012)
		Iho670	5KYH	2016	*I. hospitalis* KIN4/I	Cryo-EM	Braun et al. (2016)
Archaella-like	T4P	Pcal_1342	8FK7	2023	*Pyrobaculum calidifontis* DSM 21063	Cryo-EM	Kreutzberger et al. (2023)

Appendages	Type	Subunit	PDB ID	Year	Organism	Method	References
Adhesion pili	T4P	AapB	N	2012	*S. acidocaldarius* DSM639	Cryo-EM	Henche et al. (2012a)
		AapA-AapB	6NAV	2019	*Sa. islandicus* LAL14/1	Cryo-EM	Wang et al. (2019a)
		AapB	6W8X	2020	*Sa. solfataricus* POZ149	Cryo-EM	Wang et al. (2020)
		AapB	8FJS	2023	*Sa. solfataricus* POZ149	Cryo-EM	Kreutzberger et al. (2023)
		AapB	8TIB, 8TIF	2024	*Sa. islandicus* REY15A	Cryo-EM	Liu et al. (2024)
		AapB	8Q30	2024	*S. acidocaldarius* DSM639	Cryo-EM	Gaines et al. (2024b)
		AapB	9ETS	2024	*S. acidocaldarius* DSM639	Cryo-EM	Gaines et al. (2024a)
Other T4Ps	T4P	Pars_0116	6W8U	2020	*P. arsenaticum* 2GA	Cryo-EM	Wang et al. (2020)
Ced pili	T4SS pili	CedC (CedA1)	8DFU	2023	*A. pernix* K1	Cryo-EM	Beltran et al. (2023)
Ted pili	T4SS pili	TedC	8DFT	2023	*P. calidifontis* DSM 21063	Cryo-EM	Beltran et al. (2023)
Threads	Protein filament	Saci_0406	7PNB	2022	*S. acidocaldarius* DSM639	Cryo-EM	Gaines et al. (2022)
		Saci_0406	8RZL	2024	*S. acidocaldarius* DSM639	Cryo-EM	Gaines et al. (2024a)
Bundling pili	Protein filament	AbpA	7UEG	2022	*P. calidifontis* DSM 21063	Cryo-EM	Wang et al. (2022)

(continued)

Table 7.1 (continued)

Appendages	Type	Subunit	PDB ID	Year	Organism	Method	References
Cannulae	Tubular protein filaments	CanA	7UII	2024	*P. calidifontis* TAG11	Cryo-EM	Sleutel et al. (2024)
		CanA	N	2024	*P. calidifontis* TAG11	X-Ray	Sleutel et al. (2024)
		CanX	9H8B	2024	*P. calidifontis* AV2	Cryo-EM	Sleutel et al. (2024)
		Hyper2	9BAB	2024	*Hyperthermus* sp. MAG130	Cryo-EM	Sleutel et al. (2024)
		Hyper2	9BAC	2024	*Hyperthermus* sp. MAG130	Cryo-EM	Sleutel et al. (2024)
Extracellular cytochrome nanowires	Cytochrome filament	Pcal_1490	8E5F	2023	*P. calidifontis* DSM 21063	Cryo-EM	Baquero et al. (2023)

motility. However, the mechanisms underlying energy delivery for archaella rotation and archaellin assembly remain unresolved.

Aap were found to mediate surface adherence and are composed of assemble components AapE/AapF, pilin subunits AapA/AapB, and functional unknown protein AapX (Henche et al., 2012a; Zolghadr et al., 2010). Initially, Aap was thought to be nonretractable because of the lack of PilT homologs, an ATPase disassemble bacteria T4Ps (Chlebek et al., 2019). However, recent studies show that Aap exhibit twitching motility driven by retractation (Charles-Orszag et al., 2024), suggesting the presence of a unique molecular mechanism of pilus retraction in Sulfolobales. Similar to archaella, the pilin subunits of Aap vary among species. In *S. acidocaldarius* DSM639, AapB is the major pilin whereas AapA is a minor pilin, and deletion of AapA inhibits motility but not assembly (Charles-Orszag et al., 2024). In *Sa. islandicus* LAL14/1, both AapA and AapB are major pilin that form hetero filaments (Wang et al., 2019a). In *Sa. islandicus* REY15A and *Sa. solfataricus* POZ149, the sole adhesion pilin AapB can form a tri-pilus with subunits in three distinct conformations but can also form a mono-pilus in which all subunits are equivalent, as observed in the strain *Sa. islandicus* REY15A (Liu et al., 2024; Wang et al., 2020). In addition to mediating adhesion, Aap also serves as a target for viral attachment and infection, although the underlying mechanism remains unclear (Rowland et al., 2020; Wang et al., 2020).

Ups mediate species-specific aggregation upon exposure to UV irradiation or other DNA damages reagents and are essential for chromosomal DNA exchange (Ajon et al., 2011; Frols et al., 2008; van Wolferen et al., 2013). Ups components are encoded by a conserved *ups* operon (Van Laer et al., 2016; van Wolferen et al., 2013). UpsX is a membrane protein with unclarified function (van Wolferen et al., 2013). UpsE and UpsF function as pilin assembly ATPase and platform, respectively, while UpsA and UpsB are pilin subunits forming pilus which mediates species-specific aggregation by recognizing different glycosylation patterns (van Wolferen et al., 2013, 2020). Recently, two independent studies identified an additional pilin subunit, UpsC, which is likely a minor pilin essential for pilus formation and cell aggregation (Recalde et al., 2025; Wu et al., 2025). For species-specific recognition, a poorly conserved region within UpsA may interact with unique glycosylation patterns on cell surface (van Wolferen et al., 2020).

In Sulfolobales, archaella and Ups are regulated in response to stress conditions. Archaella operon is regulated by Arn system, which comprises transcriptional activator ArnR and ArnR1, suppressor ArnA and ArnB, kinases ArnC and ArnD, and phosphatase PP2A (Bischof et al., 2019, 2018; Duan & He, 2011; Hoffmann et al., 2017b; Reimann et al., 2012). Additionally, archaella are also regulated at the transcriptional and protein levels by the Lrs14/TrmB family protein AbfR1 and the GPN-loop GTPase *Sa*GPN, respectively (Korf et al., 2023; Orell et al., 2013). Ups pili is regulated by DNA damage-response (DDR) transcriptional activator Orc1-2 and TFB3 (Feng et al., 2018; Schult et al., 2018; Sun et al., 2018). Interestingly, deletion of *aapF* leads to an approximately 20-fold upregulation of ArlB in *S. acidocaldarius*, while it eliminates archaella in *Sa. islandicus* REY15A (Lassak et al.,

2012; Liu et al., 2024). The complex regulatory mechanisms suggest that T4Ps in Sulfolobales are subject to stringent control to save energy at extreme environment.

7.7.2 Structure and Functions of T4SS in Sulfolobales

Bacterial T4SS pili, such as F-pili, form a conduit connecting close or distant cells for conjugative DNA transfer (Beltran et al., 2024; Goldlust et al., 2023). The archaeal Type IV secretion system (T4SS) was initially discovered on the Sulfolobales conjugative plasmid pNOB8 approximately three decades ago. She et al. (1998) reported the formation of mating pairs consisting of two cells or cell aggregates during the conjugation process of pNOB8 (She et al., 1998). However, conjugative T4SS pili have not yet been observed in Sulfolobales. Through the analysis of AlphaFold-predicted structures, putative VirB6 and VirB2 homologs on pNOB8 were identified, suggesting the potential formation of T4SS pili during conjugation. Furthermore, a separate T4SS encoded by the chromosome is found in the majority of Crenarchaea. Known as the Ced system in Sulfolobales, Acidilobales, Desulfurococcales, and Fervidicoccales, and as the Ted system in Thermoproteales and Thermofilales, this system was identified through phylogenetic analysis of CedA/TedA (Beltran et al., 2023; van Wolferen et al., 2016; Wu et al., 2025). The Ced system, consisting of the pilus assembly platform CedA and CedE, DNA import ATPases CedB and CedD, pilin subunit CedC, and pseudo-pilin CedA2, collaborates with Ups to import chromosomal DNA (Recalde et al., 2025; van Wolferen et al., 2016; Wu et al., 2025). Ced pili have been exclusively characterized in *Aeropyrum pernix* K1. These pili are comprised of CedC subunits and exhibit structural resemblance to Ted pili and bacterial conjugative pili (Beltran et al., 2023). However, the mechanisms by which DNA is exported and subsequently incorporated into the DNA import apparatus remain ambiguous. Of particular importance is the inquiry into how DNA is exported via conjugative plasmids in Sulfolobales, a group lacking relaxosome homologs for DNA preprocessing.

7.7.3 Other Protein Filaments in Sulfolobales

Following the removal of archaella, Aap, and Ups, or the deletion of the T4P prepilin signal peptidase PibD, an additional filament termed a "thread" has been identified on the cellular surface of *S. acidocaldarius* (Gaines et al., 2022; Henche et al., 2012b). Upon thorough analysis of the high-quality Cryo-EM map, Gaines et al. have revealed that the thread consists of Saci_0406, a protein characterized by high glycosylation and rich in β-sheets (Gaines et al., 2022). The function and assembly mechanism of Saci_0406 have yet to be elucidated. It is important to note that Saci_0406 is not conserved among species in the order Sulfolobales. However, a recent study conducted by Sleutel et al. has shed light on the structure of Cannulae in Thermoproteales. Specifically, Cannulae in Thermoproteales are composed of CanA or CanX, in conjunction with a TasA-like protein (Sleutel et al., 2024). The authors have also identified three proteins from *Sa. solfataricus* and *Sa. islandicus* as structural homologs of Saci_0406 (Sleutel et al., 2024), suggesting that these proteins may perform similar functions, such as biofilm formation.

7.7.4 The Relationship of the Membrane Lipid GDGT Cyclization and Cellular Appendages

As other archaea, the membrane lipid is ether-based, which is synthesized by a complex pathway. Yang et al. reported that the GDGT ring synthases mutant, deficient in GDGT cyclization, inhibited archaellum formation and reduced cell motility in *S. acidocaldarius*. This inhibition was caused by decreased transcription of the archaellum operon, likely due to cleavage of the C-terminal domains in transmembrane proteins ArnRs, the transcription factors that regulate archaellum operon expression. The transcriptomic and proteomic analysis showed deficiency of GDGT cyclization broadly impacted the expression of membrane associate proteins, including respiratory chain proteins, and decreased cellular ATP concentration. Moreover, phylogenetic analysis demonstrated that the correlation between GDGT cyclization and archaellum formation is widespread among (hyper) thermophilic archaea. The findings suggested that archaea modify their membrane lipids to profoundly alter cellular appendages and cell physiology to adapt to environmental fluctuations (Yang et al., 2025).

7.8 Protein Post-Translational Modification in Sulfolobales

Protein post-translational modification offers a mechanism to intricately regulate protein functions independent of de novo protein synthesis, as highlighted by Walsh et al. (2005). Within Sulfolobales, a variety of post-translational modifications have been reported, encompassing phosphorylation, acetylation, methylation, and ubiquitin-like modifications, although their functional mechanisms have not been fully elucidated (Bell et al., 2002; Cao et al., 2018, 2019; Lewis et al., 2021).

7.8.1 Protein Phosphorylation

Protein phosphorylation in the Sulfolobales has been extensively investigated, with a focus on the role of Ser/Thr/Tyr protein kinases in mediating this process. Species within Sulfolobales typically encode between 11 and 15 Hanks-type protein kinases, which include both group one canonical Hanks-type protein kinases (eukaryotic-like protein kinase, ePK) and group two atypical protein kinases (aPK) (Hanks & Hunter, 1995), and two protein phosphatases (PP) (Kennelly, 2014; LaRonde, 2014; Lewis et al., 2021). Unexpectedly, the ratios of Tyr phosphorylation in all the phosphoproteomic data were much higher than those in eukaryotes (10.9–53.6% in Sulfolobales vs. <1% in eukaryotes) (Esser et al., 2012; Huang et al., 2020; Reimann et al., 2013). Recently, the only protein kinase Sut1 displaying tyrosine phosphorylation activity was reported (Maklad, 2024).

Protein phosphorylation was proved to play roles in transcription regulation (Haurat et al., 2017; Hoffmann et al., 2019, 2017a; Li et al., 2017; Orell et al., 2013; Reimann et al., 2012; Ye et al., 2020). In a recent study by Jiang et al. (2023), it was demonstrated that ArnA and ArnB proteins in *Sa. islandicus* are implicated in the DDR pathway. Specifically, the deletion of either ArnA or ArnB gene led to an increased resistance to the DNA damaging agent NQO. Moreover, the interaction between ArnA and ArnB was shown to be modulated by ePK1 phosphorylation in vitro, while being suppressed by NQO treatment in vivo (Jiang et al., 2023). In addition, NQO stimulated the interaction of ArnA with another partner, ArnE, whose gene deletion led to higher sensitivity to NQO than wild type (Jiang et al., 2023).

Archaea employ Orc1-2, an Orc1/Cdc6 family protein, to mediate DNA damage-responsive regulation, which orchestrates a series of cellular responses to DNA damage. Recently, Qunxin She's group found that Orc1-2 proteins present in untreated cells are phosphorylated at Thr356 (Liu et al., in revision). Characterization of T356A, the phosphor-ablative form and T356D, the phosphor-mimetic form of Orc1-2 revealed that only the phosphor-ablative form retained the specific DNA binding. Genetic characterization and RNA-seq analyses further revealed that their corresponding mutants also exhibited expected phenotypes: Orc1-2T356D no longer exhibited DNA damage responses upon NQO treatment, while the phosphor-ablative mutant Orc1-2T356A is not only more tolerant to DNA damage agents but also prolongs the window of the DDR. Taken together, these results indicated that the Thr356 phosphorylation deactivates Orc1-2, thereby attenuating the archaeal DDR (Liu et al., in revision).

In addition to transcription regulation, several processes of nucleic acids metabolism were also influenced by protein phosphorylation/kinases in Sulfolobales (Huang et al., 2019; Knuppel et al., 2018). Noticeably, we recently identified phosphorylation of the α subunit of the proteasome by ePK2 in *Sa. islandicus* in vivo and in vitro (Huang, 2025). It was shown that overexpression of ePK2 and phosphorylation-mimetic mutants of the α subunit impaired cell division by inhibiting the assembly of proteasome and its degradation on cell division proteins (Huang, 2025). Moreover, another aPK Bud32, a subunit of KEOPS complex which is conserved in both archaea and eukaryotes, plays an essential role in N6-threonylcarbamoyladenosine (t^6A) modification of tRNA as that in eukaryotes and may also be involved in DNA homologous recombination repair (Wu et al., 2023). Eukaryotic protein phosphorylation extensively regulates cell cycle regulation and DNA damage response. These findings suggest that archaeal protein phosphorylation serves similar functional roles to those observed in eukaryotes, supporting the hypothesis that archaea and eukaryotes share a common ancestral origin. Furthermore, the impact of protein phosphorylation on central metabolism was also reported, highlighting its physiological significance in cellular processes (Ray et al., 2005; Reimann et al., 2013; Wang et al., 2019b).

7.8.2 *Ubiquitin-Like Modification: Urmylation*

The ubiquitin-related modification protein in achaea (Urm1) was first discovered in a halophile and later discovered in Sulfolobales (Anjum et al., 2015; Humbard et al., 2010). Urm1 shares structural similarities with ubiquitin, possessing a beta-grasp fold and a C-terminal diglycine motif. E1-like homolog has also been identified in archaea, as an analog to the E1 enzymes found in the ubiquitin-proteasome system of eukaryotes. However, homolog of either E2 or E3 enzyme has not yet been identified in archaea (Anjum et al., 2015). Sulfolobales Urm1 and E1 were proposed to be involved in protein degradation by proteasome. *S. acidocaldarius* E1 was able to activate and transfer Urm1 to the tested substrate thermosome in the presence of ATP in vitro (Anjum et al., 2015). In addition, Urm1-conjugated GFP can be degraded by reconstituted *S. acidocaldarius* proteasome. Urmylated proteins identified by pull-down and mass spectrometry included proteasome-related proteins, ribosomal proteins, tRNA synthases, and metabolic enzymes (Anjum et al., 2015), suggesting a link between urmylation and the proteasome degradation pathway, as well as multiple cellular processes. Our recent data showed that Sulfolobales only have one copy of JAMM/MPN+ metalloprotease (JAMM) gene supposed to be involved in de-urmylation. All the genes of E1, Urm1, and JAMM could not be deleted in *Sa. islandicus* indicating that urmylation is essential for cell viability (unpublished).

Recently, a proteomic analysis of Urm1 modification in *Sa. islandicus* has been reported (Cao et al., 2025, in press). Extensive protein urmylation was observed, with a total of 783 Urm1 conjugation sites, mapped to 330 proteins, identified in the cell. Among the seven lysine residues in Urm1, six were the modification sites, of which K7 and K37 were preferentially modified. Treatment with the proteasome inhibitor bortezomib resulted in K37-linked chains being the sole major modification species, suggesting that K37 linkage served as a primary trigger of proteasomal degradation. The modified proteins were involved in a number of cellular processes, such as cell division, chromosomal organization, DNA replication, translation, proteasomal protein degradation, and sulfur relay. Protein urmylation was dynamic and influenced by growth conditions and stress treatments. As in our study, attempts to delete *urm1* was unsuccessful. The knockdown of *urm1* resulted in substantial growth delay, during which a drastic reduction in cellular concentration of cell division proteins (CdvB, CdvB1, CdvB2) occurred. The results shed significant light on the landscape and potential roles of protein urmylation in Sulfolobales (Cao et al.,

2025, in press). However, the role and mechanism of urmylation of the target proteins remain to be explored.

7.9 Stress Responses

Cells of Sulfolobales archaea appear to have developed a sophisticated network in response to external and internal environmental stimuli. The response of *Sa. islandicus* REY15A to reactive oxygen species (ROS) was investigated by our group (Zhang et al., 2025c). Using a multidisciplinary approach, we revealed that SisPerR, the homolog of bacterial PerR, is responsible for the transcriptional regulation to deal with ROS. We show that the transcription of proteins involved in ion homeostasis including Dps, NirD, VIT1/CCC1, and MntH, is significantly upregulated, while expression of ROS-scavenging enzymes remains basically unaffected. Notably, the genes coding for Dps, NirD, and MntH are direct targets of SisPerR. Three novel residues critical for ferrous ion binding and one novel residue for zinc ion binding were identified. This study has established that SisPerR is a repressive redox-sensing transcription factor regulating intracellular metal ion homeostasis in *Sa. islandicus* for oxidative stress defense.

Benninghoff et al. (2021a, b) examined the response of *S. acidocaldarius* toward the industrially relevant organic solvent 1-butanol (Benninghoff et al., 2021a). The results showed biofilm formation as a primary cellular response to 1-butanol. Upon exposure to butanol, biofilm formation of *S. acidocaldarius* was enhanced and occurred at up to 1.5% (vol/vol) 1-butanol, while planktonic growth was observed at up to 1% (vol/vol) 1-butanol (Benninghoff et al., 2021b). Furthermore, the triggered cell/membrane stress led to significant changes in culture heterogeneity accompanied by changes in central cellular processes, such as cell division and cellular defense systems, thus suggesting a global response for the protection at the population level.

Toxin-antitoxin (TA) systems are important for stress adaptation in prokaryotes, including persistence, antibiotic resistance, pathogenicity, and biofilm formation. During heat stress, *S. acidocaldarius* exhibited an increase in the expression of several bicistronic type II VapBC TA systems, with the highest expression observed in the VapBC4 system. A recent study found that VapBC4 could potentially signal a preference for persister cell formation over biofilm growth (Bhowmick et al., 2024) through the various functions of the VapC4 toxin, including inhibiting translation, inducing persister-like cell formation, and regulating biofilm formation (Bhowmick et al., 2024).

7.10 Biomass and Other Carbon Source Degradation and Utilization

Sulfolobales species are potentially useful due to their adaptation toward high temperature (75–80 °C) and low pH (pH 2–5) extreme that matches harsh process conditions in industrial applications and possess unique enzymes and enzyme cascades. The ease of growth of heterotrophic species such as *S. acidocaldarious*, *Sa. solfataricus*, and *Sa. islandicus* under aerobic or microaerophilic conditions and established fermentation technologies promote these Sulfolobales species as microbial resources for extremozymes and as valuable novel platform organisms for industrial biotechnology (Schocke et al., 2019). Recent studies have increased our understanding of the mechanisms on activity and regulation of the endogenous cellulases (Li et al., 2025), and glycerol and phenol utilization (Schmerling et al., 2025; Wolf et al., 2021), and on the ethanol production activity of the oxygen insensitive bifunctional PORs/PDCs (Pyruvate decarboxylase/pyruvate:ferredoxin oxidoreductase) in vitro (Alharbi et al., 2022b).

7.10.1 The Roles and Transcriptional Regulation of the Cellulases in Saccharolobus

Lignocellulose is the most widely distributed carbohydrate in nature and represents an important renewable biomass resource with great application potentials (Ashokkumar et al., 2022; Reshmy et al., 2022). Efficient biomass utilization faces several technical challenges, particularly in the pretreatment and conversion processes. Common pretreatment methods often combine physical, chemical, and biological approaches. Acid-thermal pretreatment, as an effective method, can disrupt the crystalline structure and physical barriers of lignocellulose (Guo et al., 2022; Zhang et al., 2023). However, it typically requires subsequent cooling and neutralization, etc. The use of heterotrophic Sulfolobales such *Saccharolobus* and *Sulfolobus* could combine biomass degradation and fermentation into one step without need of cooling and neutralization (Schocke et al., 2019).

Cellulases are required for biomass degradation. In *Sa. solfataricus*, four cellulases SsoCelS (*sso2534*), SSO1354 (*sso1354*), SSO1949 (*sso1949*), and SsoLacS (*sso3019*) were predicted (She et al., 2001). The colony of *Sa. solfataricus* on carboxymethylcellulose sodium (CMC)-containing plates formed a hydrolysis zone and *sso2534* transcript was highly induced by CMC, demonstrating that SsoCelS may have endoglucanase activity in vivo (Limauro et al., 2001). Recombinant SSO1949 protein exhibited endoglucanase activity in vitro (Huang et al., 2005). It was suggested that either SSO1354 or SSO1949 originated from a duplication/insertion event mediated by insertion sequences (Iss) (Girfoglio et al., 2012). SSO1949 has an optimal pH of 1.8 and an optimal temperature of 80 °C (Huang et al., 2005), while SSO1354 undergoes glycosylation, with an optimal pH of 3.5

and an optimal temperature of 95 °C (Maurelli et al., 2008). Additionally, it is likely that SSO1354 functions by anchoring to the cell membrane (Girfoglio et al., 2012; Maurelli et al., 2008). SsoLacS, a reporter gene often used in genetic analysis, exhibits β-galactosidase activity and can hydrolyze β-D-gluco-, fuco-, and galactosides. It is also capable of degrading G2-G5 oligosaccharides. SsoLacS can synergistically interact with ABC transporter substrate-binding protein SSO3053 (which can bind cellodextrin from G2 to G8, forming an efficient cellodextrin degradation system (Lalithambika et al., 2012; Nucci et al., 1993; Wang et al., 2023; Zheng et al., 2012). Genomic annotation identified three predicted cellulases in *Sa. islandicus* REY15A, Cel1, Cel2A, and LacS, homologs to SsoCelS (*sso2534*), SSO1354 (*sso1354*)/SSO1949 (*sso1949*), and SsoLacS (*sso3019*), respectively (Guo et al., 2011). Nevertheless, it is still obscure about how the cellulases and other proteins work in cellulose degradation in vivo and our understanding of the cellulolytic mechanisms of these microbes remain limited (Cowan et al., 2024; De Lise et al., 2023; Lewis et al., 2021). One of the standing questions is that the production and cellulase activity in the wild-type cells are very low, and our initial experiments showed that it is difficult to express the endogenous cellulases (except the LacS), despite having well-developed expression system in *Sa. islandicus* REY15A, highlighting that there exists an unknown mechanism for cellulase production, translocation, and regulation.

The roles and transcriptional regulation of two endogenous cellulases were investigated by our group using *Sa. islandicus* REY15A (Li et al., 2025). To analyze the functions of the three cellulases in the archaeon, we developed a constitutive expression vector pSeAP having one of the strongest promoters, P_{alba}, and constructed strains overexpressing each of the three cellulases. At the same time, we also constructed single and double deletion mutants of the cellulase genes. Using the overexpression strains based on the pSeAP vector, we revealed that Cel1 and Cel2A are membrane-associated, and a higher proportion of Cel2A is secreted into the medium than Cel1. The sizes of the two proteins detected were larger than their theoretical sizes. This may be attributed to glycosylation modifications in the N-terminal transmembrane regions of these proteins, similar to what has been reported for the homologous protein SSO1354 of Cel2A in *Sa. solfataricus* (Girfoglio et al., 2012; Maurelli et al., 2008). It was clarified that Cel2A is the only active endoglucanase in *Sa. islandicus* by gene knockout and gene expression. The results demonstrated that Cel1 does not possess endoglucanase activity, which is inconsistent with the result of the homolog SSO2534 in *Sa. solfataricus* (Limauro et al., 2001), but in agreement with the study by Girfoglio et al. (2012). The result supports that the activity observed in colony of *Sa. solfataricus* is produced by SSO1354 and SSO1949 (Girfoglio et al., 2012; Huang et al., 2005). Previous studies have investigated the protein expression and activity of the endoglucanase Cel2A and β-glucosidase LacS (Girfoglio et al., 2012; Maurelli et al., 2008; Nucci et al., 1993), yet the specific interactions and induction mechanisms among cellulases remain unclear.

We found that the expression of all the three cellulases is induced by carboxymethylcellulose sodium (CMC). Furthermore, we found that the transcriptional levels of *cel1*, *cel2A*, and *lacS* are interdependent and *cel1* and *cel2A* levels apparently increase with *lacS* deletion and overexpression of an ABC transporter, suggesting an intracellular oligosaccharide-dependent regulatory mechanism. One of the interesting features in the cellulase transcriptional regulation is that *cel1* is regulated by aCcr1, a cell cycle regulator in Sulfolobales (Yang et al., 2023). This result suggests that cellulase gene transcription is coupled with cell cycle progression. Another accidental but interesting finding is that in the presence of *pyrEF*, transcription of all cellulase genes was upregulated. Still another, we found that deletion of the cellulase genes facilitates cell growth in CMC-containing medium. The mechanisms underlining these phenomena are unclear. This study reveals an ingenious and intricated mechanism for cellulose utilization to adapt to extreme environment in archaea and provide insights for engineering Sulfolobales archaea as biomass-degrading and utilization platforms (Li et al., 2025).

Based on these findings, we proposed a model for cellulose utilization in *Sa. islandicus* (Fig. 7.1). Upon biomass being available, the cellulase genes are induced by oligosaccharides degraded by cellulases. The endoglucanase Cel2A, either in its membrane-anchored or secreted form, degrades extracellular long-chain cellulose into oligosaccharides, which are then transported into the cell via the ABC transporter system and subsequently hydrolyzed into monosaccharides by intracellular LacS. The transcriptional regulator aCcr1 directly regulates the transcription of *cel1*. Additionally, the transcription of *cel1*, *cel2A*, and *lacS* is influenced by intracellular oligosaccharide levels and interplay among the cellulases. Furthermore, uracil synthesis by *pyrEF* enhances cellulase gene transcription by an unknown mechanism. Once biomass is depleted, cellular oligosaccharide level decreases, leading to reduced cellulase transcription to save energy for cell propagation. By this ingenious and intricate mechanism, the cells are adapted to the nutrient-limited extreme environments.

7.10.2 Glycerol/Phenol Utilization and Oxygen-Insensitive Ethanol Fermentation Activity of PORs/PDCs

Glycerol is highly abundant in natural ecosystems and serves as both an important carbon source for microorganisms as well as a promising feedstock for industrial applications. Schmerling et al. (2025) investigated the pathways involved in glycerol degradation in archaea (Schmerling et al., 2025). They found that *S. acidocaldarius* can grow with glycerol as its sole carbon source. After uptake involving facilitated diffusion, glycerol is phosphorylated to glycerol-3-phosphate by glycerol kinase (GK), followed by oxidation to dihydroxyacetone phosphate catalyzed by an unusual glycerol-3-phosphate dehydrogenase (G3PDH) with a previously undescribed type of membrane anchoring via a CoxG-like protein. Furthermore, they showed that while *S. acidocaldarius* has two paralogous GK/G3PDH copies (*saci_1117-1119*, *saci_2031-2033*) with similar biochemical activity, only

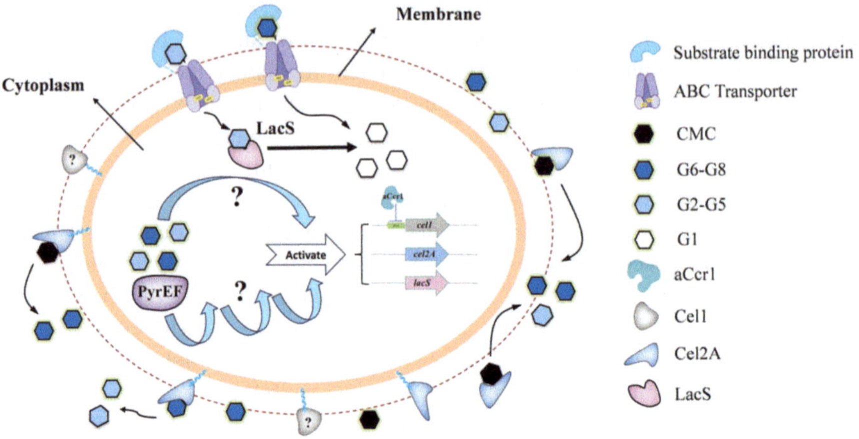

Fig. 7.1 A model for cellulose utilization in Sulfolobales archaea (from Li et al., 2025). Upon biomass being available, the cellulase genes are induced by oligosaccharides degraded by cellulases. The endoglucanase Cel2A, either in its membrane-anchored or secreted form, degrades extracellular long-chain CMC into oligosaccharides, which are then transported into the cell via the ABC transporter system and subsequently hydrolyzed into monosaccharides by intracellular LacS. The transcriptional regulator aCcr1 directly regulates the transcription of *cel1*. Additionally, the transcription of *cel1*, *cel2A*, and *lacS* are influenced by intracellular oligosaccharide levels. Furthermore, uracil synthesis by *pyrEF* enhances cellulase gene transcription by an unknown mechanism

saci_2031-2033 is highly upregulated and essential for glycerol, suggesting that distinct enzyme pairs may be regulated by different environmental conditions. They also revealed a high versatility of G3PDHs with respect to interacting proteins, electron transfer mechanisms, and modes of membrane anchoring. These findings help to elucidate the mechanisms involved in glycerol utilization in Archaea, highlighting unique evolutionary strategies that likely enabled adaptation to different lifestyles.

Wolf et al. (2021) studied phenol degradation in *Sa. solfataricus* P2 under extreme conditions (Wolf et al., 2021). Although *Sa. solfataricus* is able to utilize phenol for biomass production, the carbon source induces profound stress reactions, including genome rearrangement as well as a strong intracellular accumulation of polyamines.

Two main pathways in microbes have been employed to produce ethanol, the bi-functional alcohol/aldehyde dehydrogenase (AdhE) pathway and the bifunctional pyruvate decarboxylase/pyruvate:ferredoxin oxidoreductase (PORs/PDCs) pathway (Guterl et al., 2012; Keller et al., 2017; Pony et al., 2020). PORs/PDCs is a key enzyme involved in ethanol fermentation, catalyzing the decarboxylation of pyruvate to acetaldehyde and CO_2. Thermostable and oxygen-stable PDC activity is highly desirable for biotechnological applications at high temperatures. The enzymes from the thermoacidophiles *Sa. solfataricus* and *S. acidocaldarius* were purified and characterized (Alharbi et al., 2022a). The purified enzymes were

CoA-dependent and thermostable. There was no loss of activity in the presence of oxygen. The enzymes showed both PDC and POR activities, which were CoA-dependent, and they were thermostable with half-life times of 2.9 +/− 1 and 1.1 +/− 1 h at 80 °C, respectively. Optimal pH values for their PDC and POR activities were determined to be 7.9 and 8.6, respectively. The results may have important implication in metabolic engineering for ethanal fermentation using heterotrophic Sulfolobales as chassis microbes.

7.11 Genetic Systems

Efficient genetic manipulation systems are currently available only in a few Sulfolobales strains, namely, *S. acidocaldarius*, *Sa. Islandicus*, and *Sa. solfataricus* (Lewis et al., 2021). Several studies have reported progress in further improvement of the genetic systems for these systems. Kuschmierz et al. (2024) reported that protein production could be increased by the integration of *S. acidocaldarius* 5′-UTRs into the Para expression plasmids (Kuschmierz et al., 2024). Using the esterase Saci_1116 as a reporter protein, the authors observed a fourfold increase in soluble and active protein yield upon insertion of the *saci_1322* (*alba*) 5′-UTR. The 5′-UTR optimized expression vector improved the expression of various proteins. The type IA CRISPR-Cas was employed to establish a genome editing system in *Sa. islandicus* REY15A (Li et al., 2016). In another study, applying a similar strategy, Bost et al. (2023) developed an endogenous CRISPR-Cas type I-D-based system for the genetic engineering in *S. acidocaldarius*. The authors developed a novel plasmid, which incorporates a mini-CRISPR array that can be induced using xylose, resulting in targeted DNA cleavage. They experimentally confirmed the functionality of three different PAMs (CCA, GTA, and TCA) for gene targeting (Bost et al., 2023).

An activator XylR (Saci_2116) responsible for the transcriptional regulation of the pentose transporter and pentose metabolizing genes was identified in *S. acidocaldarius* (van der Kolk et al., 2020). Two XylR-dependent promoters next to the long-established maltose inducible promotor were used to construct a high-throughput expression vector system for *S. acidocaldarius*. A *xylR* deletion mutant showed growth retardation on D-xylose/L-arabinose containing media and the lack of transcription of the respective ABC transporter. In contrast to the so far used promoters for expression in *S. acidocaldarius*, the *xylR* responsive promoters have a very low background activity.

Dual-vector system is superior to single-vector genetic system in several respects including usefulness for studying protein–protein interaction in vivo. Dual-vector systems are available in *E. coli* and budding yeast but were not developed in archaea. Qunxin She's group has recently established a dual-plasmid system in *Sa. islandicus* (Zhao et al., 2024). Previously, the genetic system in *Sa. islandicus* is based solely on the pRN2 cryptic plasmid. Although this plasmid coexists with pRN1 in its original host, early attempts to test pRN1-based vectors consistently failed to yield any stable host-vector system. The authors hypothesized that this failure could be due to the occurrence of CRISPR immunity against pRN1. They identified a putative target sequence in orf904 encoding a putative replicase on pRN1 (target N1). Mutated targets were then designed and tested for their capability to escape the host CRISPR immunity. The results revealed that the original target triggered CRISPR immunity in this archaeon, whereas all three mutated targets did not, indicating that all the designed target mutations evaded host immunity. These mutated targets were then incorporated into orf904 individually, yielding corresponding mutated pRN1 backbones with which shuttle plasmids were constructed. They further showed that pRN1-based and pRN2-based vectors were stably maintained in the archaeal cells either alone or in combination (Zhao et al., 2024).

7.12 Summary and Future Perspectives

In recent years, significant advancements have been made in the research of Sulfolobales. This section reviewed interesting and exciting advances in the study of the biology of Sulfolobales and for the utilization of heterotrophic Sulfolobales. However, due to constraints in space, this chapter was unable to delve into the progress made in areas such as basic transcription, protein translation mechanism, immunity systems, virus–host interactions, and the physiology of chemolithotrophic ion and sulfur oxidizer Sulfolobales. In addition, we apologize for the omission of citations in the aforementioned areas.

Moving forward, we anticipate further exciting developments in the study of Sulfolobales biology and applications, particularly in understanding cell cycle control mechanisms, responses to internal and external stimuli, and the establishment of genetic manipulation systems for chemolithotrophic ion and sulfur oxidizers. Addressing both fundamental and technical inquiries will unlock the full potential of these promising thermoacidophilic archaea for evolutionary study and industrial biotechnology.

References

Abella, M., Rodriguez, S., Paytubi, S., Campoy, S., White, M. F., & Barbe, J. (2007). The Sulfolobus solfataricus radA paralogue sso0777 is DNA damage inducible and positively regulated by the Sta1 protein. *Nucleic Acids Research, 35*, 6788–6797.

Ahmad, M. U. D., Waege, I., Hausner, W., Thomm, M., Boos, W., Diederichs, K., & Welte, W. (2015). Structural insights into nonspecific binding of DNA by TrmBL2, an archaeal chromatin protein. *Journal of Molecular Biology, 427*, 3216–3229.

Ajon, M., Frols, S., van Wolferen, M., Stoecker, K., Teichmann, D., Driessen, A. J., Grogan, D. W., Albers, S. V., & Schleper, C. (2011). UV-inducible DNA exchange in hyperthermophilic archaea mediated by type IV pili. *Molecular Microbiology, 82*, 807–817.

Alharbi, F., Knura, T., Siebers, B., & Ma, K. (2022a). Thermostable and O-Insensitive pyruvate decarboxylases from thermoacidophilic archaea catalyzing the production of acetaldehyde. *Biology (Basel), 11*, 1247.

Alharbi, F., Knura, T., Siebers, B., & Ma, K. (2022b). Thermostable and O(2)-insensitive pyruvate decarboxylases from thermoacidophilic archaea catalyzing the production of acetaldehyde. *Biology (Basel), 11*(8), 1247.

Anjum, R. S., Bray, S. M., Blackwood, J. K., Kilkenny, M. L., Coelho, M. A., Foster, B. M., Li, S., Howard, J. A., Pellegrini, L., Albers, S. V., et al. (2015). Involvement of a eukaryotic-like ubiquitin-related modifier in the proteasome pathway of the archaeon Sulfolobus acidocaldarius. *Nature Communications, 6*, 8163.

Ashokkumar, V., Venkatkarthick, R., Jayashree, S., Chuetor, S., Dharmaraj, S., Kumar, G., Chen, W. H., & Ngamcharussrivichai, C. (2022). Recent advances in lignocellulosic biomass for biofuels and value-added bioproducts - a critical review. *Bioresource Technology, 344*, 126195.

Badel, C., Samson, R. Y., & Bell, S. D. (2022). Chromosome organization affects genome evolution in Sulfolobus archaea. *Nature Microbiology, 7*, 820–830.

Baes, R., Grunberger, F., Pyr Dit Ruys, S., Couturier, M., De Keulenaer, S., Skevin, S., Van Nieuwerburgh, F., Vertommen, D., Grohmann, D., Ferreira-Cerca, S., et al. (2023). Transcriptional and translational dynamics underlying heat shock response in the thermophilic crenarchaeon Sulfolobus acidocaldarius. *MBio, 14*, e0359322.

Banerjee, A., Ghosh, A., Mills, D. J., Kahnt, J., Vonck, J., & Albers, S. V. (2012). FlaX, a unique component of the crenarchaeal archaellum, forms oligomeric ring-shaped structures and interacts with the motor ATPase FlaI. *The Journal of Biological Chemistry, 287*, 43322–43330.

Banerjee, A., Tsai, C. L., Chaudhury, P., Tripp, P., Arvai, A. S., Ishida, J. P., Tainer, J. A., & Albers, S. V. (2015). FlaF is a beta-sandwich protein that anchors the archaellum in the archaeal cell envelope by binding the S-layer protein. *Structure, 23*, 863–872.

Baquero, D. P., Cvirkaite-Krupovic, V., Hu, S. S., Fields, J. L., Liu, X., Rensing, C., Egelman, E. H., Krupovic, M., & Wang, F. (2023). Extracellular cytochrome nanowires appear to be ubiquitous in prokaryotes. *Cell, 186*, 2853–2864 e2858.

Bell, S. D., Botting, C. H., Wardleworth, B. N., Jackson, S. P., & White, M. F. (2002). The interaction of Alba, a conserved archaeal, chromatin protein, with Sir2 and its regulation by acetylation. *Science, 296*, 148–151.

Beltran, L. C., Cvirkaite-Krupovic, V., Miller, J., Wang, F., Kreutzberger, M. A. B., Patkowski, J. B., Costa, T. R. D., Schouten, S., Levental, I., Conticello, V. P., et al. (2023). Archaeal DNA-import apparatus is homologous to bacterial conjugation machinery. *Nature Communications, 14*, 666.

Beltran, L., Torsilieri, H., Patkowski, J. B., Yang, J. E., Casanova, J., Costa, T. R. D., Wright, E. R., & Egelman, E. H. (2024). The mating pilus of E. coli pED208 acts as a conduit for ssDNA during horizontal gene transfer. *MBio, 15*, e0285723.

Benninghoff, J. C., Kuschmierz, L., Zhou, X., Albersmeier, A., Pham, T. K., Busche, T., Wright, P. C., Kalinowski, J., Makarova, K. S., Brasen, C., et al. (2021a). Exposure to 1-butanol exemplifies the response of the Thermoacidophilic archaeon Sulfolobus acidocaldarius to solvent stress. *Applied and Environmental Microbiology, 87*.

Benninghoff, J. C., Kuschmierz, L., Zhou, X. X., Albersmeier, A., Pham, T. K., Busche, T., Wright, P. C., Kalinowski, J., Makarova, K. S., Bräsen, C., et al. (2021b). Exposure to 1-butanol exemplifies the response of the thermoacidophilic archaeon to solvent stress. *Applied and Environmental Microbiology, 87*(11), e02988–e02920.

Bhowmick, A., Recalde, A., Bhattacharyya, C., Banerjee, A., Das, J., Rodriguez-Cruz, U. E., Albers, S. V., & Ghosh, A. (2024). Role of VapBC4 toxin-antitoxin system of Sulfolobus acidocaldarius in heat stress adaptation. *MBio, 15*, e0275324.

Bischof, L. F., Haurat, M. F., Hoffmann, L., Albersmeier, A., Wolf, J., Neu, A., Pham, T. K., Albaum, S. P., Jakobi, T., Schouten, S., et al. (2018). Early response of Sulfolobus acidocaldarius to nutrient limitation. *Frontiers in Microbiology, 9*, 3201.

Bischof, L. F., Haurat, M. F., & Albers, S. V. (2019). Two membrane-bound transcription factors regulate expression of various type-IV-pili surface structures in Sulfolobus acidocaldarius. *PeerJ, 7*, e6459.

Blombach, F., & Werner, F. (2025). Chromatin and gene regulation in archaea. *Molecular Microbiology, 123*, 218–231.

Blombach, F., Sykora, M., Case, J., Feng, X., Baquero, D. P., Fouqueau, T., Phung, D. K., Barker, D., Krupovic, M., She, Q., et al. (2024). Cbp1 and Cren7 form chromatin-like structures that ensure efficient transcription of long CRISPR arrays. *Nature Communications, 15*, 1620.

Bost, J., Recalde, A., Wassmer, B., Wagner, A., Siebers, B., & Albers, S. V. (2023). Application of the endogenous CRISPR-Cas type I-D system for genetic engineering in the thermoacidophilic archaeon Sulfolobus acidocaldarius. *Frontiers in Microbiology, 14*, 1254891.

Braun, T., Vos, M. R., Kalisman, N., Sherman, N. E., Rachel, R., Wirth, R., Schroder, G. F., & Egelman, E. H. (2016). Archaeal flagellin combines a bacterial type IV pilin domain with an Ig-like domain. *Proceedings of the National Academy of Sciences of the United States of America, 113*, 10352–10357.

Brock, T. D., Brock, K. M., Belly, R. T., & Weiss, R. L. (1972). Sulfolobus: A new genus of sulfur-oxidizing bacteria living at low pH and high temperature. *Archiv für Mikrobiologie, 84*, 54–68.

Cajili, M. K. M., & Prieto, E. I. (2022). Interplay between Alba and Cren7 regulates chromatin compaction in Sulfolobus solfataricus. *Biomolecules, 12*.

Cao, J., Wang, Q., Liu, T., Peng, N., & Huang, L. (2018). Insights into the post-translational modifications of archaeal Sis10b (Alba): Lysine-16 is methylated, not acetylated, and this does not regulate transcription or growth. *Molecular Microbiology, 109*, 192–208.

Cao, J. J., Wang, T. K., Wang, Q., Zheng, X. W., & Huang, L. (2019). Functional insights into protein acetylation in the hyperthermophilic archaeon. *Molecular & Cellular Proteomics, 18*, 1572–1587.

Cao, J., Xiong, D., Zheng, X., Yuan, W., & Huang, L. (2025). Protein modification by a eukaryotic-like ubiquitinrelated modifier in the hyperthermophilic archaeon Saccharolobus islandicus. *mSystems 10*, e0058025

Cezanne, A., Foo, S., Kuo, Y. W., & Baum, B. (2024). The archaeal cell cycle. *Annual Review of Cell and Development Biology, 40*, 1–23.

Charles-Orszag, A., van Wolferen, M., Lord, S. J., Albers, S. V., & Mullins, R. D. (2024). Adhesion pilus retraction powers twitching motility in the thermoacidophilic crenarchaeon Sulfolobus acidocaldarius. *Nature Communications, 15*, 5051.

Charles-Orszag, A., Lord, S. J., Herrera, N., Strauskulage, L., Bhowmick, A., Goddard, T., Wassmer, B., van Wolferen, M., Asper, G., Flis, A., et al. (2025). Archaeal SegAB forms a bipolar structure that promotes chromosome segregation in spherical cells. *bioRxiv*. https://doi.org/10.1101/2025.04.15.649018

Chaudhury, P., Neiner, T., D'Imprima, E., Banerjee, A., Reindl, S., Ghosh, A., Arvai, A. S., Mills, D. J., van der Does, C., Tainer, J. A., et al. (2016). The nucleotide-dependent interaction of FlaH and FlaI is essential for assembly and function of the archaellum motor. *Molecular Microbiology, 99*, 674–685.

Chlebek, J. L., Hughes, H. Q., Ratkiewicz, A. S., Rayyan, R., Wang, J. C., Herrin, B. E., Dalia, T. N., Biais, N., & Dalia, A. B. (2019). PilT and PilU are homohexameric ATPases that coordinate to retract type IVa pili. *PLoS Genetics, 15*, e1008448.

Constantinescu-Aruxandei, D., Petrovic-Stojanovska, B., Penedo, J. C., White, M. F., & Naismith, J. H. (2016). Mechanism of DNA loading by the DNA repair helicase XPD. *Nucleic Acids Research, 44*, 2806–2815.

Counts, J. A., Willard, D. J., & Kelly, R. M. (2021). Life in hot acid: A genome-based reassessment of the archaeal order Sulfolobales. *Environmental Microbiology, 23*, 3568–3584.

Cowan, D. A., Albers, S. V., Antranikian, G., Atomi, H., Averhoff, B., Basen, M., Driessen, A. J. M., Jebbar, M., Kelman, Z., Kerou, M., et al. (2024). Extremophiles in a changing world. *Extremophiles, 28*, 26.

De Kock, V., Peeters, E., & Baes, R. (2025). The Lrs14 family of DNA-binding proteins as nucleoid-associated proteins in the Crenarchaeal order Sulfolobales. *Molecular Microbiology, 123*, 132–142.

De Lise, F., Iacono, R., Moracci, M., Strazzulli, A., & Cobucci-Ponzano, B. (2023). Archaea as a model system for molecular biology and biotechnology. *Biomolecules, 13*.

Dhanaraju, R., Samson, R. Y., Feng, X., Costa, A., Gonzalez-Gutierrez, G., & Bell, S. D. (2025). An archaeal nucleoid-associated protein binds an essential motif in DNA replication origins. *Nature Communications, 16*, 5230.

Duan, X., & He, Z. G. (2011). Characterization of the specific interaction between archaeal FHA domain-containing protein and the promoter of a flagellar-like gene-cluster and its regulation by phosphorylation. *Biochemical and Biophysical Research Communications, 407*, 242–247.

Duggin, I. G., McCallum, S. A., & Bell, S. D. (2008). Chromosome replication dynamics in the archaeon. *Proc Natl Acad Sci USA, 105*, 16737–16742.

Esser, D., Pham, T. K., Reimann, J., Albers, S. V., Siebers, B., & Wright, P. C. (2012). Change of carbon source causes dramatic effects in the phospho-proteome of the archaeon Sulfolobus solfataricus. *Journal of Proteome Research, 11*, 4823–4833.

Feng, X., Sun, M., Han, W., Liang, Y. X., & She, Q. (2018). A transcriptional factor B paralog functions as an activator to DNA damage-responsive expression in archaea. *Nucleic Acids Research, 46*, 7085–7096.

Feng, X., Liu, X., Xu, R., Zhao, R., Feng, W., Liao, J., Han, W., & She, Q. (2020). A unique B-family DNA polymerase facilitating error-prone DNA damage tolerance in Crenarchaeota. *Frontiers in Microbiology, 11*, 1585.

Feng, X., Zhang, B., Gao, Z., Xu, R., Liu, X., Ishino, S., Feng, M., Shen, Y., Ishino, Y., & She, Q. (2022). A well-conserved archaeal B-family polymerase functions as an extender in translesion synthesis. *MBio, 13*, e0265921.

Foo, S., Caspy, I., Cezanne, A., Bharat, T. A. M., & Baum, B. (2025). A self-assembling surface layer flattens the cytokinetic furrow to aid cell division in an archaeon. *Proceedings of the National Academy of Sciences of the United States of America, 122*, e2501044122.

Frols, S., Ajon, M., Wagner, M., Teichmann, D., Zolghadr, B., Folea, M., Boekema, E. J., Driessen, A. J., Schleper, C., & Albers, S. V. (2008). UV-inducible cellular aggregation of the hyperthermophilic archaeon Sulfolobus solfataricus is mediated by pili formation. *Molecular Microbiology, 70*, 938–952.

Gaines, M. C., Isupov, M. N., Sivabalasarma, S., Haque, R. U., McLaren, M., Mollat, C. L., Tripp, P., Neuhaus, A., Gold, V. A. M., Albers, S. V., et al. (2022). Electron cryo-microscopy reveals the structure of the archaeal thread filament. *Nature Communications, 13*, 7411.

Gaines, M. C., Isupov, M. N., McLaren, M., Mollat, C. L., Haque, R. U., Stephenson, J. K., Sivabalasarma, S., Hanus, C., Kattnig, D., Gold, V. A. M., et al. (2024a). Towards a molecular picture of the archaeal cell surface. *Nature Communications, 15*, 10401.

Gaines, M. C., Sivabalasarma, S., Isupov, M. N., Haque, R. U., McLaren, M., Hanus, C., Gold, V. A. M., Albers, S. V., & Daum, B. (2024b). CryoEM reveals the structure of an archaeal pilus involved in twitching motility. *Nature Communications, 15*, 5050.

Girfoglio, M., Rossi, M., & Cannio, R. (2012). Cellulose degradation by Sulfolobus solfataricus requires a cell-anchored endo-β-1-4-glucanase. *Journal of Bacteriology, 194*, 5091–5100.

Goldlust, K., Ducret, A., Halte, M., Dedieu-Berne, A., Erhardt, M., & Lesterlin, C. (2023). The F pilus serves as a conduit for the DNA during conjugation between physically distant bacteria. *Proceedings of the National Academy of Sciences of the United States of America, 120*, e2310842120.

Gomez-Raya-Vilanova, M. V., Teulière, J., Medvedeva, S., Dai, Y. P., Corel, E., Lopez, P., Lapointe, F. J., Bhattacharya, D., Haraoui, L. P., Turc, E., et al. (2025). Transcriptional landscape of the cell cycle in a model thermoacidophilic archaeon reveals similarities to eukaryotes. *Nature Communications, 16*, 5697.

Grogan, D., Palm, P., & Zillig, W. (1990). Isolate B12, which harbours a virus-like element, represents a new species of the archaebacterial genus Sulfolobus, Sulfolobus shibatae, sp. nov. *Archives of Microbiology, 154*, 594–599.

Guo, L., Brugger, K., Liu, C., Shah, S. A., Zheng, H., Zhu, Y., Wang, S., Lillestol, R. K., Chen, L., Frank, J., et al. (2011). Genome analyses of Icelandic strains of Sulfolobus islandicus, model organisms for genetic and virus-host interaction studies. *Journal of Bacteriology, 193*, 1672–1680.

Guo, H., Zhao, Y., Chang, J. S., & Lee, D. J. (2022). Inhibitor formation and detoxification during lignocellulose biorefinery: A review. *Bioresource Technology, 361*, 127666.

Guterl, J. K., Garbe, D., Carsten, J., Steffler, F., Sommer, B., Reisse, S., Philipp, A., Haack, M., Ruhmann, B., Koltermann, A., et al. (2012). Cell-free metabolic engineering: Production of chemicals by minimized reaction cascades. *ChemSusChem, 5*, 2165–2172.

Hanks, S. K., & Hunter, T. (1995). Protein Kinases. 6. The eukaryotic protein-kinase superfamily - kinase (catalytic) domain-structure and classification. *The FASEB Journal, 9*, 576–596.

Haurat, M. F., Figueiredo, A. S., Hoffmann, L., Li, L. L., Herr, K., Wilson, A. J., Beeby, M., Schaber, J., & Albers, S. V. (2017). ArnS, a kinase involved in starvation-induced archaellum expression. *Molecular Microbiology, 103*, 181–194.

Henche, A. L., Ghosh, A., Yu, X., Jeske, T., Egelman, E., & Albers, S. V. (2012a). Structure and function of the adhesive type IV pilus of Sulfolobus acidocaldarius. *Environmental Microbiology, 14*, 3188–3202.

Henche, A. L., Koerdt, A., Ghosh, A., & Albers, S. V. (2012b). Influence of cell surface structures on crenarchaeal biofilm formation using a thermostable green fluorescent protein. *Environmental Microbiology, 14*, 779–793.

Hoffmann, L., Schummer, A., Reimann, J., Haurat, M. F., Wilson, A. J., Beeby, M., Warscheid, B., & Albers, S. V. (2017a). Expanding the archaellum regulatory network - the eukaryotic protein kinases ArnC and ArnD influence motility of Sulfolobus acidocaldarius. *Microbiologyopen, 6*, e00414.

Hoffmann, L., Schummer, A., Reimann, J., Haurat, M. F., Wilson, A. J., Beeby, M., Warscheid, B., & Albers, S. V. (2017b). Expanding the archaellum regulatory network - the eukaryotic protein kinases ArnC and ArnD influence motility of Sulfolobus acidocaldarius. *Microbiology, 6*, e00414.

Hoffmann, L., Anders, K., Bischof, L. F., Ye, X., Reimann, J., Khadouma, S., Pham, T. K., van der Does, C., Wright, P. C., Essen, L. O., et al. (2019). Structure and interactions of the archaeal motility repression module ArnA-ArnB that modulates archaellum gene expression in Sulfolobus acidocaldarius. *The Journal of Biological Chemistry, 294*, 7460–7471.

Huang, Y., Krauss, G., Cottaz, S., Driguez, H., & Lipps, G. (2005). A highly acid-stable and thermostable endo-beta-glucanase from the thermoacidophilic archaeon Sulfolobus solfataricus. *The Biochemical Journal, 385*, 581–588.

Huang, Q., Li, Y., Zeng, C., Song, T., Yan, Z., Ni, J., She, Q., & Shen, Y. (2015). Genetic analysis of the Holliday junction resolvases Hje and Hjc in Sulfolobus islandicus. *Extremophiles, 19*, 505–514.

Huang, Q., Mayaka, J. B., Zhong, Q., Zhang, C., Hou, G., Ni, J., & Shen, Y. (2019). Phosphorylation of the archaeal holliday junction resolvase Hjc inhibits its catalytic activity and facilitates DNA repair in Sulfolobus islandicus REY15A. *Frontiers in Microbiology, 10*, 1214.

Huang, Q., Lin, Z., Wu, P., Ni, J., & Shen, Y. (2020). Phosphoproteomic analysis reveals Rio1-related protein phosphorylation changes in response to UV irradiation in Sulfolobus islandicus REY15A. *Frontiers in Microbiology, 11*, 586025.

Huang, Q., Ya, W., Gan, Q., Liang, S., Yang, Y., Wu, P., Feng, X., She, Q., Ni, J., & Shen, Y. (2025). Phosphorylation of the α subunit inhibits proteasome assembly and regulates cell division in archaea. *biorixv preprint*. https://doi.org/10.1101/2025.05.09.653063

Huber, G., Spinnler, C., Gambacorta, A., Stetter, K. O. (1989) Metallosphaera sedula gen, and sp. nov. represents a new genus of aerobic, metal-mobilizing, thermoacidophilic archaebacteria. *Systematic and Applied Microbiology, 12*, 38–47

Humbard, M. A., Miranda, H. V., Lim, J. M., Krause, D. J., Pritz, J. R., Zhou, G. Y., Chen, S. X., Wells, L., & Maupin-Furlow, J. A. (2010). Ubiquitin-like small archaeal modifier proteins (SAMPs) in Haloferax volcanii. *Nature, 463*, 54–U60.

Ithurbide, S., Gribaldo, S., Albers, S. V., & Pende, N. (2022). Spotlight on FtsZ-based cell division in archaea. *Trends in Microbiology, 30*, 665–678.

Itoh, Y.H., Kurosawa, N., Uda, I., Sugai, A., Tanoue, S., Itoh, T., Horiuchi,T., Itoh, T. (2001). Metallosphaera sedula TA-2, a calditoglycerocaldarchaeol deletion strain of a thermoacidophilic archaeon. *Extremophiles 5*, 241-245

Jiang, Z. C., Lin, Z. J., Gan, Q., Wu, P. J., Zhang, X. M., Xiao, Y. X., She, Q. X., Ni, J. F., Shen, Y. L., & Huang, Q. H. (2023). The FHA domain protein ArnA functions as a global DNA damage response repressor in the hyperthermophilic archaeon. *MBio, 14*(4), e0094223.

Jover, A. B., & Dekker, C. (2023). The archaeal Cdv cell division system. *Trends in Microbiology, 31*, 601–615.

Kabli, A. F., Ng, I. W., Read, N., Pal, P., Reimann, J., Tran, N. T., Albers, S. V., Le, T. B. K., & Barilla, D. (2025). Coupling chromosome organization to genome segregation in archaea. *Nature Communications, 16*, 6759.

Kalichuk, V., Behar, G., Renodon-Corniere, A., Danovski, G., Obal, G., Barbet, J., Mouratou, B., & Pecorari, F. (2016). The archaeal "7 kDa DNA-binding" proteins: Extended characterization of an old gifted family. *Scientific Reports, 6*, 37274.

Kalliomaa-Sanford, A. K., Rodriguez-Castaneda, F. A., McLeod, B. N., Latorre-Rosello, V., Smith, J. H., Reimann, J., Albers, S. V., & Barilla, D. (2012). Chromosome segregation in archaea mediated by a hybrid DNA partition machine. *Proceedings of the National Academy of Sciences of the United States of America, 109*, 3754–3759.

Keller, M. W., Lipscomb, G. L., Nguyen, D. M., Crowley, A. T., Schut, G. J., Scott, I., Kelly, R. M., & Adams, M. W. W. (2017). Ethanol production by the hyperthermophilic archaeon by expression of bacterial bifunctional alcohol dehydrogenases. *Microbial Biotechnology, 10*, 1535–1545.

Kennelly, P. J. (2014). Protein Ser/Thr/Tyr phosphorylation in the archaea. *The Journal of Biological Chemistry, 289*, 9480–9487.

Kessler, A., Sezonov, G., Guijarro, J. I., Desnoues, N., Rose, T., Delepierre, M., Bell, S. D., & Prangishvili, D. (2006). A novel archaeal regulatory protein, Sta1, activates transcription from viral promoters. *Nucleic Acids Research, 34*, 4837–4845.

Knuppel, R., Christensen, R. H., Gray, F. C., Esser, D., Strauss, D., Medenbach, J., Siebers, B., MacNeill, S. A., LaRonde, N., & Ferreira-Cerca, S. (2018). Insights into the evolutionary conserved regulation of Rio ATPase activity. *Nucleic Acids Research, 46*, 1441–1456.

Koonin, E. V., Mushegian, A. R., Galperin, M. Y., & Walker, D. R. (1997). Comparison of archaeal and bacterial genomes: Computer analysis of protein sequences predicts novel functions and suggests a chimeric origin for the archaea. *Molecular Microbiology, 25*, 619–637.

Korf, L., Ye, X., Vogt, M. S., Steinchen, W., Watad, M., van der Does, C., Tourte, M., Sivabalasarma, S., Albers, S. V., & Essen, L. O. (2023). Archaeal GPN-loop GTPases involve a lock-switch-rock mechanism for GTP hydrolysis. *MBio, 14*, e0085923.

Kreutzberger, M. A. B., Sonani, R. R., Liu, J., Chatterjee, S., Wang, F., Sebastian, A. L., Biswas, P., Ewing, C., Zheng, W., Poly, F., et al. (2022). Convergent evolution in the supercoiling of prokaryotic flagellar filaments. *Cell, 185*, 3487–3500 e3414.

Kreutzberger, M. A. B., Cvirkaite-Krupovic, V., Liu, Y., Baquero, D. P., Liu, J., Sonani, R. R., Calladine, C. R., Wang, F., Krupovic, M., & Egelman, E. H. (2023). The evolution of archaeal flagellar filaments. *Proceedings of the National Academy of Sciences of the United States of America, 120*, e2304256120.

Kuo, Y. W., Traparic, J., Foo, S., & Baum, B. (2026). The mechanism of cell-cycle-dependent proteasomal degradation of archaeal ESCRT-III homolog CdvB in Sulfolobus. *The EMBO Journal*

Kurosawa, N., Itoh, Y. H., Iwai, T., Sugai, A., Uda, I., Kimura, N., Horiuchi, T., & Itoh, T. (1998). Sulfurisphaera ohwakuensis gen. nov., sp. nov., a novel extremely thermophilic acidophile of the order Sulfolobales. *International Journal of Systematic Bacteriology, 48*(Pt 2), 451–456.

Kuschmierz, L., Wagner, A., Schmerling, C., Busche, T., Kalinowski, J., Bräsen, C., & Siebers, B. (2024). 5′-untranslated region sequences enhance plasmid-based protein production in *Sulfolobus acidocaldarius*. *Frontiers in Microbiology, 15*.

Lalithambika, S., Peterson, L., Dana, K., & Blum, P. (2012). Carbohydrate hydrolysis and transport in the extreme thermoacidophile Sulfolobus solfataricus. *Applied and Environmental Microbiology, 78*, 7931–7938.

LaRonde, N. A. (2014). The ancient microbial RIO kinases. *The Journal of Biological Chemistry, 289*, 9488–9492.

Lassak, K., Neiner, T., Ghosh, A., Klingl, A., Wirth, R., & Albers, S. V. (2012). Molecular analysis of the crenarchaeal flagellum. *Molecular Microbiology, 83*, 110–124.

Lemmens, L., Wang, K., Ruykens, E., Nguyen, V. T., Lindas, A. C., Willaert, R., Couturier, M., & Peeters, E. (2022). DNA-binding properties of a novel crenarchaeal chromatin-organizing protein in Sulfolobus acidocaldarius. *Biomolecules, 12*.

Lewis, A. M., Recalde, A., Brasen, C., Counts, J. A., Nussbaum, P., Bost, J., Schocke, L., Shen, L., Willard, D. J., Quax, T. E. F., et al. (2021). The biology of thermoacidophilic archaea from the order Sulfolobales. *FEMS Microbiology Reviews, 45*.

Li, Y. J., Pan, S. F., Zhang, Y., Ren, M., Feng, M. X., Peng, N., Chen, L. M., Liang, Y. X., & She, Q. X. (2016). Harnessing type I and type III CRISPR-Cas systems for genome editing. *Nucleic Acids Research, 44*, e34.

Li, L. L., Banerjee, A., Bischof, L. F., Maklad, H. R., Hoffmann, L., Henche, A. L., Veliz, F., Bildl, W., Schulte, U., Orell, A., et al. (2017). Wing phosphorylation is a major functional determinant of the Lrs14-type biofilm and motility regulator AbfR1 in Sulfolobus acidocaldarius. *Molecular Microbiology, 105*, 777–793.

Li, X. Y., Lozano-Madueno, C., Martinez-Alvarez, L., & Peng, X. (2023). A clade of RHH proteins ubiquitous in Sulfolobales and their viruses regulates cell cycle progression. *Nucleic Acids Research, 51*, 1724.

Li, S., Huang, Q., Yang, Y., Wu, P., Li, J., Shen, Y., & Ni, J. (2025). Roles and transcriptional regulation of the endogenous cellulases in association with substrate and cellular processes in a thermoacidophilic archaeon. *Bioresource Technology, 435*, 132871.

Limauro, D., Cannio, R., Fiorentino, G., Rossi, M., & Bartolucci, S. (2001). Identification and molecular characterization of an endoglucanase gene, celS, from the extremely thermophilic archaeon Sulfolobus solfataricus. *Extremophiles, 5*, 213–219.

Lindås, A. C., & Bernander, R. (2013). The cell cycle of archaea. *Nature Reviews. Microbiology, 11*, 627–638.

Lindas, A. C., Karlsson, E. A., Lindgren, M. T., Ettema, T. J., & Bernander, R. (2008). A unique cell division machinery in the archaea. *Proceedings of the National Academy of Sciences of the United States of America, 105*, 18942–18946.

Ling, H., Boudsocq, F., Woodgate, R., & Yang, W. (2001). Crystal structure of a Y-family DNA polymerase in action: A mechanism for error-prone and lesion-bypass replication. *Cell, 107*, 91–102.

Litsios, A., Grys, B. T., Kraus, O. Z., Friesen, H., Ross, C., Masinas, M. P. D., Forster, D. T., Couvillion, M. T., Timmermann, S., Billmann, M., et al. (2024). Proteome-scale movements and compartment connectivity during the eukaryotic cell cycle. *Cell, 187*, 1490.

Liu, L. J., You, X. Y., Guo, X., Liu, S. J., & Jiang, C. Y. (2011). Metallosphaera cuprina sp. nov., an acidothermophilic, metal-mobilizing archaeon. *International Journal of Systematic and Evolutionary Microbiology, 61*, 2395–2400.

Liu, J. F., Gao, R. X., Li, C. T., Ni, J. F., Yang, Z. J., Zhang, Q., Chen, H. N., & Shen, Y. L. (2017). Functional assignment of multiple ESCRT-III homologs in cell division and budding in. *Molecular Microbiology, 105*, 540–553.

Liu, L. J., Jiang, Z., Wang, P., Qin, Y. L., Xu, W., Wang, Y., Liu, S. J., & Jiang, C. Y. (2021a). Physiology, taxonomy, and sulfur metabolism of the sulfolobales, an order of thermoacidophilic archaea. *Frontiers in Microbiology, 12*, 768283.

Liu, Y., Makarova, K. S., Huang, W. C., Wolf, Y. I., Nikolskaya, A. N., Zhang, X., Cai, M., Zhang, C. J., Xu, W., Luo, Z., et al. (2021b). Expanded diversity of Asgard archaea and their relationships with eukaryotes. *Nature, 593*, 553–557.

Liu, J., Eastep, G. N., Cvirkaite-Krupovic, V., Rich-New, S. T., Kreutzberger, M. A. B., Egelman, E. H., Krupovic, M., & Wang, F. (2024). Two distinct archaeal type IV pili structures formed by proteins with identical sequence. *Nature Communications, 15*, 5049.

Liu, J. F., Lelek, M., Yang, Y. F., Salles, A., Zimmer, C., Shen, Y. L., & Krupovic, M. (2025). A relay race of ESCRT-III paralogs drives cell division in a hyperthermophilic archaeon. *MBio, 16*, e0099124.

Lundgren, M., Andersson, A., Chen, L. M., Nilsson, P., & Bernander, R. (2004). Three replication origins in species: Synchronous initiation of chromosome replication and asynchronous termination. *Proceedings of the National Academy of Sciences of the United States of America, 101*, 7046–7051.

Lundgren, M., Malandrin, L., Eriksson, S., Huber, H., & Bernander, R. (2008). Cell cycle characteristics of crenarchaeota: Unity among diversity. *Journal of Bacteriology, 190*, 5362–5367.

Makarova, K. S., Koonin, E. V., & Kelman, Z. (2012). The CMG (CDC45/RecJ, MCM, GINS) complex is a conserved component of the DNA replication system in all archaea and eukaryotes. *Biology Direct, 7*, 7.

Makarova, K. S., Zhang, C., Wolf, Y. I., Karamycheva, S., Whitaker, R. J., & Koonin, E. V. (2024). Computational analysis of genes with lethal knockout phenotype and prediction of essential genes in archaea. *MBio, 15*, e0309223.

Maklad, H. R., Ruys, S. P. d., Bervoets, I., Vertommen, D., Gutierrez, G. J., Vranken, W. F., Siebers, B., & Peeters, E. (2024). Protein kinase Sut1 from the Crenarchaeon Sulfolobus acidocaldarius displays tyrosine phosphorylation activity. *biorixv preprint*.

Manesh, M. J. H., Willard, D. J., Lewis, A. M., & Kelly, R. M. (2024). Extremely thermoacidophilic archaea for metal bioleaching: What do their genomes tell us? *Bioresource Technology, 391*, 129988.

Maurelli, L., Giovane, A., Esposito, A., Moracci, M., Fiume, I., Rossi, M., & Morana, A. (2008). Evidence that the xylanase activity from Sulfolobus solfataricus Oalpha is encoded by the endoglucanase precursor gene (sso1354) and characterization of the associated cellulase activity. *Extremophiles, 12*, 689–700.

Nucci, R., Moracci, M., Vaccaro, C., Vespa, N., & Rossi, M. (1993). Exo-glucosidase activity and substrate specificity of the beta-glycosidase isolated from the extreme thermophile Sulfolobus solfataricus. *Biotechnology and Applied Biochemistry, 17*, 239–250.

Orell, A., Peeters, E., Vassen, V., Jachlewski, S., Schalles, S., Siebers, B., & Albers, S.-V. (2013). Lrs14 transcriptional regulators influence biofilm formation and cell motility of Crenarchaea. *The ISME Journal, 7*, 1886–1898.

Parham, J., Sorichetti, V., Cezanne, A., Hoogenberg, B., Foo, S., Kuo, Y.-W., Mawdesley, E., Gatward, L. D., Boulanger, J., Schulze, U., et al. (2025). Temporal and spatial coordination of DNA segregation and cell division in an archaeon. *Proceedings of the National Academy of Sciences of the United States of America, 122*(42), e2513939122.

Peng, T. J., Liu, L. J., Liu, C., Yang, Z. F., Liu, S. J., & Jiang, C. Y. (2015). Metallosphaera tengchongensis sp. nov., an acidothermophilic archaeon isolated from a hot spring. *International Journal of Systematic and Evolutionary Microbiology, 65*, 537–542.

Pony, P., Rapisarda, C., Terradot, L., Marza, E., & Fronzes, R. (2020). Filamentation of the bacterial bi-functional alcohol/aldehyde dehydrogenase AdhE is essential for substrate channeling and enzymatic regulation. *Nature Communications, 11*, 1426.

Quaiser, A., Constantinesco, F., White, M. F., Forterre, P., & Elie, C. (2008). The Mre11 protein interacts with both Rad50 and the HerA bipolar helicase and is recruited to DNA following gamma irradiation in the archaeon Sulfolobus acidocaldarius. *BMC Molecular Biology, 9*, 25.

Ray, W. K., Keith, S. M., DeSantis, A. M., Hunt, J. P., Larson, T. J., Helm, R. F., & Kennelly, P. J. (2005). A phosphohexomutase from the archaeon Sulfolobus solfataricus is covalently modified by phosphorylation on serine. *Journal of Bacteriology, 187*, 4270–4275.

Recalde, A., Wagner, A., Sivabalasarma, S., Yurmashava, A., Fehr, N. P., Thurm, R., Le, T. N., Koebler, C., Wassmer, B., Albers, S. V., et al. (2025). New components of the community-based DNA-repair mechanism in Sulfolobales. *Microlife, 6*, uqaf002.

Reimann, J., Lassak, K., Khadouma, S., Ettema, T. J., Yang, N., Driessen, A. J., Klingl, A., & Albers, S. V. (2012). Regulation of archaella expression by the FHA and von Willebrand domain-containing proteins ArnA and ArnB in Sulfolobus acidocaldarius. *Molecular Microbiology, 86*, 24–36.

Reimann, J., Esser, D., Orell, A., Amman, F., Pham, T. K., Noirel, J., Lindas, A. C., Bernander, R., Wright, P. C., Siebers, B., et al. (2013). Archaeal signal transduction: Impact of protein phosphatase deletions on cell size, motility, and energy metabolism in Sulfolobus acidocaldarius. *Molecular & Cellular Proteomics, 12*, 3908–3923.

Reindl, S., Ghosh, A., Williams, G. J., Lassak, K., Neiner, T., Henche, A. L., Albers, S. V., & Tainer, J. A. (2013). Insights into FlaI functions in archaeal motor assembly and motility from structures, conformations, and genetics. *Molecular Cell, 49*, 1069–1082.

Reshmy, R., Philip, E., Madhavan, A., Sirohi, R., Pugazhendhi, A., Binod, P., Kumar Awasthi, M., Vivek, N., Kumar, V., & Sindhu, R. (2022). Lignocellulose in future biorefineries: Strategies for cost-effective production of biomaterials and bioenergy. *Bioresource Technology, 344*, 126241.

Risa, G. T., Hurtig, F., Bray, S., Hafner, A. E., Harker-Kirschneck, L., Faull, P., Davis, C., Papatziamou, D., Mutavchiev, D. R., Fan, C., et al. (2020). The proteasome controls ESCRT-III-mediated cell division in an archaeon. *Science, 369*, 642.

Rouillon, C., & White, M. F. (2010). The XBP-Bax1 helicase-nuclease complex unwinds and cleaves DNA: Implications for eukaryal and archaeal nucleotide excision repair. *The Journal of Biological Chemistry, 285*, 11013–11022.

Rowland, E. F., Bautista, M. A., Zhang, C., & Whitaker, R. J. (2020). Surface resistance to SSVs and SIRVs in pilin deletions of Sulfolobus islandicus. *Molecular Microbiology, 113*, 718–727.

Sakai, H. D., & Kurosawa, N. (2018). Saccharolobus caldissimus gen. nov., sp. nov., a facultatively anaerobic iron-reducing hyperthermophilic archaeon isolated from an acidic terrestrial hot spring, and reclassification of Sulfolobus solfataricus as Saccharolobus solfataricus comb. nov. and Sulfolobus shibatae as Saccharolobus shibatae comb. nov. *International Journal of Systematic and Evolutionary Microbiology, 68*, 1271–1278.

Sakai, H. D., Nakamura, K., & Kurosawa, N. (2022). Stygiolobus caldivivus sp. nov., a facultatively anaerobic hyperthermophilic archaeon isolated from the Unzen hot spring in Japan. *International Journal of Systematic and Evolutionary Microbiology, 72*.

Samson, R. Y., Xu, Y., Gadelha, C., Stone, T. A., Faqiri, J. N., Li, D., Qin, N., Pu, F., Liang, Y. X., She, Q., et al. (2013). Specificity and function of archaeal DNA replication initiator proteins. *Cell Reports, 3*, 485–496.

Samson, R. Y., Abeyrathne, P. D., & Bell, S. D. (2016). Mechanism of archaeal MCM helicase recruitment to DNA replication origins. *Molecular Cell, 61*, 287–296.

Schmerling, C., Schroeder, C., Zhou, X. X., Bost, J., Wassmer, B., Ninck, S., Busche, T., Montero, L., Kaschani, F., Schmitz, O. J., et al. (2025). An unusual glycerol-3-phosphate dehydrogenase in elucidates the diversity of glycerol metabolism across Archaea. *Communications Biology, 8*. https://doi.org/10.1038/s42003-025-07953-9

Schocke, L., Brasen, C., & Siebers, B. (2019). Thermoacidophilic Sulfolobus species as source for extremozymes and as novel archaeal platform organisms. *Current Opinion in Biotechnology, 59*, 71–77.

Schult, F., Le, T. N., Albersmeier, A., Rauch, B., Blumenkamp, P., van der Does, C., Goesmann, A., Kalinowski, J., Albers, S. V., & Siebers, B. (2018). Effect of UV irradiation on Sulfolobus acidocaldarius and involvement of the general transcription factor TFB3 in the early UV response. *Nucleic Acids Research, 46*, 7179–7192.

Segerer, A., Neuner A., Kristjansson, J. K., Stetter, K. O. (1986). Acidianus infernus gen. nov., sp. nov., and Acidianus brierleyi comb. nov.: facultatively aerobic, extremely acidophilic thermophilic sulfur-metabolizing archaebacteria. *International Journal of Systematic and Evolutionary Microbiology, 36*, 559–564.

She, Q., Phan, H., Garrett, R. A., Albers, S. V., Stedman, K. M., & Zillig, W. (1998). Genetic profile of pNOB8 from Sulfolobus: The first conjugative plasmid from an archaeon. *Extremophiles, 2*, 417–425.

She, Q., Singh, R. K., Confalonieri, F., Zivanovic, Y., Allard, G., Awayez, M. J., Chan-Weiher, C. C., Clausen, I. G., Curtis, B. A., De Moors, A., et al. (2001). The complete genome of the crenarchaeon Sulfolobus solfataricus P2. *Proceedings of the National Academy of Sciences of the United States of America, 98*, 7835–7840.

Sivabalasarma, S., Taib, N., Mollat, C. L., Joest, M., Steimle, S., Gribaldo, S., & Albers, S.-V. (2025). Horizontal gene transfer of the functional archaellum machinery to bacteria. *bioRxiv*.

Sleutel, M., Sonani, R. R., Miller, J. G., Wang, F., Socorro, A. G., Chen, Y., Martin, R., Demeler, B., Rudolph, M. J., Alva, V., et al. (2024). Donor strand complementation and calcium ion coordination drive the chaperone-free polymerization of archaeal cannulae. *bioRxiv*. https://doi.org/10.1101/2024.12.30.630787

Sun, M., Feng, X., Liu, Z., Han, W., Liang, Y. X., & She, Q. (2018). An Orc1/Cdc6 ortholog functions as a key regulator in the DNA damage response in archaea. *Nucleic Acids Research, 46*, 6697–6711.

Suzuki, T., Iwasaki, T., Uzawa, T., Hara, K., Nemoto, N., Kon, T., Ueki, T., Yamagishi, A., & Oshima, T. (2002). Sulfolobus tokodaii sp. nov. (f. Sulfolobus sp. strain 7), a new member of the genus Sulfolobus isolated from Beppu Hot Springs, Japan. *Extremophiles, 6*, 39–44.

Takayanagi, S., Kawasaki, H., Sugimori, K., Yamada, T., Sugai, A., Ito, T., Yamasato, K., & Shioda, M. (1996). Sulfolobus hakonensis sp. nov., a novel species of acidothermophilic archaeon. *International Journal of Systematic Bacteriology, 46*, 377–382.

Takemata, N., & Bell, S. D. (2021). Multi-scale architecture of archaeal chromosomes. *Molecular Cell, 81*, 473.

Takemata, N., Samson, R. Y., & Bell, S. D. (2019). Physical and functional compartmentalization of archaeal chromosomes. *Cell, 179*, 165–179 e118.

Tang, S., Huang, C. H., Ko, T. P., Lin, K. F., Chang, Y. C., Lin, P. Y., Sun, L., & Chen, C. Y. (2024). Dual dimeric interactions in the nucleic acid-binding protein Sac10b lead to multiple bridging of double-stranded DNA. *Heliyon, 10*, e31630.

Timinskas, K., Timinskas, A., & Venclovas, C. (2025). Common themes in architecture and interactions of prokaryotic PolB2 and Pol V mutasomes inferred from in silico studies. *Computational and Structural Biotechnology Journal, 27*, 401–410.

Tsai, C. L., Tripp, P., Sivabalasarma, S., Zhang, C., Rodriguez-Franco, M., Wipfler, R. L., Chaudhury, P., Banerjee, A., Beeby, M., Whitaker, R. J., et al. (2020). The structure of the periplasmic FlaG-FlaF complex and its essential role for archaellar swimming motility. *Nature Microbiology, 5*, 216–225.

Tsuboi, K., Sakai, H. D., Nur, N., Stedman, K. M., Kurosawa, N., & Suwanto, A. (2018). Sulfurisphaera javensis sp. nov., a hyperthermophilic and acidophilic archaeon isolated from Indonesian hot spring, and reclassification of Sulfolobus tokodaii Suzuki et al. 2002 as Sulfurisphaera tokodaii comb. nov. *International Journal of Systematic and Evolutionary Microbiology, 68*, 1907–1913.

Urbieta, M. S., Toril, E. G., Giaveno, M., Bazán, A. A., Donati, E. R. (2014). Archaeal and bacterial diversity in five different hydrothermal ponds in the copahue region in Argentina. *Systematic and Applied Microbiology, 37*, 429–441.

van der Kolk, N., Wagner, A., Wagner, M., Wassmer, B., Siebers, B., & Albers, S. V. (2020). Identification of XylR, the activator of arabinose/xylose inducible regulon in and its application for homologous protein expression. *Frontiers in Microbiology, 11*, 1066.

Van Laer, B., Roovers, M., Wauters, L., Kasprzak, J. M., Dyzma, M., Deyaert, E., Kumar Singh, R., Feller, A., Bujnicki, J. M., Droogmans, L., et al. (2016). Structural and functional insights into tRNA binding and adenosine N1-methylation by an archaeal Trm10 homologue. *Nucleic Acids Research, 44*, 940–953.

van Wolferen, M., Ajon, M., Driessen, A. J., & Albers, S. V. (2013). Molecular analysis of the UV-inducible pili operon from Sulfolobus acidocaldarius. *Microbiology, 2*, 928–937.

van Wolferen, M., Wagner, A., van der Does, C., & Albers, S. V. (2016). The archaeal Ced system imports DNA. *Proceedings of the National Academy of Sciences of the United States of America, 113*, 2496–2501.

van Wolferen, M., Shajahan, A., Heinrich, K., Brenzinger, S., Black, I. M., Wagner, A., Briegel, A., Azadi, P., & Albers, S. V. (2020). Species-specific recognition of sulfolobales mediated by UV-Inducible Pili and S-Layer glycosylation patterns. *MBio, 11*.

Walsh, C. T., Garneau-Tsodikova, S., & Gatto, G. J. (2005). Protein posttranslational modifications: The chemistry of proteome diversifications. *Angewandte Chemie. International Edition, 44*, 7342–7372.

Wang, F., Cvirkaite-Krupovic, V., Kreutzberger, M. A. B., Su, Z., de Oliveira, G. A. P., Osinski, T., Sherman, N., DiMaio, F., Wall, J. S., Prangishvili, D., et al. (2019a). An extensively glycosylated archaeal pilus survives extreme conditions. *Nature Microbiology, 4*, 1401–1410.

Wang, K., Sybers, D., Maklad, H. R., Lemmens, L., Lewyllie, C., Zhou, X. X., Schult, F., Bräsen, C., Siebers, B., Valegård, K., et al. (2019b). A TetR-family transcription factor regulates fatty acid metabolism in the archaeal model organism. *Nature Communications, 10*, 1542.

Wang, F., Baquero, D. P., Su, Z., Beltran, L. C., Prangishvili, D., Krupovic, M., & Egelman, E. H. (2020). The structures of two archaeal type IV pili illuminate evolutionary relationships. *Nature Communications, 11*, 3424.

Wang, F., Cvirkaite-Krupovic, V., Krupovic, M., & Egelman, E. H. (2022). Archaeal bundling pili of Pyrobaculum calidifontis reveal similarities between archaeal and bacterial biofilms. *Proceedings of the National Academy of Sciences of the United States of America, 119*, e2207037119.

Wang, P., Tang, C., Liu, Y., Yang, J., & Fan, D. (2023). Biotransformation of high concentrations of Ginsenoside substrate into compound K by β-glycosidase from Sulfolobus solfataricus. *Genes, 14*, 897.

Wettasinghe, A. P., Seifi, M. O., Bravo, M., Adams, A. C., Patel, A., Lou, M., Kahanda, D., Peng, H. C., Stelling, A. L., Fan, L., et al. (2023). Molecular wrench activity of DNA helicases: Keys to modulation of rapid kinetics in DNA repair. *Protein Science, 32*, e4815.

White, M. F. (2003). Archaeal DNA repair: Paradigms and puzzles. *Biochemical Society Transactions, 31*, 690–693.

Wolf, J., Koblitz, J., Albersmeier, A., Kalinowski, J., Siebers, B., Schomburg, D., & Neumann-Schaal, M. (2021). Utilization of phenol as carbon source by the thermoacidophilic archaeon P2 is limited by oxygen supply and the cellular stress response. *Frontiers in Microbiology, 11*, 587032.

Wu, P. J., Gan, Q., Zhang, X. M., Yang, Y. F., Xiao, Y. X., She, Q. X., Ni, J. F., Huang, Q. H., & Shen, Y. L. (2023). The archaeal KEOPS complex possesses a functional Gon7 homolog and has an essential function independent of the cellular t6A modification level. *Mlife, 2*, 11–27.

Wu, P., Zhang, M., Kou, Y., Liang, S., Ni, J., Huang, Q., & Shen, Y. (2025). Identification of novel components of the Ced and ups systems in Saccharolobus islandicus REY15A. *mLife, 4*, 17–28.

Xiang, X., Dong, X., & Huang, L. (2003). Sulfolobus tengchongensis sp. nov., a novel thermoacidophilic archaeon isolated from a hot spring in Tengchong, China. *Extremophiles, 7*, 493–498.

Xiao, Y., Jiang, Z., Zhang, M., Zhang, X., Gan, Q., Yang, Y., Wu, P., Feng, X., Ni, J., Dong, X., et al. (2023). The canonical single-stranded DNA-binding protein is not an essential replication factor but an RNA chaperon in Saccharolobus islandicus. *IScience, 26*, 108389.

Xiong, D., Li, Z., Qi, W., Wang, S., Huang, J., Zhang, N., Zhang, Z., & Huang, L. (2025). Archaeal replicative primase mediates DNA double-strand break repair. *Nucleic Acids Research, 53*.

Yang, Y., Liu, J., Fu, X., Zhou, F., Zhang, S., Zhang, X., Huang, Q., Krupovic, M., She, Q., Ni, J., et al. (2023). A novel RHH family transcription factor aCcr1 and its viral homologs dictate cell cycle progression in archaea. *Nucleic Acids Research, 51*, 1707–1723.

Yang, W., Feng, X., Chen, H., Liman, G. L. S., Santangelo, T. J., Zhang, C., & Zeng, Z. (2025). Cyclization of archaeal membrane lipids impacts membrane protein activity and archaellum formation. *Proceedings of the National Academy of Sciences of the United States of America, 122*, e2423648122.

Ye, X., Vogt, M. S., van der Does, C., Bildl, W., Schulte, U., Essen, L. O., & Albers, S. V. (2020). The phosphatase PP2A interacts with ArnA and ArnB to regulate the Oligomeric state and the stability of the ArnA/B complex. *Frontiers in Microbiology, 11*, 1849.

Yu, X., Goforth, C., Meyer, C., Rachel, R., Wirth, R., Schroder, G. F., & Egelman, E. H. (2012). Filaments from Ignicoccus hospitalis show diversity of packing in proteins containing N-terminal type IV pilin helices. *Journal of Molecular Biology, 422*, 274–281.

Zhang, Z., Zhao, M., Chen, Y., Wang, L., Liu, Q., Dong, Y., Gong, Y., & Huang, L. (2019). Architectural roles of Cren7 in folding crenarchaeal chromatin filament. *Molecular Microbiology, 111*, 556–569.

Zhang, N., Guo, L., & Huang, L. (2020a). The Sac10b homolog from Sulfolobus islandicus is an RNA chaperone. *Nucleic Acids Research, 48*, 9273–9284.

Zhang, Z., Zhan, Z., Wang, B., Chen, Y., Chen, X., Wan, C., Fu, Y., & Huang, L. (2020b). Archaeal chromatin proteins Cren7 and Sul7d compact DNA by bending and bridging. *MBio, 11*.

Zhang, R., Gao, H., Wang, Y., He, B., Lu, J., Zhu, W., Peng, L., & Wang, Y. (2023). Challenges and perspectives of green-like lignocellulose pretreatments selectable for low-cost biofuels and high-value bioproduction. *Bioresource Technology, 369*, 128315.

Zhang, J., Feng, X., Li, M., Liu, Y., Liu, M., Hou, L. J., & Dong, H. P. (2025a). Deep origin of eukaryotes outside Heimdallarchaeia within Asgardarchaeota. *Nature, 642*, 990–998.

Zhang, L., Gao, T., Li, Z., Chen, C., Jiang, D., Yin, Y., Zheng, Y., Cao, P., Gong, Y., & Yang, Z. (2025b). Alkylated DNA repair by a novel HhH-GPD family protein from Crenarchaea. *Nucleic Acids Research, 53*.

Zhang, X., Wu, P., Bai, R., Gan, Q., Yang, Y., Li, H., Ni, J., Huang, Q., & Shen, Y. (2025c). PerR functions as a redox-sensing transcription factor regulating metal homeostasis in the thermoacidophilic archaeon Saccharolobus islandicus REY15A. *Nucleic Acids Research, 53*.

Zhao, P., Bi, X., Wang, X., Feng, X., Shen, Y., Yuan, G., & She, Q. (2024). Rational design of unrestricted pRN1 derivatives and their application in the construction of a dual plasmid vector system for Saccharolobus islandicus. *mLife, 3*, 119–128.

Zheng, T., Huang, Q., Zhang, C., Ni, J., She, Q., & Shen, Y. (2012). Development of a simvastatin selection marker for a hyperthermophilic acidophile, Sulfolobus islandicus. *Applied and Environmental Microbiology, 78*, 568–574.

Zillig, W., Stetter, K.O., Wunderl, S., Schulz, W., Priess, H., Scholz, I. (1980). The Sulfolobus-"Caldariella" group: taxonomy on the basis of the structure of DNA-dependent RNA polymerases. *Archives of Microbiology, 125*, 259–269

Zillig, W., Yeats, S., Holz, I., Bock, A., Gropp, F., Rettenberger, M., & Lutz, S. (1985). Plasmid-related anaerobic autotrophy of the novel archaebacterium Sulfolobus ambivalens. *Nature, 313*, 789–791.

Zolghadr, B., Klingl, A., Koerdt, A., Driessen, A. J., Rachel, R., & Albers, S. V. (2010). Appendage-mediated surface adherence of Sulfolobus solfataricus. *Journal of Bacteriology, 192*, 104–110.

Zolghadr, B., Klingl, A., Rachel, R., Driessen, A. J., & Albers, S. V. (2011). The bindosome is a structural component of the Sulfolobus solfataricus cell envelope. *Extremophiles, 15*, 235–244.

Part III
Production of Valuable Compounds

Chapter 8
Genome Editing Technologies and Synthetic Biology of Hyperthermophilic Archaea

Guanhua Yuan, Yang Zhang, and Qunxin She

Abstract Hyperthermophilic archaea are characterized with robust metabolic capability and outstanding enzyme stability. Since their discovery, great efforts have been devoted into the development of genetic tools for these archaea, and important progresses have been made in the past three decades including the development of conventional and novel genetic manipulation methods as well as CRISPR-based genome editing technologies. These genetic methods have enabled efficient genome engineering of these archaea, including engineering of promoter elements, highly efficient enzymes, as well as modified or new metabolic pathways for different biotechnological applications. Herein we summarize these research progresses in their genetic studies and bioprocessing efficiency and environmental sustainability of these hyperthermophilic and thermoacidophilic archaea, which provide the basis to generate environment-friendly and circular bioeconomies with these unique organisms.

Keywords CRISPR gene editing · Metabolic engineering · *Sulfolobales* · *Thermococcales* · Bioproduction

8.1 Introduction

Archaea constitute the third domain of life according to the pioneer phylogenetic analysis of 16S rRNA sequences by Woese and Fox (1977) and Woese et al. (1990). Currently, archaeal organisms are classified into four broad phyla among which a large majority of them remain to be isolated as pure cultures (Rinke et al., 2021). Archaea are also classified into thermophiles, hyperthermophiles, psychrophiles, acidophiles, alkaliphiles, halophiles, piezophiles, radiophiles, methanogens, and

G. Yuan · Y. Zhang · Q. She (✉)
CRISPR and Archaea Biology Research Center, State Key Laboratory of Microbial Technology, Microbial Technology Institute, Shandong University, Qingdao, China
e-mail: shequnxin@sdu.edu.cn

M. Ito et al. (eds.), *Extremophiles for a Sustainable Future*,
https://doi.org/10.1007/978-981-95-6970-0_8

metallophiles based on their physiological traits, and such classification is very useful for their physiological research and biotechnological application (Aparici-Carratalá et al., 2023). Some archaeal organisms were chosen for developing model systems soon after their isolation since genetic tools are essential for their basic research and application. Numerous problems have been encountered in the process of developing efficient archaeal genetic systems, and this is particularly true for hyperthermophilic archaea that thrive at a temperature of >80 °C in particular. Nevertheless, consistent efforts from international research groups have eventually yielded genetically tractable hyperthermophilic or thermoacidophilic archaea (Peng et al., 2017; Straub et al., 2018).

Hyperthermophilic archaea typically inhabit in the environments associated with volcanic activities or hydrothermal vents where temperatures can exceed 100 °C whereas thermophilic and acidophilic crenarchaea thrive in hot acidic environments such as terrestrial hotsprings. These archaea possess diverse metabolic pathways tailored to these extreme habitats, including versatile pathways for efficient carbon metabolism in high-temperature and acidic environments, thus exhibiting great potential in biocatalysis and bioindustry. However, our understanding of the biology and physiology as well as the biotech potential of these archaea is still rudimentary, and this is mainly attributed to the general lack of efficient genetic tools for these archaea until very recently. Versatile genetic tools are now available for a few model organisms, including organisms of *Sulfolobus* and *Saccharolobus* genera and those of *Pyrococcus* and *Thermococcus* genera, belonging to Crenarchaeota and Euryarchaeota, respectively. Herein, we summarize the current knowledge of the gene editing techniques in these archaeal models and research progress in the synthetic biology of these hyperthermophilic archaea.

8.2 Essential Elements for Developing Genetic Tools

Genetic tools are important for the functional characterization of genes or metabolic pathways in any microbe. These include cloning and expression vectors as well as methods of genetic manipulation for construction of gene knockout mutants. Common for these genetic tools are selection markers and methods of introducing DNA into microbial cells, which are often major bottlenecks in developing genetic tools for a new microbe.

8.2.1 Genetic Markers for Hyperthermophilic Archaea

Genetic markers fall into two distinct types: one is generally applicable in genetic studies in different microorganisms while the other is relying on complementation of a specific mutation by plasmid-borne expression of a corresponding enzyme.

Well-known examples of general selection markers are antibiotic selection widely applied in bacterial genetics. These chemicals selectively target bacterial cell wall synthesis or bacterial informational processing machineries, thereby serving as selection for any given bacterium. However, antibiotics often do not work on archaea since archaeal cell envelopes are distinct from the bacterial ones and the archaeal information-processing machineries resemble those of eukaryotes. Nevertheless, archaea possess a unique cell membrane whose synthesis requires a 3-hydroxy-3-methylglutaryl coenzyme A reductase (HMG-CoA), and some chemicals can inhibit the archaeal lipid synthesis. One of them is simvastatin, a chemical that specifically inhibits the HMG-CoA activity. Although the encoding gene is essential in archaea, overexpression of HMG-CoA in *Thermococcus kodakarensis* facilitated colony formation of transformants on selective plates containing simvastatin, and this led to the development of versatile genetic systems in this euryarchaeal model (Matsumi et al., 2007). Subsequently, genetic systems were developed with the same selection for different archaea, including *Pyrococcus furiosus* (Matsumi et al., 2007; Waege et al., 2010) and *P. yayanosii* (Li et al., 2014) as well as the hyperthermophilic acidophile, *Saccharolobus islandicus* (formerly *Sulfolobus islandicus*) (Zheng et al., 2012). Since the genetic selection is based on the elevation of the HMG-CoA activity, a subtle change in the drug concentration in selective media often greatly reduces the likelihood of obtaining colonies of transformants on the selective plates. Thus, more robust genetic selection is required for conducting robust genetic studies in these organisms.

Specific genetic selections are often based on isolation of auxotrophic mutants and testing genetic complementation of the auxotrophy using the mutants as the host. Four different types of auxotrophic mutants were reported for developing genetic selections in hyperthermophilic archaea, including uracil auxotroph, agmatine auxotroph, tryptophan auxotroph, and lactose auxotroph.

Uracil auxotrophy/*pyrEF* complementation is a common selection system in microbial genetics. This selection requires specific strains that are defective in pyrimidine synthesis such as mutants of *pyrE* or *pyrF* encoding orotate phosphoribosyltransferase and orotidine 5-phosphate decarboxylase, respectively. Plasmid transformants can be selected on complementing uracil auxotroph by plasmid-borne expression of *pyrEF*. In addition, the enzymes encoded by *pyrEF* convert 5-fluoroorotic acid (5-FOA) to 5-fluorouridine monophosphate (5-fluoro-UMP), a toxic substance that kills metabolically active cells. Thus, only *pyrEF* mutants can form colonies on 5-FOA plates. This feature facilitates plasmid-curing in the constructed mutants, thereby simplifying the procedure of mutant purification. Currently, the uracil auxotrophy/*pyrEF* complementation system has successfully been employed for developing genetic tools in several archaea, including *Sa. islandicus* (Jonuscheit et al., 2003; Deng et al., 2009; Wagner et al., 2009), *Sulfolobus acidocaldarius* (Suzuki & Kurosawa, 2016), *Saccharolobus solfataricus* (formerly *Sulfolobus solfataricus*) (Jonuscheit et al., 2003), *T. kodakarensis* (Sato et al., 2003, 2005), *T. barophilus* (Thiel et al., 2014) and *P. furiosus* (Lucas et al., 2002).

The remaining three genetic complementation selections do not enable any positive selection. In the agmatine auxotroph/*argD* gene system, *argD* encodes arginine decarboxylase and its deficiency can be complemented by addition of agmatine in the medium. This selection marker was first developed for hyperthermophilic euryarchaeon *T. kodakarensis* (Sato et al., 2003), and also applied in the genome editing tools of hyperthermophilic and acidophilic archaea (Zhang et al., 2013, 2018a; Zhao et al., 2024). The tryptophan/lactose auxotroph can be complemented by addition of the amino acid and the sugar in the selective media, respectively, and these selections were found to operate in *T. kodakarensis* (Sato et al., 2003) and in *Sa. solfataricus* (Berkner et al., 2007), respectively. However, both systems rely on chemically defined media to infer the genetic selection, and this is not applicable to many hyperthermophilic archaea since they grow very poorly on chemically defined media. In contrast, the agmatine auxotroph/*argD* selection operates well in rich media. Thus, *argD* is often the preferable second genetic marker after the *pyrEF* selection.

It is worth pointing out that many archaeal genomes are rich in insertion sequence (IS) elements, and a large majority of spontaneous mutations often carry IS-inserted mutations in an IS hop-in process. Such IS-inserted mutants are not proper hosts for developing genetic tools since revertants, i.e., the wild-type strain is generated in the IS "hop-off" process and this can occur at a high frequency. As a result, after plasmid transformation, colonies formed on uracil-free plates are predominantly wild-type cells whereas plasmid-containing colonies are hardly identifiable. Thus, it is highly recommended to employ gene deletion mutants for developing genetic tools for microorganisms.

8.2.2 Plasmid Transformation

The other essential element in microbial genetics is the methods of introducing DNA into a cell, a process that is often called transformation. Many archaea are not easy to be transformed, i.e., it is very difficult to introduce DNA into archaeal cells. For *Sulfolobales*, the thermoacidophilic archaea, electroporation was found to be efficient in getting DNA into cells although the exact experimental conditions can vary from one species to another. For organisms belonging to *Thermococcales*, many species are naturally competent in taking up DNAs from environments, such as strains of *Pyrococcus* and *Thermococcus* genera. Thus, all known hyperthermophilic archaea strains are naturally competent and they can easily take up DNA molecules from the environments.

Archaeal genetics often rely on shuttle vectors constructed with *Escherichia coli* vectors, which replicate both in *E. coli* and in an archaeon where the archaeal replicon works. The archaeal replicons on the shuttle vectors are derived from their genetic elements, carrying (putative) origins of replication from archaeal plasmids or viruses. DNA cloning is often conducted in *E. coli*, and the constructed plasmids are then introduced into archaeal cells for genetic characterization. Below we focus on genome editing methods whereas other genetic tools including reporter gene assay for assessing promoter activity, expression plasmids for recombinant protein expression, and purification are summarized in other Chapters.

8.3 Efficient Genome Editing Tools

8.3.1 CRISPR Genome Editing

One of the most important discoveries in contemporary microbiology is the discovery of the "clustered regularly interspaced palindromic repeats (CRISPR) and CRISPR-associated (Cas) systems". CRISPR-Cas system provides adaptive immunity against virus infection in archaea and bacteria: upon the first infection, the immune system selectively takes up short segments from viral genomes and inserts them into their CRISPR arrays as new spacers; upon subsequent encounters, CRISPR RNAs (crRNAs) generated from the expression of CRISPR arrays and subsequent processing bind to Cas proteins to form nucleoprotein nuclease effectors. The crRNAs in the effectors guide the Cas nuclease to specifically recognize corresponding DNA targets in viral genomes and target them for degradation (Mohanraju et al., 2022; Wang et al., 2022). Two main classes of CRISPR-Cas systems are known: those of Class 1 employ different Cas proteins to build up their nucleoprotein complexes with crRNA, whereas Class 2 CRISPR-Cas use a single Cas protein in the nucleoprotein complex assembly (Makarova et al., 2020). CRISPR-Cas systems are also classified into six different types, five of them have their own signature proteins, i.e., Cas3, Cas9, and Cas10 for Type I, II, and III whereas Cas12 and Cas13 for Type V and VI, individually. Type IV systems may represent simplified or inactivated versions of Type I and III CRISPR systems, therefore lacking any conserved signature Cas protein, and they are prevalent in hyperthermophilic archaea (Yang et al., 2024).

The discovery of the CRISPR gene editing has revolutionized the genetic manipulation in all three domains of life. After the first report on CRISPR-based genome editing with Cas9 (Jinek et al., 2012), genome-editing has been developed with different CRISPR systems, such as those based on diverse Cas9 and Cas12 proteins. Unfortunately, Cas9 and Cas12 proteins are mainly encoded in mesophilic organisms and these mesophilic enzymes do not provide efficient genome-editing activity in hyperthermophilic archaea. On the other hand, pioneering research of genome editing with endogenous CRISPR systems revealed both type I and type III CRISPR systems mediate efficient genome editing in *Sa. islandicus* REY15A, a crenarchaeon (Li et al., 2016). The experiment starts with the construction of a genome-editing plasmid carrying an artificial mini-CRISPR array and donor DNA, and crRNAs generated from plasmid-borne expression form nucleoprotein effectors that specifically recognize the target sequences for destruction, and only cells carrying

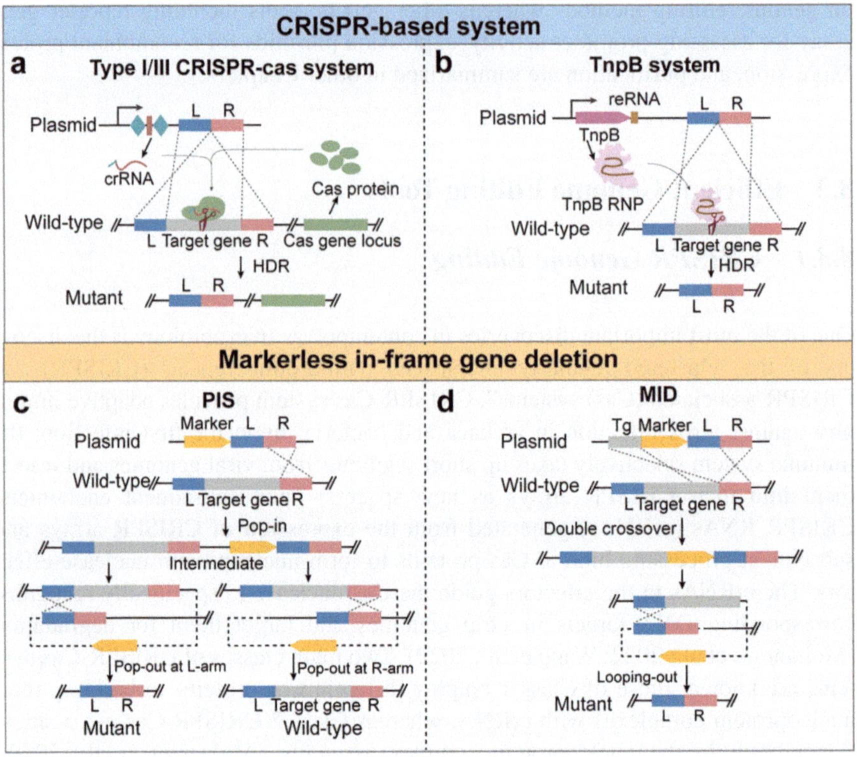

Fig. 8.1 Useful genetic manipulation methods developed for hyperthermophilic archaea. (**a**) *Type I/III CRISPR-Cas system* forms a ribonucleoprotein complex by binding the crRNA expressed from a plasmid with the endogenous CRISPR-Cas proteins of the host, thereby targeting and cleaving genomic DNA. The gene editing is achieved through HDR using the repair template carried by the plasmid. (**b**) Plasmid-borne expression of the TnpB system produces TnpB protein and reRNA, yielding TnpB-reRNA complexes that mediate targeted cleavage of genomic DNA, and completes gene editing by using the repair template carried by the plasmid through HDR. *Markerless in-frame gene deletion* involves selection for the integration of a marker into the target site, followed by counterselection to remove the marker, resulting in a markerless deletion. PIS Plasmid Integration and Segregation (**c**), MID Marker Insertion and unmarked gene Deletion (**d**). L left homologous arms of the target gene, R right homologous arms of the target gene, Tg homologous arms of the target gene, HDR homology-directed repair

the mutated gene allele can grow and form colonies on selective plates (Fig. 8.1a). This system allows for precise gene deletions, insertions, and point mutations by selectively targeting wild-type cells for degradation while retaining mutant cells. Type I-A, I-D, III-B, and III-D CRISPR-Cas systems are common in these organisms (Yu et al., 2021), all of which have been harnessed for genome editing (Li et al., 2016). *S. acidocaldarius* carries CRISPR-Cas type I-D and III-D systems. Recent research indicates that the type I-D CRISPR-Cas system can be utilized for genetic engineering in the *S. acidocaldarius* (Bost et al., 2023). In addition, gene silencing based on CRISPR type III is a widely applicable tool that can be used to study the in vivo functions of essential genes in Sulfolobales (Zink et al., 2019, 2021). More recently, TnpB proteins encoded in transposons of the IS605 family have been harnessed for genome editing in hyperthermophiles as well as in mesophiles (Feng et al., 2024; Tang et al., 2025; Luo et al., 2025) (Fig. 8.1b). Thus, these TnpB proteins enable the genome editing based on heterologously expressed ribonucleoprotein systems in hyperthermophiles.

8.3.2 Conventional Targeted Chromosome Mutagenesis

A series of genetic manipulation methods were developed for hyperthermophilic archaea before the CRISPR era. The simplest one is allelic replacement, in which a target gene is replaced with a marker gene in the double-crossover events of HR (Homologous recombination). Although the method operates well in hyperthermophilic archaea (Albers, 2008; Worthington et al., 2003; Li et al., 2011), its application has been limited by the availability of useful selection markers in these organisms, which is 2–3 for each archaeon. In this regard, markerless gene deletion methods could be more useful. Two such methods are known, one is "Plasmid Integration and Segregation" (PIS), which is also referred to as the "pop in/pop out" method, allowing the selection marker to be removed at the last step (Fig. 8.1c) (Deng et al., 2009; Zhang et al., 2013). While this method works well in *Thermococcales* (Gehring et al., 2017), doing the same in Sulfolobales can hardly recover any mutants. However, when linear DNA was employed to facilitate HR by using "Marker Replacement and Looping out," designed mutants are readily obtained (Deng et al., 2009). Strikingly, further refining the design to give the "Marker Insertion and unmarked gene Deletion" (MID) not only ensures the insertion of the *pyrEF* marker regardless of whether the target gene is essential or not, but also facilitates the deletion of any unessential gene along with the marker (Fig. 8.1d) (Zhang et al., 2010). This method operates so well in some archaea that it is still a preferred method of genetic manipulation in the CRISPR era, such as *S. acidocaldarius* and *T. kodakarensis*.

8.4 Synthetic Biology of Hyperthermophilic Archaea

Organisms of *Thermococcus*, *Pyrococcus*, and *Sulfolobus* genera have evolved thermophilic enzymes that are very stable in ambient temperatures as well as in the extremely high temperatures. They have also evolved adaptive metabolic pathways that enable them to survive in harsh conditions. In addition to their natural tolerance, these archaea also exhibit significant metabolic plasticity, capable of utilizing a wide range of substrates including polysaccharides and chitin. In the modern era of synthetic biology, these features have provided the basis for synthetic biology

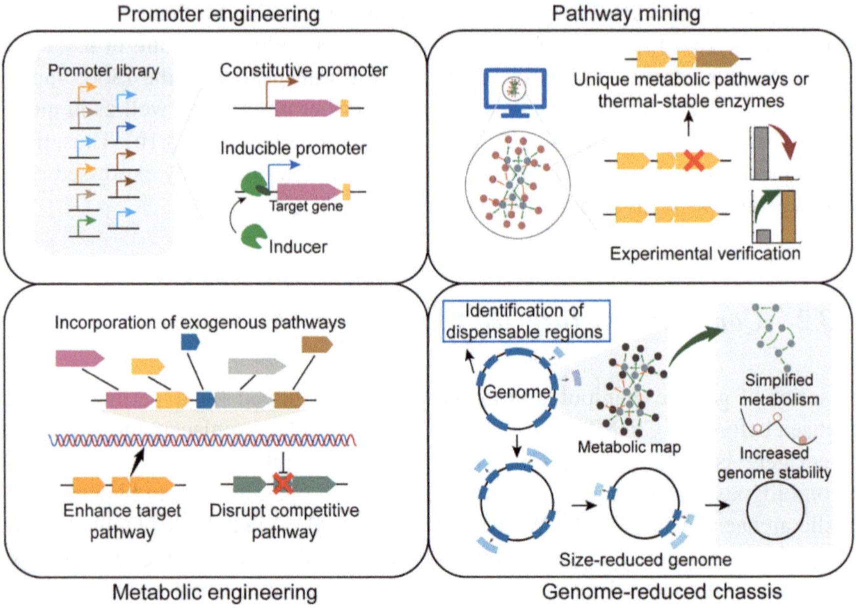

Fig. 8.2 Strategies for synthetic biology of hyperthermophilic archaea. (1) The key role of promoter engineering in regulating gene expression, including the construction of promoter libraries, the screening and optimization of inducible and constitutive promoters; (2) Metabolic engineering strategies for enhancing the synthesis of target products, achieved by disrupting competitive pathways, introducing exogenous enzyme systems and/or enhancing synthetic pathways to achieve efficient production; (3) The mining and characterization of unique metabolic pathways or thermal-stable enzymes adapted to extreme environments; (4) The construction of genome-reduced chassis, aiming to enhance cell stability and simplify complex metabolic networks by deleting redundant dispensable genes

research to explore their biotechnological applications favorable at high temperatures by promoter engineering, metabolic engineering, as well as redirecting metabolic fluxes in known pathways or incorporating new pathways in these hyperthermophilic archaea (Fig. 8.2).

8.4.1 Promoter Engineering

The aim of promoter engineering is to enable precise control over gene expression for metabolic engineering and bioproduction, and this is very important for optimizing the production of biofuels, chemicals, and other valuable products in any microbial hosts including extremophiles. Two different types of promoters are often employed for facilitating bioproduction: inducible promoters and constitutive ones. Currently, there are a few reports about pioneer research of inducible expression in these archaeal hyperthermophiles.

Inducible promoters confer the expression of target enzymes/proteins by responding to an inducer that is either an environmental factor or a chemical substance. Sugars, including arabinose, maltose, dextrin, and maltodextrin have been employed as inducers to trigger gene expression in archaea. The D-arabinose induction works in *Sa. solfataricus* and *Sa. islandicus* (Stan et al., 2006; Peng et al., 2009, 2012) whereas maltose/dextrin-inducible promoters have been identified in *S. acidocaldarius* (Silvia et al., 2010). Systematically searching for inducible promoters in *T. kodakarensis* has yielded 22 inducible promoters, including 15 maltodextrin-inducible promoters and 7 high-pressure-inducible ones, and the most effective promoter can elevate the expression of target proteins by up to eightfolds (Li & Xu, 2025). One of such inducible promoters has been employed to develop a markerless genome editing method for a piezophile, *P. yayanosii*, where high-pressure inducible expression of a toxin-antitoxin (T-A) system functions as a selection marker (Song et al., 2019).

Promoter engineering has been elegantly employed in developing a two-step process for biosynthesis of lactate in *P. furiosus*. While this archaeon does not encode lactate dehydrogenase (Ldh), *Caldicellulosiruptor bescii*, a moderately thermophile, does. But this Ldh is inactive at the physiological temperature of *P. furiosus*. To enable the production of Ldh in this hyperthermophile, the *ldh* gene was fused to the promoter of the cold-induced protein A gene encoding a membrane-bound glycoprotein, allowing the high-level cold-inducible expression of Ldh at 72 °C. The designed cell factory then synthesizes lactate in two steps: cell mass is generated at the first step of cultivation at 98 °C, and after shifting to 72 °C, the archaeal cell factory synthesizes a large amount of Ldh, and the yield of lactate increased from an undetectable level at the cultivation stage to 3 mM at the production stage (Basen et al., 2012). This has provided an excellent example for promoter engineering-based design of cell factories of hyperthermophilic archaea and their application in bioproduction.

To enable constitutively high-level expression in these hyperthermophilic archaea, promoters of S-layer protein genes are chosen since they are among the strongest promoters for these organisms. Two notable examples of their applications are reported: (a) the promoter greatly enhanced expression of an alcohol dehydrogenase gene, enabling ethanol production at the temperature of 95 °C in *P. furiosus* (Lipscomb et al., 2023), and (b) facilitating the expression of the membrane-bound hydrogenase has yielded an increase of hydrogen production by 28% in *T. kodakarensis* (Kanai et al., 2015). These represent another practice in promoter engineering.

Ribosome-binding sites (RBS) represent a distinct element in the regulation of gene expression since they constitute the sites for ribosome assembly in the 5′-leader sequence of messenger RNAs. A systematic search for efficient RBS has identified 31 effective RBS sites in *T. kadakarensis* exhibiting a fivefold difference in translation (Li & Xu, 2025). These RBS elements, along with the inducible promoter elements, provide rich resource for genetic engineering for synthetic biology studies of hyperthermophilic archaea.

8.4.2 Metabolic Engineering

Attempts have also been made to achieve efficient biotransformation of target products by introducing exogenous metabolic pathways by metabolic engineering in hyperthermophilic archaea. In an early attempt, insertion of a single gene encoding an aldehyde ferredoxin oxidoreductase (AOR) or an alcohol dehydrogenase has enabled alcohol production by *P. furiosus* from corresponding organic acids (Basen et al., 2014). Alcohol production is then optimized in this archaeon by deleting genes coding for endogenous AOR or acetyl-lactate synthase and replacing them with heterologous genes for optimized bioproduction: e.g., incorporation of the genes encoding alcohol dehydrogenase E and alcohol dehydrogenase A has facilitated ethanol production whereas utilizing those encoding a butyryl-CoA and butyraldehyde from *Thermoanaerobacter* has butanol production (Keller et al., 2015, 2017; Nguyen et al., 2016). Additionally, a hybrid pathway with both native and heterologous enzymes has been created for 1-propanol production in *P. furiosus* (O'Quinn et al., 2025).

Metabolic engineering has provided a useful means to expand the range of substrate utilization for bioproduction in hyperthermophilic archaea. Organisms of *Sulfolobus* genus lack many sugar-metabolizing pathways in contrast to those of *Sacharolobus*. However, introduction of relevant enzymes into *Sulfolobus* cells has enabled them to utilize xylan (Lee et al., 2022). Furthermore, incorporating genes of the sulfur oxidation pathway into *S. acidocaldarius* has not only enhanced its ability to utilize inorganic sulfur, but also enabled the synthesis of acetone under the mixed nutrition mode of "organic carbon + inorganic sulfur" (Willard & Kelly, 2025). In *P. furiosus*, genomic insertion of the FHL (formate hydrogen lyase) operon has led to the production of dihydrogen from formate (Lipscomb, 2014) and co-expression of a homologous alcohol dehydrogenase and a heterologous carbon monoxide dehydrogenase complex has rendered it possible to produce ethanol directly from carbon monoxide oxidation (Lipscomb et al., 2023).

Inactivating one or more enzymes by gene deletion represents another promising approach of metabolic engineering to optimize biochemical pathways for industrial applications in hyperthermophilic archaea. It has been shown that deletion of specific enzymes in the modified Embden-Meyerhof pathway altered the fermentation product profile without compromising viability (Straub et al., 2020), whereas deletion of acetyl-CoA synthetases I and II has led to an increase in the production of 3-hydroxypropionic acid (3-HP) in *P. furiosus* (Thorgersen et al., 2014). Other attempts such as disruption of the reductant recycling (Santangelo et al., 2011) and removing hydrogen-consuming pathways have been made to generate genome-streamlined strains (Li & Xu, 2025; Kanai et al., 2015), and the resulting chassis cells exhibit upregulation of hydrogen-producing enzymes and elevated levels of hydrogen production (Aslam et al., 2017).

Taken together, metabolic engineering in hyperthermophilic archaea has shown great potential in the field of industrial biotechnology.

8.4.3 Pathway Mining

Pathway mining is referred to as improving cell factories by exploration and modification of specific metabolic pathways and enzymes. One of such targeted pathways is the non-phosphorylative Entner-Doudoroff pathway first identified for Sulfolobales organisms. These archaea thrive in extreme heat and acidity, utilizing diverse sugars (e.g., arabinose, xylose) via a non-phosphorylative Entner-Doudoroff pathway to produce pyruvate and glycolate, which could be explored for lignocellulosic biomass conversion. Their thermostable trehalase (active at 85 °C/pH 4.5) and unique GT4-family trehalose-6-phosphate synthase/trehalose-6-phosphate phosphatase pathway—producing trehalose for high-salt adaptation—enhance industrial utility (Moon et al., 2016; Stracke et al., 2020).

Interestingly, heterologous expression of the *S. acidocaldarius* squalene/phytoene synthase (*Saci_1734*) in T. *kodakarensis* enabled phytoene synthesis and further optimization has attained >3-fold yield (Fuke et al., 2018). Examination of the genome annotation of *T. kodakarensis* revealed that, while it does encode farnesyl pyrophosphate /geranylgeranyl pyrophosphate prenyltransferase for synthesizing the squalene/phytoene precursors, the organism lacks key condensation enzymes (e.g., squalene synthase) (Leavell et al., 2016; Fukui et al., 2005), further emphasizing the necessity of metabolic pathway exploration of hyperthermophilic archaea.

8.4.4 Genome-Reduced Chassis Cells

In modern synthetic biology and industrial biotechnology, constructing minimal genomes has become a key strategy, aiming to create efficient and streamlined chassis cells by removing redundant genes and nonessential genomic regions. The central point of this approach is to reduce the metabolic burden on cells by simplifying genomic complexity, thereby enhancing their production efficiency and genome stability, making them more suitable for industrial applications. In *E. coli*, deleting 14.3% of nonessential genes has yielded the MDS42 strain that increased L-threonine production by ca. twofold (Lee et al., 2009). Although archaeal hyperthermophiles have a streamlined genome, genetic analysis revealed that, among the 2658 protein-coding genes in *Sa. islandicus* M.16.4, only 441 are classified as essential ones by transposition analysis (Zhang et al., 2018b), and most of these genes are conserved in organisms of the Sulfolobales family. Considering other essential genes, such as those of redundant functions or conditionally essential, cannot be identified by the transposition analysis, the list of essential genes should be (much) longer than the reported one. Nevertheless, these archaea conceivably carry a large number of nonessential or redundant genes, such as mobile genetic elements and gene clusters associated with stress responses and their regulation. These genetic units are theoretically nonessential and therefore can be removed without any negative impact on their growth. In contrast, we reason construction of genome-reduced chassis of Sulfolobales would facilitate cell growth by reducing energy consumption that otherwise occurs in the wild-type cell. It is conceivable that construction of genome-reduced chassis cells of hyperthermophilic archaea will facilitate their biotechnological applications.

8.5 Perspectives

Pioneer researches have been conducted in different archaea, and important research progress in different applications in hydrogen production, organic acids and alcohols, and in the bioindustry are listed in Table 8.1. Investigation of their environment adaptation mechanisms, and the construction of genome-reduced chassis further unlock their potential, allowing for targeted modifications to increase the yield of target products or redirect carbon flux toward high-value chemicals such as organic acids and alcohols, and expand the range of substrate utilization.

Table 8.1 Industrial applications of metabolically engineered hyperthermophilic archaea

Product	Organism	Carbon source	Description	References
Hydrogen	*Thermococcus kodakarensis*	chitin	Engineering chitin-dependent hydrogen generation strains through genome engineering modification and adaptive laboratory evolution	Aslam et al. (2017)
	T. kodakarensis	pyruvate	Promoter replacement and disruption of the pathway involved in hydrogen consumption	Kanai et al. (2015)
	T. kodakarensis	pyruvate	Replacement of promoters in genes associated with hydrogen production pathways	Li and Xu (2025)
	T. kodakarensis	pyruvate	Deletion of alternative pathways for reductant recycling	Santangelo et al. (2011)
	Pyrococcus furiosus	formate	The incorporation of the *Thermococcus. onnurineus* FHL Operon into the *P. furiosus* Chromosome	Lipscomb (2014)
Acetoin	*P. furiosus*	Maltose	The incorporation of the alcohol dehydrogenase gene (adhA) from *Thermoanaerobacter* sp. X514 and the overexpression of ALS through promoter replacement	Nguyen et al. (2016)
Lactate	*P.furiosus*	Maltose	The incorporation of the lactate dehydrogenase from the *Caldicellulosiruptor bescii*	Basen et al. 2012)
3-HP	*P. furiosus*	Maltose	The incorporation of genes from the 3-hydroxypropionate/4-hydroxybutyrate (4-HB) pathway of Metallosphaera sedula	Keller et al. (2013)
	P. furiosus	Maltose	Deletion of acetyl-CoA synthetases I and II	Thorgersen et al. (2014)

(continued)

Table 8.1 (continued)

Product	Organism	Carbon source	Description	References
Ethanol	*P. furiosus*	Maltose	The incorporation of the ethanol dehydrogenase (AdhA) gene from *Thermoanaerobacter* strain X514 into the genome of *Pyrococcus furiosus*	Basen et al. (2014)
	P. furiosus	Maltose	Deletion of AOR and incorporation of adhE and adhA from *Thermoanaerobacter*	Keller et al. (2017)
	P. furiosus	Maltose	The incorporation of Bad and Bdh from *Thermoanaerobacter* sp. X514	Keller et al. (2015)
	P. furiosus	Maltose	Deletion of ALS	Nguyen et al. (2016)
	P. furiosus	CO, maltose	Overexpress adhF by promoter replacement and integration of CO dehydrogenase complex from *Thermococcus onnurineus*	Lipscomb et al. (2023)
Butanol	*P. furiosus*	Maltose	The incorporation of Thl, Hbd, Crt from *Thermoanaerobacter tengcongensis*, Ter from *Spirochaeta thermophila* and Bad, Bdh from *Thermoanaerobacter* sp. X514	Keller et al. (2015)
1-Propanol	*P. furiosus*	Cellobiose	Construction of an engineered hybrid metabolic pathway, consisting of two natural enzymes and seven heterologous enzymes	O'Quinn et al. (2025)
Phytoene	*T. kodakarensis*	Pyruvate	The incorporation of multiple copies of phytoene synthase from *Sulfolobus solfataricus*	Fuke et al. (2018)

However, synthetic biology of these hyperthermophilic archaea has also encountered the bottleneck in the identification of the large number of unknown nonessential genes in all microbial genomes, including *E. coli* and *Bacillus subtilis*, the well-characterized bacterial workhorses. Currently, all known gene knockout strategies work only in the trail-and-error fashion. Thus, novel genome trimming methods are required to cope with the increasing demand of systematically classifying essential versus nonessential genes in microbial chassis cells. In addition, integration of systematic genome trimming and evaluation of production efficiency of cell factories will pave a new avenue for facilitating synthetic biology of different microorganisms including hyperthermophilic archaea.

References

Albers, S. V., & Driessen, A. J. (2008). Conditions for gene disruption by homologous recombination of exogenous DNA into the Sulfolobus solfataricus genome. *Archaea, 2*, 145–149. https://doi.org/10.1155/2008/948014

Aparici-Carratalá, D., Esclapez, J., Bautista, V., Bonete, M.-J., & Camacho, M. (2023). Archaea: Current and potential biotechnological applications. *Research in Microbiology, 174*. https://doi.org/10.1016/j.resmic.2023.104080

Aslam, M., et al. (2017). Engineering of the Hyperthermophilic Archaeon Thermococcus kodakarensis for Chitin-dependent hydrogen production. *Applied and Environmental Microbiology, 83*. https://doi.org/10.1128/aem.00280-17

Basen, M., Sun, J., & Adams, M. (2012). Engineering a hyperthermophilic archaeon for temperature-dependent product formation. *MBio, 3*, e00053–e00012. https://doi.org/10.1128/mBio.00053-12

Basen, M., et al. (2014). Single gene insertion drives bioalcohol production by a thermophilic archaeon. *Proceedings of the National Academy of Sciences of the United States of America, 111*, 17618–17623. https://doi.org/10.1073/pnas.1413789111

Berkner, S., Grogan, D. W., Albers, S. V., & Lipps, G. (2007). Small multicopy, non-integrative shuttle vectors based on the plasmid pRN1 for Sulfolobus acidocaldarius and Sulfolobus solfataricus, model organisms of the (cren-)archaea. *Nucleic Acids Research, 35*, e88. https://doi.org/10.1093/nar/gkm449

Bost, J., et al. (2023). Application of the endogenous CRISPR-Cas type I-D system for genetic engineering in the thermoacidophilic archaeon Sulfolobus acidocaldarius. *Frontiers in Microbiology, 14*, 1254891. https://doi.org/10.3389/fmicb.2023.1254891

Deng, L., Zhu, H., Chen, Z., Liang, Y. X., & She, Q. (2009). Unmarked gene deletion and host–vector system for the hyperthermophilic crenarchaeon Sulfolobus islandicus. *Extremophiles, 13*, 735–746. https://doi.org/10.1007/s00792-009-0254-2

Feng, X., et al. (2024). Flexible TAM requirement of TnpB enables efficient single-nucleotide editing with expanded targeting scope. *Nature Communications, 15*. https://doi.org/10.1038/s41467-024-47697-4

Fuke, T., Sato, T., Jha, S., Tansengco, M. L., & Atomi, H. (2018). Phytoene production utilizing the isoprenoid biosynthesis capacity of Thermococcus kodakarensis. *Extremophiles, 22*, 301–313. https://doi.org/10.1007/s00792-018-0998-7

Fukui, T., et al. (2005). Complete genome sequence of the hyperthermophilic archaeon Thermococcus kodakaraensis KOD1 and comparison with Pyrococcus genomes. *Genome Research, 15*, 352–363. https://doi.org/10.1101/gr.3003105

Gehring, A. M., Sanders, T. J., & Santangelo, T. J. (2017). Markerless gene editing in the Hyperthermophilic Archaeon Thermococcus kodakarensis. *Bio-Protocol, 7*. https://doi.org/10.21769/BioProtoc.2604

Jinek, M., et al. (2012). A programmable dual-RNA-guided DNA endonuclease in adaptive bacterial immunity. *Science (New York, N.Y.), 337*, 816–821. https://doi.org/10.1126/science.1225829

Jonuscheit, M., Martusewitsch, E., Stedman, K. M., & Schleper, C. (2003). A reporter gene system for the hyperthermophilic archaeon Sulfolobus solfataricus based on a selectable and integrative shuttle vector. *Molecular Microbiology, 48*, 1241–1252. https://doi.org/10.1046/j.1365-2958.2003.03509.x

Kanai, T., et al. (2015). Overproduction of the membrane-bound [NiFe]-hydrogenase in Thermococcus kodakarensis and its effect on hydrogen production. *Frontiers in Microbiology, 6*, 847. https://doi.org/10.3389/fmicb.2015.00847

Keller, M., et al. (2013). Exploiting microbial hyperthermophilicity to produce an industrial chemical, using hydrogen and carbon dioxide. *Proceedings of the National Academy of Sciences of the United States of America, 110*, 5840–5845. https://doi.org/10.1073/pnas.1222607110

Keller, M., et al. (2015). A hybrid synthetic pathway for butanol production by a hyperthermophilic microbe. *Metabolic Engineering, 27*, 101–106. https://doi.org/10.1016/j.ymben.2014.11.004

Keller, M., et al. (2017). Ethanol production by the hyperthermophilic archaeon Pyrococcus furiosus by expression of bacterial bifunctional alcohol dehydrogenases. *Microbial Biotechnology, 10*, 1535–1545. https://doi.org/10.1111/1751-7915.12486

Leavell, M. D., McPhee, D. J., & Paddon, C. J. (2016). Developing fermentative terpenoid production for commercial usage. *Current Opinion in Biotechnology, 37*, 114–119. https://doi.org/10.1016/j.copbio.2015.10.007

Lee, J., et al. (2009). Metabolic engineering of a reduced-genome strain of Escherichia coli for L-threonine production. *Microbial Cell Factories, 8*, 2. https://doi.org/10.1186/1475-2859-8-2

Lee, A., Jin, H., & Cha, J. (2022). Engineering of Sulfolobus acidocaldarius for Hemicellulosic biomass utilization. *Journal of Microbiology and Biotechnology, 32*, 663–671. https://doi.org/10.4014/jmb.2202.02016

Li, C., & Xu, J. (2025). Construction of promoter-ribosome binding site library for manipulating gene expression in the Hyperthermophilic Archaeon Thermococcus kodakarensis. *ACS Synthetic Biology, 14*, 2181–2197. https://doi.org/10.1021/acssynbio.5c00086

Li, X., et al. (2011). Deletion of the topoisomerase III gene in the hyperthermophilic archaeon Sulfolobus islandicus results in slow growth and defects in cell cycle control. *Journal of Genetics and Genomics, 38*, 253–259. https://doi.org/10.1016/j.jgg.2011.05.001

Li, X., et al. (2014). Genetic tools for the piezophilic hyperthermophilic archaeon Pyrococcus yayanosii. *Extremophiles, 19*, 59–67. https://doi.org/10.1007/s00792-014-0705-2

Li, Y., et al. (2016). Harnessing Type I and Type III CRISPR-Cas systems for genome editing. *Nucleic Acids Research, 44*, e34. https://doi.org/10.1093/nar/gkv1044

Lipscomb, G., et al. (2014). Engineering hydrogen gas production from formate in a hyperthermophile by heterologous production of an 18-subunit membrane-bound complex. *The Journal of Biological Chemistry, 289*, 2873–2879. https://doi.org/10.1074/jbc.M113.530725

Lipscomb, G., et al. (2023). Manipulating fermentation pathways in the Hyperthermophilic Archaeon Pyrococcus furiosus for ethanol production up to 95°C driven by carbon monoxide oxidation. *Applied and Environmental Microbiology, 89*, e0001223. https://doi.org/10.1128/aem.00012-23

Lucas, S., et al. (2002). Construction of a shuttle vector for, and spheroplast transformation of, the hyperthermophilic archaeon Pyrococcus abyssi. *Applied and Environmental Microbiology, 68*, 5528–5536. https://doi.org/10.1128/aem.68.11.5528-5536.2002

Luo, J., et al. (2025). STAGE: A compact and versatile TnpB-based genome editing toolkit for Streptomyces. *Proceedings. National Academy of Sciences. United States of America, 122*, e2509146122. https://doi.org/10.1073/pnas

Makarova, K. S., et al. (2020). Evolutionary classification of CRISPR-Cas systems: A burst of class 2 and derived variants. *Nature Reviews. Microbiology, 18*, 67–83. https://doi.org/10.1038/s41579-019-0299-x

Matsumi, R., Manabe, K., Fukui, T., Atomi, H., & Imanaka, T. (2007). Disruption of a sugar transporter gene cluster in a Hyperthermophilic Archaeon using a host-marker system based on antibiotic resistance. *Journal of Bacteriology, 189*, 2683–2691. https://doi.org/10.1128/jb.01692-06

Mohanraju, P., et al. (2022). Alternative functions of CRISPR-Cas systems in the evolutionary arms race. *Nature Reviews. Microbiology*. https://doi.org/10.1038/s41579-021-00663-z

Moon, J. H., Lee, W., Park, J., Choi, K. H., & Cha, J. (2016). Characterization of a trehalose-degrading enzyme from the hyperthermophilic archaeon Sulfolobus acidocaldarius. *Journal of Bioscience and Bioengineering, 122*, 47–51. https://doi.org/10.1016/j.jbiosc.2015.12.011

Nguyen, D., et al. (2016). Temperature-dependent acetoin production by Pyrococcus furiosus is catalyzed by a biosynthetic acetolactate synthase and its deletion improves ethanol production. *Metabolic Engineering, 34*, 71–79. https://doi.org/10.1016/j.ymben.2015.12.006

O'Quinn, H., et al. (2025). Engineering the hyperthermophilic archaeon Pyrococcus furiosus for 1-propanol production. *Applied and Environmental Microbiology, 91*, e0047125. https://doi.org/10.1128/aem.00471-25

Peng, N., Qiu, X., Chen, Z., Liang, Y. X., & She, Q. (2009). An upstream activation element exerting differential transcriptional activation on an archaeal promoter. *Molecular Microbiology*. https://doi.org/10.1111/j.1365-2958.2009.06908.x

Peng, N., et al. (2012). A synthetic arabinose-inducible promoter confers high levels of recombinant protein expression in hyperthermophilic archaeon Sulfolobus islandicus. *Applied and Environmental Microbiology, 78*, 5630–5637. https://doi.org/10.1128/AEM.00855-12

Peng, N., Han, W., Li, Y., Liang, Y., & She, Q. (2017). Genetic technologies for extremely thermophilic microorganisms of Sulfolobus, the only genetically tractable genus of crenarchaea. *Science China. Life Sciences, 60*, 370–385. https://doi.org/10.1007/s11427-016-0355-8

Rinke, C., et al. (2021). A standardized archaeal taxonomy for the Genome Taxonomy Database. *Nature Microbiology, 6*, 946–959. https://doi.org/10.1038/s41564-021-00918-8

Santangelo, T., Cuboňová, L., & Reeve, J. (2011). Deletion of alternative pathways for reductant recycling in Thermococcus kodakarensis increases hydrogen production. *Molecular Microbiology, 81*, 897–911. https://doi.org/10.1111/j.1365-2958.2011.07734.x

Sato, T., Fukui, T., Atomi, H., & Imanaka, T. (2003). Targeted gene disruption by homologous recombination in the hyperthermophilic archaeon Thermococcus kodakaraensis KOD1. *Journal of Bacteriology, 185*, 210–220. https://doi.org/10.1128/jb.185.1.210-220.2003

Sato, T., Fukui, T., Atomi, H., & Imanaka, T. (2005). Improved and versatile transformation system allowing multiple genetic manipulations of the hyperthermophilic archaeon Thermococcus kodakaraensis. *Applied and Environmental Microbiology, 71*, 3889–3899. https://doi.org/10.1128/aem.71.7.3889-3899.2005

Silvia, B., Alexander, W., Sonja-Verena, A., & Georg, L. (2010). Inducible and constitutive promoters for genetic systems in Sulfolobus acidocaldarius. *Extremophiles: Life Under Extreme Conditions*. https://doi.org/10.1007/s00792-010-0304-9

Song, Q., et al. (2019). Induction of a toxin-antitoxin gene cassette under high hydrostatic pressure enables markerless gene disruption in the Hyperthermophilic Archaeon Pyrococcus yayanosii. *Applied and Environmental Microbiology, 85*. https://doi.org/10.1128/aem.02662-18

Stan, J. J. B., et al. (2006). Identification of the missing links in prokaryotic pentose oxidation pathways: Evidence for enzyme recruitment. *The Journal of Biological Chemistry*. https://doi.org/10.1074/jbc.M605549200

Stracke, C., et al. (2020). Salt stress response of Sulfolobus acidocaldarius involves complex Trehalose metabolism utilizing a novel Trehalose-6-phosphate synthase (TPS)/Trehalose-6-Phosphate Phosphatase (TPP) pathway. *Applied and Environmental Microbiology, 86*. https://doi.org/10.1128/aem.01565-20

Straub, C. T., et al. (2018). Biotechnology of extremely thermophilic archaea. *FEMS Microbiology Reviews, 42*, 543–578. https://doi.org/10.1093/femsre/fuy012

Straub, C., et al. (2020). Modification of the glycolytic pathway in Pyrococcus furiosus and the implications for metabolic engineering. *Extremophiles: Life Under Extreme Conditions, 24*, 511–518. https://doi.org/10.1007/s00792-020-01172-2

Suzuki, S., & Kurosawa, N. (2016). Disruption of the gene encoding restriction endonuclease SuaI and development of a host-vector system for the thermoacidophilic archaeon Sulfolobus acidocaldarius. *Extremophiles: Life Under Extreme Conditions, 20*, 139–148. https://doi.org/10.1007/s00792-016-0807-0

Tang, H., et al. (2025). Characterization of the genome editing with miniature DNA nucleases TnpB and IscB in Escherichia coli strains. *Communications Biology, 8*, 261. https://doi.org/10.1038/s42003-025-07521-1

Thiel, A., Michoud, G., Moalic, Y., Flament, D., & Jebbar, M. (2014). Genetic manipulations of the hyperthermophilic piezophilic archaeon Thermococcus barophilus. *Applied and Environmental Microbiology, 80*, 2299–2306. https://doi.org/10.1128/aem.00084-14

Thorgersen, M., Lipscomb, G., Schut, G., Kelly, R., & Adams, M. (2014). Deletion of acetyl-CoA synthetases I and II increases production of 3-hydroxypropionate by the metabolically-engineered hyperthermophile Pyrococcus furiosus. *Metabolic Engineering, 22*, 83–88. https://doi.org/10.1016/j.ymben.2013.12.006

Waege, I., Schmid, G., Thumann, S., Thomm, M., & Hausner, W. (2010). Shuttle vector-based transformation system for Pyrococcus furiosus. *Applied and Environmental Microbiology, 76*, 3308–3313. https://doi.org/10.1128/aem.01951-09

Wagner, M., et al. (2009). Expanding and understanding the genetic toolbox of the hyperthermophilic genus Sulfolobus. *Biochemical Society Transactions, 37*, 97–101. https://doi.org/10.1042/bst0370097

Wang, J. Y., Pausch, P., & Doudna, J. A. (2022). Structural biology of CRISPR-Cas immunity and genome editing enzymes. *Nature Reviews. Microbiology, 20*, 641–656. https://doi.org/10.1038/s41579-022-00739-4

Willard, D. J., & Kelly, R. M. (2025). Metabolic engineering in Hot Acid: Strategies enabling chemolithotrophy in thermoacidophilic archaea. *Metabolic Engineering, 91*, 366–378. https://doi.org/10.1016/j.ymben.2025.06.005

Woese, C. R., & Fox, G. E. (1977). Phylogenetic structure of the prokaryotic domain: The primary kingdoms. *Proceedings of the National Academy of Sciences of the United States of America, 74*, 5088–5090.

Woese, C. R., Kandler, O., & Wheelis, M. L. (1990). Towards a natural system of organisms: Proposal for the domains Archaea, Bacteria, and Eucarya. *Proceedings of the National Academy of Sciences of the United States of America, 87*, 4576–4579.

Worthington, P., Hoang, V., Perez-Pomares, F., & Blum, P. (2003). Targeted disruption of the alpha-amylase gene in the hyperthermophilic archaeon Sulfolobus solfataricus. *Journal of Bacteriology, 185*, 482–488. https://doi.org/10.1128/jb.185.2.482-488.2003

Yang, P., et al. (2024). Research Progress on the mechanism and application of the type I CRISPR-Cas system. *International Journal of Molecular Sciences, 25*. https://doi.org/10.3390/ijms252312544

Yu, Z., et al. (2021). CRISPR-Cas adaptive immune systems in Sulfolobales: Genetic studies and molecular mechanisms. *Science China. Life Sciences, 64*, 678–696. https://doi.org/10.1007/s11427-020-1745-0

Zhang, C., et al. (2010). Revealing the essentiality of multiple archaeal pcna genes using a mutant propagation assay based on an improved knockout method. *Microbiology, 156*, 3386–3397. https://doi.org/10.1099/mic.0.042523-0

Zhang, C., Cooper, T., Krause, D., & Whitaker, R. (2013). Augmenting the genetic toolbox for Sulfolobus islandicus with a stringent positive selectable marker for agmatine prototrophy. *Applied and Environmental Microbiology, 79*, 5539–5549. https://doi.org/10.1128/aem.01608-13

Zhang, C., Whitaker, R. J., & Kivisaar, M. (2018a). Microhomology-mediated high-throughput gene inactivation strategy for the Hyperthermophilic Crenarchaeon Sulfolobus islandicus. *Applied and Environmental Microbiology, 84*, e02167–e02117. https://doi.org/10.1128/aem.02167-17

Zhang, C., Phillips, A. P. R., Wipfler, R. L., Olsen, G. J., & Whitaker, R. J. (2018b). The essential genome of the crenarchaeal model Sulfolobus islandicus. *Nature Communications, 9*, 4908. https://doi.org/10.1038/s41467-018-07379-4

Zhao, P., et al. (2024). Rational design of unrestricted pRN1 derivatives and their application in the construction of a dual plasmid vector system for Saccharolobus islandicus. *mLife, 0*. https://doi.org/10.1002/mlf2.12107

Zheng, T., et al. (2012). Development of a Simvastatin selection marker for a Hyperthermophilic Acidophile, Sulfolobus islandicus. *Applied and Environmental Microbiology, 78*, 568–574. https://doi.org/10.1128/aem.06095-11

Zink, I. A., et al. (2019). CRISPR-mediated gene silencing reveals involvement of the archaeal S-layer in cell division and virus infection. *Nature Communications, 10*, 4797. https://doi.org/10.1038/s41467-019-12745-x

Zink, I. A., et al. (2021). Comparative CRISPR type III-based knockdown of essential genes in hyperthermophilic Sulfolobales and the evasion of lethal gene silencing. *RNA Biology, 18*, 421–434. https://doi.org/10.1080/15476286.2020.1813411

Chapter 9
Future Perspectives on Sustainable Bio-Based Production by Extreme Halophilic Archaea

Rie Yatsunami and Satoshi Nakamura

Abstract Extremely halophilic archaea inhabit hypersaline environments and have evolved unique metabolic strategies to withstand osmotic and oxidative stress. This chapter highlights their potential as sustainable platforms for bio-based production. Recent advances in carotenoid biosynthesis are summarized, including novel bifunctional enzymes that expand current knowledge of archaeal metabolism. The physiological and biomedical roles of carotenoids, such as antioxidant and anticancer activities, are discussed alongside strategies for enhancing production in native and engineered hosts. In addition, other valuable products, including retinal proteins, biopolymers, biosurfactants, and halocins, are explored as emerging applications.

Keywords Extremely halophilic archaea · Carotenoid biosynthesis · Bacterioruberin · Retinal-binding membrane proteins · Biopolymers

9.1 Introduction

Extremely halophilic archaea are microorganisms that thrive in hypersaline environments such as salt lakes, salterns, and brine pools (Oren, 2024). These habitats impose high osmotic pressure, intense solar irradiation, and ionic stress, driving the evolution of unique metabolic and physiological adaptations. As a result, extremely halophilic archaea are capable of synthesizing a wide array of bio-based products, many of which are gaining attention for their potential in sustainable biotechnology.

R. Yatsunami (✉) · S. Nakamura
School of Life Science and Technology, Institute of Science Tokyo, Yokohama, Japan
e-mail: ryatsuna@life.isct.ac.jp

M. Ito et al. (eds.), *Extremophiles for a Sustainable Future*,
https://doi.org/10.1007/978-981-95-6970-0_9

This chapter provides an overview of bio-based products synthesized by extremely halophilic archaea, highlighting carotenoids as a key area of study (Yatsunami et al., 2014, 2024; Yang et al., 2015). Their biosynthetic pathways offer new insights into pigment production under hypersaline conditions. In addition to these pigments, extreme halophilic archaea are capable of producing a diverse range of bio-based compounds, including retinal-binding membrane proteins; biopolymers such as poly-hydroxy-alkanoates (PHA), poly-γ-D-glutamic acid (PGA), and exopolysaccharides; and other functional metabolites such as biosurfactants and halocins (Verma et al., 2025; Hezayen et al., 2000). This chapter also provides an overview of the microbial production of these compounds and explores the future potential of sustainable bio-based product development using extremely halophilic archaea.

9.2 Extremely Halophilic Archaea

9.2.1 Classification and Morphological Diversity of Extremely Halophilic Archaea

Microorganisms can be broadly classified into non-halophilic and halophilic types (Table 9.1) based on the NaCl concentration optimal for their growth (Kushner, 1978). Non-halophilic microorganisms thrive at NaCl concentrations below 0.2 M and include species such as *Escherichia coli* and most soil bacteria. In contrast, halophilic microorganisms grow well at NaCl concentrations above 0.2 M and are considered a type of extremophiles due to their preference for high-salt environments. Halophiles are further subdivided according to their NaCl requirements: slight halophiles grow optimally at 0.2–0.5 M, moderate halophiles at 0.5–2.5 M, and extreme halophiles at 2.5–5.0 M NaCl. While slight and moderate halophiles are typically bacterial, most extreme halophiles belong to the domain archaea.

Within the archaeal domain, halophilic representatives have also been identified in the phylum Methanobacteriota (kingdom Methanobacteriati) and in the "Candidatus" phylum Nanohalarchaeota (kingdom Nanobdellati) (Oren, 2024). Members of Nanohalarchaeota have not yet been cultivated in pure culture, and their existence was first inferred through metagenomic assembly of DNA libraries from hypersaline environments such as Lake Tyrrell in Australia. Consequently, the

Table 9.1 Classification of microorganisms based on optimal NaCl concentration for growth

Categories	Optimal NaCl concentration	Examples of microorganisms
Non-halophilic microorganisms	0–0.2 M	*Escherichia coli* and soil bacteria
Halophilic microorganisms		
Slight halophiles	0.2–0.5 M	Marine bacteria
Moderate halophiles	0.5–0.2.5 M	Bacteria from saline samples
Extreme halophiles	2.5–5.2 M	Archaea

majority of experimentally characterized halophilic archaea—including those involved in bio-based compound production—belong to the kingdom *Methanobacteriati*, particularly within the phylum Halobacteriota. This phylum includes the class Halobacteria, which comprises many of the best-characterized halophilic archaea. The number of taxa recognized in the class Halobacteria, the halophiles *par excellence*, has greatly increased in recent years. As of December 2023, this class encompassed nine families (Haladaptataceae, Haloarculaceae, Halobacteriaceae, Halococcaceae, Haloferacaceae, Halorubellaceae, Natronoarchaeaceae, Natrialbaceae, and Halorutilaceae), 82 genera, and 357 species with validly published names (Cui et al., 2024).

Among extremely halophilic archaea, *Halobacterium salinarum* (Madigan et al., 1999) and *Haloferax volcanii* (Torreblanca et al., 1986) are the most extensively studied species, both originally isolated from the Dead Sea in Israel. *Hbt. salinarum* typically exhibits a rod-shaped morphology, which remains relatively stable throughout its growth cycle. In contrast, *Hfx. volcanii* undergoes a notable morphological transition during growth, shifting from rod-shaped cells in early phases to a plate-shaped disk form in later stages. *Haloarcula japonica* strain TR-1 (Takashina et al., 1990; Otozai et al., 1991), recovered from salt field soil in the Noto Peninsula, Ishikawa Prefecture, Japan, represents the first extremely halophilic archaeon to be isolated from a domestic environment in Japan. This species exhibits a distinctive triangular, flat morphology (Horikoshi et al., 2011). In addition to rod-shaped, plate-shaped disk, and triangular-shaped disk forms, certain species such as *Haloquadratum walsbyi*, originally isolated from saltern crystallizers in Australia and Spain, exhibit nearly perfect square-shaped disk form (Dyall-Smith et al., 2011), further underscoring the structural uniqueness of this group. These examples highlight the remarkable morphological diversity observed among extremely halophilic archaea.

9.2.2 *Triangular-Shaped Extremely Halophilic Archaeon* **Haloarcula japonica** *Strain TR-1*

Haloarcula japonica strain TR-1 was motile by flagella, non-spore forming, and had red flat cells that are predominantly triangular in shape although rhomboidal cells were also observed (Nakamura et al., 1992). The cells were typically 1–2 μm × 0.2–0.3 μm in size in liquid media. *Har. japonica* was extremely halophilic, and requires high concentrations of NaCl (1.7–4.3 M) and Mg^{2+} ions (41–650 mM) for growth. The concentrations of NaCl and Mg^{2+} were critical for the formation of the triangular shape, and optimal concentrations of NaCl and Mg^{2+} were 3.4 M and 160 mM, respectively. The temperature range for growth was 24–45 °C, with an optimal temperature of 42 °C. Growth of the microbe occurred within a pH range of 6.0–8.0, with an optimal range of 7.0–7.5. Electron micrographs of thin sections of *Har. japonica* cells were prepared using the rapid freezing and substitution fixation method (Nishiyama et al., 1992, 1995) The cells had predominantly triangular shape, and a thick surface (S) layer was clearly seen outside of the cells (Fig. 9.1a). Atomic force microscopic images of typical triangular disc-shaped cells of *Har. japonica* were also observed using rapid membrane filtration method (Fig. 9.1b) (Umemura et al., 1998).

The genome of *Har. japonica* has also been sequenced (Nakamura et al., 2011). The 4,294,979-base pairs (bp) genome of *Har. japonica* is divided into relatively high and low G+C content replicons, including the largest chromosome I, a 3,028,012-bp replicon with a G+C content of 62.7%, and four smaller replicons

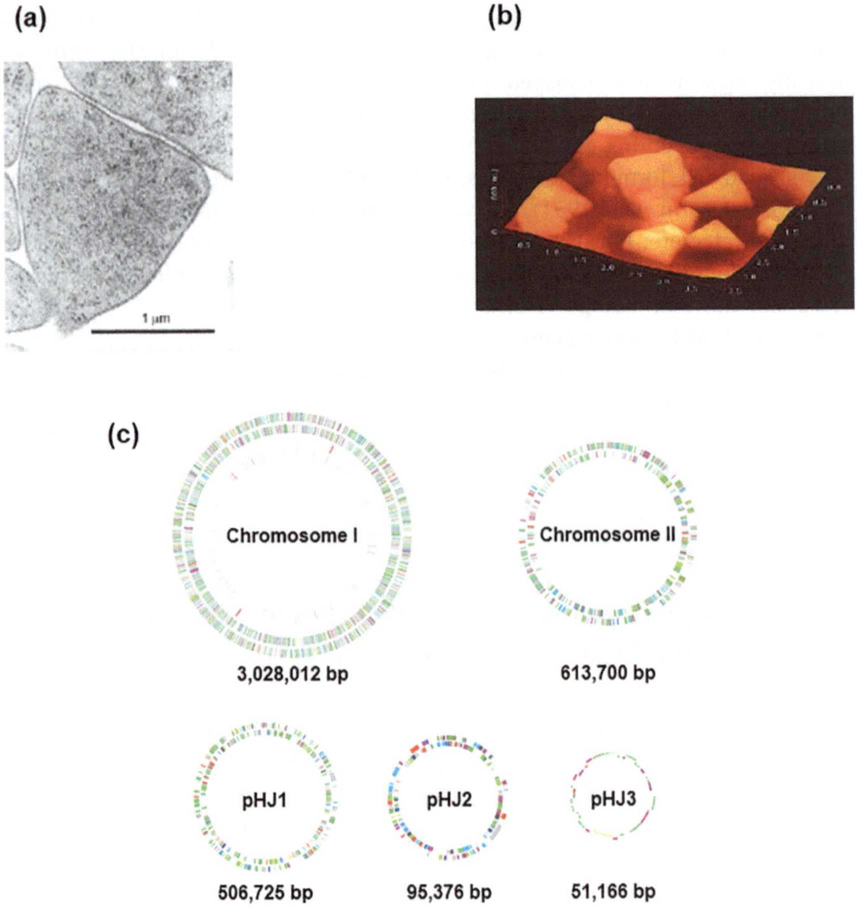

Fig. 9.1 Extremely halophilic archaeaon *Haloarucula japonica* TR-1. (**a**) Electron micrograph of thin section of *Har. japonica* cells prepared by the freeze-substitution fixation method. (**b**) Atomic force microscopic image of *Ha. japonica* cells on a membrane filter (contact mode in air) (**c**) Graphical circular maps of the genome of *Har. japonica*. From outside to the center: genes on forward strand [color by Clusters of Orthologous Groups of proteins (COG) categories], genes on reverse strand (color by COG categories). rRNA and tRNA genes were shown in chromosome I

ranging from 51,166 to 613,700 bp with G+C contents ranging from 54.7 to 58.7%. The total number of predicted protein coding sequences (CDSs) were 4057, and 2 *rrn* operons were found only in chromosome I (Fig. 9.1c).

The cell surfaces of many extremely halophilic archaea are covered by a surface (S) layer (Sleytr & Messner, 1983; Sumper, 1987). *Har. japonica* have glycoprotein on the cell surface (Nakamura et al., 1995; Wakai et al., 1997) as well as *Halobacterium salinarum* and *Haloferax volcanii* (Mescher et al., 1974; Wieland et al., 1980; Kessel et al., 1988). The cell envelope fraction of *Har. japonica* was prepared and examined by sodium dodecylsulfate-polyacrylamide gel electrophoresis (SDS-PAGE). A high molecular mass protein was detected as a major band at 170 kDa by staining with Coomassie brilliant blue (CBB) (Nakamura et al., 1992). This protein was also visualized by the periodate-Schiff (PAS) method. A similar phenomenon had been observed with the CSG of *Hbt. salinarumm* (Mescher et al., 1974). These results suggested that the cell envelope of *Har. japonica* had a high molecular mass, carbohydrate-containing protein as the major envelope protein.

9.3 Biosynthesis of Carotenoids in Extremely Halophilic Archaea

9.3.1 Carotenoids Produced by Extremely Halophilic Archaea

Carotenoids, an important class of natural isoprenoid-derived pigments, are synthesized by all photosynthetic organisms, and some non-photosynthetic bacteria, archaea, and fungi. They act as light-harvesting pigments in photosynthesis (Holt et al., 2005), and also as antioxidants (Miller et al., 1996; Naguib, 2000), and light-protecting pigments in halophilic archaea (Dundas & Larsen, 1963; Shahmohammadi et al., 1998). The basic structure of carotenoids consists of a C_{40} backbone formed by the condensation of eight isoprene (C_5H_8) units. In contrast, many extremely halophilic archaea are known to produce C_{50} carotenoids. Carotenoids produced by extremely halophilic archaea have been primarily studied in species such as *Halobacterium salinarum*, *Haloferax volcanii*, and *Har. japonica* (Rodrigo-Baños et al., 2015; Yatsunami et al., 2014). These organisms synthesize not only C_{40} carotenoids such as *cis*- and *trans*-phytoene and lycopene, but also C_{50} carotenoids including isopentenyl dehydrorhodopin (IDR), bisanhydrobacterioruberin (BABR), monoanhydrobacterioruberin (MABA), and bacterioruberin (BR) (Fig. 9.2). In *Har. japonica*, BR accounts for approximately 70% of the total carotenoids. Therefore, it was the main one responsible for the red color of this organism. The BR content in *Har. japonica* was similar to those in other halophilic archaea (Mandelli et al., 2012). Other major pigments were MABR (22.5%) and BABR (9.3%), and IDR was found at a lower level (<0.1%). These results suggest that BR is produced as a final product in *Har. japonica* and is synthesized from other C_{50} carotenoids, such as IDR, BABR, and MABR as well as compounds commonly detected in carotenoid

Fig. 9.2 Structure of carotenoids produced by *Har. japonica*

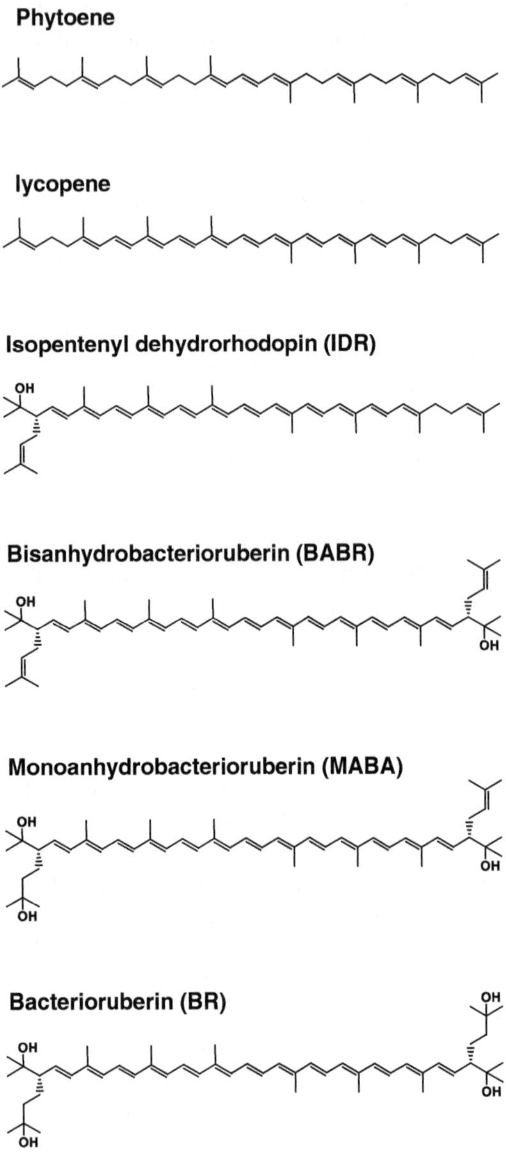

producing extremely halophilic archaea. Additional carotenoids have been identified at very low concentrations in extremely halophilic archaea: lycopersene, *cis*- and *trans*-phytofluene, neo-β-carotene and neo-α-carotene. Their low abundance implies a possible role as precursors in the biosynthesis of carotenoids such as lycopene, retinal, and members of the bacterioruberin group. Some species may also produce the ketocarotenoid canthaxanthin in addition to other carotenoids (Naziri et al., 2014). Although this is the general carotenoid profile exhibited for most of the haloarchaeal species, it is important to note that some of them can produce high amounts of canthaxanthin, β-carotene and *trans*-astaxanthin (Asker et al., 2002).

9.3.2 *Biosynthetic Pathway of Carotenoids in Extremely Halophilic Archaea*

9.3.2.1 Early Steps of Carotenoid Biosynthesis: From Acetyl-CoA to Phytoene

The initial stage of carotenoid biosynthesis in bacteria involves the synthesis of isoprenoid units. The key building block, isopentenyl pyrophosphate (IPP), is synthesized via either the mevalonate (MVA) pathway or the methylerythritol phosphate (MEP) pathway. The MEP pathway, also known as the non-mevalonate pathway, is utilized by many bacteria and plastids. However, it is known that extremely halophilic archaea lack the MEP pathway and instead rely exclusively on the MVA pathway for IPP biosynthesis. This pathway begins with acetyl-CoA and proceeds through a series of reactions analogous to those found in bacteria, including the formation of 3-hydroxy-3-methylglutaryl-CoA (HMG-CoA) and mevalonate, ultimately yielding the C5 compound IPP (Moise et al., 2014).

The synthesized IPP is then isomerized to dimethylallyl pyrophosphate (DMAPP, C5) by isopentenyl diphosphate isomerase (Idi). Subsequently, geranylgeranyl pyrophosphate synthase (CrtE) catalyzes sequential condensations of IPP units to form geranyl pyrophosphate (GPP, C10), farnesyl pyrophosphate (FPP, C15), and geranylgeranyl pyrophosphate (GGPP, C20). Finally, phytoene synthase (CrtB) utilizes adenosine triphosphate (ATP) to condense two molecules of GGPP, resulting in the formation of 15-*cis*-phytoene, the first C40 carotenoid composed of eight isoprene units (Fig. 9.3) (Alcaino et al., 2014). The biosynthetic pathway from phytoene to bacterioruberin (BR) has been identified in *Haloarcula japonica* in recent years (Yatsunami et al., 2024; Yang et al., 2015).

The following sections outline this pathway in two parts: the conversion from phytoene to lycopene, and the subsequent biosynthetic steps leading from lycopene to BR. In addition, a separate route from lycopene to retinal via β-carotene is briefly described to provide a broader context for carotenoid metabolism.

9.3.2.2 From Phytoene to Lycopene

In the biosynthetic conversion from phytoene to lycopene, phytoene desaturase (CrtI) is known to play a central role in the carotenoid biosynthesis pathways of certain bacteria. A genomic search of *Har. japonica* revealed two gene homologs, *c0507* and *d1086*, showing sequence similarity to the well-characterized *crtI* (Sandmann, 1994) from *Pantoea ananatis*. At the amino acid level, C0507 and D1086 showed 30% and 27% identity with *P. ananatis* CrtI (PaCrtI), respectively,

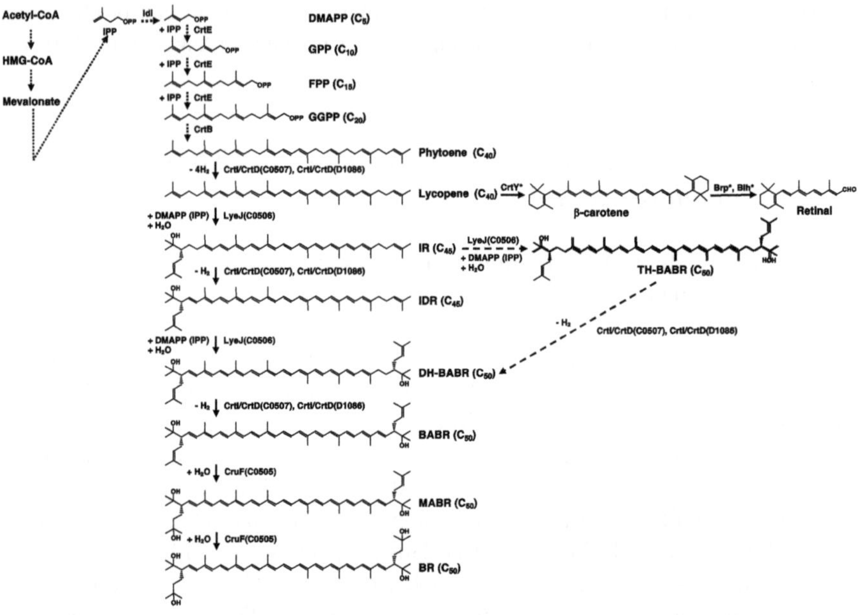

Fig. 9.3 Overview of BR and retinal biosynthetic pathways in *Har. japonica.* Solid arrows indicate the main BR biosynthetic steps. Dashed arrows indicate other possible BR biosynthetic steps. Among the intermediates from phytoene to BR, the locations of structural changes in each step are circled in red. Asterisks indicate unidentified BR and retinal biosynthetic enzymes

with E-values of 6e−70 and 9e−52. Additionally, D1086 exhibited 60% sequence homology with C0507. Accordingly, single-gene knockout strains (Δ*c0507* and Δ*d1086*) as well as a double knockout strain (Δ*c0507*Δ*d1086*) were constructed, and the carotenoid species produced by each mutant were analyzed. In addition, complementation strains for each mutant were generated and subjected to the same analysis to confirm the functional roles of the respective genes. Analysis of these mutants revealed that both C0507 and D1086 are involved in the conversion of phytoene to lycopene, functioning as phytoene desaturases (CrtI) in *Har. japonica.* Furthermore, transcript levels of *c0507* and *d1086* were quantified by RT-PCR using the wild-type strain, revealing that *c0507* was transcribed at higher levels than *d1086*. These results suggest that although both genes contribute to phytoene desaturation, *c0507* serves as the primary phytoene desaturase in this biosynthetic pathway.

9.3.2.3 From Lycopene to BR

Detailed analysis was performed on the genomic region surrounding the *c0507* in *Har. japonica*. This revealed that the downstream gene, *c0506*, shares sequence similarity with *lye*, a gene originally identified in *Hbt. salinarum* that encodes a lycopene elongase known as Lye (Dummer et al., 2011). The predicted protein encoded by *c0506* exhibited 60% sequence identity to Lye, suggesting that C0506 may function as the lycopene elongase in *Har. japonica*. Immediately downstream of *c0506* lies the *c0505*, which encodes a homolog of CruF. C0505 showed 30% and 31% amino acid identity to the CruF proteins of *Synechococcus* sp. PCC 7002 (Maresca et al., 2008) and *Deinococcus radiodurans* R1(Sun et al., 2009), respectively. In addition, *c0507*, *c0506,* and *c0505* cluster together on the genome of *Har. japonica,* in this order, and RT-PCR analysis revealed that these genes were co-transcribed. Since carotenoid synthesis-related genes in some bacteria are assembled in clusters or in neighborhoods (Misawa et al., 1990; Giraud et al., 2004), these genes were predicted to be involved in the carotenoid synthesis of *Har. japonica*. To investigate the individual roles of *c0507*, *c0506*, and *c0505*, single-gene knockout mutants were generated for each gene in *Har. japonica*, along with their respective complementation strains. The carotenoid profiles of these strains were subsequently analyzed. The results demonstrated that *c0507* encodes a 3,4-desaturase (CrtD), *c0506* encodes a bifunctional enzyme with lycopene elongase and 1,2-hydratase activities (designated LyeJ, based on ~60% sequence identity to Lye from *Hbt. salinarum*), and *c0505* encodes a hydratase (CruF). In the biosynthetic pathway from lycopene to BR, lycopene is first modified by Lye, which adds an IPP unit and a water molecule to one terminal end, forming isopentenylrhodopin (IR). This intermediate is then desaturated by CrtD to produce IDR. A similar reaction occurs at the opposite end, yielding bacteriorhodopsin (BABR) via DH-BABR. Subsequently, the added IPP moieties are hydrated by CrtF, forming monoanhydrobacterioruberin (MABR), and further hydration at the opposite end leads to the final product, BR. Further analysis of the knockout strains revealed a minor alternative pathway originating from IR, in which LyeJ acts on IR to produce tetrahydro-BABR (TH-BABR), followed by desaturation by CrtD to yield BABR.

Notably, C0507 functions as phytoene desaturase (CrtI) in the phytoene-to-lycopene pathway and as CrtD (3,4-desaturase) in the lycopene-to-BR pathway, indicating that it encodes a bifunctional enzyme, hereafter referred to as CrtI/CrtD (c0507). Subsequent analysis identified another gene, D1086, which also participates in phytoene desaturation and exhibits CrtD activity in the lycopene-to-BR pathway. This gene was similarly designated CrtI/CrtD (D1086). These two bifunctional enzymes appear to complement each other in BR biosynthesis. To date, no bifunctional enzyme possessing both CrtI and CrtD activities has been reported, making CrtI/CrtD (c0507) and CrtI/CrtD (D1086) from *Har. japonica* the first known examples.

Phylogenetic tree analysis revealed that CrtDs originated from CrtIs via gene duplication at an early stage, resulting in the divergence of CrtIs and CrtDs with unique functions. The tree shows that CrtI/CrtD(C0507) and CrtI/CrtD(D1086) were closely related members of the CrtI family. The *Har. japonica crtI/crtD* (*d1086*) was located upstream of *c1158*. This gene (*c1158*) has already been cloned (Yatsunami et al., 2003) and is homologous to the *brp* gene of *Hbt. salinarum*. The *brp* gene encodes a β-carotene cleavage dioxygenase (Peck et al., 2001) involved in retinal biosynthesis, which will be described later in this chapter. Consequently,

CrtI/CrtD (D1086) is suggested to be involved in retinal biosynthesis, whereas CrtI/CrtD (C0507) functions in BR biosynthesis. Since both CrtI and CrtD activities are essential for BR production, the ancestral CrtI (C0507) in the *Har. japonica* genome likely expanded its substrate specificity during evolution, acquiring CrtD activity and becoming bifunctional. The acquisition of CrtI/CrtD (D1086) occurred later, likely through gene duplication of the already bifunctional BR biosynthetic gene. As a result, CrtI/CrtD (D1086) also exhibits dual functionality. *Hbt. salinarum* also produces both BR and retinal, and possesses two CrtIs, [HsCrtI(1) and HsCrtI(2)]. Closely related *Hfx. volcanii* produces BR, but not retinal, with only one CrtI (HvCrtI). Both CrtI and CrtD activities are essential for BR biosynthesis, and CrtIs in these archaea may also be bifunctional enzymes. Microbes possessing bifunctional CrtI(s) are not necessarily limited to extremely halophilic archaea, such as *Har. japonica*.

9.3.3 Physiological Roles and Applications of BR

9.3.3.1 Physiological Role of BR in Extremely Halophilic Archaea

There have been several reports concerning the function of BR in extremely halophilic archaea. One such function is the scavenging of reactive oxygen species (ROS) and free radicals—a role typically attributed to carotenoids—which BR is suggested to perform. ROS generated under photo-oxidative stress can interact with macromolecules such as lipids and proteins, potentially leading to cellular damage. The antioxidant capacity of BR extracted from *Haloarcula japonica* was evaluated using 1,1-diphenyl-2-picrylhydrazyl (DPPH) radical scavenging activity (Yatsunami et al., 2014). DPPH is a stable free radical donor, which is widely used to test the free radical scavenging effect of natural antioxidants. DPPH method involves the scavenging of a performed stable radical by an electron transfer mechanism from the carotenoid to the radical, generating a carotenoid radical cation (Guo et al., 1999). In this method, a stable radical is scavenged via electron transfer from the carotenoid, resulting in the formation of a carotenoid radical cation (Huang et al., 2005). The radical scavenging activity was found to be dependent on the concentration of the carotenoid. As shown in Fig. 9.4, the scavenging capacity of BR was significantly higher than that of β-carotene. Redox potential appears to be influenced by both the length of the conjugated double bond (c.d.b.) system and the presence of functional groups. The antioxidant activity of carotenoids increases with the extension and maximal overlap of the c.d.b. molecular orbitals (Albrecht et al., 2000; Tian et al., 2007). The BR molecule contains 13 conjugated double bonds, compared to 9 in β-carotene, suggesting that BR is a highly effective DPPH radical scavenger. Such antioxidant activity is presumed to contribute to the oxidative stress tolerance of extremely halophilic archaea inhabiting environments such as salt lakes, where they are exposed to intense ultraviolet radiation.

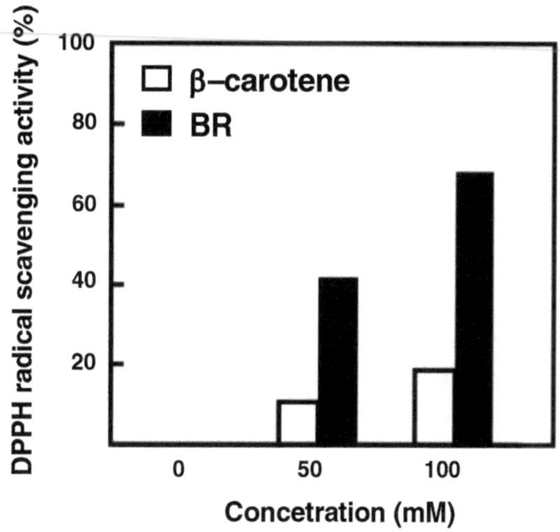

Fig. 9.4 Radical scavenging activities of BR and β-carotene against DPPH radicals. Closed bars, β-carotene; Open bars, BR

The second hypothesis is that BR, a polar carotenoid with four hydroxyl groups at its termini, functions as a membrane rivet (D'Souza et al., 1997). In *Haloferax mediterranei*, a marked increase in BR production has been observed under low-salt conditions, suggesting that BR acts as a rivet to stabilize the membrane and prevent cell lysis in hypotonic environments. Furthermore, BR has also been reported to stabilize light-driven proton pumps, such as bacteriorhodopsin. *Halorubrum* sp. aus-2 possesses a proton pump known as archaerhodopsin-2 (aR-2) (See Sect. 9.4.1). Crystallographic analysis of aR-2 revealed that BR binds within the inter-subunit gaps of the aR-2 trimer (Yoshimura & Kouyama, 2008).

9.3.3.2 Applications of BR

Recent studies have highlighted the multifaceted bioactivity of BR, particularly in contexts relevant to human health and disease (Giani et al., 2024). BR exhibits strong immunomodulatory and anti-inflammatory properties, including the down-regulation of pro-inflammatory cytokines such as TNF-α and IL-6, and the activation of cytoprotective pathways involving Nrf2 and HO-1 (Ávila-Román et al., 2023; Higa et al., 2020). In murine models of colitis, BR-loaded nanoparticles have been shown to restore intestinal barrier integrity and reduce inflammatory responses, suggesting potential applications in treating immune-mediated disorders (Higa et al., 2020). BR also demonstrates antitumor effects across various human cancer cell lines, including HepG2, MCF-7, Caco-2, and HeLa. Its proapoptotic activity is associated with the upregulation of key genes such as CASP3, CASP8, and BAX, indicating its promise as a complementary agent in cancer therapeutics (Giani et al., 2023; Shahbazi et al., 2023; Hegazy et al., 2020). These results underscore the therapeutic relevance of BR as a supportive agent in cancer therapy, particularly through its ability to trigger apoptosis in diverse tumor cell lines. While current evidence is largely in vitro, the consistency across models suggests promising biomedical significance. Future studies should clarify its mechanisms and evaluate efficacy in vivo to support clinical translation.

9.3.4 *Strategies for Enhancing BR Production in Extremely Halophilic Archaea and Non-halophilic Hosts*

Various strategies have been explored to enhance the microbial production of BR, a C_{50} carotenoid with strong antioxidant properties and industrial relevance. Recent studies have focused on optimizing cultivation parameters and investigating metabolic characteristics that influence BR biosynthesis. Among natural producers, *Haloferax marinum* MBLA0078 has demonstrated the highest reported volumetric titer of 2.80 mg/L under fed-batch fermentation with high salinity and controlled aeration (Cho et al., 2024). The cultivation conditions were optimized using one-factor-at-a-time (OFAT) experiments, in which individual parameters were varied independently, such as salt concentration, temperature, and pH. This approach is simple and effective for identifying key factors, though it does not consider how multiple factors might interact. Similarly, *Halorubrum ruber* MBLA0099 produced 1.996 mg/L in batch culture under hypersaline conditions (Hwang et al., 2024). Enhanced productivity was observed under diauxic growth induced by glucose and starch, and further improvement was achieved through osmotic stress. Although no exact BR titer was reported, *Hrr.* sp. HRM-150 was found to constitute approximately 85% of total carotenoids under optimized conditions (Ma et al., 2024). Productivity was increased by 54% when glucose and starch were used together as carbon sources, and further enhancement was observed under osmotic stress.

Extreme halophiles have been studied for carotenoid biosynthesis under hypersaline conditions. Their industrial application, however, is limited by the use of high-salt media, which can cause equipment corrosion, along with slow growth and restricted genetic tools. As an alternative strategy, *Corynebacterium glutamicum* was engineered to express three codon-optimized genes from *Hfx. marinum*, *lyeJ*, *crtD*, and *cruF* (as shown in Fig. 9.3), enabling carotenoid biosynthesis under salt-free conditions (Cho et al., 2025). The resulting strain produced BR at a final titer of 9.74 mg/L in fed-batch fermentation, whereas *Hfx. marinum* MBLA0078 achieved titers of 0.954–2.80 mg/L under high-salt conditions (Cho et al., 2024). This demonstrates the feasibility of reconstructing halophilic biosynthetic pathways in non-halophilic hosts, offering a scalable and less corrosive platform for carotenoid production.

9.4 Biosynthesis of Some Other Bio-Based Products in Extremely Halophilic Archaea

9.4.1 Biosynthesis of Retinal-Binding Membrane Proteins

Several strains of extremely halophilic archaea possess retinal-binding membrane proteins (retinal proteins), which function as photoreceptive pigment-protein complexes. The most extensively studied species, *Halobacterium salinarum*, contains four distinct retinal proteins: bacteriorhodopsin (bR), halorhodopsin (hR), sensory rhodopsin I (sRI), and sensory rhodopsin II (sRII) (Mukohata et al., 1999; Oesterhelt & Stoeckenius, 1971). Among these, bR functions as a light-driven proton pump, while hR acts as a light-driven chloride ion pump. In contrast, sRI and sRII serve as photoreceptors that mediate phototactic responses, with sRI promoting positive phototaxis and sRII triggering negative phototaxis. bR, hR, and sRs play distinct physiological roles in haloarchaeal adaptation to light and salinity. While sharing a common seven-transmembrane architecture and retinal chromophore, these proteins differ markedly in their photochemical reaction cycles and downstream effects. bR undergoes a well-characterized photocycle involving multiple intermediate states to facilitate proton translocation across the membrane. The resulting proton gradient is harnessed by ATP synthase to drive ATP production, providing energy for cellular processes under anaerobic conditions. hR responds to light by transporting chloride ions inward, stabilizing internal osmotic pressure (Schobert & Lanyi, 1982). In contrast, sRI and sRII initiate signal transduction cascades via associated transducer proteins, enabling directional movement in response to light stimuli.

All of bR, hR, and sRs incorporate retinal as their chromophore, the biosynthesis of which in *Hbt. salinarum* has been previously described. In retinal biosynthesis, retinal is synthesized from lycopene, a precursor of BR. Lycopene is first converted to β-carotene by lycopene β-cyclase (CrtY) (as shown in Fig. 9.3) (Peck et al., 2002). Subsequently, β-carotene is cleaved by β-carotene cleavage dioxygenase (Brp) (Peck et al., 2001) to produce retinal (C_{20}). In the case of bR, this retinal binds to the apoprotein bacterio-opsin to form the functional photoreceptor. Various ion-pumping rhodopsins have been identified in halophilic archaea. Among these, several function as light-driven proton pumps, such as archaerhodopsin (aR) (Mukohata et al., 1988) from *Halorubrum* sp., cruxrhodopsin (cR) (Yatsunami et al., 2000) from *Har. japonica*, and deltarhodopsin (dR) (Mukohata et al., 1999) from *Hbt.* sp. arg-4, now reclassified as *Haloterrigena* sp. strain arg-4. Notably, *Halorubrum* sp. aus-1 and aus-2 express distinct variants—archaerhodopsin-1 (aR-1) and archaerhodopsin-2 (aR-2), respectively.

Of the haloarchaeal rhodopsins identified to date, bR has been most extensively applied in bioelectronic devices, such as optical data storage, photoelectric sensors, and bio-sensitized solar cells, owing to its stable photochemical cycle and efficient light-driven proton-pumping activity (Li et al., 2018). These applications highlight the promise of haloarchaeal rhodopsins as versatile components in future biotechnological innovation.

9.4.2 Biopolymer Production: PHA, PGA, and EPS

Haloferax mediterranei has emerged as a leading extremely halophilic archaeon for polyhydroxyalkanoate (PHA) production, owing to its ability to accumulate high levels of biopolymer under non-sterile, hypersaline conditions. When cultivated using enzymatically treated extruded starch, *Hfx. mediterranei* reached a cell concentration of 39.4 g/L (cell dry weight, CDW) and produced 20 g/L of PHA, corresponding to 50.8 wt.% of CDW (Chen et al., 2006). Recent studies have demonstrated its industrial potential using low-cost substrates such as food waste, achieving PHA contents up to 52.3% of CDW and enabling simplified downstream processing through water-based extraction methods (Atarés et al., 2024). These features make *Hfx. mediterranei* particularly attractive for scalable bioplastic production, especially in applications such as biodegradable food packaging. *Cupriavidus necator* (formerly *Ralstonia eutropha*) utilizes a well-characterized class I PHA synthase (PhaC) for PHB production (Rehm, 2003). In contrast, *Hfx. mediterranei* employs a distinct haloarchaeal-type synthase that differs both structurally and mechanistically. Its PHA synthase belongs to a noncanonical class, more closely related to class III enzymes but adapted to extreme halophilic conditions (Han et al., 2010). This divergence in enzyme class may underlie its ability to produce PHBV copolymers under hypersaline environments, and highlights the evolutionary diversity of PHA biosynthetic pathways.

In addition to polyhydroxyalkanoates (PHAs), extremely halophilic archaea are also capable of synthesizing other valuable biopolymers, such as poly-γ-glutamic acid (PGA). *Natrialba aegyptiaca*, formerly referred to as strain 40, is an extremely halophilic archaeon capable of producing poly-γ-glutamic acid (PGA) under hypersaline conditions (Hezayen et al., 2000). On solid medium containing 30% (w/v) NaCl, the strain achieved a maximum PGA yield of 0.42 g/g-CDW. The exopolymer isolated from these cultures contained approximately 85% (w/v) PGA, indicating high purity. In liquid medium under the same salt concentration, the PGA yield decreased to 0.2 g/g-CDW, and the PGA content of the exopolymer dropped to 55% (w/v). PGA formation was observed only at NaCl concentrations of 25% (w/v) or higher, while colonies lost their mucous morphology below 20% (w/v) NaCl. In addition to its structural and industrial relevance, L-form PGA derived from *Natrialba aegyptiaca* has demonstrated extremolyte-like properties. Yamasaki et al. (2010) reported that this archaeal exopolymer significantly improved the stability of DNA ligase against freeze-thaw cycles, proteolytic degradation, heat, and alkaline conditions (Yamasaki et al., 2010). These findings suggest that L-PGA may serve as a bio-based stabilizer for enzymes and other labile biomolecules under harsh environmental conditions.

Archaeal exopolysaccharides (EPSs) also represent a distinct class of biopolymers with structural and functional diversity. *Halorubrum* sp. TBZ112, an extremely halophilic archaeon isolated from Urmia Lake, Iran, has been reported to produce a novel exopolysaccharide (EPS) under hypersaline cultivation conditions (Hamidi et al., 2019). The strain secreted approximately 480 mg/L of EPS, which was subsequently purified and structurally characterized. The EPS exhibited a relatively low molecular weight of 5.052 kDa and was identified as a heteropolysaccharide comprising ten sugar moieties, including mannose (19.95%), glucosamine (15.55%), galacturonic acid (15.43%), arabinose (12.24%), and glucuronic acid (12.05%), as determined by high-pressure size exclusion chromatography (HPSEC) with multi-angle laser light scattering (MALLS), high-pressure anion exchange chromatography (HPAEC), and FT-IR analyses. Functional assays revealed significant antiproliferative activity against human gastric cancer cells (MKN-45), while maintaining biocompatibility with human dermal fibroblasts. These findings suggest promising biomedical applications for archaeal EPSs with defined structural and functional properties.

9.4.3 Biosynthesis of Bioactive Compounds: Biosurfactants and Halocins

Natrialba sp. M6 was reported to produce a biosurfactant under highly alkaline and hypersaline conditions (Hegazy et al., 2022). Optimal production was achieved at pH 12, 20.8% NaCl, 3% glycerol, 37 °C, and 150 rpm, as determined by statistical optimization. Structural characterization by GC-MS, compositional analysis, and Zetasizer measurements suggested that the biosurfactant is a cationic lipoprotein. It exhibited potent antiviral activity against Hepatitis C virus (HCV) and Herpes simplex virus (HSV), primarily via deactivating viral envelopes, thus preventing viral entry, and also inhibited viral replication, as supported by molecular docking analysis. These findings highlight the biomedical potential of archaeal biosurfactants with defined structural and functional properties.

Halocin C8 is a proteinaceous antimicrobial peptide that was initially purified from *Natrinema* sp. AS7092 (formerly *Halobacterium* sp. AS7092) (Li et al., 2003). The peptide is encoded by the *halC8*, which has since been identified in multiple extreme halophilic archaeal genera, including *Natrinema*, *Haloterrigena*, *Haloferax*, and *Halobacterium* (Besse et al., 2017). Its antimicrobial activity was experimentally demonstrated by inhibition halos formed on lawns of *Halorubrum chaoviator* ETD3, highlighting its role in inter-archaeal antagonism. These findings support its potential utility as a natural antimicrobial agent in biotechnological processes operating under extreme environmental conditions.

9.5 Future Perspectives on Sustainable Bio-Based Production by Extremely Halophilic Archaea

Extremely halophilic archaea are increasingly recognized as versatile platforms for sustainable bioproduction. Beyond biopolymers and bioactive compounds, these organisms have demonstrated potential in producing biofuels such as biodiesel and bioethanol (Kasirajan et al., 2021), as well as high-value compounds like squalene, a precursor for pharmaceuticals and cosmetics (Kramer et al., 1972; Falb et al., 2008). Their remarkable tolerance to hypersaline, nutrient-poor, and metal-contaminated environments enables growth in unconventional substrates, including industrial wastewater, salt lakes, and seawater, reducing reliance on freshwater and costly sterilization processes (Kasirajan et al., 2021). Recent transcriptomic studies have revealed molecular adaptations that confer resistance to chaotropic agents like perchlorate, further expanding their utility in extreme bioprocessing conditions (Díaz-Rullo et al., 2025). These traits position extremely halophilic archaea as ideal candidates for low-input, high-resilience biomanufacturing systems. Future research is encouraged to consider improving metabolic yields, developing genetically tractable strains, and integrating haloarchaeal bioprocesses into circular bioeconomy frameworks.

References

Albrecht, M., Takaichi, S., Steiger, S., Wang, Z., & Sandmann, G. (2000). Novel hydroxycarotenoids with improved antioxidative properties produced by gene combination in *Escherichia coli*. *Nature Biotechnology, 18*, 843–846.

Alcaino, J., Romero, I., Niklitschek, M., Sepúlveda, D., Rojas, M. C., Baeza, M., & Cifuente, V. (2014). *PLoS One, 9*, e96626.

Asker, D., Awad, T., & Ohta, Y. (2002). Lipids of Haloferax. Alexandrinus strain TMT: An extremely halophilic canthaxanthin-producing archaeon. *Journal of Bioscience and Bioengineering, 93*, 37–43.

Atarés, L., Chiralt, A., González-Martínez, C., & Vargas, M. (2024). Production of polyhydroxyalkanoates for biodegradable food packaging applications using *Haloferax mediterranei*. *Food, 13*, 950.

Ávila-Román, J., Talero, E., & Motilva, V. (2023). Bacterioruberin-rich extracts reduce oxidative stress and inflammation in LPS-stimulated macrophages. *Antioxidants, 12*, 1080.

Besse, A., Vandervennet, M., Goulard, C., Peduzzi, J., Isaac, S., Rebuffat, S., & Carré-Mlouka, A. (2017). Halocin C8: An antimicrobial peptide distributed among four halophilic archaeal genera: *Natrinema*, *Haloterrigena*, *Haloferax*, and *Halobacterium*. *Extremophiles, 21*, 623–638.

Chen, C. W., Don, T. M., & Yen, H. F. (2006). Enzymatic extruded starch as a carbon source for the production of poly(3-hydroxybutyrate-co-3-hydroxyvalerate) by *Haloferax mediterranei*. *Process Biochemistry, 41*, 2289–2296.

Cho, Y., Lee, J., & Kim, S. (2024). Optimized production of bacterioruberin from *Haloferax marinum* using one-factor-at-a-time and central composite design approaches. *Bioresources and Bioprocessing, 11*, 15.

Cho, E.-S., Hwang, C. Y., & Seo, M.-J. (2025). Enhanced production of C_{50} carotenoid bacterioruberin by metabolically engineered *Corynebacterium glutamicum*. *Bioresource Technology, 391*, 132670.

Cui, H.-L., Hou, J., Amoozegar, M. A., Dyall-Smith, M. L., de la Haba, R. R., Minegishi, H., Montalvo-Rodriguez, R., Oren, A., Sanchez-Porro, C., Ventosa, A., & Vreeland, R. H. (2024). Proposed minimal standards for description of new taxa of the class Halobacteria. *International Journal of Systematic and Evolutionary Microbiology, 74*, 006290.

D'Souza, S. E., Altekar, W., & D'Souza, S. F. (1997). Lipid composition of *Halobacterium salinarum*: A study of its adaptability to salt stress. *Archives of Microbiology, 168*, 68–75.

Díaz-Rullo, J., & González-Pastor, J. E. (2025). Molecular adaptations specific to extreme halophilic archaea could promote high perchlorate tolerance. *Applied and Environmental Microbiology, 91*, e00512–e00525.

Dummer, A. M., Bonsall, J. C., Cihla, J. B., Lawry, S. M., Johnson, G. C., & Peck, R. F. (2011). Bacterioopsin-mediated regulation of bacterioruberin biosynthesis in *Halobacterium salinarum*. *Journal of Bacteriology, 193*, 5658–5667.

Dundas, I. D., & Larsen, H. (1963). A study on the killing by light of photosensitized cells of *Halobacterium salinarium*. *Archives of Microbiology, 46*, 19–28.

Dyall-Smith, M. L., Pfeiffer, F., Klee, K., Palm, P., Gross, K., Schuster, S. C., Rampp, M., & Oesterhelt, D. (2011). *Haloquadratum walsbyi*: Limited diversity in a global pond. *PLoS One, 6*(6), e20968.

Falb, M., Müller, K., Königsmaier, L., Oberwinkler, T., Horn, P., von Gronau, S., Gonzalez, O., Pfeiffer, F., Bornberg-Bauer, E., & Oesterhelt, D. (2008). Metabolism of halophilic archaea. *Extremophiles, 12*, 177–196.

Giani, M., Serino, S., Simioni, C., Montoyo-Pujol, Y. G., Peiró, G., & Martínez-Espinosa, R. M. (2023). Antiproliferative effects of bacterioruberin-rich extracts on breast cancer cells. *Scientific Reports, 13*, 7148.

Giani, M., Pire, C., & Martínez-Espinosa, R. M. (2024). Bacterioruberin: Biosynthesis, antioxidant activity, and therapeutic applications in cancer and immune pathologies. *Marine Drugs, 22*, 167.

Giraud, E., Hannibal, L., Fardoux, J., Jaubert, M., Jourand, P., Dreyfus, B., Sturgis, J. N., & Vermeglio, A. (2004). Two distinct *crt* gene clusters for two different functional classes of carotenoid in *Bradyrhizobium*. *Journal of Biological Chemistry, 279*, 15076–15083.

Guo, Q., Zhao, B., Shen, S., Hou, J., & Xin, W. (1999). ESR study on the structure-antioxidant activity relationship of tea catechins and their epimers. *Biochimica et Biophysica Acta, 1427*, 13–23.

Hamidi, M., Mirzaei, R., Delattre, C., Khanaki, K., Pierre, G., Gardarin, C., Petit, E., Karimitabar, F., & Faezi, S. (2019). Characterization of a new exopolysaccharide produced by *Halorubrum* sp. TBZ112 and evaluation of its anti-proliferative effect on gastric cancer cells. *3 Biotech, 9*, Article 1.

Han, J., Li, M., Hou, J., Wu, L., Zhou, J., & Xiang, H. (2010). Comparison of four *phaC* genes from *Haloferax mediterranei* and their function in different PHBV copolymer biosyntheses in *Haloarcula hispanica*. *Saline Systems, 6*, 9.

Hegazy, M. E., El-Baz, A. F., El-Ghandour, H., Abu-Serie, M. M., Abo-Elela, G. M., Ghozlan, H., Sabry, S. A., Soliman, N. A., & Abdel-Fattah, Y. R. (2020). Dual anticancer and antiviral activity of bacterioruberin-rich extracts from *Natrialba* sp. M6. *Scientific Reports, 10*, 5986.

Hegazy, G. E., Abu-Serie, M. M., Abou-Elela, G. M., Ghozlan, H., Sabry, S. A., Soliman, N. A., & Abdel-Fattah, Y. R. (2022). Bioprocess development for biosurfactant production by *Natrialba* sp. M6 with effective direct virucidal and anti-replicative potential against HCV and HSV. *Scientific Reports, 12*, 16577.

Hezayen, F. F., Rehm, B. H. A., Eberhardt, R., & Steinbüchel, A. (2000). Polymer production by two newly isolated extremely halophilic archaea: Application of a novel corrosion-resistant bioreactor. *Applied Microbiology and Biotechnology, 54*, 319–325.

Higa, L. H., Schilrreff, P., de Oliveira, C. I., Briski, A. M., Jerez, H. E., de Farias, M. A., Villares Portugal, R., Romero, E. L., & Morilla, M. J. (2020). Bacterioruberin and dexamethasone co-loaded nanoparticles restore intestinal barrier function in colitis models. *Colloids and Surfaces B: Biointerfaces, 191*, 110961.

Holt, N. E., Zigmantas, D., Valkunas, L., Li, X. P., Niyogi, K. K., & Fleming, G. R. (2005). Carotenoid cation formation and the regulation of photosynthetic light harvesting. *Science, 307*, 433–436.

Horikoshi, K., Aono, R., & Nakamura, S. (2011). The Triangular halophilic archaeabacteria *Haloarcuka japonica* strain TR-1. *Experientia, 49*, 497–502.

Huang, D., Ou, B., & Prior, R. L. (2005). The chemistry behind antioxidant capacity assays. *Journal of Agricultural and Food Chemistry, 53*, 1841–1856.

Hwang, C. Y., Cho, E.-S., & Seo, M.-J. (2024). Enhanced production of C_{50} carotenoid bacterioruberin by *Halorubrum ruber* MBLA0099 through fermentation optimization. *Microbial Cell Factories, 23*, 88.

Kasirajan, L., & Maupin-Furlow, J. A. (2021). Halophilic archaea and their potential to generate renewable fuels and chemicals. *Biotechnology and Bioengineering, 118*, 1066–1090.

Kessel, M., Wildhaber, I., Cohen, S., & Baumeister, W. (1988). Three-dimensional structure of the regular surface glycoprotein layer of *Halobacterium volcanii* from the Dead Sea. *The EMBO Journal, 7*, 1549–1554.

Kramer, J. K. G., Kushwaha, S. C., & Kates, M. (1972). Structure determination of the squalene, dihydrosqualene and tetrahydrosqualene in *Halobacterium cutirubrum*. *Biochimica et Biophysica Acta, 270*, 103–110.

Kushner, D. J. (1978). Microbial life in extreme environments. In D. J. Kushner (Ed.), *Microbial life in extreme environments* (p. 317). Academic Press.

Li, Y., Xiang, H., Liu, J., Zhou, M., & Tan, H. (2003). Purification and bio- logical characterization of halocin C8, a novel peptide antibiotic from *Halobacterium* strain AS7092. *Extremophiles, 7*, 401–407.

Li, Y.-T., Tian, Y., Tian, H., Tu, T., Gou, G.-Y., Wang, Q., Qiao, Y.-C., Yang, Y., & Ren, T.-L. (2018). A review on bacteriorhodopsin-based bioelectronic devices. *Sensors, 18*, 1368.

Ma, Y., Sun, Z., Yang, H., Xie, W., Song, M., Zhang, B., & Sui, L. (2024). The biosynthesis mechanism of bacterioruberin in halophilic archaea revealed by genome and transcriptome analysis. *Applied and Environmental Microbiology, 90*, e00540–e00524.

Madigan, M. T., & Oren, A. (1999). Thermophilic and halophilic extremophiles. *Current Opinion in Microbiology, 2*, 265–269.

Mandelli, F., Miranda, S. V., Rodrigus, E., & Marcadante, A. A. (2012). Identification of carotenoids with high antioxidant capacity produced by extremophile microorganisms. *World Journal of Microbiology and Biotechnology, 28*, 1781–1790.

Maresca, J. A., Graham, J. E., & Bryant, D. A. (2008). The biochemical basis for structural diversity in the carotenoids of chlorophototrophic bacteria. *Photosynthesis Research, 97*, 121–140.

Mescher, M. F., Strominger, J. L., & Watson, S. W. (1974). Protein and carbohydrate composition of the cell envelope of *Halobacterium salinarium*. *Journal of Bacteriology, 120*, 945–954.

Miller, N. J., Sampson, J., Candeias, L. P., Brameley, P. M., & Rice-Evans, C. A. (1996). Antioxidant activities of carotenes and xanthophylls. *FEBS Letter, 384*, 240–242.

Misawa, N., Nakagawa, M., Kobayashi, K., Yamano, S., Izawa, Y., Nakamura, K., & Harashima, K. (1990). Elucidation of the *Erwinia uredovora* carotenoid biosynthetic pathway by functional analysis of gene products expressed in *Escherichia coli*. *Journal of Bacteriology, 172*, 6704–6712.

Moise, A. R., Al-Babili, S., & Wurtzel, E. T. (2014). Mechanistic aspects of carotenoid biosynthesis. *Chemical Reviews, 114*, 164–193.

Mukohata, Y., Sugiyama, Y., Ihara, K., & Yoshida, M. (1988). An Australian *Halobacterium* contains a novel protein pump retinal protein: Archaerhodopsin. *Biochemical and Biophysical Research Communications, 151*, 1339–13445.

Mukohata, Y., Ihara, K., Tamura, T., & Sugiyama, Y. (1999). Halobacterial rhodopsins. *The Journal of Biochemistry, 125*, 649–657.

Naguib, Y. M. (2000). Antioxidant activities of astaxanthin and related carotenoids. *Journal of Agricultural and Food Chemistry, 48*, 1150–1154.

Nakamura, S., Aono, R., Mizutani, S., Takashina, T., Grant, W. D., & Horikoshi, K. (1992). The cell surface glycoprotein of *Haloarcula japonica* TR-1. *Bioscience, Biotechnology, and Biochemistry, 56*, 996–998.

Nakamura, S., Mizutani, S., Wakai, H., Kawasaki, H., Aono, R., & Horikoshi, K. (1995). Purification and partial characterization of cell surface glycoprotein from extremely halophilic archaeon *Haloarcula japonica* strain TR-1. *Biotechnology Letters, 17*, 705–706.

Nakamura, S., Nakasone, K., & Takashina, T. (2011). Haloarcula japonica. In K. Horikoshi, G. Antranikian, A. T. Bull, F. T. Robb, & K. O. Stetter (Eds.), *Extremophiles handbook* (pp. 364–381). Springer.

Naziri, D., Hamidi, M., Hassanzadeh, S., Tarhriz, V., Maleki Zanjani, B., Nazemyieh, H., Hejazi, M. A., & Hejazi, M. S. (2014). Analysis of carotenoid production by *Halorubrum* sp. TBZ126: An extremely halophilic archaeon from Urmia Lake. *Advanced Pharmaceutical Bulletin, 4*, 61–67.

Nishiyama, Y., Takashina, T., Grant, W. D., & Horikoshi, K. (1992). Ultrastructure of the cell wall of the triangular halophilic archaebacterium *Haloarcula japonica* strain TR-1. *FEMS Microbiology Letter, 9*, 43–48.

Nishiyama, Y., Nakamura, S., Aono, R., & Horikoshi, K. (1995). Electron microscopy of halobacteria. In F. T. Robb, K. R. Sowers, S. DasSarma, A. P. Place, H. J. Schreier, & E. M. Fleischmann (Eds.), *Archaea: A laboratory manual, halophiles* (pp. 29–33). Cold Spring Harbor Laboratory Press.

Oesterhelt, D., & Stoeckenius, W. (1971). Rhodopsin-like protein from the purple membrane of *Halobacterium halobium. Nature: New Biology, 233*, 149–152.

Oren, A. (2024). Novel insights into the diversity of halophilic microorganisms and their functioning in hypersaline ecosystems. *npj Biodiversity, 3*, 50.

Otozai, K., Takashina, T., & Grant, W. D. (1991). A novel triangular archaebacterium, *Haloarcula japonica*. In K. Horikoshi & W. D. Grant (Eds.), *Superbugs: Microorganisms in extreme environments* (pp. 63–75). Japan Scientific Societies Press.

Peck, R. F., Echavarri-Erasun, C., Johnson, E. A., Ng, W. V., Kennedy, S. P., Hood, L., DasSarma, S., & Krebs, M. P. (2001). *brp* and *blh* are required for synthesis of the retinal cofactor of bacteriorhodopsin in *Halobacterium salinarum. Journal of Biological Chemistry, 276*, 5739–5744.

Peck, R. F., Johonson, E. A., & Krebs, M. P. (2002). Identification of a lycopene β-cyclase required for bacteriorhodopsin biogenesis in the archaeon *Halobacterium salinarum. Journal of Bacteriology, 184*, 2889–2897.

Rehm, B. H. A. (2003). Polyester synthases: Natural catalysts for plastics. *Biochemical Journal, 376*, 15–33.

Rodrigo-Baños, M., Garbayo, I., Vílchez, C., Bonete, M. J., & Martínez-Espinosa, R. M. (2015). Carotenoids from haloarchaea and their potential in biotechnology. *Marine Drugs, 13*, 5508–5532.

Sandmann, G. (1994). Phytoene desaturase: Genes, enzymes and phylogenetic aspects. *Journal of Plant Physiology, 143*, 444–447.

Schobert, B., & Lanyi, J. K. (1982). Halorhodopsin is a light-driven chloride pump. *Journal of Biological Chemistry, 257*, 10306–10313.

Shahbazi, M., Ghanbarzadeh, S., Hamishehkar, H., Mohammadi, M., Khazaei, M., & Ebrahimi, M. (2023). Apoptotic effects of bacterioruberin from *Haloarcula* sp. A15 on MCF-7 cells. *Cell Biochemistry and Function, 41*, 344–354.

Shahmohammadi, H. R., Asgarani, E., Terato, H., Saito, T., Ohyama, Y., Gekko, K., Yamamoto, O., & Ide, H. (1998). Protective roles of bacterioruberin and intracellular KCl in the resistance of *Halobacterium salinarium* against DNA-damaging agents. *Journal of Radiation Research, 39*, 251–262.

Sleytr, U. B., & Messner, P. (1983). Crystalline surface layers on bacteria. *Annual Review of Microbiology, 37*, 311–339.
Sumper, M. (1987). Halobacterial glycoprotein biosynthesis. *Biochimica et Biophysica Acta, 906*, 69–79.
Sun, Z. T., Shen, S. C., Wang, C., Wang, H., Hu, Y. P., Jiao, J. D., Ma, T. T., Tian, B., & Hua, Y. J. (2009). A novel carotenoid 1,2-hydratase (CruF) from two species of the non-photosynthetic bacterium *Deinococcus*. *Microbiology, 155*(8), 2775–2783.
Takashina, T., Hamamoto, T., Otozai, K., Grant, W. D., & Horikoshi, K. (1990). *Haloarcula japonica* sp. nov., a new triangular halophilic archaebacterium. *Systematic and Applied Microbiology, 13*(2), 177–181.
Tian, B., Xu, Z., Sun, Z., Lin, J., & Hua, Y. (2007). Evaluation of the antioxidant effects of carotenoids from *Deinococcus radiodurans* through targeted mutagenesis, chemiluminescence, and DNA damage analysis. *Biochimica et Biophysica Acta, 1770*, 902–911.
Torreblanca, M., Rodriguez-Valera, F., Juez, G., Ventosa, A., Kamekura, M., & Kates, M. (1986). Classification of non-alkaliphilic halobacteria based on numerical taxonomy and polar lipid composition, and description of *Haloarcula* gen. nov. and *Haloferax* gen. nov. *Systematic and Applied Microbiology, 8*, 89–99.
Umemura, K., Wakai, H., Takada, K., Nakamura, S., & Hara. (1998). Atomic force microscopy of an extremely halophilic archaeon using rapid membrane filtration method. *Bioimages, 6*, 77–81.
Verma, M., Devi, B. P., Singh, S. P., Reddy, G. B., Kavitake, D., & Shetty, P. H. (2025). Microbial poly-glutamic acid: Production, biosynthesis, properties, and their applications in food, environment, and biomedicals. *Fermentation, 11*, 208.
Wakai, H., Nakamura, S., Kawasaki, H., Takada, K., Mizutani, S., Aono, R., & Horikoshi, K. (1997). Cloning and sequencing of the gene encoding the cell surface glycoprotein of *Haloarcula japonica* strain TR-1. *Extremophiles, 1*, 29–35.
Wieland, F., Dompert, W., Bernhardt, G., & Sumper, M. (1980). Halobacterial glycoprotein saccharides contain covalently linked sulphate. *FEBS Letter, 120*, 110–114.
Yamasaki, D., Minouchi, Y., & Ashiuchi, M. (2010). Extremolyte-like applicability of an archaeal exopolymer, poly-γ-L-glutamate. *Environmental Technology, 31*, 1129–1134.
Yang, Y., Yatsunami, R., Ando, A., Miyoko, N., Fukui, T., Takaichi, S., & Nakamura, S. (2015). Complete biosynthetic pathway of the C50 carotenoid bacterioruberin from lycopene in the extremely halophilic archaeon *Haloarcula japonica*. *Journal of Bacteriology, 197*, 1614–1623.
Yatsunami, R., Kawakami, T., Ohtani, H., & Nakamura, S. (2000). A novel bacteriorhodopsin-like protein from *Haloarcula japonica* strain TR-1: Gene cloning, sequencing and transcript analysis. *Extremophiles, 4*, 109–144.
Yatsunami, Y., Iwamoto, M., Ito, S., & Nakamura, S. (2003). A *brp* homolog of *Haloarcula japonica*: Gene cloning and transcript analysis. *Nucleic Acids Research: Special Publication, 3*, 267–268.
Yatsunami, R., Ando, A., Yang, Y., Takaichi, S., Kohno, M., Matsumura, Y., Ikeda, H., Fukui, T., Nakasone, K., & Fujita, N. (2014). Identification of carotenoids from the extremely halophilic archaeon *Haloarcula japonica*. *Frontiers in Microbiology, 5*, 100.
Yatsunami, R., Ando, A., Miyoko, N., Yang, Y., Takaichi, S., & Nakamura, S. (2024). Two distinct enzymes have both phytoene desaturase and 3,4-desaturase activities involved in carotenoid biosynthesis of extremely halophilic archaeon *Haloarcula japonica*. *Microbes and Environments, 39*, ME24004.
Yoshimura, K., & Kouyama, T. (2008). Structural role of water molecules in bacteriorhodopsin. *Journal of Molecular Biology, 375*, 1267–1278.

Chapter 10
Industrial Applications and Challenges of Antarctic-Derived Fungi

Masaharu Tsuji

Abstract Antarctica, renowned for its extreme harshness, unexpectedly harbors over 1000 species of fungi. These fungi have recently garnered significant attention due to their remarkable cold-adapted characteristics, making them potential microbial resources. This chapter explores the history of fungi research conducted during the Japanese Antarctic Regional Expedition (JARE). Subsequently, it presents the author's applied studies, which investigate the utilization of Antarctic fungi for wastewater treatment and brewing purposes.

Keywords Antarctic fungi · Dairy-wastewater treatment · Alcohol fermentation

10.1 Introduction

Antarctica, characterized by the Antarctic Continent covered in ice and snow, is part of the Polar Regions. Antarctica is one of the planet's extreme environments, exposed to cold and dry conditions with the lowest temperature ever recorded at −90 °C. Approximately 98% of Antarctica's surface is covered in snow and ice, while the remaining 2% is an ice-free area located in coastal areas and high mountains. In this ice-free area, snow melts in the summer, exposing the ground, and most of Antarctica's terrestrial ecosystems are distributed there (Onofri et al., 2007).

The first fungus recorded from Antarctica was *Scleotium antarcticum*, discovered during the SY Belgica expedition in 1900 at Danco Land, Antarctic Peninsula (Boomer & Rousseau, 1900). Since then, over 1000 fungal species from 421 genera have been isolated and documented in Antarctica. The list of known species,

M. Tsuji (✉)
Department of Science of Technology Innovation, Nagaoka University of Technology, Nagaoka, Niigata, Japan

National Institute of Polar Research (NIPR), Tokyo, Japan

M. Ito et al. (eds.), *Extremophiles for a Sustainable Future*,
https://doi.org/10.1007/978-981-95-6970-0_10

obtained through culturing and collection, comprises 68% ascomycetes, 23% basidiomycetes, 5% zygomycetes, and the remaining 4% includes various other lineages (Bridge & Spooner, 2012).

10.2 History of Fungal Research Near the Syowa Station

The inaugural report on fungi discovered in proximity to Syowa Station, the administrative hub of the Japanese Antarctic Research Expedition (JARE), was published in 1961 (Tubaki, 1961a). Subsequently, three additional articles were released, detailing the discovery of twelve ascomycetes and four basidiomycetes (Tubaki, 1961b; Soneda, 1961; Tubaki & Asano, 1965). Despite the suspension of the JARE in the 1960s, *Moesziomyces antarcticus* was isolated from Ross Island in 1969 (Goto et al., 1969). This species stands out for its ability to secrete a stable lipase and produce biosurfactants (Kitamoto et al., 1993).

From 1965 to 2012, the Japan Aerospace Exploration Agency (JARE) did not report any further fungal species in the vicinity of Syowa Station. However, mycologists Dr. Tamotsu Hoshino and Dr. Takashi Osono conducted a study that revealed partial diversities of fungi in the area. They participated in the 48th JARE (2006–2007) and the 51st JARE (2009–2010), respectively (Tsuji et al., 2013a; Hirose et al., 2013).

Samples collected during JARE-48, JARE-49, and JARE-56 provided a comprehensive survey of fungal diversity in the Skarvsnes ice-free area, East Ongul Island, and Inhovde area (Fig. 10.1). The species names were verified and updated using the Index Fungorum (http://www.indexfungorum.org). While some isolates or strains were not identified at the species level in the original article, their DNA sequences were deposited in a DNA database, enabling an attempt at reclassification. In instances where reclassification at the species level was not feasible, the isolates or strains were presented at the genus level.

In the Skarvsnes ice-free region, a comprehensive study yielded the isolation and classification of 97 fungal strains. These strains were categorized into five Ascomycota genera and five species, while eight Basidiomycota genera and ten species were identified. Notably, the genus *Mrakia* dominated the list, comprising 53% of the isolates. Among the most prevalent species were *Mrakia blollopis* (23%), *M. gelida* (21%), and *M. robertii* (9%). *Mrakia*, a basidiomycete yeast, has been successfully isolated from diverse cold environments across the globe (Tsuji et al., 2013a).

On East Ongul Island, a comprehensive study yielded the isolation and classification of 196 fungal strains. These strains were categorized into eight Ascomycota genera and nine species, while 10 Basidiomycota genera and 16 species were identified. Notably, *Thelebolus microspores* dominated the island's fungal flora, comprising 27% of the total. *Vishniacozyma victoriae* represented 11% of the island's fungal diversity, while *Naganishia friedmannii* accounted for 9% (Tsuji, 2018a).

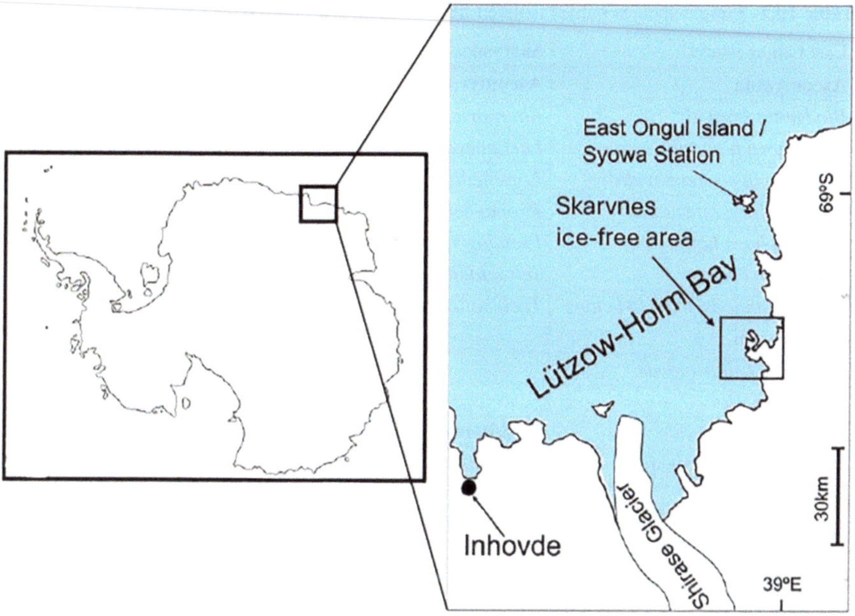

Fig 10.1 Antarctic map and Syowa Station area. From left to right: Antarctic Map, Syowa Station Area Map (with Inhovde, Skarvesnes ice-free area, and East Ongul Island)

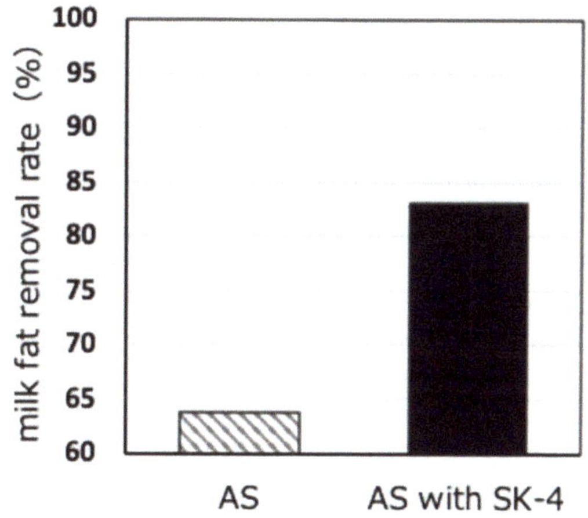

Fig. 10.2 Milk fat degradation ability using batch model dairy wastewater by activated sludge method (Tsuji et al., 2013c). Abbreviations for Fig. 10.2 are activated sludge (AS) and activated sludge containing *Mrakia blollopis* SK-4 (AS with Sk-4)

The Inhovde region, situated in East Antarctica, is approximately 130 kilometers from Syowa Station. A comprehensive study yielded the isolation of 148 fungal strains, with 132 successfully analyzed through DNA sequencing. These strains were subsequently classified into four genera belonging to the Ascomycota phylum and four genera belonging to the Basidiomycota phylum, along with 132 and 10 species, respectively. Notably, *Vishniacozyma victoriae* (35.6%), *Glaciozyma*

Table 10.1 Fungal species names isolated from three sites around Syowa Station

East Ongul Island	Skarvsnes ice-free area	Inhovde area
Ascomycota	**Ascomycota**	**Ascomycota**
Blodgettia bornetii	*Alternaria embellisia*	*Aotearoamyces* sp.
Cadophora luteo-olivacea	*Laetinaevia carneoflavida*	*Leuconeurospora* sp.
Laetinaevia carneoflavida	*Lecanicillium attenuatum*	*Pseudeurotium hygrophilum*
Lecanicillium attenuatum	*Phoma herbarum*	*Symbiotaphrina microtheca*
Lecythophora hoffmannii	*Pseudogymnoascus pannorum*	
Phoma herbarum	*Tetracladium* sp.	
Pseudogymnoascus destructans	*Thelebolus microsporus*	
P. pannorum		
Thelebolus ellipsoideus		
T. microsporus		
Basidiomycota	**Basidiomycota**	**Basidiomycota**
Cystobasidium lysinophilum	*Cystobasidium laryngis*	*C. ongulense*
C. ongulense	*Dioszegia fristingensis*	*Cystobasidium* sp.
C. tubakii	*Glaciozyma watsonii*	*Glaciozyma watsonii*
Glaciozyma antarctica	*Goffeauzyma gastrica*	*Leucosporidium yakuticum*
G. martinii	*Mrakia blollopis*	*Mrakia gelida*
Goffeauzyma gilvescens	*M. gelida*	*M. robertii*
Holtermanniella wattica	*M. roberti*	*Naganishia adeliensis*
Mrakia gelida	*Phenoliferia glacialis*	*N. albidosimilis*
Naganishia adeliensis	*Vishniacozyma victoriae*	*N. friedmannii*
N. albidosimilis		*Vishniacozyma. victoriae*
N. friedmannii		
Phenoliferia glacialis		
Tausonia pullulans		
Udeniomyces puniceus		
Vishniacozyma carnescens		
V. victoriae		

watsonii (28.0%), and *Cystobasidium ongulense* (15.2%) emerged as the most frequently isolated fungal species (Tsuji et al., 2024).

A comprehensive list of fungal species isolated from the three sites is presented in Table 10.1. Notably, three genera and two species of basidiomycetous fungi, specifically *M. gelida* and *V. victoriae*, were prevalent across all three locations. East Ongul Island, home to the Syowa Station, is situated approximately 60 kilometers from the Skarvsnes ice-free region, while Inhovde is approximately 130 kilometers away. The fungal diversity in these areas surrounding the Syowa Station exhibited significant variations. East Ongul Island exhibited the highest level of diversity, followed by Skarvsnes and Inhovde, which demonstrated lower levels of diversity. These findings strongly suggest that human activities exert a notable influence on the fungal diversity in the vicinity of the Syowa Station.

During the fungal surveys near Syowa Station, two novel fungal species, *C. tubakii* and *C. ongulense*, were discovered for the first time in JARE's history (Tsuji et al., 2017).

Consequently, the number of fungal species reported in the vicinity of Syowa Station experienced a significant surge, increasing from 16 to an astonishing 92 species over a 12-year period. This remarkable rise represents an approximate 5.8-fold increase (Tsuji, 2018b).

The National Institute of Polar Research (NIPR) in Japan has amassed an extensive collection of over 3000 strains of Antarctic fungi. This comprehensive database encompasses species names, marker gene sequences, and detailed information regarding the isolation sites of each strain. These strains are readily available for collaboration with NIPR. For further information, please visit the NIPR Biosphere Research Group website at https://www.nipr.ac.jp/biology/.

Additionally, we are collaborating with the National Institute of Genetics, Japan and Chiba University to establish a freely accessible genome database for a substantial portion of Antarctic fungi. This database encompasses complete genome sequences, gene sequences, and gene function predictions (https://antarcfungi.annotation.jp).

10.3 Industrial Applications of Antarctic Fungi

Low-Temperature Dairy Wastewater Treatment Utilizing Antarctic Fungi Presents a Substantial Challenge

Sewage containing milk fat curdle necessitates active sludge treatment, but this becomes challenging in cold temperatures. Dairy wastewater discharge, often untreated or partially treated, contaminates rivers and groundwater, potentially containing detergents, bactericides, mucus, lactose, and milk fat (Healy et al., 2007).

Various treatment systems encounter operational difficulties due to high costs and space requirements. Activated sludge systems are widely adopted due to their ease of maintenance and lower costs, but they face limitations during winter due to low temperatures (Ying et al., 2010).

Recent research indicates that microorganisms isolated from polar regions can remove nitrogen and phosphorus compounds from wastewater under low-temperature conditions (Chevalier et al., 2000; Hirayama-Katayama et al., 2003). However, their application to milk fat remains unexplored.

To identify an appropriate agent for bioremediation, fungi were collected from various locations and tested for their ability to decompose milk fat on fresh cream agar at 10 °C. Only Antarctic basidiomycetous yeast of the genus *Mrakia* exhibited the capacity to degrade milk fat. Among the *Mrakia* strains, *M. blollopis* strain SK-4 demonstrated the highest ability to degrade milk fat at both 4 °C and 10 °C (Shimohara et al., 2012; Tsuji et al., 2013b).

Mrakia blollopis SK-4 exhibits a maximum growth temperature of 22 °C, whereas other species have a maximum growth temperature of less than 20 °C. Furthermore, *M. blollopis* SK-4 can thrive on lactose, D-arabinose, and inositol as carbon sources. These findings suggest that it could be a potential candidate for decomposing milk parlor wastewater sugars under low temperatures (Tsuji et al., 2013c).

To test this hypothesis, activated sludge (AS) from dairy facilities in Hokkaido, Japan, and AS containing SK-4 were utilized to treat model dairy wastewater using milk as a substrate in a batch system. The wastewater underwent aeration, sedimentation, and the treated water was discharged. The treatment was conducted at 10 °C for 24 h. The results indicated that AS containing *M. blollopis* SK-4 removed 1.25 times more milk fat compared to the control.

A continuous model test was conducted in a 3.5 L wastewater treatment tank with a 2.9 L aeration tank and a 0.6 L settling tank for 35 weeks. Temperature varied between 3 °C and 25 °C. Milk fat removal rate and SK-4 survival rate were measured at each sampling time.

Under all temperatures, the average milk fat removal rate exceeded 80%. At 10 °C, it was 86.0%, at 20 °C, it was 96.5%, at 25 °C, it was 97.3%, and at 3 °C, it was 83.5%. SK-4 survival gradually decreased between the start of the experiment and the 23rd week, with an 80% decrease occurring after 1 week. After 4 weeks, the cells experienced a slight increase, but they decreased when the temperature was raised to 20 °C–25 °C between the 8th and 23rd weeks. After the 23rd week, the temperature was lowered to 3 °C, and SK-4 cells gradually increased through the 30th week (Fig. 10.3).

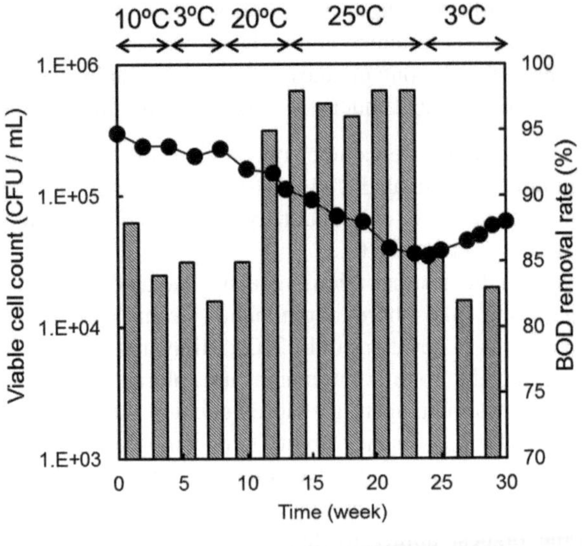

Fig. 10.3 Continuous model wastewater treatment by activated sludge containing *Mrakia blollopis* SK-4 (Tsuji et al., 2019). The left-hand side of the vertical axis shows the viable cell counts (CFU/mL) and right-hand side of the vertical axis indicates the milk fat removal rate (%). The horizontal axis indicates time (weeks).

The water temperature of actual dairy wastewater in Hokkaido, Japan, ranges from 3 °C in winter to 25 °C in summer. Since the maximum growth temperature of SK-4 is 22 °C, we assessed its milk fat removal rate in activated sludge (AS) and its survival at 20 °C and 25 °C. Model experiments showed that SK-4's colony counts decreased over 23 weeks but increased when the temperature was changed to 3 °C. Therefore, SK-4 can survive at 20–25 °C for 15 weeks and fix the AS (Tsuji et al., 2019). These results suggest efficient dairy wastewater treatment with SK-4 in AS under low-temperature conditions. SK-4 meets biosafety level-l criteria, being harmless to animals, plants, and humans, and presenting the lowest environmental risk. Hence, it's a promising biological agent for dairy wastewater treatment, even in cold regions.

The SK-4 strain's lipase, capable of decomposing milk fat at low temperatures, was purified and analyzed. The results showed an optimal activity temperature of 60–65 °C, thermal stability up to 65 °C, and an optimal activity pH of 8–9, showing stability in the pH range of 3–10. These research results, registered in Japan as Patent No. 5867954, "Basidiomycetous yeast from the south pole having milk fat decomposition ability and use of the same," are currently sold by private companies in Japan as a dairy wastewater treatment system. This wastewater treatment technology using Antarctic microbes can be used in the future for treating domestic and industrial wastewater in cold regions, as well as at Syowa Station. The highly stable lipase produced by SK-4 has potential for industrial use.

10.4 Application of Antarctic Fungi to Alcoholic Beverages

The harsh environment of continental Antarctica is home to a high concentration of single-celled fungi called basidiomycetes yeast. *Mrakia* is the only fungus capable of ethanol fermentation in Antarctica (Tsuji et al., 2016).

Ethanol fermentation by *Mrakia* spp. has been known since the 1960s (Sinclair & Stokes, 1965). They can ferment more efficiently than common yeast at low temperatures. Thomas-Hall et al. (2010) reported homemade beers made with *Mrakia* species native to Antarctica. *M. frigida*, *M. blollopis*, *M. gelida*, and *M. robertii* fermented sucrose at 6 °C and 15 °C, but cell growth ceased when ethanol levels exceeded 2% (v/v) (Thomas-Hall et al., 2010). These reports didn't investigate ethanol production in detail.

We tested ethanol production by 27 strains isolated from Skarvsnes ice-free area, East Antarctica. Eight strains produced over 2% (v/v) ethanol in a medium with 40 g/L glucose at 10 °C (Tsuji et al., 2016). We didn't know the ethanol concentration limit for *Mrakia*. We conducted ethanol fermentation in a medium with a high glucose concentration (100 g/L or 120 g/L) using six *Mrakia* strains selected based on their ethanol productivity in a medium with 40 g/L glucose. Two strains (*M. blollopis* SK-4 and NIR-4) produced over 4% (v/v) ethanol at 10 °C. Strain SK-4 produced approximately 4.8% (v/v) ethanol in a medium with 120 g/L glucose at 10 °C (Tsuji et al., 2013d).

Mrakia blollopis SK-4 was isolated from an algal mat in the lake sediment of Naga-ike, a lake located in the Skarvsnes ice-free area of East Antarctica. SK-4 possesses the ability to produce ethanol at temperatures ranging from −1 to 20 °C, with an optimal temperature range of 10 to 15 °C. SK-4 can produce ethanol within a pH range of 5.0–10.0, although it does not completely convert glucose to ethanol at pH levels below 4.0 or above 10.5. The optimal pH ranges for maximum ethanol productivity are 4.5 to 5.0 and 8.0 to 10.0 (Tsuji et al., 2013e).

Under low temperatures, SK-4 secretes extracellular enzymes like cellulase, β-glucosidase, catalase, amylase, and lipase. Notably, SK-4's lipase is stable against metal ions and organic solvents, and its optimum pH range is 8.5–9.0 (Tsuji et al., 2013c). Naga-ike, an oligotrophic lake, has a pH of 8.5 (Tsuji et al., 2013a).

Studies have documented ethanol tolerance and productivity of conventional yeast species *Saccharomyces cerevisiae* (Beaven et al., 1981; You et al., 2003). In contrast, sake yeasts produce high ethanol but have low stress tolerance. Within *S. cerevisiae*, Rim15p, a PAS kinase, is upstream of stress-responsive genes Msn2P and Msn4P. In sake yeast strains, Rim15p is truncated due to a mutation. Deleting the full-length Rim15p gene enhances fermentation rate but compromises stress tolerance, including ethanol tolerance. Consequently, Rim15p deletion mutants of sake yeasts exhibit low ethanol tolerance but robust ethanol fermentation (Watanabe et al., 2012).

We assessed ethanol tolerance of six *Mrakia* strains near the Syowa Station in media with varying ethanol concentrations (2%, 4%, and 8%). Our findings showed a correlation between ethanol tolerance and ethanol productivity in *Mrakia* strains (Tsuji et al., 2016). High ethanol-producing *Mrakia* strains exhibit enhanced ethanol tolerance. Notably, *Mrakia* strains from the Syowa Station site exhibit high osmotic pressure tolerance, such as 10% NaCl or 50% glucose. Ethanol is cytotoxic, and wild-type fungi typically cannot produce more than 4% ethanol due to feedback repression. In *S. cerevisiae*, ethanol tolerance is inversely proportional to ethanol productivity. However, *Mrakia* strains exhibit distinct characteristics, suggesting a distinct ethanol fermentation system compared to conventional sake yeast.

Antarctic basidiomycetous yeast *M. blollopis* SK-4 can produce 5% ethanol, the highest among its genus.

We attempted to brew sake using only SK-4 at 15 °C, but the sake mash (moromi) had less than 2% ethanol (Fig. 10.4). Moromi is a starter mixture of steamed rice, water, and koji fungus. The rice starch is digested into glucose by the koji fungus, but the glucose production may exceed SK-4's growth rate. This high glucose concentration and osmotic pressure likely contributed to the low yield.

To address this, we conducted co-fermentation experiments with SK-4 and sake yeast strain K 701. We varied the initial cell density ratios of SK-4 and K701.

After evaluating various conditions, we found that an initial inoculum ratio of 40% SK-4 and 60% K701 yielded the same total decrement value as brewing sake with K701 alone (Fig. 10.4). Using 40% SK-4 with 60% K701, we estimated a final ethanol concentration of 19.7% (v/v), while using K701 alone, we estimated 19.5% (v/v).

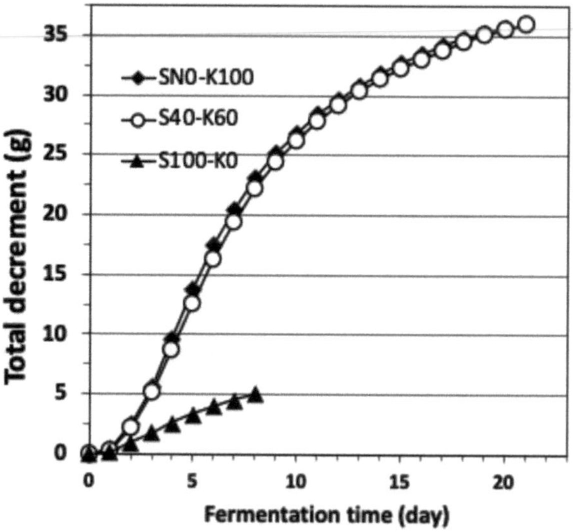

Fig. 10.4 Total decrement in sake mash (Tsuji & Hoshino, 2019). SN0-K100 indicates sake brewed by 100% K701. S40-K60 shows sake brewed by 40% SK-4 and 60% K701 used as initial inoculum. S100-K0 indicates sake brewed by 100% SK-4

Addition of SK4 to sake mash increased isoamyl acetate content by 1.25-fold compared to K701 alone. Iso-amyl acetate, a ginjoko, imparts a banana-like fragrance to sake. Co-fermentation with SK4 and K701 reduced amino acid content in sake compared to K701 alone. After yeast cell death, proteases digest peptides and proteins in steamed rice, increasing amino acid content. Cell death ratios after brewing with SK4 and K701 were 43.9% and 32.2%, respectively. SK4 significantly reduced cell death, suggesting the reduction in amino acid content is due to improved cell death. Consequently, ethanol concentration in co-fermentation was likely higher than with K701 alone (Tsuji & Hoshino, 2019).

Co-fermentation with SK4 and K701 resulted in quality sake similar to K701 alone.

Mrakia blollopis, a Basidiomycota and Tremellales member, has high β-glucan concentrations in its cell wall, which have immunostimulating and allergy-preventing properties. Strain SK4 meets Biosafety Level 1 criteria, making it safe for animals, plants, and humans. Therefore, SK4 is a strong candidate for alcoholic beverage production and amazake, a traditional Japanese sweet made from sake mash.

References

Beaven, M. J., Charpentier, C., & Rose, A. H. (1981). Production and tolerance of ethanol in relation to phospholipid fatty-acyl composition in *Saccharomyces cerevisiae* NCYC 431. *Microbiology, 128*(7), 1447–1455. https://doi.org/10.1099/00221287-128-7-1447

Boomer, E., & Rousseau, M. (1900). Note préliminaire sur les champignons recueillis par l'Expedition Antarctique Belge. *Bulletin de l'Académie Royale des Sciences de Belgique Classe des Sciences, 8*, 640–646.

Bridge, P. D., & Spooner, B. M. (2012). Non-lichenized Antarctic fungi: Transient visitors or members of a cryptic ecosystem? *Fungal Ecology, 5*, 381–394. https://doi.org/10.1016/j.funeco.2012.01.007

Chevalier, P., Proulx, D., Lessard, P., Vincent, W. F., & de la Noue, J. (2000). Nitrogen and phosphorus removal by high latitude mat-forming cyanobacteria for potential use in tertiary wastewater treatment. *Journal of Applied Phycology, 12*, 105–112. https://doi.org/10.1023/A:1008168128654

Goto, S., Sugiyama, J., & Iizuka, H. (1969). A taxonomic study of Antarctic yeasts. *Mycologia, 61*, 784–774. https://doi.org/10.1080/00275514.1969.12018794

Healy, M. G., Rodgers, M., & Mulqueen, J. (2007). Treatment of dairy wastewater using constructed wetland and intermittent sand filter. *Bioresource Technology, 98*(12), 2268–2281. https://doi.org/10.1016/j.biortech.2006.07.036

Hirayama-Katayama, K., Koike, Y., Kobayashi, K., & Hirayama, K. (2003). Removal of nitrogen by Antarctic yeast cells at low temperature. *Polar Bioscience, 16*, 43–48.

Hirose, D., Tanabe, Y., Uchida, M., Kudoh, S., & Osono, T. (2013). Microfungi associated with withering willow wood in ground contact near Syowa Station, East Antarctica for 40 years. *Polar Biology, 36*, 919–924. https://doi.org/10.1007/s00300-013-1320-x

Kitamoto, D., Yanagishita, H., Shinbo, T., Nakane, T., Kamisawa, C., & Nakahara, T. (1993). Surface active properties and antimicrobial activities of mannosylerythritol lipids as biosurfactants produced by *Candida Antarctica*. *Journal of Biotechnology, 29*(1), 91–96. https://doi.org/10.1016/0168-1656(93)90042-L

Onofri, S., Zucconi, L., & Tosi, S. (2007). *Continental Antarctic Fungi*. IHW Verlag.

Shimohara, K., Fujiu, S., Tsuji, M., Kudoh, S., Hoshino, T., & Yumoto, I. (2012). Lipolytic activities and their thermal dependence of Mrakia species, basidiomycetous yeast from Antarctica. *Journal of Water and Waste, 54*, 691–696. (Japnese).

Sinclair, N. A., & Stokes, J. L. (1965). Obligately psychrophilic yeasts from the polar regions. *Canadian Journal of Microbiology, 11*(2), 259–269. https://doi.org/10.1139/m65-032

Soneda, M. (1961). On some yeasts from the Antarctic region. In *Biological results of the Japanese Antarctic Research Expedition* (Vol. 15, pp. 3–10).

Thomas-Hall, S. R., Turchetti, B., Buzzini, P., Branda, E., Boekhout, T., Threelen, B., & Watson, K. (2010). Cold-adapted yeasts from Antarctica and Italian Alps-description of three novel species: *Mrakia robertii* sp. nov., *Mrakia blollopis* sp. nov. and *Mrakiella niccombsii* sp. nov. *Extremophiles, 14*, 47–59. https://doi.org/10.1007/s00792-009-0286-7

Tsuji, M. (2018a). Genetic diversity of yeasts from East Ongul Island, East Antarctica and their extracellular enzymes secretion. *Polar Biology, 41*, 249–258. https://doi.org/10.1007/s00300-017-2185-1

Tsuji, M. (2018b). A catalog of fungi recorded from the vicinity of Syowa Station. *Mycoscience, 59*(4), 319–324. https://doi.org/10.1016/j.myc.2017.08.007

Tsuji, M., & Hoshino, T. (2019). Ethanol fermentation by the Basidiomycetous yeast *Mrakia blollopis* under low temperature conditions. In M. Tsuji & T. Hoshino (Eds.), *Fungi in polar regions* (pp. 110–119). Taylor & Francis.

Tsuji, M., Fujiu, S., Xiao, N., Hanada, Y., Kudoh, S., Kondo, H., Tsuda, S., & Hoshino, T. (2013a). Cold adaptation of fungi obtained from soil and lake sediment in the Skarvsnes ice-free area, Antarctic. *FEMS Microbiology Letters, 346*(2), 121–130. https://doi.org/10.1111/1574-6968.12217

Tsuji, M., Yokota, Y., Kudoh, S., & Hoshino, T. (2013b). Comparative analysis of milk fat decomposition activity by *Mrakia* spp. isolated from Skarvsnes ice-free area, East Antarctica. *Cryobiology, 70*(3), 293–299. https://doi.org/10.1016/j.cryobiol.2015.04.002

Tsuji, M., Yokota, Y., Shimohara, K., Kudoh, S., & Hoshino, T. (2013c). An application of wastewater treatment in a cold environment and stable lipase production of Antarctic basidiomycetous yeast *Mrakia blollopis*. *PLoS One, 8*(3), e59376. https://doi.org/10.1371/journal.pone.0059376

Tsuji, M., Goshima, T., Matsushika, A., Kudoh, S., & Hoshino, T. (2013d). Direct ethanol fermentation from lignocellulosic biomass by Antarctic Basidiomycetous yeast *Mrakia blollopis* under a low temperature condition. *Cryobiology, 67*(2), 241–243. https://doi.org/10.1016/j.cryobiol.2013.06.003

Tsuji, M., Singh, S. M., Yokota, Y., Kudoh, S., & Hoshino, T. (2013e). Influence of initial pH on ethanol production by the Antarctic Basidiomycetous Yeast *Mrakia blollopis*. *Bioscience, Biotechnology, and Biochemistry, 77*(12), 2483–2485.

Tsuji, M., Kudoh, S., & Hoshino, T. (2016). Ethanol productivity of cryophilic basidiomycetous yeast *Mrakia* spp. correlates with ethanol tolerance. *Mycoscience, 57*(1), 42–50. https://doi.org/10.1016/j.myc.2015.08.002

Tsuji, M., Tsujimoto, M., & Imura, S. (2017). *Cystobasidium tubakii* and *Cystobasidium ongulense*, new basidiomycetous yeast species isolated from East Ongul Island, East Antarctica. *Mycoscience, 58*(2), 103–107. https://doi.org/10.1016/j.myc.2016.11.002

Tsuji, M., Kudoh, S., & Hoshino, T. (2019). Dairy wastewater treatment under low-temperature condition by an Antarctic Basidiomycetous Yeast. In M. Tsuji & T. Hoshino (Eds.), *Fungi in polar regions* (pp. 120–132). Taylor & Francis.

Tsuji, M., Kobayashi, T., & Fujikawa, T. (2024). *Fungal diversity in the Inhovde Area, East Antarctica. Preprint* 202407. https://www.preprints.org/manuscript/202407.1884/v1

Tubaki, K. (1961a). On some fungi isolated from the Antarctic materials. In *Biological results of the Japanese Antarctic Research Expedition* (Vol. 14, pp. 1–9). https://doi.org/10.5134/176444

Tubaki, K. (1961b). Note on some fungi and yeasts from Antarctica. *Antarctic Record (Tokyo), 11*, 161–162.

Tubaki, K., & Asano, I. (1965). Additional species of fungi isolated from the Antarctic materials. *JARE Scientific Reports. Series E: Biology, 27*, 1–27.

Watanabe, D., Araki, Y., Zhou, Y., Maeya, N., Akao, T., & Shimoi, H. (2012). A loss-of-function mutation in the PAS kinase Rim15p is related to defective quiescence entry and high fermentation rates in *Saccharomyces cerevisiae* sake yeast strains. *Applied and Environmental Microbiology, 78*(11), 4008–4016. https://doi.org/10.1128/AEM.00165-12

Ying, C., Umetsu, K., Ihara, I., Sakai, Y., & Yamashiro, T. (2010). Simultaneous removal of organic matter and nitrogen form milking parlor wastewater by a magnetic activated sludge (MAS) process. *Bioresource Technology, 101*(12), 4349–4353. https://doi.org/10.1016/j.biortech.2010.01.087

You, K. M., Rosenfield, C. L., & Knipple, D. C. (2003). Ethanol tolerance in the yeast Saccharomyces cerevisiae is dependent on cellular oleic acid content. *Applied and Environmental Microbiology, 69*(3), 1499–1503. https://doi.org/10.1128/aem.69.3.1499-1503.2003

Chapter 11
Frontiers in Ectoine Production Using Halophiles and Future Perspectives

Masahiro Ito

Abstract Ectoine, a natural osmolyte that stabilizes proteins, membranes, and whole cells, is used in cosmetics, foods, and biopharmaceutical formulations. Traditional production relies on salt-rich batch fermentation of native halophiles, which limits scalability and sustainability. This chapter reviews the recent breakthroughs in reshaping the field. Engineered *Halomonas* now achieves >20 g L^{-1} ectoine in non-sterile, low-salt air-lift reactors. In parallel, multi-layered pathway balancing enables *Escherichia coli* and *Corynebacterium glutamicum* to produce >130 g L^{-1} from renewable feedstocks. Advances in *Pichia pastoris* and cell-free systems have further reduced salinity constraints, simplifying downstream purification. Techno-economic and life-cycle assessments indicate the potential to lower the cost of goods below US$10 kg^{-1}, positioning ectoine as a flagship molecule for circular, seawater-based fermentation systems.

Keywords Ectoine · Halophiles · Metabolic engineering · Non-sterile fermentation · Circular bioeconomy

11.1 Introduction

Ectoine (1,4,5,6-tetrahydro-2-methyl-4-pyrimidinecarboxylic acid) is a cyclic amino acid ($C_6H_{10}N_2O_2$) with a zwitterionic structure. It functions as a compatible solute that protects proteins and cellular structures under stresses, such as high osmolarity, heat, and oxidative conditions (Guynn et al., 2023). In addition to its ecological role, ectoine is a sought-after ingredient in dermatological, biopharmaceutical, and food formulations owing to its hydration, anti-inflammatory, and DNA-protective properties. Demand has reached a multibillion-dollar market with

M. Ito (✉)
Faculty of Life Sciences, and Bio-Resilience Research Project (BRRP), Toyo University, Asaka, Saitama, Japan
e-mail: masahiro.ito@toyo.jp

M. Ito et al. (eds.), *Extremophiles for a Sustainable Future*,
https://doi.org/10.1007/978-981-95-6970-0_11

retail prices near US\$1000 kg^{-1} (Daga-Quisbert et al., 2024), creating strong incentives to improve manufacturing economics.

Industrial production began with *Halomonas elongata* via the "bacterial-milking" process (Sauer & Galinski, 1998). Cells accumulate ectoine in 2.5–4 M NaCl and release it upon hypo-osmotic shock, achieving up to 7.4 g L^{-1} day^{-1}. However, high salinity can cause equipment corrosion, generate hypersaline effluents, and increase downstream costs (Pastor et al., 2010; Daga-Quisbert et al., 2024). Similar issues affect the continuous culture of *Chromohalobacter salexigens* and *Halomonas* spp. (>15% NaCl). High salinity also complicates crystallization and increases water use.

Two strategies have emerged to address these bottlenecks:

1. Upgrading native halophiles via genetic and process engineering, e.g., *Halomonas hydrothermalis* Y2 with Δ*doeA* and transporter overexpression achieves >10 g L^{-1} at one-third of the traditional NaCl (Xingrong et al., 2025); *Vreelandella boliviensis* in non-sterile air-lift reactors reaches 1.76 g L^{-1} day^{-1} at 8.5% NaCl (Daga-Quisbert et al., 2024).
2. Transplanting the ectABC(D) pathway into low-salt hosts, enabling *E. coli* and *C. glutamicum* to exceed 100 g L^{-1} with <10 g L^{-1} NaCl and without sterile water or corrosion-resistant vessels (Ma et al., 2024).

Thus, ectoine is positioned as a flagship molecule for the circular bioeconomy (Wang et al., 2025). This chapter reviews biosynthetic/regulatory networks, compares engineering tactics in native and heterologous hosts, examines low-salt and nonsterile process innovations, surveys new applications, and outlines future research directions for AI-assisted enzyme design and marine-integrated biorefineries.

11.2 Biosynthetic Pathway and Regulatory Mechanisms of Ectoine

Ectoine biosynthesis is a tightly regulated process embedded in the central metabolism of aspartate (Fig. 11.1). In natural halophiles, such as *H. elongata*, *C. salexigens*, and *V. boliviensis*, ectoine acts as a compatible solute, enabling adaptation to osmotic stress. Its biosynthesis is controlled by a conserved genetic cluster, precursor flux management, stress-responsive regulation, and the repression of competing pathways.

11.2.1 Genetic Organization: ectABC(D) *Operon*

Canonical ectoine biosynthetic genes are organized in the *ectABC* operon and are often extended with an adjacent *ectD* gene encoding ectoine hydroxylase, which converts ectoine to hydroxyectoine. EctB converts L-aspartate-β-semialdehyde (ASA) to L-2,4-diaminobutyric acid (DABA), EctA acetylates DABA to N-γ-acetyldiaminobutyric acid, and EctC cyclizes this intermediate to form ectoine. In

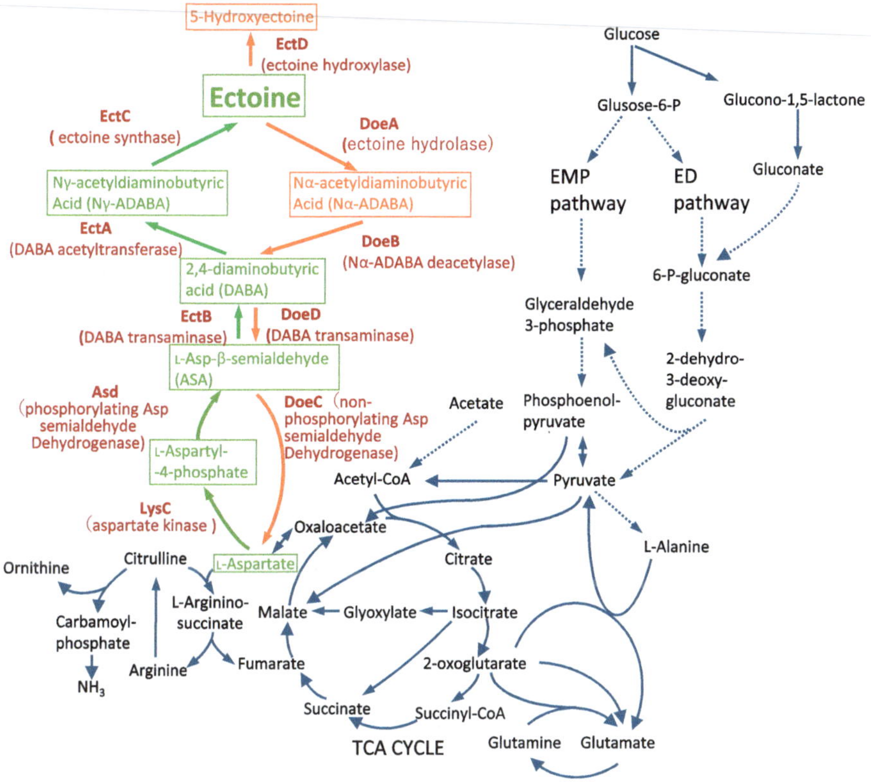

Fig. 11.1 Overview of metabolic pathways in *H. elongata*. The diagram summarizes glycolysis (EMP Embden–Meyerhof–Parnas pathway, ED Entner–Doudoroff pathway), the TCA cycle, and the glutamate biosynthetic route from the TCA intermediate 2-oxoglutarate, ectoine biosynthetic branch, and arginine catabolic pathway via ornithine and carbamoyl phosphate (carbamoyl-P). For the ectoine branch from L-aspartate, enzyme names are indicated for both the biosynthetic (green) and degradation/recycling (orange) routes

some species, *ectD* is colocalized and produces hydroxyectoine, which enhances stress tolerance. Figure 11.2 shows that, under high osmolarity, the pathway for *H. elongata* originates from L-aspartate via aspartate kinase (LysC) and canonical aspartate semialdehyde dehydrogenase to form ASA, followed by the *ectABC(D)* steps, resulting in the intracellular accumulation of ectoine/hydroxyectoine. Tight expression balance among *ectA*, *ectB*, and *ectC* is critical; in *E. coli*, ribosome binding site tuning or 5′-UTR modification improves titers (Lei et al., 2025).

11.2.2 Precursor Supply: Rerouting Central Metabolism

ASA availability limits ectoine flux. In *E. coli* and *C. glutamicum*, ASA is mainly directed toward lysine/threonine biosynthesis. Feedback-resistant aspartokinase variants (e.g., S301Y and T311I) and knockdowns/knockouts of *thrA* and *dapA*

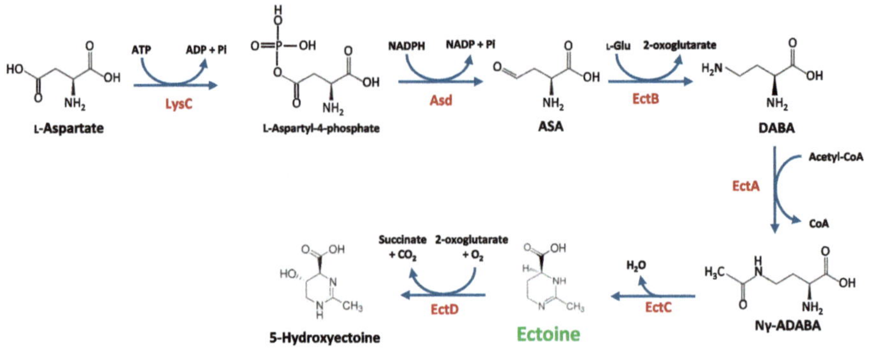

Fig. 11.2 Ectoine and 5-hydroxyectoine biosynthesis in *H. elongata*. Under high osmolarity, *H. elongata* synthesizes ectoine from L-aspartate, initiating the pathway by activating the aspartate kinase (LysC) and canonical aspartate semialdehyde dehydrogenase (Asd) enzymes to produce ASA. Subsequently, the *ectABC* cluster is activated, where EctB converts ASA to DABA, EctA acetylates DABA to Nγ-ADABA, and EctC cyclizes Nγ-ADABA to ectoine; EctD optionally hydroxylates ectoine to 5-hydroxyectoine. These compatible solutes accumulate intracellularly, thereby ensuring osmoprotection. Abbreviations: ASA L-aspartate-β-semialdehyde, DABA L-2,4-diaminobutyric acid, Nγ-ADABA N-γ-acetyldiaminobutyric acid

redirect ASA to ectoine (Ohnishi & Ikeda, 2006; Jiang et al., 2022). Dynamic regulation of *lysA*, coupled with degron tags, maintains growth while limiting lysine overflow, achieving >160 g L^{-1} titers in fed-batch fermentation (Lei et al., 2025). Overexpression of TCA cycle genes, such as *gltA*, increases the oxaloacetate supply and improves yield (Daga-Quisbert et al., 2024). In *H. elongata*, enhanced phosphoenolpyruvate (PEP) carboxylase expression under high-salt conditions converts PEP into oxaloacetate, supported by the suppression of malate decarboxylation (Yu et al., 2025). The deletion of *crr* or *ptsG* and the introduction of non-PTS transporters (*galP* and *glf*) conserve PEP for oxaloacetate synthesis (Lei et al., 2025).

11.2.3 Blocking Ectoine Degradation: The Doe *Operon*

Ectoine catabolism proceeds via the *doeA–D* gene cluster, converting ectoine to aspartate. The deletion of *doeA* or the entire operon in *H. hydrothermalis* Y2 prevents degradation and enhances accumulation (Zhao et al., 2019). In chemostats, blocking catabolic fluxes redirects carbon from maintenance respiration toward ectoine synthesis, particularly under high salinity (Yu et al., 2025).

11.2.4 Salt Stress and Gene Expression Control

Ectoine biosynthesis is upregulated under hyperosmotic stress. In many halophiles, the ect operon is induced by NaCl or KCl via regulators such as EctR1, σ^{38}, and two-component systems (EnvZ/OmpR) (Saum & Müller, 2008; Schwibbert et al., 2011). Common features include increased ectABC transcription, mRNA stabilization, and activation of retention transporters (e.g., TeaABC) (Grammann et al., 2002). Multi-omics studies in *H. elongata* have shown that NaCl elevation to 13% enhances ED pathway flux, suppresses TCA activity, and increases *ectABC* expression (Yu et al., 2025). Some species release ectoine adaptively, aiding in downstream recovery (*V. boliviensis*) (Daga-Quisbert et al., 2024). Engineered regulatory circuits in *E. coli* use growth-phase promoters and degron tags for dynamic control (Lei et al., 2025).

11.2.5 Native Halophilic Hosts and Metabolic Engineering Strategies for Ectoine Production

Members of the Halomonadaceae family are natural ectoine producers with a strong potential for sustainable bioprocesses (Liu et al., 2021). Their salinity tolerance allows non-sterile cultivation and reduces sterilization and corrosion-control costs (Hu et al., 2024).

11.2.6 H. elongata

Prototype organisms used for ectoine production. The strain accumulates ectoine at 2.5–4.0 M NaCl. "Bacterial milking" achieves up to 7.4 g L^{-1} day^{-1} (Sauer & Galinski, 1998) but suffers from corrosion and brine disposal costs. Chemostat studies in 13% NaCl show TCA cycle flux redirected toward ectoine via the ED pathway and PEP carboxylation (Yu et al., 2025). Engineering has targeted *pepCK* deletion (↑OAA yield), *teaBC* overexpression (↑export), LysC deregulation (feedback resistance), and ectABC promoter/UTR optimization, reaching 22 g L^{-1} at ~1 g L^{-1} h^{-1} under ≤6% NaCl (Pérez-García et al., 2017; Hobmeier et al., 2022).

11.2.7 H. hydrothermalis *Y2*

Δ*doeA* + Δ*ectD* mutants accumulate ectoine as the sole solute, achieving >10.5 g L^{-1} at 6% NaCl (Zhao et al., 2019). Promoter engineering raised this to ~11.5 g L^{-1} (Li et al., 2021). Genome sequence availability facilitates chassis optimization (Zhao et al., 2018).

11.2.8 Halomonas bluephagenesis

Originally developed for PHA production (Tan et al., 2011; Chen & Jiang, 2018), it grows in unsterilized seawater-based media. For ectoine, *ectABC* expression, Δ*doeA*, and NADPH supply enhancement yield up to 39.2 g L^{-1} day^{-1} in the non-sterile fed-batch (Hu et al., 2024).

Table 11.1 Comparative summary of native halophilic hosts for ectoine production

Host strain	Key engineering strategies	Max titer (g L^{-1}) or productivity (g L^{-1} day^{-1})	Special features	Reference
H. elongata	Δ*pepCK* (↑ OAA), *teaBC* overexpression, LysC feedback resistance, *ectABC* promoter tuning	22 g L^{-1}; ~1 g L^{-1} h^{-1}	Prototype halophile; reduced-salt, non-sterile fermentation	Pérez-García et al. (2017) and Hobmeier et al. (2022)
H. hydrothermalis Y2	Δ*doeA* + Δ*ectD*, *ectABC* promoter optimization	11.5 g L^{-1}; 0.32 g L^{-1} h^{-1}	Low-salt (6% NaCl) tolerance; genome-sequenced	Zhao et al. (2019) and Li et al. (2021)
H. bluephagenesis	*ectABC* expression, Δ*doeA*, ↑ NADPH supply	39.2 g L^{-1}/day	Seawater-based, fully open cultivation; scalable	Hu et al. (2024) and Chen and Jiang (2018)
Haloarchaea	CRISPR-based editing, extremophile synthetic biology	Not yet optimized	Thrives in >3.5 M NaCl; niche for extreme environments	Becker et al. (2014), Oren (2013), and Bonnaud et al. (2024)

11.2.9 Prospects in Haloarchaea

Haloarchaea (*Haloferax mediterranei* and *Natrialba asiatica*) can tolerate >3.5 M NaCl and produce ectoine (Oren, 2013; Becker et al., 2014). Adoption is limited by slow growth, limited availability of genetic tools, and downstream challenges. Advances in CRISPR tools for *Haloferax* spp. and extremophile synthetic biology (Bonnaud et al., 2024) may unlock their potential (Table 11.1).

11.3 Heterologous Hosts for High-Level Ectoine Production

11.3.1 Rationale and Design Philosophy

Native halophiles possess inherent osmotolerance; however, their ≥10% (w/v) NaCl requirement complicates downstream processing and causes equipment corrosion (Chen & Jiang, 2018). A parallel strategy redirects the ectoine biosynthetic pathway into a low-salt, genetically tractable chassis with tunable metabolism, stress responses, and product recovery (Hobmeier et al., 2022). Two complementary engineering logics emerged.

- Demand-driven modules eliminate feedback inhibition and create a strong sink, for example, by overexpressing feedback-insensitive aspartokinase (LysC S301Y) or deleting competing branches (Hyung-Min et al., 2006; Li et al., 2025).

- *Supply-driven modules* expand precursor pools (OAA, glutamate, and NADPH) via PTS repression, anaplerotic rewiring, or transporter engineering (Becker et al., 2014; Naeem et al., 2024).

The first committed step, Asp → L-aspartyl phosphate via LysC, is tightly feedback-inhibited. Site-directed LysC mutants (S301Y in *C. glutamicum* and T311I in *Bacillus licheniformis*) abolish allosteric lysine binding and increase the ASA pools (Hyung-Min et al., 2006; Li et al., 2025). In *E. coli*, S301Y raises titers by ~0.4 g L^{-1} per gene copy. Heterologous overexpression of LysC under strong promoters doubles the ectoine levels in *C. glutamicum* and *Halomonas* hybrids (Becker et al., 2013; Hobmeier et al., 2022).

LysC deregulation often requires *lysA* attenuation or dynamic control to prevent lysine overflow (Liu et al., 2021). Even in the presence of abundant precursors, ectoine titers plateaued when *ectABC* expression was unbalanced. Promoter/UTR tuning, modular operon assembly (e.g., *B. pseudofirmus* under P_{Lac}–P_{Tac}–P_{Tac}), and plasmid copy number modulation enhance yields (Hobmeier et al., 2022; Naeem et al., 2024).

11.3.2 E. coli: *A Platform for High-Titer Ectoine Production*

Metabolic engineering of *E. coli* has advanced; ectoine titers can surpass 130 g L^{-1}, with the highest value of 134 g L^{-1} obtained from wheat straw hydrolysate at a productivity of 3.7 g L^{-1} h^{-1} (Feng et al., 2025). These achievements rely on modifications such as deletion of *crr* and introduction of feedback-resistant LysC variants (S301Y), which redirect the carbon flux to the aspartate semialdehyde node. The ability of *E. coli* to function at NaCl concentrations below 10 g L^{-1} is advantageous, allowing the use of stainless-steel fermenters, nanofiltration for product recovery, and mixed C_5/C_6 sugar substrates. In addition, *E. coli* has benefitted from long-standing regulatory acceptance in industrial biotechnology, strengthening its position as a versatile host for sustainable ectoine production.

11.3.3 C. glutamicum: *Generally Recognized As Safe (GRAS) Chassis for High-Titer Ectoine Production*

C. glutamicum is a GRAS organism widely used for amino acid production, making it an attractive platform for ectoine biosynthesis. Through the introduction of *ectABC* from *H. elongata*, combined with targeted deletions (*ΔlysE*, *Δpck*) and incorporation of a feedback-insensitive LysC (S301Y), titers of 115.9 g L^{-1} have been achieved at 5% NaCl (Hobmeier et al., 2022; Li et al., 2025). Notably, *C. glutamicum* tolerates crude molasses and acidic feedstocks, thereby extending its potential for use in cost-effective industrial processes. Its well-established use in large-scale

L-glutamate production offers valuable process knowledge and infrastructure that can be leveraged for ectoine manufacturing, reinforcing its role as a safe and productive host (Becker et al., 2013).

11.3.4 **B. licheniformis***: A Robust Industrial Host for Ectoine Production*

The spore-forming bacterium *B. licheniformis* has also been explored as a chassis for ectoine production, building on its industrial background in enzyme secretion. Engineering of strain DW2 with *ectABC*, promoter and UTR optimization, introduction of *E. coli ppc*, and attenuation of by-product pathways enabled production of 3.29 g L^{-1} ectoine at low salinity in a 5 L fed-batch system (Li et al., 2025). Although the titers are lower than those in gram-negative hosts, *B. licheniformis* offers intrinsic robustness, GRAS potential, and suitability for large-scale operations. Its ability to secrete heterologous products and withstand harsh conditions suggests that further optimization can be used as a complementary platform for ectoine biosynthesis.

11.3.5 Yeast Systems: Toward GRAS Eukaryotic Factories

Research on yeast-based ectoine production has gained momentum recently, in line with increasing interest in GRAS eukaryotic hosts for food and pharmaceutical applications. In engineered eukaryotic hosts, such as *Saccharomyces cerevisiae* and *P. pastoris*, introduction of the *ectABC* gene cluster enables growth at <1% NaCl, with the additional advantage of producing endotoxin-free products that are secreted efficiently, setting them apart from bacterial counterparts. Nevertheless, these strains face several bottlenecks, including limited flux through the aspartate semialdehyde (ASA) node and the absence of native ectoine exporters. Recent strategies for addressing these challenges include enhancing mitochondrial acetyl-CoA recycling and introducing heterologous TRAP transporters. Although still in the proof-of-concept stage, titers of 2–4 g L^{-1} have been reported, highlighting their future potential (Ma et al., 2022; Zhang et al., 2025).

11.3.6 Emerging Non-conventional Chassis

In addition to conventional model hosts, several nontraditional microorganisms have been explored for ectoine production. For example, the methanotroph *Methylomicrobium alcaliphilum* 20Z, engineered to block ectoine degradation, has achieved titers of approximately 1.5 g L^{-1} from methane as a carbon source (Lim et al., 2024). The electroactive *Salinivibrio* sp. EAGSL couples ectoine biosynthesis with anodic electron flow, representing a unique approach for bioelectrochemical production (Guynn et al., 2023). Moreover, the moderate halophile *Vibrio boliviensis* has produced ~0.9 g L^{-1} day^{-1} in nonaseptic air-lift reactors, illustrating the potential of low-maintenance, environmentally compatible cultivation systems (Daga-Quisbert et al., 2024).

11.3.7 Process Integration and Techno-Economic Outlook

The path toward the industrial deployment of ectoine production depends on metabolic engineering and on integrated process optimization and techno-economic validation. For instance, switching *E. coli* from glucose to wheat straw hydrolysate as a carbon source reduced raw material costs by >50%, without compromising titers (Feng et al., 2025). Furthermore, all known heterologous hosts operate below 6% (w/v) NaCl, often under 10 g L^{-1}, which significantly reduces costs associated with brine management and effluent disposal (Chen & Jiang, 2018). Adaptive evolution has introduced halophile-derived traits into nonhalophilic hosts, thereby enabling nonsterile fermentation systems (Ma et al., 2020). Techno-economic analyses, including Monte Carlo simulations, suggest that achieving titers >80 g L^{-1} at NaCl concentrations <3% could result in operating expenditures of approximately €7 per kilogram, underscoring the promise of these strategies for future large-scale implementation (Hu et al., 2024) (Table 11.2).

11.3.8 Process Engineering and Feedstock Diversification

Industrial deployment of ectoine biotechnology must address the "three S bottleneck"—sterility, salinity, and substrate cost. Recent process-engineered platforms have demonstrated that halophiles and heterologous hosts can operate in open, low-salt, and low-cost regimes without sacrificing space–time yield (Chen & Jiang, 2018; Hu et al., 2024).

11.3.9 Operating Without Sterile Walls

Early ectoine fermentation mirrored classical amino acid plants: stainless-steel tanks, full steam-in-place, and strict asepsis. Halophiles overcome this paradigm because hypersaline broth naturally suppresses contaminants. *H. bluephagenesis* reached 39.2 g L^{-1} day^{-1} ectoine in a 1000 m^3 nonaseptic, seawater-based process (Chen & Jiang, 2018).

Table 11.2 Comparative summary of native halophiles and heterologous hosts for ectoine production

Category	Halomonas (native halophile)	References	Heterologous hosts (*E. coli, C. glutamicum*)	References
NaCl requirement	6–13% (w/v) for the production phase	Yu et al. (2025) and Zhao et al. (2019)	<1–5% (w/v)	Becker et al. (2013) and Li et al. (2025)
Typical titer range	10–22 g L^{-1}	Pérez-García et al. (2017)	2–134 g L^{-1}	*E. coli*: Lei et al. (2025) and *C. glutamicum*: Becker et al. (2013)
Max reported the titer	22 g L^{-1} (*H. elongata* variants)	Pérez-García et al. (2017)	134 g L^{-1} (*E. coli* from wheat straw hydrolysate)	Lei et al. (2025)
Key demand-side strategies	Δ*doeA*, *ectABC* promoter tuning, *teaBC* overexpression	Hobmeier et al. (2022) and Li et al. (2021)	LysC (S301Y/T311I), competing branch removal	Hyung-Min et al. (2006) and Li et al. (2025)
Key supply-side strategies	*pepCK* deletion (↑ OAA), NADPH supply balancing	Hobmeier et al. (2022) and Yu et al. (2025)	PTS repression (*crr* knockout), anaplerotic reinforcement, NADPH engineering	Becker et al. (2013) and Li et al. (2025)
Process salinity advantage	High salinity limits corrosion savings; low-salt variants are emerging	Zhao et al. (2019) and Li et al. (2021)	Major—enables stainless steel, nanofiltration	Becker et al. (2013) and Lei et al. (2025)
Open fermentation capability	Yes—inherent contamination resistance	Chen and Jiang (2018) and Hu et al. (2024)	Partial—some strains adapted for low-salt contamination resistance	Adaptive laboratory evolution studies (various)
Industrial readiness	Proven at pilot and demo scale (PHA and ectoine)	Chen and Jiang (2018) and Ma et al. (2020)	Commercial readiness for food/cosmetic-grade production	Becker et al. (2013) and Lei et al. (2025)

Airlift reactors have been used to extend this concept. *V. boliviensis* in a 70 L air-lift delivered 4.37 g L^{-1} and 1.76 g L^{-1} day^{-1} in repeated-batch, remaining contamination-free for 60 h; continuous mode sustained ~1.8 g L^{-1} day^{-1} after 15 residence times (Daga-Quisbert et al., 2024). Pneumatic mixing prevents salt-crystal fouling and eliminates the need for rotating seals.

Continuous bioreactors with *C. salexigens*, notably a two-stage cascade, achieved ~2.1 g L^{-1} h^{-1} (≈50.4 g L^{-1} day^{-1}) under ~15% NaCl (Fallet et al., 2010). Airlift data showed moderate halophiles could match these fluxes at <10% NaCl, reducing corrosion and cleaning costs.

11.3.10 Salinity Reduction and Wastewater Burden

Lowering the salt content reduces the corrosion and brine volume. *H. hydrothermalis* Y2 (Δ*ectD*/Δ*doeA*) produced 10.5 g L^{-1} ectoine at 6% (w/v) NaCl, one-third of "bacterial milking" salinity (Zhao et al., 2019). Chemostat fluxomics of *H. elongata* showed that, at 6% vs. 13% NaCl, carbon flux shifted from respiration to ectoine without growth loss (Yu et al., 2024).

ISPR combined with low-salt milking prevents NaCl shock. Mild vacuum evaporation after continuous culture precipitates ectoine from 5% to 8% brine, removing the need for high-salt crystallizers and reducing freshwater demand by >40% (Hobmeier et al., 2022). Non-sterile operations plus ISPR can save 25–30% of utilities (Hu et al., 2024).

11.3.11 Feedstock Flexibility: From Glucose to Gas

Lignocellulosic hydrolysate: *E. coli* (feedback-insensitive mutant LysC + Δ*crr*) achieved 115 g L^{-1} from glucose; switching to wheat straw hydrolysate achieved 134 g L^{-1} (3.7 g L^{-1} h^{-1}) with sustained productivity, highlighting nutrient richness (Feng et al., 2025); continuous feeding kept residual sugars at <5 g L^{-1}, minimizing overflow metabolism.

Molasses and desalinated seawater: NGIB with *H. bluephagenesis* used cane molasses without refined sugars. A high K^+ content buffered osmotic stress, allowing ~80 g L^{-1} NaCl while maintaining yield (Chen & Jiang, 2018).

Gaseous C sources: Methane- or methanol-oxidizing bacteria, such as *M. alcaliphilum* 20Z, can produce ectoine (Rodero et al., 2023; Lim et al., 2024). Here, "gaseous C" includes CH_4, CO_2, CO, CH_3OH vapor, or syngas (CO + H_2) from natural gas, biogas, flue gas, or gasified biomass. Valorizing C_1 streams lowers costs and emissions (Strong et al., 2015). Trickle-bed trials show 2–3 g L^{-1} proof-of-principle titers.

Electro-active set-ups: *Salinivibrio* sp. EAGSL couples ectoine synthesis with the current generation in microbial fuel cells, valorizes hypersaline wastewater, and eliminates aeration energy (Guynn et al., 2023).

11.3.12 Integrated Process Improvements

Repeated-batch with medium recycling reduces downtime; in *V. boliviensis*, a three-step recycling tripled biomass and doubled titer vs. single batch (Daga-Quisbert et al., 2024). Hybrid feed + milking, continuous growth at 5–8% NaCl with periodic hypo-osmotic extraction, saves ~30% salt per kg product vs. classical two-tank milking of *H. elongata* (7.4 g L^{-1} day^{-1}) (Pastor et al., 2010).

In situ product crystallization with selective ethanol at low temperatures recovers >90% ectoine from low-salt broth, producing <4% NaCl recycling streams suitable for reuse, halving water discharge (Hobmeier et al., 2022).

11.3.13 Future Outlook

Converging processes and metabolic engineering allow low-salt strains to simplify the equipment and expand feedstock options (Hu et al., 2024). Fully continuous, membrane-based bleed/feed with nanofiltration could exceed 60 g L^{-1} day^{-1} at seawater salinity, supplying "green" extremolytes with minimal footprint.

11.4 Future Perspectives

Ectoine biotechnology is transitioning from single-product sugar-fed batches to integrated climate-positive biorefineries. Three trends will be defined in the next decade:

(i) valorization of polysaccharide waste streams,
(ii) one-pot synthesis via halophilic consortia, and
(iii) regulatory/LCA alignment for net-zero carbon.

11.4.1 Fully Integrated Biorefineries Based on Polysaccharide Wastes

Lignocellulosic residues exceed 1 billion t $year^{-1}$. Detoxified wheat straw hydrolysate can replace glucose without yield loss: *E. coli* (Δcrr + LysC*) reached 134 g L^{-1}, 3.7 g L^{-1} h^{-1} in a 5 L fed-batch. Co-processing can use cellulose for ectoine and hemicellulose/lignin gases for onsite energy, approaching a zero cradle-to-gate CO_2 footprint. At <US\$60 t^{-1} biomass cost, feedstock is <15% of OPEX vs. 35–40% for food-grade dextrose. Continuous low-salt fermentation plus solid-state pretreatment could halve COG to ≈€5 kg^{-1}.

11.4.2 One-Pot Synthesis via Halophilic Consortia

Natural saline ecosystems rely on phototroph–heterotroph consortia that recycle C/N/reducing power (Oren, 2011, 2013). Halophilic cyanobacteria (*Euhalothece* sp.) fix CO_2 and excrete sucrose, feeding on engineered *H. bluephagenesis* without desalting or sterilization (Chen & Jiang, 2018). Pilot mixed cultures at 7% (w/v) NaCl under light yielded 1.2 g L^{-1} day^{-1} using only seawater and flue gas CO_2 (TRL 4).

Division-of-labor chemistry: *V. boliviensis* with a glutamate-supplying halophile increased ectoine 30% in an airlift without genetic modification; helper species can also stabilize pH, remove inhibitors, or regenerate NADPH, as in electroactive *Salinivibrio* (Daga-Quisbert et al., 2024). Needs: Robust population control (e.g., quorum-sensing lysis) and computational tools for cross-feeding under variable osmolarity (Brenner et al., 2008).

11.4.3 Regulatory Alignment, LCA, and Carbon Neutrality

Ectoine is GRAS; however, its scale-up requires environmental scrutiny (Li et al., 2021). Reducing NaCl from 15% to <6% reduces wastewater chloride by >40% and corrosion by ~70% (Chen & Jiang, 2018). LCA of seawater-based *H. bluephagenesis* shows 0.8 kg CO_2-eq kg^{-1} ectoine, 55% lower than petrochemical analogs; lignocellulose/methane plus CO_2 credits could go net-negative (Hu et al., 2024). Anaerobic digestion of high-salt biomass can recover ~12% process energy as biogas (Zhao et al., 2019). The EU Green Deal and ISO 14067 require traceable feedstocks and brine/water reuse metrics; ag-residue processes fare better than virgin sugarcane (ISO, 2018; European_Commission, 2020).

11.4.4 Research and Policy Roadmap (2025 → 2035)

2026—CRISPR multiplex editing of *C. glutamicum* + low-pH pretreatment liquor: ≥150 g L^{-1} ectoine from hydrolysates in non-sterile STRs; COG <€7 kg^{-1} (Becker et al., 2013; Li et al., 2025).

2028—Phototroph–*Halomonas* co-cultures at seawater salinity sustain 4 g L^{-1} day^{-1}; net-negative CO_2 with biogenic accounting (Pérez-García et al., 2017; Zhao et al., 2019).

2030—Third-party LCA (ISO 14040/14044, ISO 14067) + EU ETS crediting; LCA <0 kg CO_2-eq kg^{-1} product, chloride <20 kg t^{-1}; tier-1 pharma market access (Chen & Jiang, 2018; Hu et al., 2024).

2035—Offshore ectoine factories with solar photobioreactors + seawater cooling, decentralized near biomass ports (Guynn et al., 2023; Daga-Quisbert et al., 2024). By 2035, fully marine-based ectoine factories situated in offshore vessels are envisioned, integrating solar-powered photobioreactors with seawater cooling to enable decentralized production near biomass ports (Guynn et al., 2023; Daga-Quisbert et al., 2024).

11.4.5 Concluding Outlook

Ectoine's trajectory parallels the rise of lactic acid from a specialty to a bioeconomy pillar. Integrating waste polysaccharides, halophilic ecology, and strict LCA can deliver gigaliter-scale and climate-positive capacities. The remaining challenges, such as complex substrates, consortium stability, and salt–energy trade-offs, are solvable within one innovation cycle (Naeem et al., 2024; Feng et al., 2025), enabling an "extremophile playbook" for bioplastics, nutraceuticals, and climate-resilient agriculture.

References

Becker, J., Schäfer, R., Kohlstedt, M., Harder, B. J., Borchert, N. S., Stöveken, N., et al. (2013). Systems metabolic engineering of Corynebacterium glutamicum for production of the chemical chaperone ectoine. *Microbial Cell Factories, 12*(1), 110. https://doi.org/10.1186/1475-2859-12-110

Becker, E. A., Seitzer, P. M., Tritt, A., Larsen, D., Krusor, M., Yao, A. I., et al. (2014). Phylogenetically driven sequencing of extremely halophilic archaea reveals strategies for static and dynamic osmo-response. *PLoS Genetics, 10*(11), e1004784. https://doi.org/10.1371/journal.pgen.1004784

Bonnaud, E., Oger, P. M., Ohayon, A., & Louis, Y. (2024). Haloarchaea as promising chassis to green chemistry. *Microorganisms, 12*(8), 1738.

Brenner, K., You, L., & Arnold, F. H. (2008). Engineering microbial consortia: A new frontier in synthetic biology. *Trends in Biotechnology, 26*(9), 483–489. https://doi.org/10.1016/j.tibtech.2008.05.004

Chen, G.-Q., & Jiang, X.-R. (2018). Next generation industrial biotechnology based on extremophilic bacteria. *Current Opinion in Biotechnology, 50*, 94–100. https://doi.org/10.1016/j.copbio.2017.11.016

Daga-Quisbert, J., Mendieta, D., Rajarao, G. K., van Maris, A. J. A., & Quillaguamán, J. (2024). Production of ectoine by Vreelandella boliviensis using non-aseptic repeated-batch and continuous cultivations in an air-lift bioreactor. *International Microbiology*. https://doi.org/10.1007/s10123-024-00626-3

European_Commission. (2020). *A European green deal*. European Commission.

Fallet, C., Rohe, P., & Franco-Lara, E. (2010). Process optimization of the integrated synthesis and secretion of ectoine and hydroxyectoine under hyper/hypo-osmotic stress. *Biotechnology and Bioengineering, 107*(1), 124–133. https://doi.org/10.1002/bit.22750

Feng, Y., Xiao, W., Li, X., Cao, W., Jiang, Y., Zhang, W., et al. (2025). High Ectoine production from Lignocellulosic hydrolysate by Escherichia coli through metabolic and fermentation engineering. *ACS Synthetic Biology, 14*(2), 609–620. https://doi.org/10.1021/acssynbio.4c00899

Grammann, K., Volke, A., & Kunte Hans, J. (2002). New type of Osmoregulated solute transporter identified in halophilic members of the bacteria domain: TRAP transporter TeaABC mediates uptake of Ectoine and Hydroxyectoine in Halomonas elongata DSM 2581T. *Journal of Bacteriology, 184*(11), 3078–3085. https://doi.org/10.1128/jb.184.11.3078-3085.2002

Guynn, I. P. A., Beaver, K., Gaffney, E. M., Zani, A. B., Dantanarayana, A., & Minteer, S. D. (2023). Salinivibrio sp. EAGSL as a halophilic and ectoine-producing bacteria for broad microbial electrochemistry applications. *Cell Reports Physical Science, 4*(6), 101420. https://doi.org/10.1016/j.xcrp.2023.101420

Hobmeier, K., Oppermann, M., Stasinski, N., Kremling, A., Pflüger-Grau, K., Kunte, H. J., et al. (2022). Metabolic engineering of Halomonas elongata: Ectoine secretion is increased by demand and supply driven approaches. *Frontiers in Microbiology, 13*(1664-302X (Print)), 968983. https://doi.org/10.3389/fmicb.2022.968983

Hu, Q., Sun, S., Zhang, Z., Liu, W., Yi, X., He, H., et al. (2024). Ectoine hyperproduction by engineered Halomonas bluephagenesis. *Metabolic Engineering, 82*, 238–249. https://doi.org/10.1016/j.ymben.2024.02.010

Hyung-Min, K., Elmar, H., & Christoph, W. (2006). Deregulation of Aspartokinase by single nucleotide exchange leads to global flux rearrangement in the central metabolism of Corynebacterium glutamicum. *Journal of Microbiology and Biotechnology, 16*(8), 1174–1179.

International Organization for Standardization (ISO). (2018). ISO 14067:2018 Greenhouse gases — Carbon footprint of products — Requirements and guidelines for quantification. Geneva, Switzerland. Available at: https://www.iso.org/standard/71206.html

Jiang, A., Song, Y., You, J., Zhang, X., Xu, M., & Rao, Z. (2022). High-yield ectoine production in engineered Corynebacterium glutamicum by fine metabolic regulation via plug-in repressor library. *Bioresource Technology, 362*, 127802. https://doi.org/10.1016/j.biortech.2022.127802

Li, S. N., Shang, Y., Zhao, Q., Liu, Y. X., Dong, X. Q., Wang, W. K., et al. (2021). Promoter engineering for high ectoine production in a lower saline medium by Y2. *Biotechnology Letters, 43*(4), 825–834. https://doi.org/10.1007/s10529-021-03084-3

Li, X., Dong, A., Yang, J., Zhu, J., Zhan, Y., Ma, X., et al. (2025). Metabolic engineering of Bacillus licheniformis DW2 for ectoine production. *World Journal of Microbiology and Biotechnology, 41*(1), 23. https://doi.org/10.1007/s11274-024-04238-x

Lei, Z., Chen, X., Yuan, L., Wu, J., & Yao, J. (2025). Metabolic control for high-efficiency Ectoine synthesis in engineered Escherichia coli. *ACS Synthetic Biology*. https://doi.org/10.1021/acssynbio.5c00330

Lim, S. E., Cho, S., Choi, Y., Na, J. G., & Lee, J. (2024). High production of ectoine from methane in genetically engineered Methylomicrobium alcaliphilum 20Z by preventing ectoine degradation. *Microbial Cell Factories, 23*(1), 127. https://doi.org/10.1186/s12934-024-02404-2

Liu, M., Liu, H., Shi, M., Jiang, M., Li, L., & Zheng, Y. (2021). Microbial production of ectoine and hydroxyectoine as high-value chemicals. *Microbial Cell Factories, 20*(1), 76. https://doi.org/10.1186/s12934-021-01567-6

Ma, H., Zhao, Y., Huang, W., Zhang, L., Wu, F., Ye, J., et al. (2020). Rational flux-tuning of Halomonas bluephagenesis for co-production of bioplastic PHB and ectoine. *Nature Communications, 11*(1), 3313. https://doi.org/10.1038/s41467-020-17223-3

Ma, Z., Wu, C., Zhu, L., Chang, R., Ma, W., Deng, Y., et al. (2022). Bioactivity profiling of the extremolyte ectoine as a promising protectant and its heterologous production. *3 Biotech, 12*(12), 331. https://doi.org/10.1007/s13205-022-03370-5

Ma, Z., Chang, R., Zhu, L., Zhu, D., Deng, Y., Guo, X., et al. (2024). Metabolic engineering of Corynebacterium glutamicum for highly efficient production of Ectoine. *ACS Synthetic Biology, 13*(7), 2081–2090. https://doi.org/10.1021/acssynbio.4c00019

Naeem, M., Yuan, H., Luo, S., Zhang, S., Wei, X., He, G., et al. (2024). Efficient biosynthesis of Ectoine in recombinant Escherichia coli by biobrick method. *Fermentation, 10*(9), 450.

Ohnishi, J., & Ikeda, M. (2006). Comparisons of potentials for L-lysine production among different Corynebacterium glutamicum strains. *Bioscience, Biotechnology, and Biochemistry, 70*(4), 1017–1020. https://doi.org/10.1271/bbb.70.1017

Oren, A. (2011). Thermodynamic limits to microbial life at high salt concentrations. *Environmental Microbiology, 13*(8), 1908–1923. https://doi.org/10.1111/j.1462-2920.2010.02365.x

Oren, A. (2013). Life at high salt concentrations. In E. Rosenberg, E. F. DeLong, S. Lory, E. Stackebrandt, & F. Thompson (Eds.), *The prokaryotes: Prokaryotic communities and ecophysiology* (pp. 421–440). Springer Berlin Heidelberg.

Pastor, J. M., Salvador, M., Argandona, M., Bernal, V., Reina-Bueno, M., Csonka, L. N., et al. (2010). Ectoines in cell stress protection: Uses and biotechnological production. *Biotechnology Advances, 28*(6), 782–801. https://doi.org/10.1016/j.biotechadv.2010.06.005

Pérez-García, F., Ziert, C., Risse, J. M., & Wendisch, V. F. (2017). Improved fermentative production of the compatible solute ectoine by Corynebacterium glutamicum from glucose and alternative carbon sources. *Journal of Biotechnology, 258*, 59–68. https://doi.org/10.1016/j.jbiotec.2017.04.039

Rodero, M. D. R., Carmona-Martínez, A. A., Martínez-Fraile, C., Herrero-Lobo, R., Rodríguez, E., García-Encina, P. A., et al. (2023). Ectoines production from biogas in pilot bubble column bioreactors and their subsequent extraction via bio-milking. *Water Research, 245*, 120665. https://doi.org/10.1016/j.watres.2023.120665

Sauer, T., & Galinski, E. A. (1998). Bacterial milking: A novel bioprocess for production of compatible solutes. *Biotechnology and Bioengineering, 59*(1), 128. https://doi.org/10.1002/(sici)1097-0290(19980705)59:1<128::aid-bit17>3.0.co;2-e

Saum, S. H., & Müller, V. (2008). Regulation of osmoadaptation in the moderate halophile Halobacillus halophilus: Chloride, glutamate and switching osmolyte strategies. *Saline Systems, 4*(1), 4. https://doi.org/10.1186/1746-1448-4-4

Schwibbert, K., Marin-Sanguino, A., Bagyan, I., Heidrich, G., Lentzen, G., Seitz, H., et al. (2011). A blueprint of ectoine metabolism from the genome of the industrial producer Halomonas elongata DSM 2581 T. *Environmental Microbiology, 13*(8), 1973–1994. https://doi.org/10.1111/j.1462-2920.2010.02336.x

Strong, P. J., Xie, S., & Clarke, W. P. (2015). Methane as a resource: Can the methanotrophs add value? *Environmental Science & Technology, 49*(7), 4001–4018. https://doi.org/10.1021/es504242n

Tan, D., Xue, Y.-S., Aibaidula, G., & Chen, G.-Q. (2011). Unsterile and continuous production of polyhydroxybutyrate by Halomonas TD01. *Bioresource Technology, 102*(17), 8130–8136. https://doi.org/10.1016/j.biortech.2011.05.068

Wang, K., Cui, B., Wang, Y., & Luo, W. (2025). Microbial production of Ectoine: A review. *ACS Synthetic Biology, 14*(2), 332–342. https://doi.org/10.1021/acssynbio.4c00490

Xingrong, L., Gorish, B. M. T., Qaria, M. A., Hussain, A., Abdelmula, W. I. Y., & Zhu, D. (2025). Unlocking Ectoine's Postbiotic therapeutic promise: Mechanisms, applications, and future directions. In *Probiotics and antimicrobial proteins*. https://doi.org/10.1007/s12602-025-10506-5

Yu, J., Zhang, Y., Liu, H., Liu, Y., Mohsin, A., Liu, Z., et al. (2024). Temporal dynamics of stress response in Halomonas elongata to NaCl shock: Physiological, metabolomic, and transcriptomic insights. *Microbial Cell Factories, 23*(1), 88. https://doi.org/10.1186/s12934-024-02358-5

Yu, J., Liu, H., Wang, Y., Ju, R., Zhang, Y., Moshin, A., et al. (2025). Decoding metabolic trade-offs in Halomonas elongata: Chemostat-based flux remodeling for industrial ectoine biosynthesis. *Synthetic and Systems Biotechnology*. https://doi.org/10.1016/j.synbio.2025.07.004

Zhang, S., Cheng, B., Liao, Q., Huang, X., Mi, M., Huang, M., et al. (2025). Multi-gene metabolic engineering of Pichia pastoris to synthesize ectoine. *Journal of Bioscience and Bioengineering, 139*(5), 347–353. https://doi.org/10.1016/j.jbiosc.2025.02.006

Zhao, Q., Meng, Y. W., Li, S. N., Lv, P. W., Xu, P., & Yang, C. Y. (2018). Genome sequence of Y2, an efficient ectoine-producer isolated from pulp mill wastewater. *Journal of Biotechnology, 285*, 38–41. https://doi.org/10.1016/j.jbiotec.2018.08.017

Zhao, Q., Li, S., Lv, P., Sun, S., Ma, C., Xu, P., et al. (2019). High ectoine production by an engineered Halomonas hydrothermalis Y2 in a reduced salinity medium. *Microbial Cell Factories, 18*(1), 184. https://doi.org/10.1186/s12934-019-1230-x

Chapter 12
Peptide Syntheses Using an Organic Solvent-Tolerant Enzyme

Hiroyasu Ogino

Abstract An organic solvent-tolerant bacterium that produced an organic solvent-tolerant protease was isolated from soil. The protease secreted from the organic solvent-tolerant bacteria had very high tolerance in the presence of various organic solvents. The organic solvent-tolerant protease was characterized, and its conformational transition in the presence and absence of an organic solvent was monitored. By using site-directed mutagenesis and random mutagenesis, amino acid residues that influenced organic solvent tolerance were investigated. And the organic solvent-tolerant protease was used as a catalyst of peptide synthesis in the presence of various organic solvents. Furthermore, the organic solvent-tolerant protease, which had high peptide synthetic activity, was developed by protein engineering.

Keywords Protease · Peptide synthesis · Organic solvent · Tolerance · Stability

12.1 Introduction

Enzymes catalyze various biochemical reactions in aqueous solutions, but they are easily denatured and inactivated in the presence of organic solvents, resulting in the loss of their catalytic function. However, enzymatic reactions in the presence of organic solvents have several advantages: (1) improved solubility of poorly water-soluble substrates, resulting in faster reaction rates; (2) hydrolases can catalyze synthetic reverse reactions; and (3) reduced microbial contamination (Ogino & Ishikawa, 2001; Doukyu & Ogino, 2010). Therefore, the development of organic solvent-tolerant enzymes that function in the presence of organic solvents was performed. Organic solvent tolerant enzymes have high activity and tolerance not only in aqueous solutions but also in reaction systems containing organic solvents. They

H. Ogino (✉)
Department of Chemical Engineering, Osaka Metropolitan University, Naka-ku, Sakai, Osaka, Japan
e-mail: ogino@omu.ac.jp

M. Ito et al. (eds.), *Extremophiles for a Sustainable Future*,
https://doi.org/10.1007/978-981-95-6970-0_12

can be used in aqueous, organic, and organic-containing systems; reaction systems with high substrate concentrations where the substrate itself is the solvent; and solvent-free systems.

Proteases are among the most valuable catalysts used in the food, pharmaceutical, and detergent industries because they hydrolyze peptide bonds in aqueous environments and synthesize peptide bonds in microaqueous environments. There is also current interest in the application of proteases to the production of certain oligopeptides without producing by-products.

So far several well-known proteases, such as thermolysis, papain, and chymotrypsin, have been used as biocatalysts of peptide synthesis in the presence of organic solvents. However, most of these proteases did not have sufficient tolerance in the presence of organic solvents.

12.2 Organic Solvent-Tolerant Bacterium That Secretes an Organic Solvent-Tolerant Protease

An organic solvent-tolerant microorganism that produced organic solvent-tolerant protease, *Pseudomonas aeruginosa* PST-01, was isolated from soil, grew in the presence of a high concentration of several organic solvents, and secreted an organic solvent-tolerant protease (Ogino et al., 1995). It was expected that both the solvent-tolerant microorganism and the organic solvent-tolerant enzyme produced by this strain could be used as catalysts for reactions in the presence of organic solvents.

12.3 Organic Solvent-Tolerant Protease Secreted by an Organic Solvent-Tolerant Bacterium

The organic solvent-tolerant protease (PST-01 protease) in a culture broth of organic solvent-tolerant *Pseudomonas aeruginosa* PST-01 was purified by successive hydrophobic interaction chromatography. PST-01 protease had a molecular mass of 38 kDa. The optimum temperature and pH for casein hydrolysis were 55 °C and 8.5, respectively. PST-01 protease had high tolerance at pH 8–12 and below 50 °C and was determined to be a metalloprotease, which was inhibited by EDTA, 1,10-phenanthroline, and phosphoramidon. PST-01 protease inhibited by EDTA was reactivated completely by the addition of zinc or cobalt ions (Ogino et al., 1999a).

The effect of various organic solvents on the tolerance of various proteases, that is, purified PST-01 protease, subtilisin Carlsberg, thermolysis, and α-chymotrypsin, was studied. Mixtures of 3 ml of an enzyme solution (5 U/ml) and 1 ml of an organic solvent were incubated at 30 °C with shaking, and the remaining activity was measured at appropriate time intervals. Deactivation of all the proteases tested in the presence or absence of organic solvents obeyed first-order kinetics. The

Table 12.1 The half-lives of the various proteases in the presence and the absence of 25% (v/v) organic solvent

Organic solvent	Half-life [d] PST-01 protease	Subtilisin	Thermolysis	α-chymotrypsin
Ethylene glycol	>100	14.3	>50	6.9
1,4-Butanediol	>100	25.0	4.4	0.7
1,5-Pentanediol	>100	>50	1.7	0.4
Ethanol	>100	>50	3.0	27.0
1-Hexanol	>50	13.7	18.2	14.2
Methanol	>50	26.2	4.6	6.0
Dimethyl sulfoxide (DMSO)	>50	6.4	2.6	33.6
2-Propanol	>50	47.6	1.2	0.6
Triethylene glycol	>50	39.7	5.1	11.1
tert-Butanol	>50	41.6	0.8	0.5
1-Heptanol	>50	8.6	13.1	3.8
N,N-Dimethylformamide (DMF)	25.3	39.8	0.9	2.2
1-Octanol	24.2	n. t.	n. t.	n. t.
1-Butanol	24.2	47.6	0.9	0.3
Acetone	23.1	24.8	0.7	0.6
1-Decanol	19.4	n. t.	n. t.	n. t.
1,4-Dioxane	17.7	29.3	0.8	0.5
Toluene	12.0	5.7	22.5	>100
Benzene	7.8	n. t.	n. t.	n. t.
n-Heptane	4.8	n. t.	n. t.	n. t.
p-Xylene	4.4	n. t.	n. t.	n. t.
n-Hexane	3.8	n. t.	n. t.	n. t.
n-Decane	2.4	n. t.	n. t.	n. t.
Cyclohexane	2.3	n. t.	n. t.	n. t.
None	9.7	0.3	10.8	13.2

The filter-sterilized solutions (3 ml) containing various proteases, the purified PST-01 protease, subtilisin Carlsberg, thermolysin, and α-chymotrypsin, were incubated at pH 8.5 and 30 °C with shaking at 160 strokes per minute in the absence or presence of 1 ml of organic solvent for 15 days. The remaining activities were measured after 0, 1, 2, 3, 5, 7, 10, and 15 days. The half-lives were calculated from the exponential regression curves
n. t. not tested

half-lives of the activity of various proteases are summarized in Table 12.1. The half-life of PST-01 protease in the absence of organic solvent was about 9.7 d. The half-life of PST-01 protease in the presence of water-insoluble organic solvents such as toluene, benzene, *n*-heptane, *p*-xylene, *n*-hexane, *n*-decane, and cyclohexane was similar to or shorter than that in the absence of organic solvent. However, the half-life of PST-01 protease in the presence of water-soluble organic solvents or alcohols such as ethylene glycol, 1,4-butanediol, 1,5-pentanediol, ethanol, 1-hexanol, methanol, dimethyl sulfoxide (DMSO), 2-propanol, triethylene glycol, *tert*-butanol, 1-heptanol, *N,N*-dimethylformamide (DMF), 1-octanol, 1-butanol, acetone,

1-decanol, and 1,4-dioxane was longer than that in the absence of organic solvent. Inactivation of the enzyme in the presence of ethylene glycol, 1,4-butanediol, 1,5-pentanediol, ethanol, and 1-hexanol was not observed at all during the 15-day experiment. When organic solvents, of which the log *P* values (Lane et al., 1987) are between −0.23 and 3.1, were added to the filtered culture supernatant of *P. aeruginosa* PST-01 (the crude enzyme solution of PST-01 protease), the enzyme was inactivated a little more rapidly than when no organic solvents were added. When organic solvents of which the log *P* values are less than −0.23 or more than 3.1 were added, the inactivation rates of the enzyme were the same as those in the absence of organic solvent. However, in the case of the purified PST-01 protease, the relationship between the tolerance against organic solvent and the log *P* value of the added organic solvent was not found. In the presence of water-soluble organic solvents or alcohols, the tolerance of PST-01 protease was much higher than in the absence of organic solvent. The half-life of the PST-01 protease activity was longer than that of any other protease tested in the presence of water-soluble organic solvents or alcohols.

12.4 Conformational Transitions of Various Proteases in the Presence of an Organic Solvent

The conformational transitions of the proteases were monitored by measuring the circular dichroism (CD) spectra. The CD spectra of the PST-01 protease, α-chymotrypsin, thermolysin, and subtilisin in the presence and absence of 25% (v/v) methanol are shown in Fig. 12.1. The CD spectrum of the PST-01 protease in the presence of methanol was almost unchanged after 24 h of incubation, although that in the absence of methanol changed considerably. A similar phenomenon was also observed in the case of subtilisin. In the presence of methanol, the CD spectrum of the subtilisin was almost unchanged. However, the CD spectrum of subtilisin significantly changed in the absence of the methanol. On the other hand, the presence of methanol greatly enhanced the change in the CD spectra of α-chymotrypsin and thermolysin. It was shown that the conformational stabilities of the PST-01 protease and subtilisin were high in the presence of methanol (Ogino et al., 2007a).

For all four proteases in the presence and absence of methanol, the extents of the change in the CD spectrum were closely related to those of the change in the activities. Under the conditions in which conformational transitions were larger, the half-lives of the protease activity were shorter in all proteases.

Some polyamino acids form the particular conformations under the specific conditions. Poly-L-glutamic acid forms an α-helix at pH 4.3 (Blout et al., 1962; Holzwarth et al., 1962; Holzwarth & Doty, 1965; Tiffany & Krimm, 1968a), and poly-L-lysine forms a β-pleated structure at pH 7.8 in the presence of SDS (Sarkar & Doty, 1966; Li & Spector, 1969). Poly-L-proline in water after incubation for 24 h forms a peculiar helical structure (Blout et al., 1963; Tiffany & Krimm, 1968b).

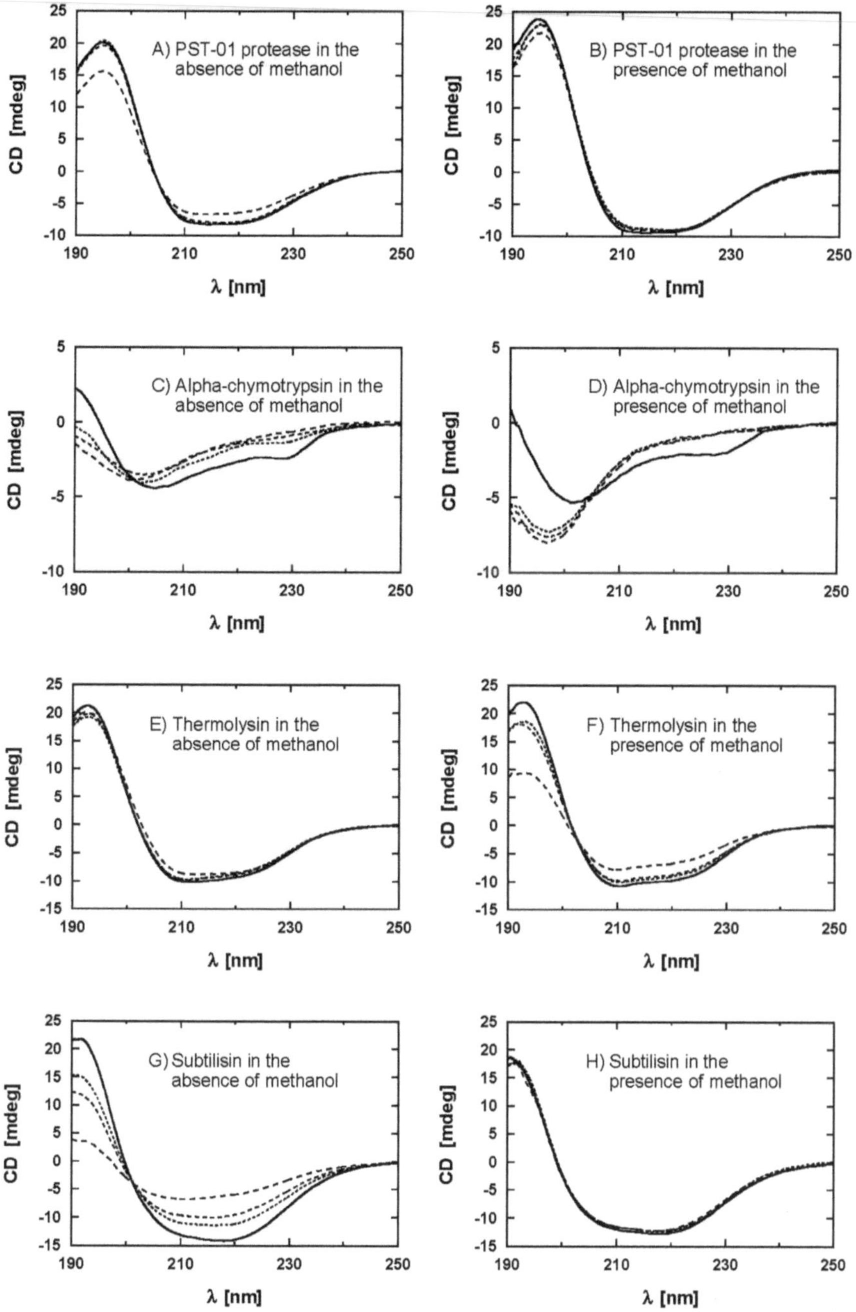

Fig. 12.1 CD spectra of various proteases in the presence and absence of methanol. The solutions containing PST-01 protease (**A**, **B**), α-chymotrypsin (**C**, **D**), thermolysin (**E**, **F**), and subtilisin (**G**, **H**) were incubated at 30 °C in the absence (**A**, **C**, **E**, **G**) and presence (**B**, **D**, **F**, **H**) of 25% (v/v) methanol. The CD spectra were measured after 0 (———), 3 (.........), 6 (------), and 24 (———) h incubations

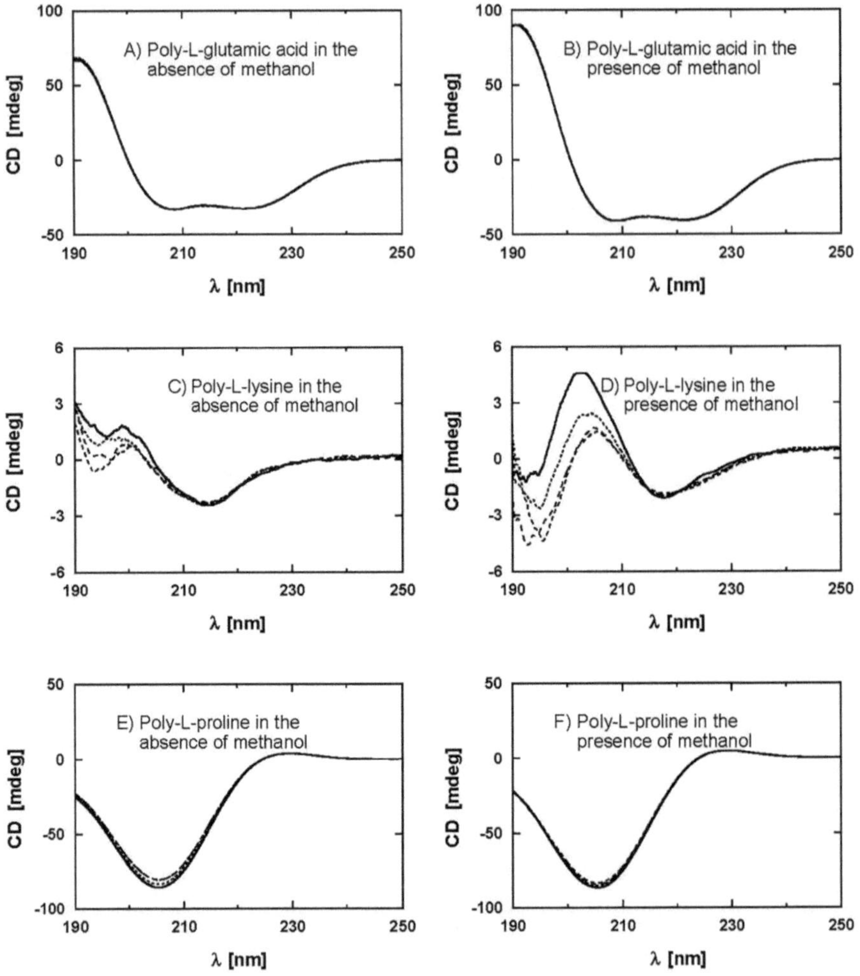

Fig. 12.2 CD spectra of various poly-amino acids in the presence and absence of methanol. The solutions containing poly-L-glutamic acid (**A**, **B**), poly-L-lysine (**C**, **D**), and poly-L-proline (**E**, **F**) were incubated at 30 °C in the absence (**A**, **C**, **E**) and presence (**B**, **D**, **F**) of 25% (v/v) methanol. The CD spectra were measured after 0 (———), 3 (.........), 6 (------), and 24 (———) h incubations

Figure 12.2 shows the CD spectra of these poly-amino acids. The CD spectra of poly-L-glutamic acid in water show typical double minima at 208 and 223 nm and a maximum around 190 nm (Fig. 12.2a, b). The CD spectra of poly-L-lysine in water show a typical minimum around 217 nm and a typical maximum around 200 nm (Fig. 12.2c, d). Furthermore, poly-L-lysine in water exhibits a typical minimum at 205 nm in the CD spectra (Fig. 12.2e, f).

To these solutions containing poly-amino acids, 25% (v/v) methanol or water was added and incubated at 30 °C. The CD spectra of poly-L-glutamic acid and poly-L-proline, which form helical structures, were unchanged by the addition of methanol and the incubation in the presence and absence of methanol. However, the CD spectra of poly-L-lysine, which forms a β-pleated structure, changed greatly by the addition of methanol and the incubation in the presence of methanol. From these conformational transitions of the poly-amino acids, it can be concluded that the conformational stability of the β-pleated structure is lower than that of the helical structures. In fact, the contents of the α-helical structure of the PST-01 protease, α-chymotrypsin, thermolysin, and subtilisin are 37, 7, 38, and 30%, respectively. The contents of the β-pleated structure of the PST-01 protease, α-chymotrypsin, thermolysin, and subtilisin are 13, 30, 17, and 16%, respectively. The content of the α-helical structure of α-chymotrypsin is the lowest, and that of the β-pleated structure of α-chymotrypsin is the highest of the four proteases. Therefore, it was suggested that the organic solvent tolerance of enzymes is closely related to the secondary structure and that when the content of the β-pleated structure is high, the organic solvent tolerance of enzymes is low.

On the other hand, although the overall conformations of the PST-01 protease and thermolysin were very similar, the tolerances of the PST-01 protease and thermolysin were quite different, as shown in Table 12.1. As will be explained later, the presence of a disulfide bond in the intermolecular of the PST-01 protease was one of the very important factors. Although the β-pleated structure was not the only decisive factor for the organic solvent tolerance of enzymes, it was found that the β-pleated structure was one of the important factors.

12.5 Role of Intermolecular Disulfide Bonds of the Organic Solvent-Tolerant PST-01 Protease

The organic solvent-tolerant PST-01 protease has two disulfide bonds (between Cys-30 and Cys-58 and between Cys-270 and Cys-278) in its molecule (Ogino et al., 2000b). Mutant PST-01 proteases in which one or both of the disulfide bonds were deleted were constructed by site-directed mutagenesis, and the effect of the disulfide bonds on the activity and the various tolerances was investigated. The disulfide bond between Cys-270 and Cys-297 in the PST-01 protease was found to be essential for its activity. The disulfide bond between Cys-30 and Cys-58 played an important role in the organic solvent tolerance of the PST-01 protease (Ogino et al., 2001).

12.6 Effect of Exchanging Amino Acid Residues of the Surface Region of the PST-01 Protease on Its Organic Solvent Tolerance

The structure gene of the PST-01 protease was amplified by the error-prone PCR method. The mutated proteases were incubated in the presence of acetonitrile. By measuring remaining activities, two kinds of mutated PST-01 proteases whose tolerances were changed were selected. The 45th serine and 201st aspartic acid were substituted for asparagine and tyrosine, respectively. Common characteristic changes between substitutions at 45th serine and 201th aspartic acid and for asparagine and tyrosine, respectively, were not found. However, both the substituted amino acid residues were located at the surface of the PST-01 protease. It is speculated that the organic solvent tolerance of the protease decreased because of the substitution of amino acid residues that may be effective for preventing the penetration of organic solvent into the inside of the protein molecule (Ogino et al., 2007b).

12.7 Peptide Synthesis Catalyzed by Organic Solvent-Tolerant Protease in Monophasic Aqueous-Organic Solvent Systems

Although the synthesis of peptide bonds was carried out in the absence of an organic solvent in a few cases, the addition of organic solvents to the reaction mixture enhanced the yield of the product because of a shift in the thermodynamic equilibrium in favor of peptide synthesis. However, enzymes are usually inactivated rapidly in the presence of organic solvents. To achieve a high yield, the tolerance of the protease is very important because enzymes are usually denatured or inactivated in the presence of organic solvents before completing the reactions. The peptide syntheses catalyzed by the PST-01 protease in water and monophasic aqueous-organic systems containing water-soluble organic solvents were investigated. The initial rates of peptide synthesis in the presence of various kinds of water-soluble organic solvents and the equilibrium yields were investigated. The initial rates of Cbz-Arg-Leu-NH_2 synthesis from 10 mM Cbz-Arg and 500 mM Leu-NH_2 in the mixture of various water-soluble organic solvents and 50 mM Tris–HCl buffer (pH 7.0) (1:1 (v/v)) at 30 °C and the equilibrium yields are shown in Fig. 12.3. In the absence of an organic solvent, the initial rate and the equilibrium yield were about 220 $\mu mol \cdot min^{-1} \cdot g\text{-}enzyme^{-1}$ and 11%, respectively. The initial rates in the presence of 50% (v/v) methanol, DMF, and DMSO were about 1.6, 2.4, and 4.6 times higher, respectively, than that in the absence of an organic solvent (Ogino et al., 2000a). Especially, the equilibrium yields in the presence of 50% (v/v) methanol, DMF, and DMSO were 53.8%, 72.1%, and 72.0%, respectively. They were about 4.7, 6.3, and 6.3 times higher than that in the absence of an organic solvent (Ogino et al., 1999b).

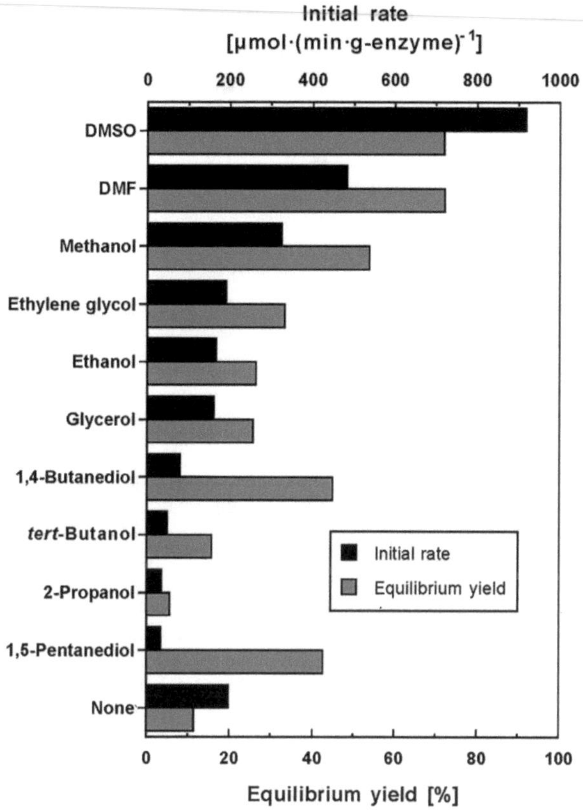

Fig. 12.3 Effect of organic solvents on the initial rate and equilibrium yield of Cbz-Arg-Leu-NH_2 synthesis catalyzed by the PST-01 protease in the presence of 50% (v/v) various organic solvents. All reactions were carried out using 10 mM Cbz-Arg, 500 mM Leu-NH_2, and 0.167–0.176 g/L of the PST-01 protease in the mixture of various organic solvents and 50 mM Tris–HCl buffer (pH 7.0) (1:1 (v/v)) at 30 °C

The effect of the DMSO concentrations on the initial rates of peptide synthesis and the equilibrium yields were further investigated. The initial rates of Cbz-Arg-Leu-NH_2 synthesis from 10 mM Cbz-Arg and 500 mM Leu-NH_2 in the mixture of various volume ratios of DMSO and 50 mM Tris–HCl buffer (pH 7.0) at 30 °C and the equilibrium yields are shown in Fig. 12.4. When the organic solvent concentration was lower than 60% (v/v), the initial rates and the equilibrium yields increased with an increase in the organic solvent concentration. The highest rates were attained in the presence of 60% (v/v) DMSO and were about 5.1 times higher than that in the absence of an organic solvent. In the presence of 60% (v/v) DMSO, the equilibrium yields of Cbz-Arg-Leu-NH_2 attained 80.2%. That was about 7.0 times higher than that in the absence of an organic solvent, respectively.

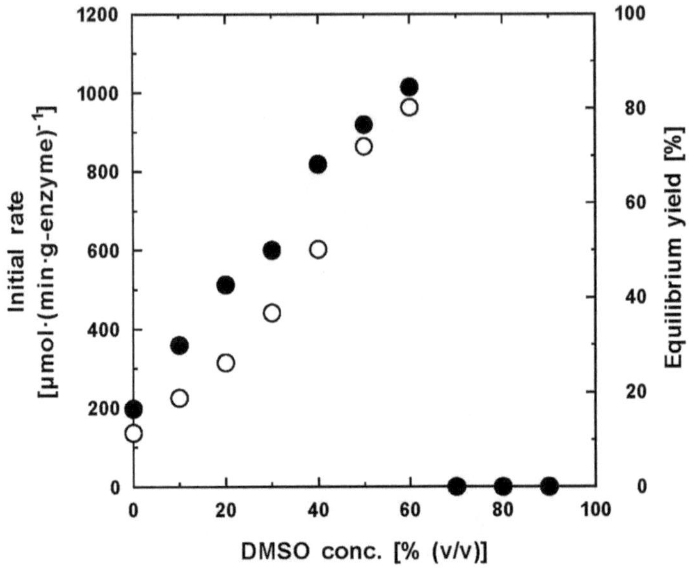

Fig. 12.4 Effect of DMSO concentration on the initial rate and equilibrium yield of Cbz-Arg-Leu-NH_2 synthesis catalyzed by the PST-01 protease. All reactions were carried out using 10 mM Cbz-Arg, 500 mM Leu-NH_2, and 0.167–0.176 g/L of the PST-01 protease in the mixture of DMSO and 50 mM Tris–HCl buffer (pH 7.0) (1:1 (v/v)) at 30 °C. Symbols: ●, initial rate; ○, equilibrium yield

12.8 Enhancement of the Aspartame Precursor Synthetic Activity of an Organic Solvent-Tolerant Protease

For example, aspartame, a low-caloric artificial sweetener, was enzymatically synthesized using a protease in the presence of organic solvents (Yagasaki & Hashimoto, 2008). Some proteases, including thermolysin, have been found to be capable of catalyzing the synthetic reaction of the aspartame precursor, N-carbobenzoxy-L-aspartyl-L-phenylalanine methyl ester (Cbz-Asp-Phe-OMe). Thermolysin has been well studied (Matthews, 1988). Although thermolysin has very high tolerance at high temperatures, it does not have high tolerance in the presence of organic solvents (Ogino et al., 1999a; Yagasaki & Hashimoto, 2008). The comparison of reaction rates of aspartame precursor synthesis using the PST-01 protease and thermolysin was performed. High-yield synthesis of an aspartame precursor was attained when the PST-01 protease was used as a catalyst in the presence of 50% (v/v) DMSO (Tsuchiyama et al., 2007). However, the synthesis rate of the aspartame precursor using the PST-01 protease was about 16.5 ± 3.0 μmol·min^{-1}·g-enzyme^{-1} and is approximately a third of that using thermolysin (50.7 ± 1.1 μmol·min^{-1}·g-enzyme^{-1}) under the same reaction conditions. Product yields at equilibrium using either the PST-01 protease or thermolysin have been shown to be almost identical when the same reaction conditions were used.

The major difference seemed to be the substrate specificity of both enzymes. The overall structure of the PST-01 protease is very similar to that of thermolysin. Although the same motif of the catalytic center was conserved in the PST-01 protease, some residues that formed the substrate binding site of the PST-01 protease were different from those of thermolysin. It was found that Phe-114 of thermolysin corresponds to Tyr-114 of the PST-01 protease, and Phe-130, Leu-133, Val-139, Ile-188, Gly-189, Val-192, and Leu-202 of thermolysin correspond to Phe-129, Leu-132, Val-137, Ile-186, Ile-190, and Leu-197 of the PST-01 protease, respectively. And Tyr-114 and Ile-190 of the PST-01 protease differ in amino acid type when compared with the corresponding residues in thermolysin. Both differences were found in the substrate binding sites of thermolysin. By comparison of the active centers of the PST-01 protease and thermolysin, the differences were found to be located at Tyr-114 and Ile-190 of the PST-01 protease. Tyr-114 and Ile-190 of the PST-01 protease correspond to Phe-114 and Val-192 in thermolysin, respectively. To examine the influence of these residues on enzymatic activity, Tyr-114 and Ile-190 of the PST-01 protease were replaced with Phe and Val, respectively. The peptide synthesis rates of the aspartame precursor were measured using the purified mutant PST-01 proteases. Although the catalytic activity of the PST-01-I190V mutant (15.3 ± 0.1 μmol·min^{-1}·g-enzyme^{-1}) was similar to the wild-type PST-01 protease (16.5 ± 3.0 μmol·min^{-1}·g-enzyme^{-1}), the activity of the PST-01-Y114F mutant (44.2 ± 0.1 μmol·min^{-1}·g-enzyme^{-1}) was greatly enhanced and similar to that of thermolysin (50.7 ± 1.1 μmol·min^{-1}·g-enzyme^{-1}). The Y114F mutation increased the PST-01 protease synthetic activity by a factor of 2.7. The synthesis rate of the aspartame precursor of the double mutated PST-01-Y114F-I190V protease (46.7 ± 7.3 μmol·min^{-1}·g-enzyme^{-1}) was similar to that using the PST-01-Y114F mutant protease and thermolysin.

The mutations of the PST-01 protease showed the importance of residue 114 for the synthesis of the aspartame precursor. Site-saturated mutagenesis of residue 114 was subsequently performed. PST-01 protease mutants, in which the 114th residue was changed from Tyr to Asn, Asp, Gln, Glu, Gly, Lys, and Pro, did not show catalytic activity. Consequently, mutation at position 114 of the PST-01 proteases to polar amino acids such as Asn, Asp, Gln, Glu, Gly, Lys, and Pro led to the loss of catalytic activity. Figure 12.5 shows the results of syntheses of the aspartame precursor using other mutant PST-01 proteases generated. It was confirmed that all reactions attained the similar equilibrium yield and the mutants having activity had enough tolerance in the presence of 50% (v/v) DMSO. Substitutions to highly non-polar amino acids, Ile, Val, and Trp, led to a decrease in catalytic activity. Mutations at this position to Leu and Thr showed moderate catalytic activity. Here, the activity was found to lie between the wild-type PST-01 protease and thermolysin. Mutations at position Tyr-114 to Met, His, Cys, Ala, Arg, and Ser led to a significant increase in catalytic activity. In particular, the activities of the PST-01-Y114R and PST-01-Y114S mutant proteases were about ten times higher than that of the wild-type PST-01 protease and about 3.4 times higher than that of thermolysin (Ogino et al., 2010).

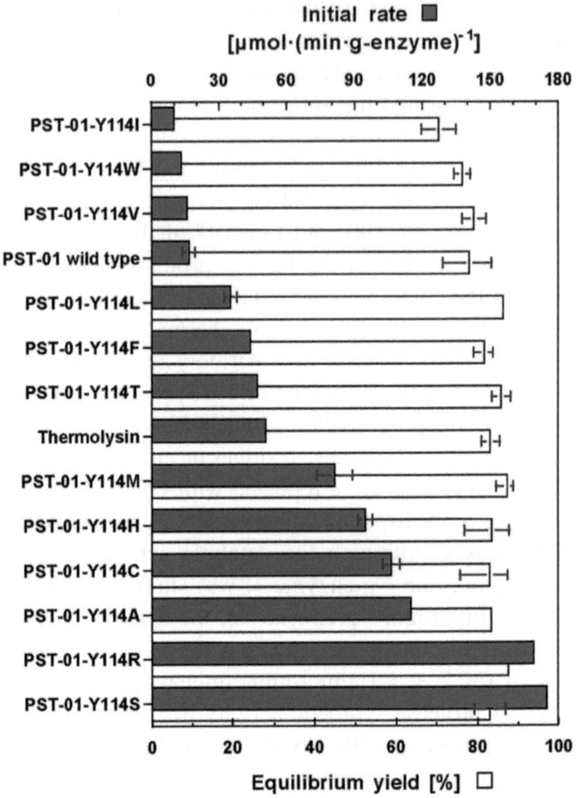

Fig. 12.5 Synthesis of Cbz-Asp-Phe-OMe using the various PST-01 protease mutants at Tyr-114. All reactions were carried out using 30 mM Cbz-Asp, 500 mM Phe-Ome, and 0.748–3.47 g·L^{-1} of the PST-01 protease in the mixture of DMSO and 50 mM Tris–HCl buffer (pH 7.0) (1:1 (v/v)) at 30 °C

References

Blout, E. R., Schmier, I., & Simmons, N. S. (1962). New cotton effects in polypeptides and proteins. *Journal of the American Chemical Society, 84*, 3193–3194.

Blout, E. R., Carver, J. P., & Gross, J. (1963). Intrinsic cotton effects in collagen and poly-L-proline. *Journal of the American Chemical Society, 85*, 644–646.

Doukyu, N., & Ogino, H. (2010). Organic solvent-tolerant enzymes. *Biochemical Engineering Journal, 48*(3), 270–282.

Holzwarth, G., & Doty, P. (1965). The ultraviolet circular dichroism of polypeptides. *Journal of the American Chemical Society, 87*, 218–228.

Holzwarth, G., Gratzer, W. B., & Doty, P. (1962). The optical activity of polypeptides in the far ultraviolet. *Journal of the American Chemical Society, 84*, 3194–3196.

Lane, C., Boeren, S., Vos, K., & Veeger, C. (1987). Rules for optimization of biocatalysis in organic solvents. *Biotechnology & Bioengineering, 30*, 81–87.

Li, L.-K., & Spector, A. (1969). Circular dichroism of β-poly-L-lysine. *Journal of the American Chemical Society, 91*, 220–222.

Matthews, B. W. (1988). Structural basis of the action of thermolysin and related zinc peptidases. *Accounts of Chemical Research, 21*, 333–340.

Ogino, H., & Ishikawa, H. (2001). Enzymes which are stable in the presence of organic solvents. *Journal of Bioscience and Bioengineering, 91*(2), 109–116.

Ogino, H., Yasui, K., Shiotani, T., Ishihara, T., & Ishikawa, H. (1995). Organic solvent-tolerant bacterium which secretes an organic solvent-stable proteolytic enzyme. *Applied and Environmental Microbiology, 61*(12), 4258–4262.

Ogino, H., Watanabe, F., Yamada, M., Nakagawa, S., Hirose, T., Noguchi, A., Yasuda, M., & Ishikawa, H. (1999a). Purification and characterization of organic solvent-stable protease from organic solvent-tolerant *Pseudomonas aeruginosa* PST-01. *Journal of Bioscience and Bioengineering, 87*(1), 61–68.

Ogino, H., Yamada, M., Watanabe, F., Ichinose, H., Yasuda, M., & Ishikawa, H. (1999b). Peptide synthesis catalyzed by organic solvent-stable protease from *Pseudomonas aeruginosa* PST-01 in monophasic aqueous-organic solvent systems. *Journal of Bioscience and Bioengineering, 88*(5), 513–518.

Ogino, H., Gemba, Y., Yamada, M., Shizuka, M., Yasuda, M., & Ishikawa, H. (2000a). The synthetic rate of dipeptide catalyzed by organic solvent-stable protease from *Pseudomonas aeruginosa* PST-01 in the presence of water-soluble organic solvents. *Biochemical Engineering Journal, 5*(3), 219–223.

Ogino, H., Yokoo, J., Watanabe, F., & Ishikawa, H. (2000b). Cloning and sequencing of a gene of organic solvent-stable protease secreted from *Pseudomonas aeruginosa* PST-01 and its expression in *Escherichia coli*. *Biochemical Engineering Journal, 5*(3), 191–200.

Ogino, H., Uchiho, T., Yokoo, J., Kobayashi, R., Ichise, R., & Ishikawa, H. (2001). Role of intermolecular disulfide bonds of the organic solvent-stable PST-01 protease in the organic solvent-stability. *Applied and Environmental Microbiology, 67*(2), 942–947.

Ogino, H., Gemba, Y., Yutori, Y., Doukyu, N., Ishimi, K., & Ishikawa, H. (2007a). Stabilities and conformational transitions of various proteases in the presence of an organic solvent. *Biotechnology Progress, 23*(1), 155–161.

Ogino, H., Uchiho, T., Doukyu, N., Yasuda, M., Ishimi, K., & Ishikawa, H. (2007b). Effect of exchange amino acid residues of the surface region of the PST-01 protease on its organic solvent-stability. *Biochemical and Biophysical Research Communications, 358*(4), 1028–1033.

Ogino, H., Tsuchiyama, S., Yasuda, M., & Noriyuki Doukyu, N. (2010). Enhancement of the aspartame precursor synthetic activity of an organic solvent-stable protease. *Protein Engineering, Design & Selection, 23*(3), 147–152.

Sarkar, P. K., & Doty, P. (1966). The optical rotatory properties of the β-configuration in polypeptides and proteins. *Proceedings of the National Academy of Sciences of the United States of America, 55*, 981–989.

Tiffany, M. L., & Krimm, S. (1968a). New chain conformations of poly(glutamic acid) and polylysine. *Biopolymers, 6*, 1379–1382.

Tiffany, M. L., & Krimm, S. (1968b). Circular dichroism of poly-L-proline in an unordered conformation. *Biopolymers, 6*, 1767–1770.

Tsuchiyama, S., Doukyu, N., Yasuda, M., Ishimi, K., & Ogino, H. (2007). Peptide synthesis of aspartame precursor using organic-solvent-stable PST-01 protease in monophasic aqueous-organic solvent systems. *Biotechnology Progress, 23*, 820–823.

Yagasaki, M., & Hashimoto, S. (2008). Synthesis and application of dipeptides; current status and perspectives. *Applied Microbiology and Biotechnology, 81*, 13–22.

Part IV
Fermentation and Biomass Applications for Renewable Energy

Chapter 13
Sustainable Biomass Composting Technology Using Extremophiles and Its Applications

Toshiyuki Moriya

Abstract Composting is gaining attention as a technology for treating solid organic waste in an environmentally friendly and economical manner, and it is also useful for producing high-quality organic fertilizers. This technology has been used since ancient times, even before the existence and role of microorganisms were recognized. As time progressed, it became clear that microorganisms played a central role in composting. Since the latter half of the twentieth century, controlled composting technologies based on microbiological knowledge have been developed. In recent years, large-scale commercial composting plants that apply this knowledge have been operated, contributing to sustainable waste disposal and resource recycling.

Composting is generally classified into four major stages: the mesophilic phase, the thermophilic phase, the cooling phase, and the maturation phase. Each stage involves various microbial communities that perform diverse reactions. In particular, the thermophilic phase (the second stage) holds significant importance in the composting process. During this phase of typical composting, the internal temperatures of the compost pile reach 60–70 °C or higher, enabling the efficient decomposition of complex biological macromolecules. This significantly reduces waste volume and promotes the extinction of pathogenic microorganisms and weed seeds, making this stage indispensable for producing hygienic and safe fertilizers. Thermophiles are responsible for reactions during the thermophilic phase, and a wide variety of these bacteria have been isolated from compost.

In the composting plant we operate, a notable feature is that the temperature at the center of the compost exceeds 80 °C, suggesting the presence of extreme thermophiles. We have isolated novel extreme thermophiles, *Calditerricola satsumensis* YMO81^T and *Calditerricola yamamurae* YMO722^T, from this compost, and it has been confirmed that these strains can grow at temperatures as high as 80 °C. These bacteria are actively involved in the early stages of the thermophilic phase but have relatively poor tolerance to lower temperatures. However, they are also present in the final product of the composting process, and how they survive the nonthermophilic

T. Moriya (✉)
Institute of Environmental Microbiology, Kyowa Kako Co., Ltd., Tokyo, Japan
e-mail: toshiyuki.moriya@kyowa-kako.co.jp

M. Ito et al. (eds.), *Extremophiles for a Sustainable Future*,
https://doi.org/10.1007/978-981-95-6970-0_13

phase of composting remains unclear. Furthermore, to evaluate the impact of high-temperature processes on the quality of the final product, we conducted experimental composting using *Alternanthera philoxeroides*, a highly reproductive aquatic plant. The final composting products were mixed with soil, and cultivation tests were conducted. The results showed that the aquatic plants did not regenerate and died. These results suggest that aquatic plants, pathogenic microorganisms, and weed seeds are inactivated in high-temperature environments. Therefore, sufficiently high temperature processes are extremely important for safe composting.

Keywords Composting · Sustainable waste management · Organic waste recycling · Thermophiles · Calditerricola

13.1 Introduction

Composting is a sustainable method for the treatment of organic waste that uses microbial activity to decompose biomass and produce organic fertilizers (Bernal et al., 2009; Cerda et al., 2018). This practice has a long history, dating back to ancient Roman times, when animal excrement and plant residues were decomposed, and the resulting product was spread on farmland to increase soil fertility and improve crop productivity. Composting has evolved over the years from traditional farming methods to a sophisticated waste treatment technology based on scientific evidence. Since the twentieth century, systematic research has been conducted on various parameters related to the composting process, such as temperature control and moisture content, and industrial-scale composting systems have been established based on such research. In recent years, against the backdrop of progressing climate change and the worsening global waste problem, the environmental and social significance of composting has come into focus.

Traditionally, organic waste has been primarily processed through incineration or landfilling. However, these methods present numerous environmental and economic problems (Farhidi et al., 2022). For example, incineration consumes large amounts of fossil fuels and emits significant amounts of carbon dioxide. Additionally, harmful substances, such as nitrogen oxides, sulfur oxides, and dioxins, are generated, necessitating advanced treatment facilities for their removal and neutralization. This results in high construction and maintenance costs. On the other hand, landfill disposal carries the risk of soil and groundwater pollution from leachate, making countermeasures against its environmental impacts indispensable. Furthermore, securing appropriate landfill sites is becoming increasingly difficult, and due to land resource constraints, this method cannot be considered a sustainable option in the long term.

In contrast, composting significantly reduces the overall emissions of greenhouse gases and harmful substances during processing, despite requiring fuel or electricity for some machinery, such as wheel loaders and aeration equipment (Hermann et al., 2011; Nordahl et al., 2020). Another major advantage is that it does not require expensive facilities for the neutralization of harmful substances and can be readily implemented at low cost. Therefore, it proves to be a highly flexible technology suitable for a variety of regions, including not only developed countries but also developing countries and small communities in rural areas. Furthermore, the final product obtained through composting is an important resource that contributes to sustainable agriculture. Compost is rich in humic substances such as humic acid, fulvic acid, and humin, and applying it to the soil improves water retention and aeration, contributing to enhanced soil fertility (Trevisan et al., 2010; Canellas & Olivares, 2014; Tahoun et al., 2022). Unlike chemical fertilizers, which release nutrients quickly, compost supplies nutrients slowly over time, helping to maintain the health and nutritional balance of the soil in the long term (Ahmed et al., 2023; Joseph & Babu, 2024). Besides, phosphate rock, the main material for chemical fertilizers, is used in the manufacture of batteries for eco-friendly cars, leading to a continuous increase in its global demand. As a result, the price of chemical fertilizers has been rising year by year, becoming a significant burden on agricultural operations. Against this backdrop, the value of compost as an organic fertilizer that is stable and utilizes local resources is expected to increase further in the future. In addition, composting is not merely a waste disposal technology but is also gaining attention as a means to contribute to international climate policies and sustainable development goals. For example, under international frameworks such as the Paris Agreement and the Sustainable Development Goals (SDGs), many countries are aiming to reduce greenhouse gas emissions and build a resource-circulating society, and composting can be positioned as a concrete and feasible option within this framework.

Despite its benefits, several challenges have impeded the widespread implementation of composting. First, composting is limited to biodegradable organic wastes, necessitating strict source separation. In urban environments, it is common for food waste to be contaminated with non-biodegradable materials, such as plastics and metals, which can significantly degrade compost quality. Secondly, composting typically requires more time and space than incineration, which poses logistical constraints for large-scale facilities in densely populated areas with limited land availability. However, recent technological innovations have begun to address these issues. For example, the development and adoption of biodegradable plastics have made it possible to compost certain packaging materials and garbage bags along with organic waste. This simplifies the sorting process. Furthermore, the introduction of high-temperature aerobic composting technology can reduce the composting time from months to weeks and significantly reduce the volume of organic waste. These advances have improved the processing efficiency and capacity of composting facilities.

In this paper, I first provide an overview of the general theory of composting and its microbiological basis, then discuss in detail the characteristics and distribution of thermophiles central to the thermophilic phase. I then introduce my ongoing investigation into high-temperature composting, examining the characteristics and ecological significance of isolated thermophilic bacteria. Furthermore, I present an experiment in which highly reproductive aquatic plants were inactivated via high-temperature composting, highlighting the process's utility in ensuring the safety of composting products. Through these discussions, I aim to demonstrate composting as a promising sustainable technology for waste treatment and resource recycling.

13.2 Fundamentals of General Composting

Composting is a complex and dynamic biochemical process in which organic materials, such as agricultural residues, food waste, cattle farming waste, pruning branches, and waste sludge, are decomposed by microbial activity and eventually converted into stable humic acid and other substances. This transformation occurs through a succession of microbial communities that change over time in response to evolving environmental conditions, including temperature, oxygen concentration, water content, and pH (Richard et al., 2006). During this process, a significant portion of the organic matter is mineralized into water, carbon dioxide, and gaseous nitrogen compounds such as ammonia. Simultaneously, a fraction of the organic constituents is transformed into stable humified substances, including humic and fulvic acids, which contribute to soil fertility and structure.

To ensure efficient and hygienic composting, it is essential to create and maintain an optimal environment for microbial activity. This includes sustaining aerobic conditions by ensuring adequate aeration, maintaining appropriate moisture levels (ideally between 50% and 60%), and carefully regulating the temperature to support microbial metabolism. When these factors are properly managed, the decomposition rate of organic matter is enhanced, pathogenic organisms are suppressed, and nutrient mineralization is facilitated. These processes collectively lead to the production of a high-quality, biologically stable compost.

Although the composting process can vary depending on the structural configuration of the composting facilities and the operational conditions under which it is carried out, it is generally divided into four stages: initial mesophilic, thermophilic, cooling, and maturation phases. Each stage is characterized by distinct microbial communities, biochemical reactions, and changes in physicochemical parameters. Understanding each stage is crucial to improve compost quality, shorten the composting period, and ensure the hygienic safety of the final product as a fertilizer (Du et al., 2019; Wainaina et al., 2020).

In the mesophilic phase of composting, the easily degradable organic compounds—such as simple sugars, amino acids, and lipids—present in the organic waste (composting materials) are rapidly metabolized by mesophilic microorganisms. Key microbial taxa active during this stage include members of the genera *Bacillus*, *Pseudomonas*, *Flavobacterium*, various species of *Actinomycetes,* and so on. The active metabolism of these microorganisms generates large amounts of heat. Because compost contains numerous microscopic spaces, it is highly insulating. Therefore, the heat generated by microorganisms is retained inside the compost pile, and as a result, the temperature inside the pile rises rapidly, often reaching 50–70 °C within a few days. The microbial activity during this stage is highly dependent on the availability of oxygen and moisture. An optimal moisture content of 50–60% is generally recommended. However, the appropriate initial moisture content depends on the material to be composted; therefore, the optimal value for each should be investigated. Excessive moisture impairs oxygen diffusion, leading to anaerobic conditions and the formation of malodorous compounds and methane. Unlike anaerobic plants that actively recover methane, the methane generated in aerobic composting plants drifts through the facility and can sometimes lead to explosions; hence, care must be taken. In contrast, insufficient moisture restricts microbial metabolism and results in a loss of composting progress. Forced aeration systems and regular mechanical turning (mixing of the compost pile) are often implemented to maintain aerobic conditions and promote microbial activity. These interventions help ensure the uniform distribution of oxygen and moisture and establish the conditions necessary for an efficient composting system.

Following the rapid microbial activity observed during the initial mesophilic stage, the temperature within the composting mass increases significantly, leading to a thermophilic phase. At these high temperatures, thermophilic microbial communities become dominant as the mesophiles that dominate the initial stages gradually become less active or die. However, not all mesophilic bacteria are killed. Some microorganisms form spores and tolerate high-temperature conditions. The thermophiles produce thermostable enzymes that degrade more structurally complex and persistent organic matter, such as cellulose, hemicellulose, keratin, and collagen (Wilson, 2004; Finore et al., 2023). Common thermophiles involved in this stage include species of the genera *Thermus, Thermoactinomyces*, *Geobacillus*, and certain thermophilic *Bacillus*. During this stage, the internal temperature typically ranges from 55 °C to 70 °C. Under such high temperatures, biopolymers are thermally denatured, and their molecular conformations are destroyed. Denatured biopolymers are more susceptible to digestion by degradative enzymes than those in their natural states (Matthyssens et al., 1972).

The internal environment of compost piles is inherently heterogeneous, with significant spatial variations in temperature, moisture, and oxygen distributions. The temperatures are higher at the center and lower at the surface. Therefore, denaturation of the biopolymer is favored at the center, but the surface may have a higher population of microorganisms and more active digestion. Thus, turning during composting is important. By turning, persistent compounds that have undergone thermal denaturation in the hot central part move to the periphery. Conversely, organic matter that is still undenatured moves to the central part, where it undergoes thermal denaturation. By repeating this process, the organic waste in the composting pile can be thermally denatured and digested evenly, thereby improving the efficiency and uniformity of the composting process.

During the third stage (cooling phase), the internal temperatures of the composting piles decrease. This temperature drop is due to the depletion of available organic matter and the resulting decrease in the amount of heat produced by microorganisms. As the temperature decreases, mesophilic microorganisms become active again. A key feature of this phase is the onset of humification, wherein microbial activity leads to the formation of humic substances such as fulvic acid, humic acid, and humin. These compounds enhance the value of compost as a soil amendment, improving the soil structure, water retention, and nutrient exchange capacity.

Maturation is the final stage of composting, during which the compost reaches a biologically and chemically stable state. An important biochemical transformation in this phase involves the nitrification of ammonium (NH_4^+) into nitrate (NO_3^-), facilitated by nitrifying bacteria, such as *Nitrosomonas* and *Nitrobacter*. With some exceptions, such as rice, most crops absorb nitrogen in the form of nitrate-nitrogen. This makes nitrate-nitrogen a fast-acting fertilizer, and its conversion to nitrate significantly increases its value. On the other hand, nitrate-nitrogen is characterized by its tendency to run off during rainfall and irrigation, posing the risk of groundwater contamination. Therefore, the amount and timing of fertilizer application should be carefully considered.

The carbon/nitrogen (C/N) ratio is considered an important parameter in the composting process, as it influences the quality of the final compost, which is then used as a fertilizer. An initial C/N ratio of 25–30 is often recommended for composting (Shen et al., 2024). If the C/N ratio is higher than this value, the composting process tends to be delayed because of the lack of nitrogen. Also, when compost with a high C/N ratio (generally referred to as immature compost) is applied to agricultural land, it leads to consumption of nitrogen by soil microorganisms, resulting in nitrogen starvation, which often hinders crop growth. Conversely, if the C/N ratio in composting is low, the composting reaction proceeds rapidly but tends to generate odors such as ammonia. When this type of compost with a low C/N ratio (generally referred to as fully matured compost) is used as a fertilizer, the risk of nitrogen starvation decreases, and plants utilize the nitrogen contained in the compost, which has a positive effect on the growth of crops.

Thus, the C/N ratio is an indicator of the composting process and the characteristics of the compost used as a fertilizer, though it is dangerous to place too much importance on it. For example, pruning branches, grass clippings, and other materials that contain a large amount of cellulose, hemicellulose, and lignin also contain abundant carbon; although, they are difficult to decompose, and their usefulness as a direct energy source for microbial metabolism is limited (Yang et al., 2021). However, in many cases, the C/N ratio is analyzed using a measuring device that employs the dry combustion method (Dieckow et al., 2007), and it is not possible to distinguish whether the detected carbon is easily utilized by microorganisms. As a result, the C/N ratio of the difficult-to-decompose substances mentioned above can appear high. Still, most of these are not carbon that can be easily utilized by microorganisms. Additionally, a low C/N ratio does not necessarily indicate that the compost is fully mature. Various raw materials are available for composting, and their C/N ratios vary significantly. For instance, the C/N ratio of sludge is generally low, sometimes in the form of single digits. Therefore, the maturity of compost should not be evaluated only by the C/N ratio but should also be judged by a comprehensive evaluation of various parameters, including moisture and temperature.

13.3 High-Temperature Composting System Operated by Kyowa Kako Co., Ltd.

Kyowa Kako Co., Ltd. is engaged in the planning, construction, operation, and maintenance of composting plants in both domestic and international settings. The composting system managed by the company is characterized by extremely high temperatures, with internal temperatures exceeding 80 °C during the thermophilic phase (Oshima & Moriya, 2008; Yoshii et al., 2013). This section introduces the structural features and operational principles of this composting system, followed by an overview of extreme thermophiles isolated from the compost and their biotechnological applications.

The composting system generally comprises multiple composting bays (referred to as Bacaly bay: Bio-Aero-Carol-Lysis), each typically measuring approximately 7 m in width, 3.2 m in height, and 13 m in depth (Fig. 13.1). The floor of each Bacaly bay has perforated aeration pipes connected to a blower system that delivers intermittent forced air to sustain the aerobic conditions of the compost and enhance microbial activity.

Composting materials vary by facility and include a wide range of organic wastes such as food residues, livestock waste, pruning biomass, sewage sludge, and agricultural byproducts. The collected waste is first stored, then thoroughly mixed with matured compost (referred to as BisDoc: Biosolid-Diorganic Compost), which acts

Fig. 13.1 An example of a Bacaly composting bay. The size and number of these systems vary depending on the amount of waste the facility accepts and other factors. Pipes with holes are installed in the floor trenches for aeration, and air is pumped through them by a blower

Fig. 13.2 A scene during the turning process

as a microbial inoculum (usually moisture content is 50% ± 5%). The mixture is then loaded into the Bacaly to initiate the composting process. Turning using a wheel loader once a week ensures homogeneous aeration and thermal distribution. After approximately 6 weeks, the compost's moisture content is sufficiently reduced (usually 25–30%), and the process is considered complete.

During peak activity of the thermophilic phase, the internal temperature can surpass 80 °C, effectively inactivating pathogens and weed seeds. A portion of the final compost product (BisDoc) is sold to local farmers as fertilizer, whereas the remainder is reused as a starter material for subsequent batches. This plant is capable of completely self-sustaining composting without the need for external heating. Figure 13.2 shows the compost-turning process. The white smoke in the image represents the release of water vapor that accumulates inside.

13.4 Extreme Thermophiles *Calditerricola satsumensis* YMO81^{T} and *C. yamamurae* YMO722^{T}

Through investigation of this high-temperature composting system, we isolated two novel thermophilic bacterial strains, *Calditerricola satsumensis* YMO81^{T} and *Calditerricola yamamurae* YMO722^{T} (Moriya et al., 2011) (Fig. 13.3). Phylogenetic analysis based on 16S rRNA gene sequences revealed a 91% similarity to *Planifilum yunnanense* LA5^{T} (Zhang et al., 2007).

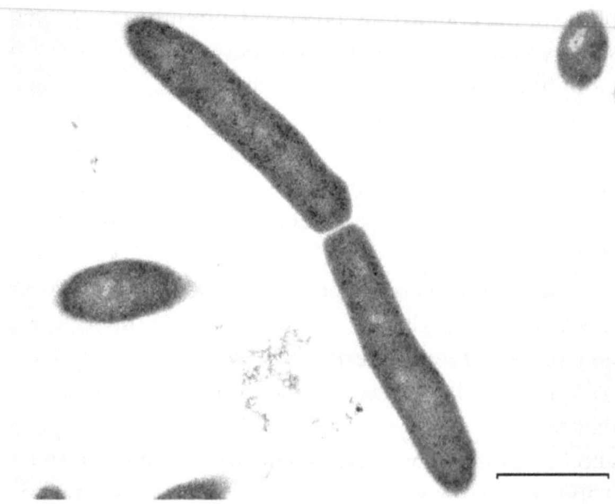

Fig. 13.3 Transmission electron microscope image of *Calditerricola satsumensis* YMO81^T. The bar represents 1 μm

C. satsumensis YMO81^T is a thermophilic, Gram-negative, non-motile, non-sporulating, rod-shaped bacterium (approx. 0.2 × 3 μm), capable of proliferating at temperatures from 65 to 83 °C. It grows vigorously in liquid CYS medium (Moriya et al., 2011) containing 2% NaCl, with a doubling time of approximately 30 min under optimal conditions. *C. yamamurae* YMO722^T has characteristics similar to *C. satsumensis* YMO81^T, but its growth temperature range is 56–81 °C and its optimal NaCl concentration is 1.6%. DNA-DNA hybridization results showed that the two species have 55% homology, suggesting that they are different species. Because agar cannot maintain a gel state at 80 °C, gellan gum was used as a gelling agent in the plate medium culture; however, the colony formation efficiency in this plate medium was very low. Specifically, 100 μL of culture medium with $OD_{660} = 0.5$ applied to CYS plate medium yields only 10 to 20 colonies.

To improve the colony formation efficiency, we adjusted the medium components and found that the addition of $FeSO_4$ (100 mg/L) and $VOSO_4$ (2 mg/L) significantly improved the colony formation efficiency by 1000-fold. Furthermore, spot application of genomic DNA (10 m g/mL) from *C. satsumensis* YMO81^T or *Escherichia coli* on the surface of CYS gellan gum plates promoted colony formation, specifically in the applied area. The DNA used in this study was dissolved in Tris-EDTA (TE) buffer, but this effect was not observed in areas where only TE buffer was applied. (unpublished data). Although the mechanism underlying this phenomenon remains unclear, it has been speculated that iron, vanadium, or extracellular DNA influence membrane stability, DNA repair processes, or the metabolism of trace metals. Further investigation is warranted to elucidate these pathways.

An interesting characteristic of these strains is their temperature sensitivity. Unlike many thermophiles, these two strains sometimes fail to grow when cultures are inoculated into room-temperature media. Pre-warming the fresh medium to an incubation temperature such as 80 °C before inoculation increases the success of the culture. This phenomenon is important for their survival during the low-temperature phase of the composting process. Currently, the mechanisms by which these strains survive composting are not understood.

In this high-temperature composting, many thermophiles other than those belonging to the genus *Calditerricola* are involved. We collected composting samples at each turn throughout the entire period from the start to the completion of composting and extracted metagenomic DNA. We then conducted a microbial flora analysis using polymerase chain reaction-denaturing gradient gel electrophoresis (PCR-DGGE) (Saiki et al., 1988; Ercolini, 2004) to investigate the dynamics of the diverse microorganisms present in the compost (Yoshii et al., 2013). The results revealed that in addition to *Calditerricola*, multiple thermophiles belonging to the genera *Planifilum*, *Thermaerobacter*, *Thermus*, *Saccharomonospora*, and so on were frequently detected. All identified microorganisms were dominant during the thermophilic phase of composting. Composting proceeds through several phases, and *C. satsumensis* is detected in large numbers during the early thermophilic phase. This composting system has a long high-temperature period; however, the microbial flora during this period is not constant. In the middle of the thermophilic phase of composting, the population of *C. satsumensis* decreases to approximately one thousandth. The early thermophilic phase is characterized by an abundance of easily degradable organic substrates, such as soluble carbohydrates and low-molecular-weight fatty acids. Therefore, *C. satsumensis* is thought to capitalize on easily degradable organic matter during this period, generating a large amount of metabolic heat, which, combined with its thermophilic properties, contributes to increasing the temperature inside the compost. In contrast, the genus *Thermus* gradually appeared approximately 10 days after the start of composting and was detected until around the 25th day. This period coincides with the middle of the thermophilic phase of composting, and the *Thermus* genus is thought to not only exhibit high thermophilicity but also produce multiple thermostable enzymes that contribute to the decomposition of refractory organic matter.

To gain a more detailed understanding of these dynamics, we used real-time PCR (Wittwer et al., 1997; Ririe et al., 1997) to quantify the number of 16S rDNA copies in each sample, targeting primers common to all bacteria and *Thermus*-specific 16S rDNA primers. The results showed that approximately 40% of the detected 16S rDNA was derived from the *Thermus* genus during days 15–20. However, 16S rDNA is a gene that exists in multiple copies in the bacterial genome, and the number of copies varies depending on the microbial species (Klappenbach et al., 2001; Větrovský & Baldrian, 2013; Louca et al., 2018). Therefore, it is difficult to directly convert the number of copies into the number of individuals. Although, this high proportion strongly indicates that the *Thermus* genus is one of the major components during the middle stage of composting. In addition, bacteria belonging to the *Thermaerobacter* genus were widely distributed during the thermophilic phase. *Thermaerobacter* has previously been isolated as a thermophilic bacterium from compost-derived samples. In this study, its band was clearly observed and remained stable throughout the thermophilic phase in DGGE analysis. Furthermore, quantitative real-time PCR indicated that *Thermaerobacter* accounted for approximately 4% of the 16S rDNA, suggesting its active role in the composting process.

These results show that the microbial community in the high-temperature composting process is not constant, but rather that various thermophilic bacteria replace each other over time, playing appropriate roles for each stage of decomposition. However, the specific functions of these microorganisms remain unclear, and a detailed analysis is necessary to further improve the composting efficiency and quality.

In this study, samples collected after each turning during composting were suspended in phosphate buffer, and the resulting supernatant was used for zymography (Heussen & Dowdle, 1980). Enzymatic reactions were performed at 70 °C, gelatin was used as a substrate, and Coomassie Brilliant Blue was used for staining. Figure 13.4 shows the zymography results. Gelatinase activity was barely detectable at the beginning of composting but gradually increased, with the strongest activity observed at the end of the process. As the enzymatic reaction was conducted at 70 °C, only thermostable gelatinase was detected. While this does not preclude gelatinase presence in the mesophilic phase, it clearly indicates that thermostable gelatinase activity began around day 20 of composting and became more pronounced toward completion. This suggests that as composting progresses, readily degradable proteins decrease, and microorganisms producing thermostable gelatinases proliferate and become active. These results, combined with changes in the microbial community based on 16S rDNA analysis, clearly demonstrate the dynamic shifts in microbial community structure and function during composting.

Fig. 13.4 Gelatin zymography results. The gel contains gelatin throughout, and areas with gelatinase activity are detected without staining. (1) Start of composting, (2) 1st turning, (3) 2nd turning, (4) 3rd turning, (5) 4th turning, (6) 5th turning, (7) 6th turning, and (8) end of composting

13.5 Application of DNA Polymerase Derived from Thermophiles and Compost

Thermophiles have various applications, with the DNA polymerase used in PCR being a typical example. We also attempted to clone the DNA polymerase gene from *C. satsumensis* YMO81^T. Given that microbial genome analysis was not as readily available then as it is today, we employed classical methods. This involved multiple-aligning known DNA polymerase amino acid sequences, designing degenerate primers for conserved regions, amplifying partial gene fragments via PCR, and subsequently obtaining full-length DNA polymerase genes through repeated inverse PCR.

Applying this method to the metagenomic DNA extracted from compost samples allowed us to retrieve polymerase genes from uncultured microbial sources as well. The isolated enzymes demonstrated both 5′→3′ polymerase activity and strand displacement activity and proved highly effective in loop-mediated isothermal amplification (Notomi et al., 2000).

Currently, these enzymes are commercially available from Nippon Gene (Tokyo, Japan) as *Csa* DNA Polymerase and 96-7 DNA Polymerase. Their high amplification efficiency, specificity, and ability to operate at a constant temperature make them ideal for rapid genetic diagnostics in clinical, food safety, and environmental applications.

13.6 Thermal Inactivation of *Alternanthera philoxeroides* and *Ludwigia grandiflora* Through High-Temperature Composting

A notable feature of the high-temperature composting system is that the internal temperature exceeds 80 °C. During composting, the high temperature in the central area is important in two ways. The first is to promote the effective decomposition of biopolymer compounds. Second, it is essential to ensure that the products obtained after composting are hygienic and biologically safe. Organic matter in compost contains high-molecular-weight proteins, such as keratin and collagen, which, in their native state, have a very stable tertiary structure and are resistant to enzymatic hydrolysis by microorganisms. However, heat treatment at 80 °C denatures these proteins. This denaturation renders them more susceptible to degradation by digestive enzymes. Furthermore, it has been reported that temperatures above 55 °C sustained for several days inactivate pathogenic microorganisms such as *E. coli* O157:H7 and *Salmonella* spp. This is a basic requirement for converting organic waste, including food residue and fecal matter, into safe agricultural materials (Singh et al., 2012). Thus, the high-temperature composting process has several advantages, including ensuring hygienic safety by inactivating pathogenic microorganisms, producing more mature compost in a shorter period by promoting the decomposition of difficult-to-decompose organic matter, and ultimately contributing to the production of safe and effective fertilizers with high nutrient content (Bernal et al., 2009).

To assess the efficacy of high-temperature composting in inactivating harmful organisms, we performed experimental composting using the invasive alien aquatic plants, *Alternanthera philoxeroides* and *Ludwigia grandiflora*, as raw materials. These species exhibit extremely high reproductive capacities and are capable of regenerating from minimal stem or leaf fragments (Dugdale et al., 2010; Genitoni et al., 2020). Consequently, they proliferate rapidly in waterways and lakes, displacing native species and causing significant ecological disruptions. Due to these impacts, many countries have imposed restrictions on the cultivation, storage, and transportation of these organisms (Chen et al., 2013; Billet et al., 2018). All procedures for the transport and disposal of these plants during the experiment were conducted with the approval of the relevant authorities.

The composting trials were conducted twice. Harvested aquatic plants were mixed with BisDoc (mature compost) and placed on Bacaly for composting. After the initial composting, additional aquatic plants were incorporated, and a second composting cycle was initiated. As described previously, air was supplied from the floor, the compost was turned weekly using a wheel loader, and the composting process was completed after 45 days. After composting, we examined whether *A. philoxeroides* and *L. grandiflora* were inactivated. Specifically, 30 g of the product obtained after two composting cycles was mixed with 300 g of soil and placed in growing pots. These pots were well watered and maintained at 20 °C for 2 months to observe the emergence of the target plant species.

Figure 13.5 illustrates the temperature profile of the compost during the experiment, which was measured using thermocouples. Although temperatures temporarily decreased during the turning process, they rapidly increased thereafter, exceeding 80 °C. Notably, during the initial composting cycle, temperatures frequently surpassed 90 °C. Figure 13.6 shows the results of plant inactivation tests. Plant growth was observed in pots where untreated plant material was mixed with soil (the two pots on the right in the figure). Conversely, no plant growth was detected in the pots containing composted material mixed with soil (the two pots on the left in the figure). These tests were repeated multiple times and consistently showed no regeneration of *A. philoxeroides* and *L. grandiflora* from the composted material. In conclusion, *A. philoxeroides* and *L. grandiflora* were effectively inactivated by the high-temperature composting process. This suggests that the entire compost pile was exposed to a high-temperature environment owing to turning, which resulted in the death of these aquatic plants. Although this study evaluated the high-temperature composting process using the fertility of highly reproductive aquatic plants as an indicator, it suggests that common weed seeds and pathogenic microorganisms may be similarly inactivated by high-temperature composting (Larney & Blackshaw, 2003). These findings demonstrate the importance of high temperatures in the composting process.

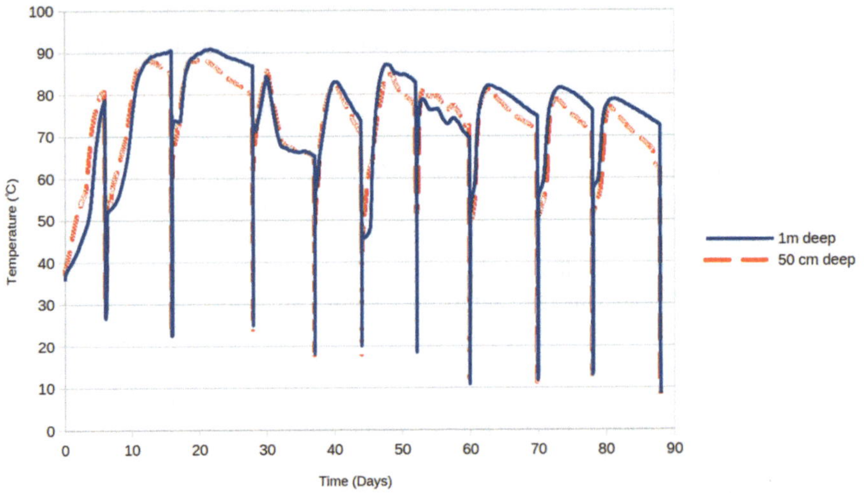

Fig. 13.5 Internal temperatures during composting. The solid line shows the temperature approximately 1 m from the top of the pile, whereas the dashed line shows it approximately 50 cm from the top. Each was recorded continuously using thermocouples. The periodic drop in temperature indicates when turning was performed

Fig. 13.6 Results of the cultivation test for aquatic-plant-composting products. The two pots on the left show the cultivation of the composted product mixed with soil. The two pots on the right show the sample immediately before the start of composting mixed with soil and cultivated. This experiment was conducted twice for each sample

13.7 Conclusion and Future Prospects

Composting is a typical organic waste treatment method that combines three major advantages: low environmental impact, low cost, and effective utilization of the resulting fertilizers. In recent years, the introduction of commercial composting plants by municipalities and companies has progressed, and the efficiency and stability of this treatment have dramatically improved. As of 2025, our company has installed and operated composting plants at 40 locations in Japan and overseas, including Davao City in the Philippines, and the number of plants is steadily increasing every year. This will enable us to further increase the ratio of organic waste treatment using composting, which is expected to contribute to the construction of a sustainable society by helping reduce CO_2 emissions.

As composting technology cannot decompose plastics, they must be separated. Therefore, the introduction of biodegradable plastics is an effective way to expand composting technology. For example, food waste is typically stored in plastic bags. When initiating composting, it is necessary to tear open these bags and separate their contents. However, it requires manpower, and depending on the season, food scraps may rot, making the work environment difficult. By replacing plastic bags with biodegradable alternatives, these can be directly composted without tearing. Additionally, using biodegradable plastics for packaging and disposable tableware (e.g., plates, cups, cutlery) at events, outdoor festivals, or on passenger planes allows for co-composting them with leftover food. Our company has confirmed that a high-temperature composting system using BisDoc effectively decomposes bioplastic materials (Soeda et al., 2024). This approach promises more efficient collection and disposal operations.

The microorganisms responsible for composting are extremely diverse, as are the enzymes they produce. Enzymes produced by thermophiles that are active during the composting process are highly thermostable and suitable for industrial use. From this perspective, compost is an important resource in the search for useful enzymes. As research on enzyme utilization progresses, it is expected to have a ripple effect in areas such as food processing, pharmaceuticals, and bioremediation.

The thermophilic phase plays an important role in composting. Under high-temperature conditions, organic matter that is normally difficult to decompose is broken down more efficiently by the action of enzymes produced by thermophiles, leading to the production of high-quality fertilizers. In addition, this high-temperature process eliminates pathogenic microorganisms, viruses, and weed seeds, resulting in safe fertilizer application. The term "pathogenic" here encompasses pathogens that affect humans as well as those that infect crops. Although plant residues remaining after the crop harvest may be used as composting materials, if pathogens are not sufficiently eliminated during the composting process, there is a risk that they may accumulate and spread through repeated composting. Therefore, high-temperature composting is indispensable for ensuring the safety of fertilizers.

Although composting is a traditional and universal technology that humans have used since ancient times, many of its aspects remain unexplored. Therefore, further research is needed to elucidate the microbial ecology and decomposition mechanisms, assess the environmental impact, and optimize the composting process. Progress in these areas can be expected with the development of new environmental technologies and the realization of a resource-recycling society.

Acknowledgments We would like to thank Editage (www.editage.jp) for English language editing.

References

Ahmed, T., Noman, M., Qi, Y., et al. (2023). Fertilization of microbial composts: A technology for improving stress resilience in plants. *Plants, 12*, 3550.

Bernal, M. P., Alburquerque, J. A., & Moral, R. (2009). Composting of animal manures and chemical criteria for compost maturity assessment. A review. *Bioresource Technology, 100*, 5444–5453.

Billet, K., Genitoni, J., Bozec, M., et al. (2018). Aquatic and terrestrial morphotypes of the aquatic invasive plant, Ludwigia grandiflora, show distinct morphological and metabolomic responses. *Ecology and Evolution, 8*, 2568–2579.

Canellas, L. P., & Olivares, F. L. (2014). Physiological responses to humic substances as plant growth promoter. *Chemical and Biological Technologies in Agriculture, 1*, 1–11.

Cerda, A., Artola, A., Font, X., et al. (2018). Composting of food wastes: Status and challenges. *Bioresource Technology, 248*, 57–67.

Chen, Y., Zhou, Y., Yin, T.-F., et al. (2013). The invasive wetland plant Alternanthera philoxeroides shows a higher tolerance to waterlogging than its native Congener Alternanthera sessilis. *PLoS One, 8*, e81456.

Dieckow, J., Mielniczuk, J., Knicker, H., et al. (2007). Comparison of carbon and nitrogen determination methods for samples of a Paleudult subjected to no-till cropping systems. *Scientia Agricola (Piracicaba, Braz), 64*, 532–540.

Du, J., Zhang, Y., Qu, M., et al. (2019). Effects of biochar on the microbial activity and community structure during sewage sludge composting. *Bioresource Technology, 272*, 171–179.

Dugdale, T. M., Clements, D., Hunt, T. D., & Butler, K. L. (2010). Alligatorweed produces viable stem fragments in response to herbicide treatment. *Journal of Aquatic Plant Management, 48*, 84–91.

Ercolini, D. (2004). PCR-DGGE fingerprinting: Novel strategies for detection of microbes in food. *Journal of Microbiological Methods, 56*, 297–314.

Farhidi, F., Madani, K., & Crichton, R. (2022). How the US economy and environment can both benefit from composting management. *Environmental Health Insights, 16*, 11786302221128454.

Finore, I., Feola, A., Russo, L., et al. (2023). Thermophilic bacteria and their thermozymes in composting processes: A review. *Chemical and Biological Technologies in Agriculture, 10*, 7.

Genitoni, J., Vassaux, D., Delaunay, A., et al. (2020). Hypomethylation of the aquatic invasive plant, *Ludwigia grandiflora* subsp. *hexapetala* mimics the adaptive transition into the terrestrial morphotype. *Physiologia Plantarum, 170*, 280–298.

Hermann, B. G., Debeer, L., De Wilde, B., et al. (2011). To compost or not to compost: Carbon and energy footprints of biodegradable materials' waste treatment. *Polymer Degradation and Stability, 96*, 1159–1171.

Heussen, C., & Dowdle, E. B. (1980). Electrophoretic analysis of plasminogen activators in polyacrylamide gels containing sodium dodecyl sulfate and copolymerized substrates. *Analytical Biochemistry, 102*, 196–202.

Joseph, B., & Babu, S. (2024). Effect of organic and chemical fertilizer on the diversity of rhizosphere and leaf microbial composition in sunflower plant. *Current Microbiology, 81*, 331.

Klappenbach, J. A., Saxman, P. R., Cole, J. R., & Schmidt, T. M. (2001). rrndb: The ribosomal RNA Operon copy number database. *Nucleic Acids Research, 29*, 181–184.

Larney, F. J., & Blackshaw, R. E. (2003). Weed seed viability in composted beef cattle feedlot manure. *Journal of Environmental Quality, 32*, 1105–1113.

Louca, S., Doebeli, M., & Parfrey, L. W. (2018). Correcting for 16S rRNA gene copy numbers in microbiome surveys remains an unsolved problem. *Microbiome, 6*, 1–12.

Matthyssens, G. E., Simons, G., & Kanarek, L. (1972). Study of the thermal-denaturation mechanism of hen egg-white lysozyme through proteolytic degradation. *European Journal of Biochemistry, 26*, 449–454.

Moriya, T., Hikota, T., Yumoto, I., et al. (2011). Calditerricola satsumensis gen. nov., sp. nov. and Calditerricola yamamurae sp. nov., extreme thermophiles isolated from a high-temperature compost. *International Journal of Systematic and Evolutionary Microbiology, 61*, 631–636.

Nordahl, S. L., Devkota, J. P., Amirebrahimi, J., et al. (2020). Life-cycle greenhouse gas emissions and human health trade-offs of organic waste management strategies. *Environmental Science & Technology, 54*, 9200–9209.

Notomi, T., Okayama, H., Masubuchi, H., et al. (2000). Loop-mediated isothermal amplification of DNA. *Nucleic Acids Research, 28*, E63.

Oshima, T., & Moriya, T. (2008). A preliminary analysis of microbial and biochemical properties of high-temperature compost. *Annals of the New York Academy of Sciences, 1125*, 338–344.

Richard, T. L., Walker, L. P., & Gossett, J. M. (2006). Effects of oxygen on aerobic solid-state biodegradation kinetics. *Biotechnology Progress, 22*, 60–69.

Ririe, K. M., Rasmussen, R. P., & Wittwer, C. T. (1997). Product differentiation by analysis of DNA melting curves during the polymerase chain reaction. *Analytical Biochemistry, 245*, 154–160.

Saiki, R. K., Gelfand, D. H., Stoffel, S., et al. (1988). Primer-directed enzymatic amplification of DNA with a thermostable DNA polymerase. *Science, 239*, 487–491.

Shen, B., Zheng, L., Zheng, X., et al. (2024). Insights from meta-analysis on carbon to nitrogen ratios in aerobic composting of agricultural residues. *Bioresource Technology, 413*, 131416.

Singh, R., Kim, J., & Jiang, X. (2012). Heat inactivation of Salmonella spp. in fresh poultry compost by simulating early phase of composting process. *Journal of Applied Microbiology, 112*, 927–935.

Soeda, S., Toshii, T., & Yoshikawa, Y. (2024). Study on degradation of biodegradable plastics and other materials by aerobic thermophilic flora used for composting. *One Health Sciences, 7*, 1–17.

Tahoun, A. M. M. A., El-Enin, M. M. A., Mancy, A. G., et al. (2022). Integrative soil application of humic acid and foliar plant growth stimulants improves soil properties and wheat yield and quality in nutrient-poor Sandy soil of a Semiarid region. *Journal of Soil Science and Plant Nutrition, 22*, 2857–2871.

Trevisan, S., Francioso, O., Quaggiotti, S., & Nardi, S. (2010). Humic substances biological activity at the plant-soil interface. *Plant Signaling & Behavior, 5*, 635–643.

Větrovský, T., & Baldrian, P. (2013). The variability of the 16S rRNA gene in bacterial genomes and its consequences for bacterial community analyses. *PLoS One, 8*, e57923.

Wainaina, S., Awasthi, M. K., Sarsaiya, S., et al. (2020). Resource recovery and circular economy from organic solid waste using aerobic and anaerobic digestion technologies. *Bioresource Technology, 301*, 122778.

Wilson, D. B. (2004). Studies of Thermobifida fusca plant cell wall degrading enzymes. *Chemical Record, 4*, 72–82.

Wittwer, C. T., Herrmann, M. G., Moss, A. A., & Rasmussen, R. P. (1997). Continuous fluorescence monitoring of rapid cycle DNA amplification. *BioTechniques, 22*, 130–131, 134–138.

Yang, H., Zhang, H., Qiu, H., et al. (2021). Effects of C/N ratio on lignocellulose degradation and enzyme activities in aerobic composting. *Horticulturae, 7*, 482.

Yoshii, T., Moriya, T., & Oshima, T. (2013). Bacterial and biochemical properties of newly invented aerobic, high-temperature compost. In T. Satyanarayana, J. Littlechild, & Y. Kawarabayasi (Eds.), *Thermophilic microbes in environmental and industrial biotechnology: Biotechnology of thermophiles* (pp. 119–135). Springer.

Zhang, Y.-X., Dong, C., & Biao, S. (2007). Planifilum yunnanense sp. nov., a thermophilic thermoactinomycete isolated from a hot spring. *International Journal of Systematic and Evolutionary Microbiology, 57*, 1851–1854.

[illegible]

Chapter 14
Social Implementation of Methanation Technology Using Methanogenic Archaea

Noriyuki Doukyu

Abstract Power-to-methane, a component of power-to-gas, is recognized globally as a key technology in the transition to a sustainable energy system. Methanation is the process of synthesizing methane (CH_4), the main component of city gas, from hydrogen (H_2) and carbon dioxide (CO_2). This process typically involves the Sabatier reaction, which uses an inorganic catalyst to react H_2 and CO_2 to produce CH_4; however, it can also be carried out by microorganisms, a process known as biomethanation. Archaean methanogens, particularly hydrogenotrophic methanogens of the genus *Methanothermobacter*, have been extensively studied in methanation research due to their capacity for mass cultivation. Understanding the methanogenic pathways of these organisms is valuable for enhancing biomethanation processes. Various methods for biomethanation, including in situ and ex situ types, have also been studied, with discussions surrounding the advantages and disadvantages of each. In addition, various reactor designs have been developed to improve the efficiency of the biomethanation reaction.

Keywords Methanation · Methanogens · Hydrogenotrophic methanogens · Methanothermobacter

14.1 Introduction

The use of fossil hydrocarbons has positively impacted the economy and energy infrastructure (Rondinelli & Berry, 2000). However, their use is also associated with adverse effects on health and the environment, as well as contributing to global warming (Hansen et al., 2000). Consequently, the decline of fossil energy sources has sparked increased interest in renewable energy alternatives. Biofuels, which are

N. Doukyu (✉)
Faculty of Life Sciences, and Bio-Resilience Research Project (BRRP), Toyo University, Asaka, Saitama, Japan
e-mail: dokyu@toyo.jp

M. Ito et al. (eds.), *Extremophiles for a Sustainable Future*,
https://doi.org/10.1007/978-981-95-6970-0_14

environmentally friendly energy carriers produced from biological resources, help recycle CO_2 and reduce greenhouse gas emissions. Additionally, wind and solar energy are being developed to bolster renewable energy initiatives. However, renewable energy production fluctuates daily and seasonally, creating a significant gap between electricity supply and demand (Rafrafi et al., 2021). It is important to estimate this gap when integrating renewable energy sources. For instance, if electricity is overproduced during the warm season, converting the excess electricity into heat may not be practical. To address these challenges, several energy storage technologies have been developed. Converting electricity to H_2 or CH_4 is one of the more effective solutions, providing sufficient storage capacity and discharge times to meet global energy needs. Methanation, the process of converting CO and CO_2 to CH_4 using H_2, plays a key role in this context. There are two methods of methanation: one employs inorganic catalysts, while the other utilizes methanogens. This chapter focuses primarily on the biological formation of methane by methanogens.

14.2 Biological Formation of Methane by Methanogens

Methanogens are strictly anaerobic microorganisms that belong to the *Euryarchaeota* and are obligate CH_4 producers. They derive all or most of their energy from methane production. The methanogenic pathway is complex and requires numerous unique cofactors and membrane-bound enzyme complexes (Hedderich & Whitman, 2013). Methanogens thrive in anaerobic habitats, particularly in low-sulfate environments such as freshwater pond sediments, the digestive tracts of animals, insects, marine sediments, and terrestrial subsurface environments, including various extreme conditions (Buan, 2018).

Methanogens can grow by reducing three main substrates: CO_2, methyl-group-containing compounds, and acetate (Table 14.1) (Liu & Whitman, 2008). Organic substances that serve as growth substrates for non-methanogenic bacteria, such as carbohydrates, long-chain fatty acids, and alcohols, are not metabolized through methanogenic pathways. Thus, non-methanogenic organisms can utilize most of the available energy for growth in most methanogenic environments.

The majority of methanogens produce CH_4 by reducing CO_2 using H_2 or formate as electron donors (Fig. 14.1). This CO_2 reduction pathway consists of seven steps and an additional branch. Formate is oxidized to CO_2 by formate dehydrogenase (Fdh). CO_2 is then successively reduced to CH_4 via formyl, methylene, and methyl levels. Methanofuran (MFR), tetrahydromethanopterin (H_4MPT), and coenzyme M (CoM) serve as the main methyl vectors. Initially, CO_2 binds to MFR and is reduced to the formyl level. The formyl group is transferred to H_4MPT, producing formyl-H_4MPT. Formyl-H_4MPT is then converted to methenyl-H_4MPT, methylene-H_4MPT, and methyl-H_4MPT in that order. Reductive coenzyme F_{420} ($F_{420}H_2$) acts as the direct electron donor in the reductions from methine-H_4MPT to methyl-H_4MPT. The methyl group is then transferred to CoM, resulting in the formation of methyl-CoM. In the final step, methyl-coenzyme M-reductase (Mcr), the main enzyme in

Table 14.1 Types of methanogenic reactions and typical methanogens

Substrate	Reaction	Methanogens
CO_2	$4\ H_2 + CO_2 \rightarrow CH_4 + 2\ H_2O$ $4\ HCOOH \rightarrow CH_4 + 3\ CO_2 + 2\ H_2O$ $CO_2 + 4$ isopropanol $\rightarrow CH_4 +$ 4 acetone $+ 2\ H_2O$ $4\ CO + 2H_2O \rightarrow CH_4 + 3\ CO_2$	Methanogens in general Many H_2/CO_2-utilizing methanogens Some H_2/CO_2-utilizing methanogens *Methanothermobacter* and *Methanosarcina*
C1 compounds	$4\ CH_3OH \rightarrow 3\ CH_4 + CO_2 + 2\ H_2O$ $CH_3OH + H_2 \rightarrow CH_4 + H_2O$ $2\ (CH_3)_2\text{-}S + 2\ H_2O \rightarrow 3\ CH_4 + CO_2 + 2\ H_2S$ $4\ CH_3\text{-}NH_2 + 2\ H_2O \rightarrow 3\ CH_4 + CO_2 + 4\ NH_3$ $2\ (CH_3)_2\text{-}NH + 2\ H_2O \rightarrow 3\ CH_4 + CO_2 + 2\ NH_3$ $4\ (CH_3)_3\text{-}N + 6\ H_2O \rightarrow 9\ CH_4 + 3\ CO_2 + 4\ NH_3$ $4\ CH_3NH_3Cl + 2\ H_2O \rightarrow 3\ CH_4 + CO_2 + 4\ NH_4Cl$	*Methanosarcina* and other methylotrophic methanogens *Methanomicrococcus blatticola* and *Methanosphaera* Some methylotrophic methanogens Some methylotrophic methanogens Some methylotrophic methanogens Some methylotrophic methanogens Some methylotrophic methanogens
Acetate	$CH_3COOH \rightarrow CH_4 + CO_2$	*Methanosarcina* and *Methanosaeta*

methanogenesis, reduces methyl-CoM to CH_4. In this reduction step, coenzyme B (CoB) serves as the direct electron donor. The oxidized CoB forms a heterodisulfide with CoM (CoM–S–S–CoB). Finally, the thiol is regenerated by heterodisulfide reduction. A membrane-bound complex, methyl-H_4MPT:HS–CoM methyltransferase (Mtr), catalyzes the methyl group transfer reaction. The methyl transfer from H_4MPT to CoM and the reduction of the heterodisulfide are exergonic and contribute to energy conservation. In the additional branch, heterodisulfide reductase (Hdr) catalyzes the reversible reduction of the heterodisulfide (CoM–S–S–CoB).

In some heterotrophic methanogens, secondary alcohols, such as 2-propanol, 2-butanol, cyclopentanol, and ethanol, can also serve as electron donors (Widdel, 1986; Widdel et al., 1988; Bleicher et al., 1989; Frimmer & Widdel, 1989). F_{420}-dependent alcohol dehydrogenase (ADF) has been shown to catalyze the oxidation of secondary alcohols (Widdel & Wolfe, 1989), while the oxidation of ethanol is catalyzed by NADP-dependent alcohol dehydrogenase (Berk & Thauer, 1997). Generally, the growth of methanogens on alcohol is less effective than their growth on H_2.

M. thermoautotrophicus and *M. barkeri* have been demonstrated to utilize CO as a reductant for methanogenesis from CO_2 (Daniels et al., 1977; O'Brien et al., 1984). In these methanogens, CO is oxidized to CO_2 by CO dehydrogenase, with H_2 serving as the direct electron donor for the reduction of CO_2. The degree of growth on CO is lower than that on CO_2 in these methanogens.

Compounds containing methyl groups (C1 compounds), such as methanol, methylated amines, and methylated sulfides, can also act as substrates. Methylotrophic methanogens are primarily affiliated with the order *Methanosarcinales*, except for *Methanosphaera* species, which belong to the order

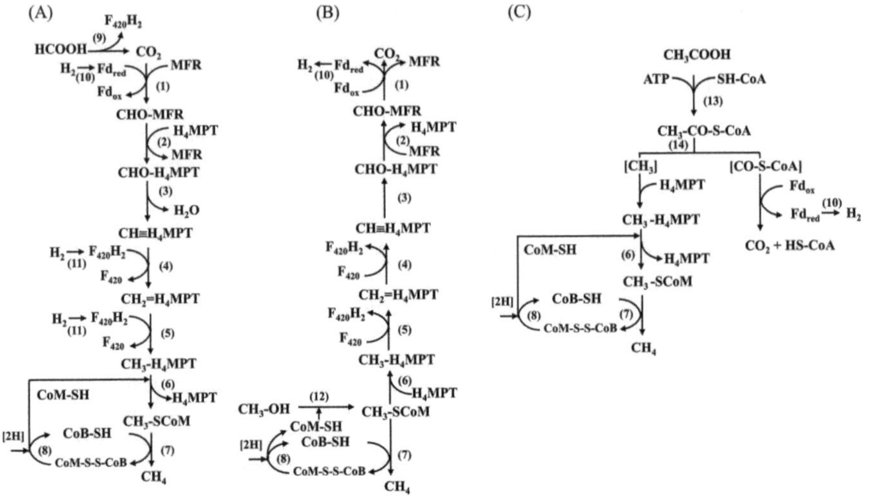

Fig. 14.1 Methanogenesis pathways. Methanogenesis occurs from H_2/CO_2 or formate (**a**), methanol (**b**), and acetate (**c**). Abbreviations: Fd_{red}, reduced ferredoxin; Fd_o, oxidized ferredoxin; $F_{420}H_2$, reduced coenzyme F_{420}; MFR, methanofuran; H_4MPT, tetrahydromethanopterin; CoM-SH, coenzyme M; CoB-SH, coenzyme B; CoM–S–S–CoB, heterodisulfide of CoM and CoB; CoASH, coenzyme A. Enzymes: (1) formyl-MFR dehydrogenase (Fmd); (2) formyl-MFR:H_4MPT formyltransferase (Ftr); (3) methenyl-H_4MPT cyclohydrolase (Mch); (4) methylene-H_4MPT dehydrogenase (Hmd); (5) methylene-H_4MPT reductase (Mer); (6) methyl-H_4MPT:HS–CoM methyltransferase (Mtr); (7) methyl-CoM reductase (Mcr); (8) heterodisulfide reductase (Hdr); (9) formate dehydrogenase (Fdh); (10) energy-conserving hydrogenase (Ech); (11) F_{420}-reducing hydrogenases; (12) methyltransferase; (13) acetate kinase (AK)-phosphotransacetylase (PTA) system in *Methanosarcina*; AMP-forming acetyl-CoA synthetase in *Methanosaeta*; (14) CO dehydrogenase/acetyl-CoA synthase (CODH/ACS)

Methanobacteriales. In methylotrophic methanogenesis, the methyl groups from methyl-containing compounds are transferred to a cognate corrinoid protein and then to CoM, generating methyl-CoM (Fig. 14.2b) (Burke & Krzycki, 1997; Sauer et al., 1997; Ferguson Jr. et al., 2000). Methyl-CoM is subsequently reduced to CH_4 in the methanogenic pathway. Substrate-specific methyltransferases facilitate the activation and transfer of methyl groups. In most methylotrophic methanogens, the electrons needed to reduce methyl groups to CH_4 are derived from the further oxidation of the methyl groups to CO_2, proceeding as the reverse of hydrogenotrophic methanogenesis. In this process, three methyl groups are reduced to CH_4, resulting in the formation of one molecule of CO_2. The methylotrophic growth of *Methanomicrococcus blatticola* and *Methanosphaera* species is H_2-dependent and differs from the previously described growth mode (Miller & Wolin, 1985; Biavati et al., 1988; Sprenger et al., 2000, 2005, 2007; Fricke et al., 2006).

Acetate can serve as a methanogenesis substrate for methanogens of the genera *Methanosarcina* and *Methanosaeta*. CH_4 is produced from acetate via aceticlastic pathways. In this process, the C–C bond of acetate is cleaved, the methyl group is reduced to CH_4, and the carboxyl group is oxidized to CO_2 (Fig. 14.2c).

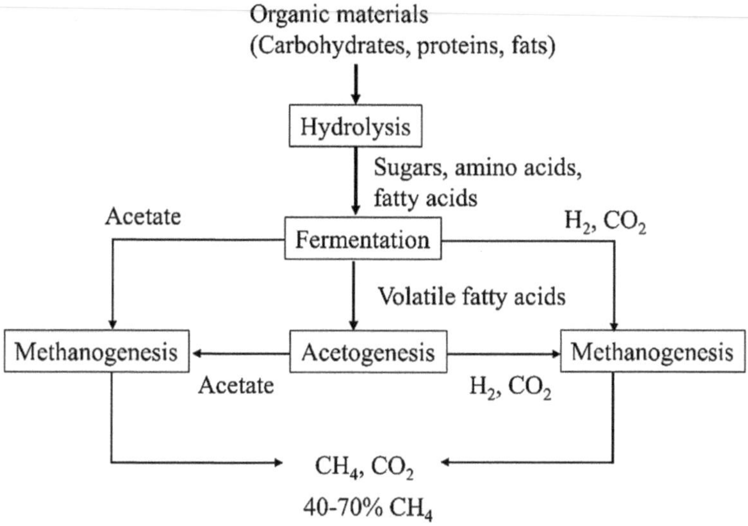

Fig. 14.2 Degradation of organic material through anaerobic digestion

Methanosarcina is a metabolically versatile methanogen capable of utilizing methanol, methylamine, and acetate, while *Methanosaeta* specifically utilizes acetate. The first step involves the generation of acetyl-CoA from acetate by acetate kinase and phosphotransacetylase. The C–C bond in acetyl-CoA and the C–S bond between coenzyme A and the acetyl group are cleaved by the carbon monoxide dehydrogenase/acetyl-CoA synthase (CODH/ACS) complex (Gong et al., 2008). Ferredoxin then converts the carbonyl group to CO_2. The methyl group is transferred to H_4MPT and subsequently to HS–CoM, forming CH_3–SCoM through the action of methyltransferase (Mtr). Finally, CH_3–SCoM is converted to CH_4, resulting in the transfer of electrons to CoM–S–S–CoB. The CoM–S–S–CoB acts as the terminal electron acceptor in the energy-conserving electron transport chain and is cleaved to HS–CoB and HS–CoM by heterodisulfide reductase (Hdr).

14.3 Taxonomic Group of Methanogens

Methanogenic archaea have been classified into seven orders based on *mcrA* and 16S rRNA gene phylogeny (Table 14.2) (Adam et al., 2017). These orders include Methanobacteriales, Methanococcales, Methanomicrobiales, Methanosarcinales, Me thanopyrales, Methanocellales, and Methanomassiliicoccales. Five of these orders, namely Methanobacteriales, Methanococcales, Methanomicrobiales, Methanopyrales, and Methanocellales, contain hydrogenotrophic methanogens. Members of the Methanobacteriales generally use H_2 as the electron donor and CO_2 as the electron acceptor to produce CH_4. Some species within this order also utilize

Table 14.2 Taxonomic groups and characteristics of methanogens

Order	Family	Some of the selected organisms	Substrates for methanogenesis	Growth temp. (°C)	References
Methanobacteriales	Methanobacteriaceae	*Methanobacterium ruminantium*	H_2	37–45	Smith and Hungate (1958)
Methanococcales	Methanothermaceae Methanococcaceae	*Methanobrevibacter arboriphilicus*	H_2, HCOOH	35–40 (Opt.)	Asakawa et al. (1993)
	Methanocaldococcaceae	*Methanothermobacter thermautotrophicus*	H_2	65–70 (Opt.)	Zeikus and Wolfe (1972)
Methanomicrobiales	Methanomicrobiaceae	*Methanothermus fervidus*	H_2	65–97	Stetter et al. (1981)
		Methanococcus maripaludis	H_2, HCOOH	18–47	Jones et al. (1983a)
		Methanocaldococcus jannaschii	H_2	50–86	Jones et al. (1983b)
		Methanomicrobium antiquum	H_2, HCOOH	15–40	Mochimaru et al. (2016)
		Methanoculleus chikugoensis	H_2, HCOOH, 2-propanol, 2-butanol, cyclopentanol	25–30 (Opt.)	Dianou et al. (2001)
	Methanospirillaceae	*Methanofollis ethanolicus*	H_2, HCOOH, ethanol, 1-propanol, 1-butanol	15–40	Imachi et al. (2009)
	Methanocorpusculaceae	*Methanogenium cariaci*	H_2, HCOOH	20–25 (Opt.)	Romesser et al. (1979)
		Methanoplanus petrolearius	H_2, HCOOH, 1-propanol	25–45	Ollivier et al. (1997)
Methanosarcinales	Methanosarcinaceae	*Methanospirillum hungatii*	H_2, HCOOH	30–37	Ferry et al. (1974)
		Methanocorpusculum labreanum	H_2, HCOOH	25–40	Zhao et al. (1989)

		Methanocalculus chunghsingensis	H_2, HCOOH	20–45	Lai et al. (2004)
		Methanosarcina barkeri	H_2, acetate, methanol, methylamines	30–40 (Opt.)	Bryant and Boone (1987)
Methanosarcinales	Methanosarcinaceae	*Methanococcoides burtonii*	Methanol, methylamines	23.4 (Opt.)	Franzmann et al. (1992)
		Methanohalophilus halophilus	Methylamines	26–36 (Opt.)	Wilharm et al. (1991)
		Methanolobus halotolerans	Methanol	4–40	Shen et al. (2020)
		Methanomethylovorans thermophila	Methanol, methylamines	42–58	Jiang et al. (2005)
		Methanosalsum natronophilum	Methanol, methylamines, dimethyl sulfide	43 (Opt.)	Sorokin et al. (2015)
Methanopyrales	Methanosaetaceae	*Methanosaeta pelagica*	Acetate	30 (Opt.)	Mori et al. (2012)
Methanocellales	Methanopyraceae	*Methanopyrus kandleri*	H_2	84–110	Kurr et al. (1991)
Methanomassiliicoccale	Methanocellaceae	*Methanocella paludicola*	H_2, HCOOH	25–40	Sakai et al. (2008)
	Methanomassiliicoccaceae	*Methanomassiliicoccus luminyensis*	Methanol	37 (Opt.)	Dridi et al. (2012)

Abbreviation: *Opt*. Optimum temperature

formate, CO, or secondary alcohols as electron donors. Methanobacteriales have been isolated from a wide variety of anaerobic habitats, including marine and freshwater sediments, soil, animal gastrointestinal tracts, anaerobic sewage digesters, and geothermal environments (Liu & Whitman, 2008). Members of the Methanococcales, which are cocci, use H_2 and CO_2 to produce CH_4. Most of these methanogens have been isolated from marine habitats and require sea salts for optimal growth (Liu & Whitman, 2008). The Methanomicrobiales also utilize H_2 and CO_2 for CH_4 production, with many species using formate or secondary alcohols as electron donors. They have been found in various anaerobic environments, such as marine and freshwater sediments, anaerobic sewage digesters, and animal gastrointestinal tracts (Liu & Whitman, 2008). The Methanosarcinales have a broader substrate range than other orders of methanogens and are the only order where aceticlastic methanogens have been identified. Some species in this order can produce CH_4 from H_2 and CO_2 and have been isolated from various anaerobic environments, including marine and freshwater sediments, anaerobic sewage digesters, and animal gastrointestinal tracts (Liu & Whitman, 2008).

Members of the Methanopyrales can reduce CO_2 with H_2 for methanogenesis. The only member of this order is the hyperthermophilic methanogen *Methanopyrus kandleri*, which was isolated from the base of a deep-sea 'black smoker' chimney in the Gulf of California and can grow at high temperatures ranging from 84 to 110 °C (Kurr et al., 1991). The Methanocellales include *Methanocella paludicola*, which was isolated from rice paddy soil. Paddy fields are significant sources of CH_4, a greenhouse gas (Sakai et al., 2008). *Methanocella paludicola* can utilize H_2 and CO_2 for CH_4 production, and formate can also serve as a substrate for methanogenesis in this strain. The Methanomassiliicoccales contain obligate methyl-respiring methylotrophic methanogens.

The members of the Methanobacteriales, Methanomicrobiales, Methanococcales, and Methanopyrales lack cytochromes and methanophenazine as components of their electron transport (Jussofie & Gottschalk, 1986; Abken et al., 1998). The Methanococcales and Methanopyrales possess several selenocysteine-containing enzymes, such as [NiFeSe]-hydrogenases, which are involved in methanogenesis from H_2 and CO_2 and/or from formate (Thauer et al., 2010). In contrast, the Methanobacteriales and Methanomicrobiales have cysteine-containing enzymes. Thus, the growth of Methanococcales and Methanopyrales is promoted by selenium, whereas the Methanosarcinales and Methanocellales contain cytochromes and methanophenazine and do not require selenium for growth.

14.4 Anaerobic Digestion

Anaerobic digestion (AD) is a series of processes in which microorganisms break down biodegradable organic materials in the absence of oxygen (Fig. 14.2). In the first stage of AD, organic polymers, including proteins, lipids, and carbohydrate polymers, are hydrolyzed by microbial enzymes to produce amino acids, long-chain fatty acids, and sugars, respectively. The resulting compounds are metabolized by fermentative bacteria, leading to the production of a mixture of short-chain volatile fatty acids (VFAs) and minor products such as CO_2, H_2, and acetate. Acetogens then convert these organic acids into acetate, CO_2, and/or H_2, which serve as direct substrates for methanogenesis. Methanogens produce CH_4 from H_2/CO_2 and acetate. H_2/CO_2 can also be converted to acetate via the Wood–Ljungdahl pathway by acetogens such as *Acetobacterium woodii* (Diekert & Wohlfarth, 1994). This reaction competes with the conversion to CH_4 by hydrogenotrophic methanogens.

Anaerobic digestion is widely recognized for its effectiveness in organic waste remediation, as it combines energy production (in the form of biogas) with nutrient recovery (digestate) (Karki et al., 2021). This process converts waste into CH_4, which can significantly reduce greenhouse gas emissions associated with CH_4 and increase the share of renewable energy in the form of green gas (Rusmanis et al., 2019). Biomass resources, including livestock manure and slurry, as well as non-food cellulosic terrestrial biomass, can be utilized in anaerobic digestion systems. By replacing open slurry storage, which is associated with substantial CH_4 leakage, anaerobic digestion can further decrease greenhouse gas emissions while simultaneously generating low-carbon energy in the form of biogas. Biogas consists primarily of CH_4 (50–70% v/v) and CO_2 (30–50% v/v) (Angelidaki et al., 2018). Hydrogenotrophic methanogens require a substantial amount of H_2 for CO_2 consumption, leading to competition and equilibrium between H_2 producers (acetogens) and consumers (methanogens). This interaction results in a very low dissolved H_2 partial pressure, maintaining a balanced microbiological community. To enhance CH_4 concentration, CO_2 must be removed from the biogas. In recent years, methanation has been explored as a method for upgrading biogas (Angelidaki et al., 2018; Voelklein et al., 2019). Biological upgrading using heterotrophic methanogens for the conversion of CO_2 and H_2 into CH_4 has gained attention due to its simplicity and high technical potential (Antukh et al., 2022). The biogas upgrading process aims to remove and separate CO_2 and other impurities, with the objective of increasing the CH_4 content to 95% (Sun et al., 2015). This ensures that the upgraded biogas meets natural gas standards, making it suitable for use as fuel. The overarching goal of biogas upgrading is to establish biogas as a stable energy source and a viable alternative to fossil fuels (Lecker et al., 2017). Once upgraded, biogas can be fed directly into the existing gas grid without the need for additional processes.

14.5 CO_2 Methanation

There are two distinct systems for methanation: catalytic chemical methanation and biomethanation (Lecker et al., 2017). Both pathways are based on the Sabatier reaction (Eq. 14.1), in which one molecule of CO_2 reacts with four molecules of H_2 to produce one molecule of CH_4 and two molecules of liquid water.

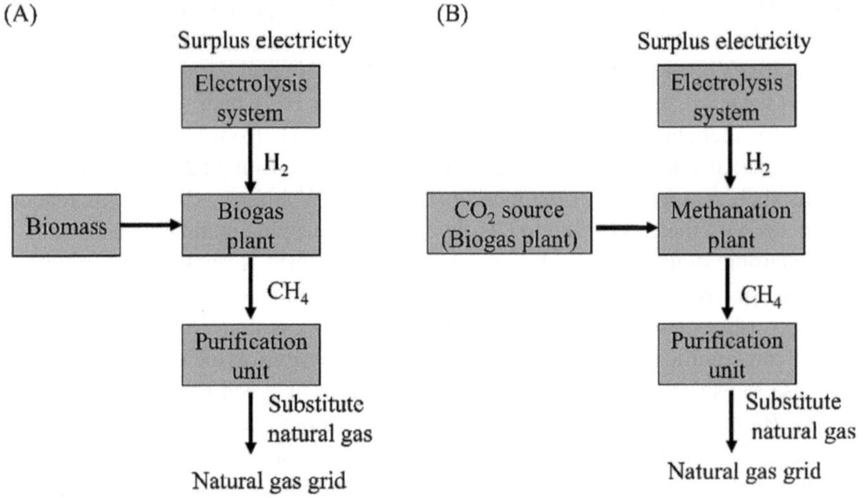

Fig. 14.3 Two technical pathways for biomethanation. (**a**) In situ. (**b**) Ex situ

$$CO_2 + 4H_2 \rightarrow CH_4 + 2H_2O \quad \Delta H = -165\,\text{kJ/mol} \tag{14.1}$$

Metal-based catalysts are employed in methanation, which occurs at temperatures between 300 and 550 °C and under high pressures (Götz et al., 2016). Noble metal or nickel-based catalysts are typically used in CO_2 methanation (Strucks et al., 2021). Although noble metals such as rhodium and palladium are highly active, their cost makes them impractical for widespread use. Therefore, non-noble metals like nickel are favored due to their lower cost and high efficiency. The biomethanation process operates at lower temperatures and pressures than the chemical methanation process, and it shows improved tolerance to chemical contaminants such as hydrogen sulfide (H_2S), organic acids, and ammonia (Burkhardt et al., 2015). Hydrogenotrophic methanogens serve as key catalysts, utilizing H_2 as a reducing agent to convert CO_2 into CH_4. These methanogens are active within a temperature range of 0–122 °C (Cavicchioli, 2011; Lecker et al., 2017). Some hydrogenotrophic methanogens have been used mainly as model organisms (Biderre-Petit et al., 2024). Several nonbiological factors, including temperature, pH, H_2/CO_2 ratio, and H_2 partial pressure, can influence CH_4 content and the microbial community (Wahid et al., 2019). Biological methanation can be performed in situ within a biogas digester or ex situ in an adjacent external reactor (Fig. 14.3) (Lecker et al., 2017; Angelidaki et al., 2018). In the in-situ method, H_2 is introduced directly into the reactor where organic waste is anaerobically digested, in proportion to the CO_2 produced. In contrast, the ex-situ method involves a bioreactor that operates separately from the reaction tank for CH_4 fermentation, focusing solely on the biomethanation reaction. As with the in-situ method, hydrogen is introduced in proportion to the amount of CO_2 flowing into the bioreactor.

14.6 Methanogens Used in Biomethanation

In the in-situ method, genera such as *Methanoculleus*, *Methanothermobacter*, *Methanobacterium*, and *Methanosarcina* play crucial roles in the methanation processes. The relative abundance of these genera in biogas upgrading reactors varies with factors such as temperature, pH, and CO (Thema et al., 2021). *Methanothermobacter* is predominant under thermophilic conditions due to its favorable growth at higher temperatures (Kaster et al., 2011). Specifically, *Methanothermobacter thermautotrophicus* and *Methanothermobacter marburgensis* demonstrate robust growth, achieving high cell densities and exhibiting significant CH_4 production rates. Thus, they have been utilized as biocatalysts in ex-situ biomethanization and large-scale industrial processes (Pfeifer et al., 2021; Thema et al., 2021; Kaul et al., 2022).

Methanothermobacter thermautotrophicus (DSM 1053), formerly known as *Methanobacterium thermoautotrophicum* strain ΔH, was isolated from sludge at an anaerobic sewage digestion plant in Urbana, Illinois, USA (Zeikus & Wolfe, 1972). Strain ΔH uses H_2 and CO_2 as its sole energy source and CO_2 as its sole carbon source; it has a doubling time of 3 h and a specific growth rate of 0.231 (Zeikus & Wolfe, 1972; Rusmanis et al., 2019). *Methanothermobacter marburgensis*, formerly *Methanobacterium thermoautotrophicum* strain Marburg, was isolated from anaerobic sewage sludge in Marburg, Germany (Fuchs et al., 1978). Strain Marburg grows on H_2 and CO_2 even faster than strain ΔH, with a doubling time of 0.288 h and a specific growth rate of 0.1 (Rusmanis et al., 2019). *Methanothermobacter marburgensis* differs from *Methanothermobacter thermautotrophicus* in the composition of the sacculus polymer (pseudomurein), the size of the subunit O of DNA-dependent RNA polymerase, and the membrane-associated ATPase activity (Kaster et al., 2011). The optimal growth temperatures for these strains are similar, around 65 °C. Like all *Methanothermobacter* species, these two strains require Na^+, K^+, Fe^{2+}, Co^{2+}, Ni^{2+}, Zn^{2+}, MoO_2^{-4}, and/or WO_2^{-4} (and possibly Ca^{2+}) for growth. They can assimilate formate, acetate, propionate, pyruvate, isobutyrate, isovalerate, phenylacetate, *p*-hydroxyphenylacetate, indoleacetate, succinate, δ-aminolevulinate, methionine, guanine, and biotin (Kaster et al., 2011). However, the addition of organic compounds does not stimulate their growth. The ability of these strains to assimilate formate does not imply that they can grow on formate as an energy source.

14.7 Factors Affecting Biomethanation

Temperature, pH, H_2 supply, and H_2/CO_2 ratio are microbial factors that can influence CH_4 production (Rachbauer et al., 2017; Wahid et al., 2019). During in-situ biomethanation, H_2 is directly injected into the reactor digesting organic matter to increase the CH_4 content in the produced biogas. In this process, the amount of H_2 injected into the reactor must be carefully controlled. From a thermodynamic perspective, high H_2 partial pressure can directly inhibit hydrolysis and acetate formation (Rafrafi et al., 2021). Acetogenic bacteria require a higher H_2 partial pressure than methanogens (Cord-Ruwisch et al., 1988). Some studies have shown that H_2 addition positively affects CH_4 production but has no effect on acetogenesis (Luo & Angelidaki, 2013; Rachbauer et al., 2016).

The optimum growth temperature for many methanogens that play key roles in CH_4 fermentation ranges from 35 to 65 °C. Although homoacetogens and methanogens compete during in-situ biomethanation, methanogens typically dominate under thermophilic conditions (Grimalt-Alemany et al., 2020). Acetoclastic methanogens exhibit lower growth rates at 35 °C; however, they are abundant during in-situ biomethanation because this mild temperature is effective for acetogenic bacteria. In contrast, ex-situ biomethanation can be conducted under conditions favorable for hydrogenotrophic methanogens, such as high temperature and high H_2 pressure. Several studies have investigated the effect of temperature on biomethanation performance (Luo & Angelidaki, 2012; Guneratnam et al., 2017), reporting that the efficiency of the methanation reaction increases with temperature.

pH affects the activity of hydrolytic enzymes in anaerobic biodegradation. The optimal pH for methanogenesis has been studied in anaerobic digesters (Rafrafi et al., 2021), with neutral pH values found to be optimal for the reaction. During in-situ biomethanation, the pH is selected to favor the entire process, including hydrolysis, acid formation, acetate formation, and methanogenic reactions. Conversely, in ex-situ systems, the reaction pH is optimized to promote only methanogenic reactions.

The methanogenic pathway is considered one of the most metal-rich enzymatic pathways in biology (Glass & Orphan, 2012). The required metals vary depending on the pathway but generally include Fe, Ni, Co, Mo, W, and Zn. Fe exists primarily as Fe-S clusters, used for electron transport and catalysis. Ni is bound to Fe-S clusters or to the porphyrins of coenzyme F_{430}. Co is present in cobamides that facilitate the transfer of methyl groups, while Zn serves as a structural atom in some enzymes. Mo or W binds to the pterin cofactor to form "molybdopterin" or "tungstopterin," catalyzing two-electron redox reactions.

Several studies have reported that the gas–liquid mass transfer rate is the most significant limiting factor for H_2 metabolism (Bassani et al., 2016; Lecker et al., 2017). The volumetric gas–liquid mass transfer coefficient (kLa) is widely used to quantify the rate of gas transfer from the gas phase to the liquid phase in bioreactors and other gas–liquid systems. The following equation represents the relationship between gas–liquid mass transfer rates:

$$r_t = 22.4 k_L a \left(H_{2g} - H_{2l} \right)$$

where r_t (L/L/h) is the gas–liquid mass transfer rate, 22.4 (L/mol) is the molar volume, k_La (/h) is the gas transfer coefficient, H_{2g} is the H_2 concentration in the gas phase, and H_{2l} is the H_2 concentration in the liquid phase. k_La can be controlled by adjusting parameters such as mixing rate, gas recirculation, and H_2 diffuser.

14.8 Reactors Used in Biomethanation

Three main types of reactors are known to provide good conversion efficiency and capability, including continuously stirred tank reactors (CSTRs), fixed-bed reactors (FBRs), and trickle-bed reactors (TBRs) (Fig. 14.4) (Rusmanis et al., 2019). CSTRs are the most commonly employed bioreactors in anaerobic digestion. The reactor is continuously agitated by an impeller or by recirculating the reactor contents, thus maintaining homogeneity. Agitated diffusion effectively allows hydrogenotrophic methanogenic archaea to contact H_2. FBRs consist of a cylindrical column filled with biocatalyst pellets, where microorganisms adhere to a packing material. This reactor minimizes interference between liquid and gas to the greatest extent possible. The gas diffuses from the bottom of the reactor and passes through the packed bed layer. In TBRs, the reaction chamber is not filled with liquid; instead, a small amount of nutrient solution continually wets the surface of the microorganisms, enhancing the phase interface for mass transfer. The biocarriers are surrounded by a gas phase, allowing independent control of gas surface velocities, unlike liquid-filled reactors such as CSTRs.

14.9 Conclusions

Methanation is gaining increasing attention as a key technology in the transition toward a sustainable and circular energy system. With the global push to reduce carbon emissions and enhance renewable energy utilization, methanation offers a promising route to store excess renewable electricity in the form of synthetic natural gas through the Power-to-Gas concept. This stored methane can be integrated into existing natural gas infrastructure, facilitating flexible energy distribution and

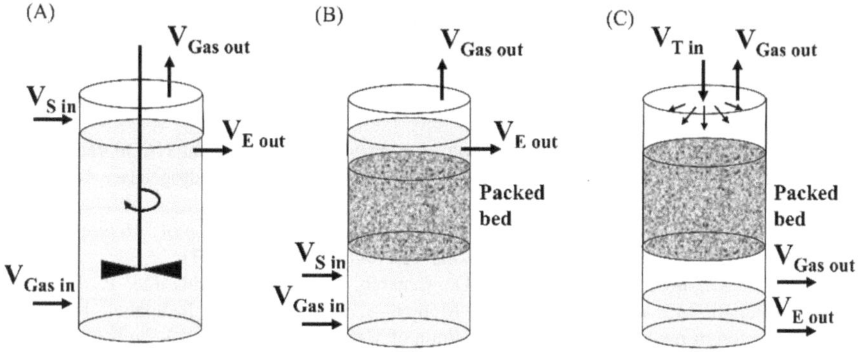

Fig. 14.4 Reactors used in biomethanation. (**a**) CSTR, (**b**) FBR, and (**c**) TBR. Abbreviations: $V_{\text{Gas in}}$, volume flow of substrate gas; $V_{\text{Gas out}}$, volume flow of biogas; $V_{\text{S in}}$, volume flow of liquid substrate; $V_{\text{E out}}$, volume flow of effluent; $V_{\text{T in}}$, volume flow of trace element

long-term storage. Numerous studies have demonstrated that biomethanation is an effective method for CO_2 recycling and carbon neutrality. While various methanogens have been identified, those of the genus *Methanothermobacter* are widely utilized for methanation. Researchers are also analyzing the methanogenic pathways of these organisms, with the potential to develop methanogens optimized specifically for methanation. In addition, the discovery of novel methanogens suitable for this process could expand the applications of methanation. There are two types of methanation methods: in situ and ex situ, each suited for different applications. Notably, the use of microbial consortia, as seen in the in situ method, requires careful control of various parameters to optimize the desired reaction. The contact efficiency of microorganisms with the substrates, H_2 and CO_2, is crucial for reaction efficiency, and several reactor designs have been proposed to enhance these reactions. The commercialization of industrial processes for biomethanation is under way, and further efficiency improvements in these processes are anticipated.

References

Abken, H. J., Tietze, M., Brodersen, J., Bäumer, S., Beifuss, U., & Deppenmeier, U. (1998). Isolation and characterization of methanophenazine and function of phenazines in membrane-bound electron transport of *Methanosarcina mazei* Göl. *Journal of Bacteriology, 180*, 2027–2032.

Adam, P. S., et al. (2017). The growing tree of Archaea: New perspectives on their diversity, evolution and ecology. *The ISME Journal, 11*, 2407–2425.

Angelidaki, I., Treu, L., Tsapekos, P., Luo, G., Campanaro, S., Wenzel, H., & Kougias, P. G. (2018). Biogas upgrading and utilization: Current status and perspectives. *Biotechnology Advances, 36*, 452–466.

Antukh, T., Lee, I., Joo, S., & Kim, H. (2022). Hydrogenotrophs-based biological biogas upgrading technologies. *Frontiers in Bioengineering and Biotechnology, 10*, 833482.

Asakawa, S., Morii, H., Akagawa-Matsushita, M., Koga, Y., & Hayano, K. (1993). Characterization of Methanobrevibacter arboriphilicus SA isolated from a Paddy field soil and DNA-DNA hybridization among M. arboriphilicus strains. *International Journal of Systematic and Evolutionary Microbiology, 43*, 683–686.

Bassani, I., Kougias, P. G., & Angelidaki, I. (2016). In-situ biogas upgrading in thermophilic granular UASB reactor: Key factors affecting the hydrogen mass transfer rate. *Bioresource Technology, 221*, 485–491.

Berk, H., & Thauer, R. K. (1997). Function of coenzyme F_{420}-dependent NADP reductase in methanogenic archaea containing an NADP-dependent alcohol dehydrogenase. *Archives of Microbiology, 168*, 396–402.

Biavati, B., Vasta, M., & Ferry, J. G. (1988). Isolation and characterization of "*Methanosphaera cuniculi*" sp. nov. *Applied and Environmental Microbiology, 54*, 768–771.

Biderre-Petit, C., Mbarki, M., Courtine, D., Benarab, Y., Vial, C., Fontanille, P., Dubessay, P., Keramati, M., Jouan-Dufournel, I., Monjot, A., Guez, J. S., & Fadhlaoui, K. (2024). Comparison of methane yield of a novel strain of *Methanothermobacter marburgensis* in pure and mixed adapted culture derived from a methanation bubble column bioreactor. *Bioresource Technology, 406*, 131021.

Bleicher, K., Zellner, G., & Winter, J. (1989). Growth of methanogens on cyclopentanol/CO_2 and specificity of alcohol dehydrogenase. *FEMS Microbiology Letters, 59*, 307–312.

Bryant, M. P., & Boone, D. R. (1987). Emended description of strain MST (DSM 800T), the type strain of *Methanosarcina barkeri. International Journal of Systematic and Evolutionary Microbiology, 37*, 169–170.

Buan, N. R. (2018). Methanogens: Pushing the boundaries of biology. *Emerging Topics in Life Sciences, 2*, 629–646.

Burke, S. A., & Krzycki, J. A. (1997). Reconstitution of Monomethylamine:Coenzyme M methyl transfer with a corrinoid protein and two methyltransferases purified from *Methanosarcina barkeri. The Journal of Biological Chemistry, 272*, 16570–16577.

Burkhardt, M., Koschack, T., & Busch, G. (2015). Biocatalytic methanation of hydrogen and carbon dioxide in an anaerobic three-phase system. *Bioresource Technology, 178*, 330–333.

Cavicchioli, R. (2011). Archaea — Timeline of the third domain. *Nature Reviews Microbiology, 9*, 51–61.

Cord-Ruwisch, R., Seitz, H.-J., & Conrad, R. (1988). The capacity of hydrogenotrophic anaerobic bacteria to compete for traces of hydrogen depends on the redox potential of the terminal electron acceptor. *Archives of Microbiology, 149*, 350–357.

Daniels, L., Fuchs, G., Thauer, R. K., & Zeikus, J. G. (1977). Carbon monoxide oxidation by methanogenic bacteria. *Journal of Bacteriology, 132*, 118–126.

Dianou, D., Miyaki, T., Asakawa, S., Morii, H., Nagaoka, K., Oyaizu, H., & Matsumoto, S. (2001). *Methanoculleus chikugoensis* sp. nov., a novel methanogenic archaeon isolated from paddy field soil in Japan, and DNA-DNA hybridization among Methanoculleus species. *International Journal of Systematic and Evolutionary Microbiology, 51*, 1663–1669.

Diekert, G., & Wohlfarth, G. (1994). Metabolism of homoacetogens. *Antonie Van Leeuwenhoek, 66*, 209–221.

Dridi, B., Fardeau, M.-L., Ollivier, B., Raoult, D., & Drancourt, M. (2012). *Methanomassiliicoccus luminyensis* gen. nov., sp. nov., a methanogenic archaeon isolated from human faeces. *International Journal of Systematic and Evolutionary Microbiology, 62*, 1902–1907.

Ferguson, D. J., Jr., Gorlatova, N., Grahame, D. A., & Krzycki, J. A. (2000). Reconstitution of dimethylamine: Coenzyme M methyl transfer with a discrete corrinoid protein and two methyltransferases purified from *Methanosarcina barkeri. The Journal of Biological Chemistry, 275*, 29053–29060.

Ferry, J. G., Smith, P. H., & Wolfe, R. S. (1974). Methanospirillum, a new genus of methanogenic bacteria, and characterization of *Methanospirillum hungatii* sp.nov. *International Journal of Systematic and Evolutionary Microbiology, 24*, 465–469.

Franzmann, P. D., Springer, N., Ludwig, W., Conway De Macario, E., & Rohde, M. (1992). A methanogenic Archaeon from Ace Lake, Antarctica: *Methanococcoides burtonii* sp. nov. *Systematic and Applied Microbiology, 15*, 573–581.

Fricke, W. F., Seedorf, H., Henne, A., Krüer, M., Liesegang, H., Hedderich, R., Gottschalk, G., & Thauer, R. K. (2006). The genome sequence of Methanosphaera stadtmanae reveals why this human intestinal archaeon is restricted to methanol and H_2 for methane formation and ATP synthesis. *Journal of Bacteriology, 188*, 642–658.

Frimmer, U., & Widdel, F. (1989). Oxidation of ethanol by methanogenic bacteria. *Archives of Microbiology, 152*, 479–483.

Fuchs, G., Stupperich, E., & Thauer, R. K. (1978). Acetate assimilation and the synthesis of alanine, aspartate and glutamate in *Methanobacterium thermoautotrophicum. Archives of Microbiology, 117*, 61–66.

Glass, J. B., & Orphan, V. J. (2012). Trace metal requirements for microbial enzymes involved in the production and consumption of methane and nitrous oxide. *Frontiers in Microbiology, 3*, 61.

Gong, W., Hao, B., Wei, Z., Ferguson, D. J., Tallant, T., Krzycki, J. A., & Chan, M. K. (2008). Structure of the alpha2epsilon2 Ni-dependent CO dehydrogenase component of the *Methanosarcina barkeri* acetyl-CoA decarbonylase/synthase complex. *Proceedings of the National Academy of Sciences, 105*, 9558–9563.

Götz, M., Lefebvre, J., Mörs, F., McDaniel Koch, A., Graf, F., Bajohr, S., Reimert, R., & Kolb, T. (2016). Renewable power-to-gas: A technological and economic review. *Renewable Energy, 85*, 1371–1390.

Grimalt-Alemany, A., Łężyk, M., Kennes-Veiga, D. M., Skiadas, I. V., & Gavala, H. N. (2020). Enrichment of Mesophilic and Thermophilic mixed microbial Consortia for Syngas Biomethanation: The role of kinetic and thermodynamic competition. *Waste and Biomass Valorization, 11*, 465–481.

Guneratnam, A. J., Ahern, E., FitzGerald, J. A., Jackson, S. A., Xia, A., Dobson, A. D. W., & Murphy, J. D. (2017). Study of the performance of a thermophilic biological methanation system. *Bioresource Technology, 225*, 308–315.

Hansen, J., Sato, M., Ruedy, R., Lacis, A., & Oinas, V. (2000). Global warming in the twenty-first century: An alternative scenario. *Proceedings of the National Academy of Sciences, 97*, 9875–9880.

Hedderich, R., & Whitman, W. B. (2013). In E. Rosenberg, E. F. DeLong, S. Lory, E. Stackebrandt, & F. Thompson (Eds.), *Physiology and biochemistry of the methane-producing Archaea. The Prokaryotes: Prokaryotic physiology and biochemistry* (pp. 635–662). Springer.

Imachi, H., Sakai, S., Nagai, H., Yamaguchi, T., & Takai, K. (2009). *Methanofollis ethanolicus* sp. nov., an ethanol-utilizing methanogen isolated from a lotus field. *International Journal of Systematic and Evolutionary Microbiology, 59*, 800–805.

Jiang, B., Parshina, S. N., van Doesburg, W., Lomans, B. P., & Stams, A. J. M. (2005). *Methanomethylovorans thermophila* sp. nov., a thermophilic, methylotrophic methanogen from an anaerobic reactor fed with methanol. *International Journal of Systematic and Evolutionary Microbiology, 55*, 2465–2470.

Jones, W. J., Paynter, M. J. B., & Gupta, R. (1983a). Characterization of *Methanococcus maripaludis* sp. nov., a new methanogen isolated from salt marsh sediment. *Archives of Microbiology, 135*, 91–97.

Jones, W. J., Leigh, J. A., Mayer, F., Woese, C. R., & Wolfe, R. S. (1983b). *Methanococcus jannaschii* sp. nov., an extremely thermophilic methanogen from a submarine hydrothermal vent. *Archives of Microbiology, 136*, 254–261.

Jussofie, A., & Gottschalk, G. (1986). Further studies on the distribution of cytochromes in methanogenic bacteria. *FEMS Microbiology Letters, 37*, 15–18.

Karki, R., Chuenchart, W., Surendra, K. C., Shrestha, S., Raskin, L., Sung, S., Hashimoto, A., & Kumar Khanal, S. (2021). Anaerobic co-digestion: Current status and perspectives. *Bioresource Technology, 330*, 125001.

Kaster, A.-K., Goenrich, M., Seedorf, H., Liesegang, H., Wollherr, A., Gottschalk, G., & Thauer, R. K. (2011). More than 200 genes required for methane formation from H2 and CO2 and energy conservation are present in *Methanothermobacter marburgensis* and *Methanothermobacter thermautotrophicus*. *Archaea, 2011*, 973848.

Kaul, A., Böllmann, A., Thema, M., Kalb, L., Stöckl, R., Huber, H., Sterner, M., & Bellack, A. (2022). Combining a robust thermophilic methanogen and packing material with high liquid hold-up to optimize biological methanation in trickle-bed reactors. *Bioresource Technology, 345*, 126524.

Kurr, M., Huber, R., König, H., Jannasch, H. W., Fricke, H., Trincone, A., Kristjansson, J. K., & Stetter, K. O. (1991). *Methanopyrus kandleri*, gen. and sp. nov. represents a novel group of hyperthermophilic methanogens, growing at 110°C. *Archives of Microbiology, 156*, 239–247.

Lai, M.-C., Lin, C.-C., Yu, P.-H., Huang, Y.-F., & Chen, S.-C. (2004). *Methanocalculus chunghsingensis* sp. nov., isolated from an estuary and a marine fishpond in Taiwan. *International Journal of Systematic and Evolutionary Microbiology, 54*, 183–189.

Lecker, B., Illi, L., Lemmer, A., & Oechsner, H. (2017). Biological hydrogen methanation – A review. *Bioresource Technology, 245*, 1220–1228.

Liu, Y., & Whitman, W. B. (2008). Metabolic, phylogenetic, and ecological diversity of the methanogenic archaea. *Annals of the New York Academy of Sciences, 1125*, 171–189.

Luo, G., & Angelidaki, I. (2012). Integrated biogas upgrading and hydrogen utilization in an anaerobic reactor containing enriched hydrogenotrophic methanogenic culture. *Biotechnology and Bioengineering, 109*, 2729–2736.

Luo, G., & Angelidaki, I. (2013). Co-digestion of manure and whey for in situ biogas upgrading by the addition of H_2: Process performance and microbial insights. *Applied Microbiology and Biotechnology, 97*, 1373–1381.

Miller, T. L., & Wolin, M. J. (1985). *Methanosphaera stadtmaniae* gen. nov., sp. nov.: A species that forms methane by reducing methanol with hydrogen. *Archives of Microbiology, 141*, 116–122.

Mochimaru, H., Tamaki, H., Katayama, T., Imachi, H., Sakata, S., & Kamagata, Y. (2016). *Methanomicrobium antiquum* sp. nov., a hydrogenotrophic methanogen isolated from deep sedimentary aquifers in a natural gas field. *International Journal of Systematic and Evolutionary Microbiology, 66*, 4873–4877.

Mori, K., Iino, T., Suzuki, K.-I., Yamaguchi, K., & Kamagata, Y. (2012). Aceticlastic and NaCl-requiring methanogen "*Methanosaeta pelagica*" sp. nov., isolated from marine tidal flat sediment. *Applied and Environmental Microbiology, 78*, 3416–3423.

O'Brien, J. M., Wolkin, R. H., Moench, T. T., Morgan, J. B., & Zeikus, J. G. (1984). Association of hydrogen metabolism with unitrophic or mixotrophic growth of *Methanosarcina barkeri* on carbon monoxide. *Journal of Bacteriology, 158*, 373–375.

Ollivier, B., Cayol, J.-L., Patel, B. K. C., Magot, M., Fardeau, M.-L., & Garcia, J.-L. (1997). *Methanoplanus petrolearius* sp. nov., a novel methanogenic bacterium from an oil-producing well. *FEMS Microbiology Letters, 147*, 51–56.

Pfeifer, K., Ergal, İ., Koller, M., Basen, M., Schuster, B., & Rittmann, S. K. M. R. (2021). Archaea biotechnology. *Biotechnology Advances, 47*, 107668.

Rachbauer, L., Voitl, G., Bochmann, G., & Fuchs, W. (2016). Biological biogas upgrading capacity of a hydrogenotrophic community in a trickle-bed reactor. *Applied Energy, 180*, 483–490.

Rachbauer, L., Beyer, R., Bochmann, G., & Fuchs, W. (2017). Characteristics of adapted hydrogenotrophic community during biomethanation. *Science of the Total Environment, 595*, 912–919.

Rafrafi, Y., Laguillaumie, L., & Dumas, C. (2021). Biological Methanation of H_2 and CO_2 with mixed cultures: Current advances, hurdles and challenges. *Waste and Biomass Valorization, 12*, 5259–5282.

Romesser, J. A., Wolfe, R. S., Mayer, F., Spiess, E., & Walther-Mauruschat, A. (1979). *Methanogenium*, a new genus of marine methanogenic bacteria, and characterization of *Methanogenium cariaci* sp. nov. and *Methanogenium marisnigri* sp. nov. *Archives of Microbiology, 121*, 147–153.

Rondinelli, D., & Berry, M. (2000). Multimodal transportation, logistics, and the environment: Managing interactions in a global economy. *European Management Journal, 18*, 398–410.

Rusmanis, D., O'Shea, R., Wall, D. M., & Murphy, J. D. (2019). Biological hydrogen methanation systems – an overview of design and efficiency. *Bioengineered, 10*, 604–634.

Sakai, S., Imachi, H., Hanada, S., Ohashi, A., Harada, H., & Kamagata, Y. (2008). *Methanocella paludicola* gen. nov., sp. nov., a methane-producing archaeon, the first isolate of the lineage 'Rice Cluster I', and proposal of the new archaeal order *Methanocellales ord.* nov. *International Journal of Systematic and Evolutionary Microbiology, 58*, 929–936.

Sauer, K., Harms, U., & Thauer, R. K. (1997). Methanol: Coenzyme M methyltransferase from *Methanosarcina barkeri*. Purification, properties and encoding genes of the corrinoid protein MT1. *European Journal of Biochemistry, 243*, 670–677.

Shen, Y., Chen, S.-C., Lai, M.-C., Huang, H.-H., Chiu, H.-H., Tang, S.-L., Rogozin, D. Y., & Degermendzhy, A. G. (2020). *Methanolobus halotolerans* sp. nov., isolated from the saline Lake Tus in Siberia. *International Journal of Systematic and Evolutionary Microbiology, 70*, 5586–5593.

Smith, P. H., & Hungate, R. (1958). Isolation and characterization of *Methanobacterium ruminantium* n. sp. *Journal of Bacteriology, 75*, 713–718.

Sorokin, D. Y., Abbas, B., Merkel, A. Y., Rijpstra, W. I. C., Damsté, J. S. S., Sukhacheva, M. V., & van Loosdrecht, M. C. M. (2015). *Methanosalsum natronophilum* sp. nov., and *Methanocalculus alkaliphilus* sp. nov., haloalkaliphilic methanogens from hypersaline soda lakes. *International Journal of Systematic and Evolutionary Microbiology, 65*, 3739–3745.

Sprenger, W. W., van Belzen, M. C., Rosenberg, J., Hackstein, J. H., & Keltjens, J. T. (2000). *Methanomicrococcus blatticola* gen. nov., sp. nov., a methanol- and methylamine-reducing methanogen from the hindgut of the cockroach Periplaneta americana. *International Journal of Systematic and Evolutionary Microbiology, 50*, 1989–1999.

Sprenger, W. W., Hackstein, J. H., & Keltjens, J. T. (2005). The energy metabolism of *Methanomicrococcus blatticola*: Physiological and biochemical aspects. *Antonie Van Leeuwenhoek, 87*, 289–299.

Sprenger, W. W., Hackstein, J. H., & Keltjens, J. T. (2007). The competitive success of *Methanomicrococcus blatticola*, a dominant methylotrophic methanogen in the cockroach hindgut, is supported by high substrate affinities and favorable thermodynamics. *FEMS Microbiology Ecology, 60*, 266–275.

Stetter, K. O., Thomm, M., Winter, J., Wildgruber, G., Huber, H., Zillig, W., Jané-Covic, D., König, H., Palm, P., & Wunderl, S. (1981). *Methanothermus fervidus*, sp. nov., a novel extremely thermophilic methanogen isolated from an Icelandic hot spring. *Zentralblatt für Bakteriologie Mikrobiologie und Hygiene: I Abt Originale C: Allgemeine, angewandte und ökologische Mikrobiologie, 2*, 166–178.

Strucks, P., Failing, L., & Kaluza, S. (2021). A short review on Ni-catalyzed Methanation of CO_2: Reaction mechanism, catalyst deactivation, dynamic operation. *Chemie Ingenieur Technik, 93*, 1526–1536.

Sun, Q., Li, H., Yan, J., Liu, L., Yu, Z., & Yu, X. (2015). Selection of appropriate biogas upgrading technology-a review of biogas cleaning, upgrading and utilisation. *Renewable and Sustainable Energy Reviews, 51*, 521–532.

Thauer, R. K., Kaster, A. K., Goenrich, M., Schick, M., Hiromoto, T., & Shima, S. (2010). Hydrogenases from methanogenic archaea, nickel, a novel cofactor, and H2 storage. *Annual Review of Biochemistry, 79*, 507–536.

Thema, M., Weidlich, T., Kaul, A., Böllmann, A., Huber, H., Bellack, A., Karl, J., & Sterner, M. (2021). Optimized biological CO_2-methanation with a pure culture of thermophilic methanogenic archaea in a trickle-bed reactor. *Bioresource Technology, 333*, 125135.

Voelklein, M. A., Rusmanis, D., & Murphy, J. D. (2019). Biological methanation: Strategies for in-situ and ex-situ upgrading in anaerobic digestion. *Applied Energy, 235*, 1061–1071.

Wahid, R., Mulat, D. G., Gaby, J. C., & Horn, S. J. (2019). Effects of H_2:CO_2 ratio and H2 supply fluctuation on methane content and microbial community composition during in-situ biological biogas upgrading. *Biotechnology for Biofuels, 12*, 104.

Widdel, F. (1986). Growth of methanogenic bacteria in pure culture with 2-propanol and other alcohols as hydrogen donors. *Applied and Environmental Microbiology, 51*, 1056–1062.

Widdel, F., & Wolfe, R. S. (1989). Expression of secondary alcohol dehydrogenase in methanogenic bacteria and purification of the F_{420}-specific enzyme from Methanogenium thermophilum strain TCI. *Archives of Microbiology, 152*, 322–328.

Widdel, F., Rouvière, P. E., & Wolfe, R. S. (1988). Classification of secondary alcohol-utilizing methanogens including a new thermophilic isolate. *Archives of Microbiology, 150*, 477–481.

Wilharm, T., Zhilina, T. N., & Hummel, P. (1991). DNA-DNA hybridization of methylotrophic halophilic methanogenic bacteria and transfer of *Methanococcus halophilusvp* to the genus *Methanohalophilus* as *Methanohalophilus halophilus* comb. nov. *International Journal of Systematic and Evolutionary Microbiology, 41*, 558–562.

Zeikus, J. G., & Wolfe, R. S. (1972). *Methanobacterium thermoautotrophicus* sp. n., an anaerobic, autotrophic, extreme thermophile. *Journal of Bacteriology, 109*, 707–715.

Zhao, Y., Boone, D. R., Mah, R. A., Boone, J. E., & Xun, L. (1989). Isolation and characterization of *Methanocorpusculum labreanum* sp. nov. from the LaBrea Tar Pits. *International Journal of Systematic and Evolutionary Microbiology, 39*, 10–13.

Part V
Contribution of Biohydrogen Production Using Extremophiles to the Renewable Energy Industry

Chapter 15
Advancements in One-Carbon Feedstock-Based Biohydrogen Production Using the Hyperthermophilic Archaeon *Thermococcus onnurineus* NA1

Hyun Sook Lee and Sung Gyun Kang

Abstract Hydrogen (H_2) is increasingly recognized as a promising solution for a sustainable energy future. The production of biohydrogen using microbes, combined with renewable, low-cost, and abundant resources, presents an economically attractive and environmentally friendly approach. One-carbon compounds, such as carbon monoxide (CO) and formate, are considered promising feedstocks. These compounds can be derived from CO-rich industrial by-products or waste gases or produced through the electrochemical reduction of greenhouse gas carbon dioxide, providing a cost-effective and practical solution. Researchers have identified a variety of microbes capable of utilizing CO or formate, and significant efforts are underway to enhance H_2 production through strain improvement and bioprocess optimization. *Thermococcus onnurineus* NA1, isolated from a deep-sea hydrothermal vent, has been engineered to become an efficient H_2 producer. Enhancements in its performance have involved addressing CO toxicity and improving enzymatic substrate conversion through promoter engineering and adaptive evolution techniques. In this context, various strategies that have been implemented are outlined, along with explanations of the underlying molecular mechanisms based on genomic and transcriptomic analyses. The developed technology was successfully scaled up to large bioreactors, demonstrating the feasibility of using industrial by-product gases as feedstock for biohydrogen production.

Keywords Hyperthermophilic archaeon · *Thermococcus onnurineus* NA1 · One-carbon substrate · H_2 production · Process engineering · Industrial gas

H. S. Lee · S. G. Kang (✉)
Korea Institute of Ocean Science and Technology, Busan, Republic of Korea

Department of Marine Biotechnology, University of Science and Technology, Daejeon, Republic of Korea
e-mail: leeh522@kiost.ac.kr; sgkang@kiost.ac.kr

M. Ito et al. (eds.), *Extremophiles for a Sustainable Future*,
https://doi.org/10.1007/978-981-95-6970-0_15

15.1 Introduction

Concerns about climate change and the energy crisis are driving a paradigm shift toward sustainable and renewable energy development. Hydrogen (H_2) draws attention as a viable solution for a green energy future due to its higher energy content per unit weight (~120 MJ kg^{-1}) than most other hydrocarbon fuels, lack of environmental pollution, sustainability, and so on (Hosseini & Wahid, 2016; Kovač et al., 2021; Yue et al., 2021). At present, the majority of hydrogen (H_2) on the market is produced from fossil sources (Megía et al., 2021; Nikolaidis & Poullikkas, 2017; Sharma & Ghoshal, 2015). However, there is an increasing demand for sustainable H_2 production that utilizes renewable resources. The production of biological hydrogen (biohydrogen) provides an alternative option with considerably reduced cost and environmental impact because it can be generated from organic or industrial waste and produced in bioreactors that require relatively little energy and few environmental prerequisites (Brentner et al., 2010; Das, 2009; Lu et al., 2020).

There are four known basic mechanisms for biohydrogen production: direct biophotolysis, indirect biophotolysis, photofermentation, and dark fermentation (Levin et al., 2004; Manish & Banerjee, 2008; Rittmann et al., 2014; Show et al., 2012). Dark fermentation has been extensively studied, in which microbes produce H_2 by disposing of electrons generated by renewable, cheap, and abundant resources (Das & Veziroglu, 2008). The primary advantage of dark fermentation lies in its higher H_2 production rate compared to other biological H_2 processes (Nath & Das, 2004; Rittmann & Herwig, 2012). Thermophilic bacteria and archaea, including genera such as *Thermotoga*, *Caldicellulosiruptor*, *Pyrococcus*, and *Thermococcus*, exhibit a remarkable potential for H_2 production. These microbes possess a thermodynamic advantage over mesophilic microbes, such as those in the families of Enterobacteriaceae and Clostridiaceae (Lee et al., 2011; Rittmann & Herwig, 2012; Show et al., 2012), as they can achieve close to the so-called "Thauer limit" of 4 mol H_2 produced per 1 mol of glucose (Kádár et al., 2004; Kanai et al., 2005; Levin et al., 2004; Roy et al., 2014; Schicho et al., 1993; Thauer et al., 1977; Verhagen & Adams, 2001). Significant advancements have been made in recent decades in identifying novel H_2-producing microbes and enhancing H_2 productivity through genetic and process engineering. Nonetheless, biohydrogen production is still largely limited to laboratory-scale settings (Akhtar & Jones, 2008; Davila-Vazquez et al., 2008; Maeda et al., 2008). There has been a need for improvements in H_2 yield and overall productivity (Brentner et al., 2010).

A sustainable supply of affordable feedstock is crucial for making biohydrogen production economically feasible. One-carbon compounds, such as carbon monoxide (CO) and formate, are receiving increasing attention as potential feedstocks. Formate can be produced through the electrochemical reduction of carbon dioxide (CO_2) (Fernández-Caso et al., 2023). Therefore, utilizing formate to produce valuable products directly contributes to carbon capture and utilization (CCU) strategies, reducing greenhouse gas emissions. CO is globally abundant. It was estimated that approximately 249 million tons of CO are produced annually from a variety of industrial processes, including steel manufacturing, coal power generation, and oil refining (Fatimah et al., 2024). Steel-mill gas primarily consists of H_2, CO, CO_2, nitrogen (N_2), and methane (CH_4), and its composition varies depending on the type of furnace used (Cho et al., 2009; Leibold et al., 2008). Additionally, syngas, which is mainly composed of CO, CO_2, and H_2, can also be a feasible substrate. It can be produced from various sources, including the partial oxidation or autothermal reforming of hydrocarbon-rich fossil fuels, as well as domestic and agricultural wastes and biomass. CO is a highly toxic substance for many chemical catalysts, especially those made of noble metals and certain transition metals, due to its strong and often irreversible binding to active sites. In contrast, biological processes, although typically slower than chemical reactions, have several advantages, such as higher yields (similar to direct fermentation of biomass), specificity, and reduced potential for catalyst poisoning (Fackler et al., 2021; Köpke & Simpson, 2020; Seifert et al., 2013; Spath & Dayton, 2003; Takors et al., 2018). While industrial waste gas has been explored for the production of metabolites, such as ethanol, acetone, isopropanol, and 2,3-butanediol, using acetogen (Fackler et al., 2021; Köpke & Simpson, 2020; Liew et al., 2022) and methanogen (Seifert et al., 2013), research on biohydrogen production remains quite limited (Rittmann et al., 2014). This raises the possibility of utilizing industrial gas as an inexpensive and practical substrate for biohydrogen production (Fatimah et al., 2024; Kim et al., 2013; Lee et al., 2020). However, the H_2 productivity is relatively low compared to studies utilizing other feedstocks.

In this chapter, we will conduct a comprehensive review of the research focused on H_2 production using the hyperthermophilic archaeon *Thermococcus onnurineus* NA1 (Fig. 15.1). Emphasis will be placed on elucidating the molecular mechanisms

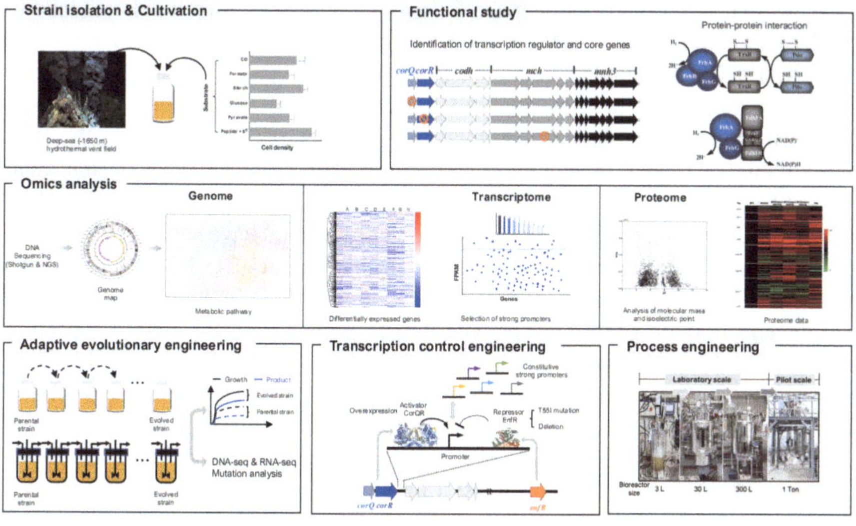

Fig. 15.1 Strategies to advance H_2 production in *T. onnurineus* NA1

underlying hydrogenases, which are involved in the metabolism of one-carbon substrates, specifically CO and formate. Additionally, we will review the integrated efforts of genetic and process engineering aimed at enhancing H_2 productivity. Finally, we will discuss recent advancements in the demonstration study of biohydrogen production using industrial by-product gas from the steel-mill process. This review aims to advance our understanding of the enzymatic processes involved and their implications for biotechnological applications in sustainable H_2 production.

15.2 Unique Hydrogenases of *T. onnurineus* NA1

The hyperthermophilic archaeon *T. onnurineus* NA1 was isolated in 2002 from a deep-sea hydrothermal vent area at a depth of 1650 m within the Papua New Guinea-Australia-Canada-Manus (PACMANUS) field. Phylogenetic analysis of the 16S rRNA gene indicated that the strain is most closely related to *Thermococcus gorgonarius*; however, no significant homology was observed among the species by DNA-DNA hybridization (Bae et al., 2006). Hyperthermophilic archaea, such as *Thermococcus* and *Pyrococcus* species, thrive in extremely high-temperature environments and utilize unique metabolic pathways. These pathways often involve hydrogenases, which play crucial roles in energy conservation and the disposal of reducing equivalents (Adams et al., 2001; Bryant & Adams, 1989; Kanai et al., 2003, 2011; Ma et al., 2000; Sapra et al., 2003; Yu et al., 2018). There are two main types of hydrogenases involved in these processes. The membrane-bound hydrogenase (MBH) couples the oxidation of ferredoxin with the evolution of H_2, along with the translocation of protons (H^+) or sodium ions (Na^+) across the membrane, vital for energy conservation. On the other hand, soluble (cytoplasmic) hydrogenase, also known as sulfhydrogenase, is involved in maintaining a delicate balance of electron disposal and hydrogen metabolism, particularly in its interactions with the $NADP^+$/NADPH pool.

The genome of *T. onnurineus* NA1 contains several conserved hydrogenases. The MBH homolog, named Mbh, is encoded by the genes TON_1583 to TON_1595, and soluble hydrogenase homologs, SulfI and SulfII, are encoded by the genes TON_0534 to TON_0537 and TON_0052 to TON_0055, respectively (Lee et al., 2008). Several additional membrane-bound hydrogenases have been identified, specifically membrane-bound formate-dependent hydrogenase 1 (Mfh1), membrane-bound formate-dependent hydrogenase 2 (Mfh2), and membrane-bound CO-dependent hydrogenase (Mch). These hydrogenases are clustered with proteins annotated as formate dehydrogenase (Fdh1 and Fdh2) and carbon monoxide dehydrogenase (CODH), along with Na^+/H^+ antiporters (Mnh1, Mnh2, and Mnh3). These associations are encoded by specific gene clusters: *fdh1-mfh1-mnh1* (TON_0266-0282), *fdh2-mfh2-mnh2* (TON_1563-1580), and *codh-mch-mnh3* (TON_1017-1032) (Lee et al., 2008; Lim et al., 2010). Since these additional hydrogenases were unique to *T. onnurineus* NA1 at the time of genome sequencing, their functionality was further explored. In addition, another soluble hydrogenase encoded by the *frhAGB* genes (TON_1559-1561) was identified near the gene cluster *fdh2-mfh2-mnh2*. The subunits of the FrhAGB complex—FrhA (the catalytic subunit with a NiFe center), FrhG (a small subunit with Fe-S clusters), and FrhB (the FAD and F_{420} binding subunit)—exhibited distinct primary sequences. These sequences showed low similarity, ranging from 33% to 35%, to the coenzyme F_{420}-reducing hydrogenase (Frh) found in *Methanococcus maripaludis*. The presence of an Frh homolog within the order Thermococcales is particularly intriguing, as *T. onnurineus* NA1 does not seem related to methanogenesis or methane utilization and lacks F_{420} biosynthesis (Jeon et al., 2015). Recently, the potential role of this hydrogenase was identified to play a role in transferring electrons to specific protein targets. Upon the oxidation of H_2, electrons are transferred from the FrhAGB complex directly to thioredoxin reductase (TrxR). This process leads to the reduction of TrxR and its redox partner, protein disulfide oxidoreductase (Pdo), without the need for an intermediate electron carrier, such as F_{420} or NAD(P)H (Jung et al., 2020). This finding suggests that H_2 can be an electron source of the thioredoxin system to help maintain redox homeostasis. Moreover, the *frhAGB*-encoded hydrogenase demonstrates a direct interaction with formate dehydrogenase 3 (Fdh3), composed of Fdh3ABG1G2, specifically, the Fdh3B subunit, which shares a similar domain structure with TrxR (Yang et al., 2023). These results broaden our understanding of the physiological roles of Frh homologs, indicating that they may be crucial in regulating diverse metabolic processes in Thermococcales, including *T. onnurineus* NA1.

15.2.1 Formate-Dependent Hydrogenase and Formate-Driven Growth

The formate hydrogenlyase (FHL) complex is a multi-protein assembly consisting of a formate dehydrogenase and a hydrogenase, along with several iron-sulfur proteins that aid in electron transfer. In some instances, it may also contain membrane-anchoring proteins and various accessory proteins that contribute to assembly and regulation. The primary function of FHL is to catalyze the disproportionation of formate into carbon dioxide (CO_2) and H_2. FHL contributes to formate detoxification and pH homeostasis by consuming formate and protons ($HCOO^- + H^+ \rightarrow CO_2 + H_2$), which deacidifies the medium and helps microbes like *Escherichia coli* to maintain a more stable intracellular pH (Sawers, 1994). In addition to its primary physiological role of formate oxidation, it can catalyze the reverse reaction under specific conditions, such as high partial pressures of H_2 and CO_2, reducing them to formate, as demonstrated by the H_2-dependent CO_2 reductase (HDCR) in acetogens (Schuchmann & Müller, 2013; Burger et al., 2022). The reaction of converting formate to H_2 ($HCOO^- + H_2O \rightarrow HCO_3^- + H_2$) occurs very close to thermodynamic equilibrium with a standard Gibbs free-energy change (ΔG^o) of +1.3 kJ mol^{-1} (Thauer et al., 1977). Therefore, the FHL of mesophiles is generally not considered to directly generate a proton motive force for ATP synthesis

(Steinhilper et al., 2022). Thus, the majority of H_2 production from formate has been carried out using resting cells of various microbes utilizing the FHL system (McDowall et al., 2014; Rittmann et al., 2014) and HDCR (Burger et al., 2022; Kottenhahn et al., 2018). In syntrophic associations, the removal of H_2 by a partner organism, commonly a methanogen or a sulfate-reducing bacterium, drives the thermodynamically unfavorable reaction converting formate to H_2 forward (Dolfing et al., 2008; McInerney et al., 2009). This process makes it energetically feasible for the formate-oxidizing bacterium to obtain a small but sufficient amount of energy for growth.

The FHL-like complex encoded by the *fdh2-mfh2-mnh2* gene cluster of *T. onnurineus* NA1 has been identified to be energy-conserving. When the temperature rises to 80 °C, the reaction for formate oxidation becomes more exergonic, with a ΔG^o of −2.6 kJ mol^{-1} at this temperature (Amend & Shock, 2001; Kim et al., 2010). The archaeon was the first known organism to utilize formate as its sole electron donor for growth, oxidizing it to produce H_2 (Bae et al., 2012; Kim et al., 2010; Lee et al., 2008; Lim et al., 2012). The architecture of the gene cluster is conserved across several strains of Thermococcales (Holden & Sistu, 2023; Kim et al., 2010). This includes the hyperthermophilic and radioresistant archaeon *Thermococcus gammatolerans* EJ3 (Zivanovic et al., 2009), the hyperthermophilic archaeon *Thermococcus paralvinellae* ES1 (Jung et al., 2014), and the obligate piezophilic hyperthermophilic archaeon *Pyrococcus yayanoshii* CH1 (Xu et al., 2011). To clarify how the strain generates ATP from the formate conversion reaction, experiments demonstrated that the reaction is coupled with ATP synthesis through a transmembrane ion current using inverted vesicles (Lim et al., 2014). Specifically, the oxidation of formate by formate dehydrogenase (Fdh2, TON_1563-1564) and hydrogenase (Mfh2, TON_1565-1571) leads to the translocation of H^+ ions across the cytoplasmic membrane. This movement of H^+ ions then drives the translocation of Na^+ ions by Na^+/H^+ antiporter (Mnh2, TON_1574-1580), which is ultimately coupled to ATP synthesis by Na^+-type ATP synthase (Kim et al., 2010) (Fig. 15.2). TON_1573 encodes formate transporter, which is homologous to the bacterial FocA (formate channel protein A), belonging to a formate/nitrite transporter (FNT) family. The *fdh2-mfh2-mnh2* gene cluster seems to be induced by formate. Cells grown on formate exhibited significant upregulation of the gene cluster compared to those grown in a sulfur-containing medium (Kim et al., 2010). In the comparative proteomic analysis, the gene cluster exhibited a significantly higher abundance in cells grown on formate compared to those grown on starch or CO (Moon et al., 2012). Although the exact mechanism for the induction of the gene cluster remains unidentified, a clue emerged from transcriptomic analysis. It was found that the expression of the gene cluster was significantly downregulated in a mutant with a deletion of the TON_1525 gene, a transcription regulator homologous to the DNA-binding protein Tfx from *Methanobacterium thermoautotrophicum* (Hochheimer et al., 1999; Lee et al., 2021). A study involving adaptive laboratory evolution of *T. onnurineus* NA1 on formate has identified mutations (A52T or A52E) in TON_1573 that promote uptake of formate (Jung et al., 2017). This strongly indicates that the efficiency of formate uptake is a rate-limiting step.

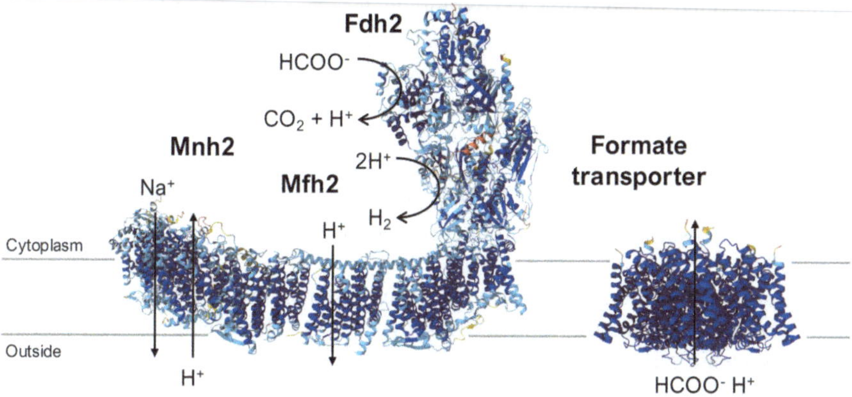

Fig. 15.2 The molecular mechanism of formate oxidation coupled with H_2 production involving the Fdh2-Mfh2-Mnh2 complex and a formate transporter in *T. onnurineus* NA1. The structures of proteins were modeled using AlphaFold3 (https://alphafoldserver.com/). In the model, colors represent a per-atom confidence based on the predicted Local Distance Difference Test (pLDDT) scores: blue for very high (pLDDT > 90), cyan for high (90 > pLDDT > 70), yellow for low (70 > pLDDT > 50), and orange for very low (pLDDT < 50)

In contrast, the levels of mRNA and protein of the *fdh2-mfh2-mnh2* gene cluster remained unchanged throughout the adaptation process.

15.2.2 Carbon Monoxide Dehydrogenase-Linked Hydrogenase and CO-Driven Growth

CO serves as a powerful electron donor, exhibiting a notably low CO_2/CO reduction potential ($E^{0\prime}$ = −520 mV), which can support the growth of diverse lithotrophic microbes. Nevertheless, the severe toxicity of CO has hindered its use by microbes (Adams, 1990; Fauque et al., 1988; Thauer et al., 1974). Despite this toxicity, some carboxydotrophic microbes can thrive on CO and generate H_2 through the following reaction (Amend & Shock, 2001): $CO + H_2O \rightarrow CO_2 + H_2$, $\Delta G^{0\prime} = -20$ kJ/mol. After a species of *Rhodopseudomonas* was shown to grow anaerobically in a CO-rich atmosphere (Uffen, 1976), CO-dependent H_2 production has been observed across various phylogenetic groups (Oelgeschläger & Rother, 2008; Sokolova et al., 2009; Techtmann et al., 2009). The biochemical capability for microbes to link CO oxidation to H_2 production and vice versa has been assessed via the actions of CODH and hydrogenase enzymes. Typically, this process occurs via indirect coupling between the two enzymes. In this mechanism, electrons generated by CODH are transferred to a soluble electron carrier, such as ferredoxin or NAD(P)H. This carrier then donates the electrons to a separate hydrogenase enzyme. For example, the acetogenic bacterium *Acetobacterium woodii* oxidizes CO using CODH, reducing ferredoxin. Reduced ferredoxin, along with NADH formed by an RnF complex, a membrane-bound ferredoxin:NAD oxidoreductase, can then be utilized to generate H_2 through an electron-bifurcating hydrogenase (Schuchmann & Müller, 2012). On the other hand, CODH and a hydrogenase can be physically linked as a complex to facilitate direct electron transfer. This would allow the electrons released from CO oxidation to be directly channeled to the hydrogenase for H_2 production. For instance, in *Carboxydothermus hydrogenoformans* Z-2901, CODH forms a functional complex with an energy-converting hydrogenase (Ech) and membrane-bound components, resulting in the production of H_2 and energy conservation (Wu et al., 2005).

The presence of the *codh-mch-mnh3* gene cluster in the genome suggested that *T. onnurineus* NA1 has the capacity to grow on CO. This metabolic potential for hydrogenogenic carboxydotrophy was experimentally validated by cultivation (Lee et al., 2008). The following gene disruption study indicated that CO is oxidized by CODH (TON_1017-1018), which then transfers electrons to an energy-converting hydrogenase (TON_1021-1025) through a ferredoxin-like protein encoded by the gene TON_1017 (Fig. 15.3) (Kim et al., 2013). The presence of genes encoding Na^+/H^+ antiporter subunits (TON_1026-1032) within the *codh-mch-mnh*3 gene cluster is distinct compared to clusters found in mesophilic or thermophilic bacteria. The antiporter is similar to the one identified in the Fdh2-Mfh2-Mnh2 complex (Kim et al., 2015; Lee et al., 2008). Therefore, it has been proposed that the CO conversion reaction is also coupled with ATP synthesis through a transmembrane ion current, similar to the formate conversion observed in *T. onnurineus* NA1 (Lim et al., 2014). The expression of the gene cluster is regulated by a novel type of

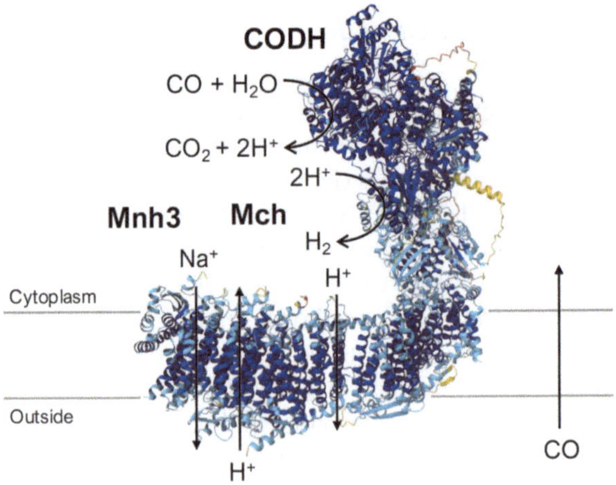

Fig. 15.3 The molecular mechanism of CO oxidation coupled with H_2 production involving the CODH-Mch-Mnh3 complex in *T. onnurineus* NA1. The structures of proteins were modeled using AlphaFold3 (https://alphafoldserver.com/). In the model, colors represent a per-atom confidence based on the predicted Local Distance Difference Test (pLDDT) scores: blue for very high (pLDDT > 90), cyan for high (90 > pLDDT > 70), yellow for low (70 > pLDDT > 50), and orange for very low (pLDDT < 50)

CO-responsive regulatory system that is distinct from previously characterized bacterial CO-sensing transcription factors, such as CooA (CO oxidation activator) and RcoM (regulator of CO metabolism), with a heme-based CO sensing mechanism (He et al., 1996; Kerby et al., 2008). In *T. onnurineus* NA1, CorQ (TON_1015-1) and CorR (TON_1016) act as a positive regulatory protein pair for the expression of the *codh-mch-mnh*3 gene cluster (Kim et al., 2015). In the *corQR* overexpression mutant, created by deleting the *corR* gene and then complementing it with integrated *corQR* genes under a strong constitutive *gdh* promoter from *T. kodakarensis*, the mRNA and protein levels of the *codh-mch-mnh*3 gene cluster were significantly increased compared to the wild-type strain (Kim et al., 2015). In contrast, EnfR (eighth note fold regulator, TON_1525) functions as a negative regulator. In a TON_1525 deletion mutant, the repression of the *codh-mch-mnh*3 gene cluster is lifted, leading to increased expression of the cluster.

15.3 Enhancement of H_2 Productivity and Scale-Up of Fermentation Processes

The production of H_2 from CO faces two major challenges: mass transfer and CO toxicity (Munasinghe & Khanal, 2010; Rittmann et al., 2014; Takors et al., 2018). *T. onnurineus* NA1 demonstrates the ability to produce H_2 under various initial CO partial pressures, reaching up to 4 bar (Kim et al., 2013), which highlights its exceptional tolerance to CO and may provide technical options to increase CO solubility. To harness the metabolic potential of CO-dependent H_2 production of *T. onnurineus* NA1, metabolic engineering was employed to enhance H_2 productivity and mitigate CO toxicity using the genetic toolbox of *T. kodakarensis* KOD1 (Matsumi et al., 2007; Sato et al., 2003), leading to a remarkable improvement in the H_2 production rate from CO. The first strategy involves bypassing the native regulation entirely and driving high-level expression of target genes by using a constitutive strong promoter, a process known as promoter engineering. The replacement of the native promoter with a stronger promoter led to a significant increase in the transcription and translation levels of the *codh-mch-mnh*3 gene cluster (Kim et al., 2013; Lee et al., 2014, 2015). In this approach, previously identified strong promoters that drive expression of genes encoding the S-layer protein and glutamate dehydrogenase, whether of endogenous or exogenous origin, were tested. A new promoter was selected based on transcriptomic analysis from *T. onnurineus* NA1. As a result of this modification, the overproduction of the CODH-hydrogenase enzyme complex resulted in a notable increase in the H_2 production rate, reaching levels that were 1.9–4.7 times greater than that observed in the wild-type strain. In all instances, the molar yields of CO consumed relative to H_2 produced were estimated to be approximately 1 mol mol^{-1}. The second strategy involves manipulating native regulatory machinery. For example, by providing a strong and persistent activation signal from CorQR through overexpression using a strong promoter, the cell was able to produce more CODH-Mch-Mnh3 enzymes necessary for CO metabolism. This approach resulted in a 5.8-fold increase in H_2 production rate compared to the wild-type strain (Kim et al., 2015). In addition, the deletion of the TON_1525 gene, which encodes EnfR and functions as a repressor of the *codh-mch-mnh3* gene cluster, resulted in a remarkable increase in H_2 production by 2.2-fold, respectively, compared to the wild-type strain (Lee et al., 2021).

Furthermore, the hydrogen productivity was significantly enhanced by laboratory adaptive evolutionary engineering. By conducting successive serial transfers in a medium containing CO or formate as a substrate, the growth and H_2 production were significantly enhanced (Jung et al., 2017; Lee et al., 2016). Particularly, the 156th transferred strains under each condition showed significantly improved H_2 production, 5.2-fold and 1.3-fold for CO- and formate-adapted strains, respectively. Genome sequencing revealed several mutations in the adapted strains. Reverse engineering indicated that specific mutations were critical in altering the phenotype of these adapted strains: the T55I mutation in TON_1525 was significant under CO condition, while the A52T or A52E mutations in TON_1573 were important under formate condition. The T55I mutation in TON_1525 was found to reduce its binding affinity to DNA, as demonstrated by electrophoretic mobility shift assay (EMSA) and X-ray crystallography (Bae et al., 2023; Lee et al., 2016). However, it is important to note that a single mutation cannot account for all the changes that occurred during adaptation.

Mutant strains developed in *T. onnurineus* NA1 showed significantly enhanced H_2 production rates and CO tolerance. Thus, industrial gases containing CO were evaluated as potential substrates for H_2 production. (Kim et al., 2013; Lee et al., 2020). Notably, one of the mutants has exhibited the capability to produce H_2 by utilizing steel mill waste gas, specifically Linz-Donawitz converter gas (LDG), which is produced as a by-product of the steel industry (Kim et al., 2013). Additionally, another strain has been shown to utilize syngas derived directly from a coal gasification pilot plant, comprised of 39.09% CO, 20.4% H_2, 1.4% CH_4, 6.95% CO_2, and 32.12% N2 (Lee et al., 2020). The maximum H_2 production rate and cell densities demonstrated a nearly linear increase with the flow rate of the waste gas. Interestingly, the maximum cell densities and the maximum H_2 production rates for 100% CO and waste gas were quite comparable when the same amount of CO supply was adjusted. In addition to LDG and coal-gasified syngas, other byproduct gases from the steel mill process, such as blast furnace gas (BFG) or syngas derived from biomass, are also suitable for H_2 production by *T. onnurineus* NA1 (unpublished data). It is worth noting that the processes did not require any pretreatment of the industry gases, which typically contain sulfur or other contaminants that can affect growth or H_2 production yields. This potential was demonstrated in a laboratory-scale bioreactor (Kim et al., 2013; Lee et al., 2020) and successfully demonstrated at a pilot plant scale of a 1 m^3 bioreactor and a 45 m^3 bioreactor system. This indicates that the process can be effectively scaled up.

15.4 Conclusions and Prospects

For decades, a variety of H_2-producing microbes that utilize CO have been isolated, and numerous efforts have been made to improve H_2 productivity. A significant breakthrough was made with the discovery of a new hyperthermophilic archaeon, *T. onnurineus* NA1. This strain was engineered to achieve an impressive H_2 production rate of 300 mmol L^{-1} h^{-1} by utilizing by-product gas from the steel mill process. This finding highlights the promising potential of converting industrial by-products or waste gases into H_2. Considering the cost of feedstocks, this result presents viable options for large-scale biohydrogen production, whether conducted on-site or off-site.

The economic and safety challenges associated with conventional H_2 storage and delivery have created a growing demand for new hydrogen storage materials. In this context, formate is getting attention as a promising hydrogen carrier (Chu et al., 2023; Jung et al., 2025; Valentini et al., 2022). The unique feature of *T. onnurineus* NA1 is its ability to grow on formate while simultaneously producing H_2, making it a valuable H_2-producing strain.

In addition to *T. onnurineus* NA1, other hyperthermophilic strains with carboxydotrophic or formatotrophic hydrogenogenic metabolism may also serve as promising candidates for H_2 production from CO and formate. This highlights the importance of studying previously unexplored microbes from various environments (Kim et al., 2010; Lim et al., 2012). Furthermore, the availability of genetic engineering tools will help realize their potential by enhancing productivity and enabling the co-production of other products as well (Jung et al., 2025; Lee et al., 2020).

Acknowledgments This work was supported by KIOST in-house programs (EA0411 and EA0413).

References

Adams, M. W. (1990). The structure and mechanism of iron-hydrogenases. *Biochimica et Biophysica Acta, 1020*, 115–145.

Adams, M. W., Holden, J. F., Menon, A. L., Schut, G. J., Grunden, A. M., Hou, C., Hutchins, A. M., Jenney, F. E., Jr., Kim, C., Ma, K., Pan, G., Roy, R., Sapra, R., Story, S. V., & Verhagen, M. F. (2001). Key role for sulfur in peptide metabolism and in regulation of three hydrogenases in the hyperthermophilic archaeon *Pyrococcus furiosus*. *Journal of Bacteriology, 183*, 716–724.

Akhtar, M. K., & Jones, P. R. (2008). Engineering of a synthetic *hydF-hydE-hydG-hydA* operon for biohydrogen production. *Analytical Biochemistry, 373*, 170–172.

Amend, J. P., & Shock, E. L. (2001). Energetics of overall metabolic reactions of thermophilic and hyperthermophilic Archaea and Bacteria. *FEMS Microbiology Reviews, 25*, 175–243.

Bae, S. S., Kim, Y. J., Kang, S. G., Yang, S. H., Lim, J. K., Jeon, J. H., et al. (2006). *Thermococcus onnurineus* sp, nov., a hyperthermophilic archaeon isolated from a deep-sea hydrothermal vent area at the PACMANUS field. *Journal of Microbiology and Biotechnology, 16*, 1826–1831.

Bae, S. S., Kim, T. W., Lee, H. S., Kwon, K. K., Kim, Y. J., Kim, M.-S., et al. (2012). H_2 production from CO, formate or starch using the hyperthermophilic archaeon, *Thermococcus onnurineus*. *Biotechnology Letters, 34*, 75–79.

Bae, D. W., Lee, S. H., Park, J. H., Son, S. Y., Lin, Y., Lee, J. H., Jang, B. R., Lee, K. H., Lee, Y. H., Lee, H. S., Kang, S. G., Kim, B. S., & Cha, S. S. (2023). An archaeal transcription factor EnfR with a novel 'eighth note' fold controls hydrogen production of a hyperthermophilic archaeon *Thermococcus onnurineus* NA1. *Nucleic Acids Research, 51*, 10026–10040.

Brentner, L. B., Peccia, J., & Zimmerman, J. B. (2010). Challenges in developing biohydrogen as a sustainable energy source: Implications for a research agenda. *Environmental Science & Technology, 44*, 2243–2254.

Bryant, F. O., & Adams, M. W. (1989). Characterization of hydrogenase from the hyperthermophilic archaebacterium, *Pyrococcus furiosus*. *The Journal of Biological Chemistry, 264*, 5070–5079.

Burger, Y., Schwarz, F. M., & Müller, V. (2022). Formate-driven H_2 production by whole cells of *Thermoanaerobacter kivui*. *Biotechnology for Biofuels and Bioproducts, 15*, 1–12.

Cho, H. C., Hong, J. G., Cho, K. W., & Kim, K. H. (2009). Oxygen-enriched combustion technology of by-product gases in iron-and steelmaking process. *RIST Research Report, 23*, 71–75.

Chu, C., Wu, K., Luo, B., Cao, Q., & Zhang, H. (2023). Hydrogen storage by liquid organic hydrogen carriers: Catalyst, renewable carrier, and technology - a review. *Carbon Resources Conversion, 6*, 334–351.

Das, D. (2009). Advances in biohydrogen production processes: An approach towards commercialization. *International Journal of Hydrogen Energy, 34*, 7349–7357.

Das, D., & Veziroglu, T. N. (2008). Advances in biological hydrogen production processes. *International Journal of Hydrogen Energy, 33*, 6046–6057.

Davila-Vazquez, G., Arriaga, S., Alatriste-Mondragón, F., León-Rodríguez, A., Rosales-Colunga, L. M., & Razo-Flores, E. (2008). Fermentative biohydrogen production: Trends and perspectives. *Reviews in Environmental Science and Biotechnology, 7*, 27–45.

Dolfing, J., Jiang, B., Henstra, A. M., Stams, A. J., & Plugge, C. M. (2008). Syntrophic growth on formate: A new microbial niche in anoxic environments. *Applied and Environmental Microbiology, 74*, 6126–6131.

Fackler, N., Heijstra, B. D., Rasor, B. J., Brown, H., Martin, J., Ni, Z., Shebek, K. M., Rosin, R. R., Simpson, S. D., Tyo, K. E., Giannone, R. J., Hettich, R. L., Tschaplinski, T. J., Leang, C., Brown, S. D., Jewett, M. C., & Köpke, M. (2021). Stepping on the gas to a circular economy: Accelerating development of carbon-negative chemical production from gas fermentation. *Annual Review of Chemical and Biomolecular Engineering, 12*, 439–470.

Fatimah, M., Qyyum, M. A., Lee, M., Alshareef, R. S., Aslam, M., Saeed, B., Dai, L., Gilani, M. A., Bazmi, A. A., Chang, I. S., et al. (2024). Industrial waste gases as a resource for sustainable hydrogen production: Resource availability, production potential, challenges, and prospects. *Carbon Capture Science & Technology, 12*, 100228.

Fauque, G., Peck, H. D., Moura, J. J. G., Huynh, B. H., Berlier, Y., DerVartanian, D. V., et al. (1988). The three classes of hydrogenases from sulfate-reducing bacteria of the genus *Desulfovibrio*. *FEMS Microbiology Letters, 54*, 299–344.

Fernández-Caso, K., Díaz-Sainz, G., Alvarez-Guerra, M., & Irabien, A. (2023). Electroreduction of CO_2: Advances in the continuous production of formic acid and formate. *ACS Energy Letters, 8*, 1992–2024.

He, Y., Shelver, D., Kerby, R. L., & Roberts, G. P. (1996). Characterization of a CO-responsive transcriptional activator from *Rhodospirillum rubrum*. *The Journal of Biological Chemistry, 271*, 120–123.

Hochheimer, A., Hedderich, R., & Thauer, R. K. (1999). The DNA binding protein Tfx from *Methanobacterium thermoautotrophicum*: Structure, DNA binding properties and transcriptional regulation. *Molecular Microbiology, 31*, 641–650.

Holden, J. F., & Sistu, H. (2023). Formate and hydrogen in hydrothermal vents and their use by extremely thermophilic methanogens and heterotrophs. *Frontiers in Microbiology, 14*, 1093018.

Hosseini, S. E., & Wahid, M. A. (2016). Hydrogen production from renewable and sustainable energy resources: Promising green energy carrier for clean development. *Renewable and Sustainable Energy Reviews, 57*, 850–866.

Jeon, J. H., Lim, J. K., Kim, M. S., Yang, T. J., Lee, S. H., Bae, S. S., Kim, Y. J., Lee, S. H., Lee, J. H., Kang, S. G., & Lee, H. S. (2015). Characterization of the *frhAGB*-encoding hydrogenase from a non-methanogenic hyperthermophilic archaeon. *Extremophiles, 19*, 109–118.

Jung, J.-H., Kim, Y.-T., Jeon, E.-J., Seo, D.-H., Hensley, S. A., Holden, J. F., Lee, J.-H., & Park, C.-S. (2014). Complete genome sequence of hyperthermophilic archaeon *Thermococcus* sp. ES1. *Journal of Biotechnology, 174*, 14–15.

Jung, H. C., Lee, S. H., Lee, S. M., An, Y. J., Lee, J. H., Lee, H. S., & Kang, S. G. (2017). Adaptive evolution of a hyperthermophilic archaeon pinpoints a formate transporter as a critical factor for the growth enhancement on formate. *Scientific Reports, 7*, 6124.

Jung, H. C., Lim, J. K., Yang, T. J., Kang, S. G., & Lee, H. S. (2020). Direct electron transfer between the *frhAGB*-encoded hydrogenase and thioredoxin reductase in the nonmethanogenic archaeon *Thermococcus onnurineus* NA1. *Applied and Environmental Microbiology, 86*, e02630–e02619.

Jung, H. C., Lee, S. M., Yang, J. I., Lee, S. H., Lee, H. S., & Kang, S. G. (2025). Exploration of formate as a liquid organic hydrogen carrier in biohydrogen production through evolutionary and process engineering of hyperthermophilic archaeon. *Bioresource Technology, 425*, 132318.

Kádár, Z., de Vrijek, T., van Noorden, G. E., Budde, M. A. W., Szengyel, Z., & Reczey, K. (2004). Yields from glucose, xylose, and paper sludge hydrolysate during hydrogen production by the extreme thermophile *Caldicellulosiruptor saccharolyticus*. *Applied Biochemistry and Biotechnology, 114*, 497–508.

Kanai, T., Ito, S., & Imanaka, T. (2003). Characterization of a cytosolic NiFe-hydrogenase from the hyperthermophilic archaeon *Thermococcus kodakaraensis* KOD1. *Journal of Bacteriology, 185*, 1705–1711.

Kanai, T., Imanaka, H., Nakajima, A., Uwamori, K., Omori, Y., Fukui, T., et al. (2005). Continuous hydrogen production by the hyperthermophilic archaeon, *Thermococcus kodakaraensis* KOD1. *Journal of Biotechnology, 116*, 271–282.

Kanai, T., Matsuoka, R., Beppu, H., Nakajima, A., Okada, Y., Atomi, H., & Imanaka, T. (2011). Distinct physiological roles of the three [NiFe]-hydrogenase orthologs in the hyperthermophilic archaeon *Thermococcus kodakarensis*. *Journal of Bacteriology, 193*, 3109–3116.

Kerby, R. L., Youn, H., & Roberts, G. P. (2008). RcoM: A new single-component transcriptional regulator of CO metabolism in bacteria. *Journal of Bacteriology, 190*, 3336–3343.

Kim, Y. J., Lee, H. S., Kim, E. S., Bae, S. S., Lim, J. K., Matsumi, R., Lebedinsky, A. V., Sokolova, T. G., Kozhevnikova, D. A., Cha, S.-S., Kim, S.-J., Kwon, K. K., Imanaka, T., Atomi, H., Bonch-Osmolovskaya, E. A., Lee, J.-H., & Kang, S. G. (2010). Formate-driven growth coupled with H_2 production. *Nature, 467*, 352–355.

Kim, M.-S., Bae, S. S., Kim, Y. J., Kim, T. W., Lim, J. K., Lee, S. H., et al. (2013). CO-dependent H2 production by genetically engineered *Thermococcus onnurineus* NA1. *Applied and Environmental Microbiology, 79*, 2048–2053.

Kim, M. S., Choi, A. R., Lee, S. H., Jung, H. C., Bae, S. S., Yang, T. J., Jeon, J. H., Lim, J. K., Youn, H., Kim, T. W., Lee, H. S., & Kang, S. G. (2015). A novel CO-responsive transcriptional regulator and enhanced H_2 production by an engineered *Thermococcus onnurineus* NA1 strain. *Applied and Environmental Microbiology, 81*, 1708–1714.

Köpke, M., & Simpson, S. D. (2020). Pollution to products: Recycling of 'above ground' carbon by gas fermentation. *Current Opinion in Biotechnology, 65*, 180–189.

Kottenhahn, P., Schuchmann, K., & Müller, V. (2018). Efficient whole cell biocatalyst for formate-based hydrogen production. *Biotechnology for Biofuels, 11*, 1–9.

Kovač, A., Paranos, M., & Marciuš, D. (2021). Hydrogen in energy transition: A review. *International Journal of Hydrogen Energy, 46*, 10016–10035.

Lee, H. S., Kang, S. G., Bae, S. S., Lim, J. K., Cho, Y., Kim, Y. J., et al. (2008). The complete genome sequence of *Thermococcus onnurineus* NA1 reveals a mixed heterotrophic and carboxydotrophic metabolism. *Journal of Bacteriology, 190*, 7491–7499.

Lee, D.-J., Show, K.-Y., & Su, A. (2011). Dark fermentation on biohydrogen production: Pure culture. *Bioresource Technology, 102*, 8393–8402.

Lee, S. H., Kim, M. S., Bae, S. S., Choi, A. R., Lee, J. W., Kim, T. W., Lee, J. H., Lee, H. S., & Kang, S. G. (2014). Comparison of CO-dependent H_2 production with strong promoters in *Thermococcus onnurineus* NA1. *Applied Microbiology and Biotechnology, 98*, 979–986.

Lee, S. H., Kim, M. S., Jung, H. C., Lee, J., Lee, J. H., Lee, H. S., & Kang, S. G. (2015). Screening of a novel strong promoter by RNA sequencing and its application to H_2 production in a hyperthermophilic archaeon. *Applied Microbiology and Biotechnology, 99*, 4085–4092.

Lee, S. H., Kim, M. S., Lee, J. H., Kim, T. W., Bae, S. S., Lee, S. M., Jung, H. C., Yang, T. J., Choi, A. R., Cho, Y. J., Lee, J. H., Kwon, K. K., Lee, H. S., & Kang, S. G. (2016). Adaptive engineering of a hyperthermophilic archaeon on CO and discovering the underlying mechanism by multi-omics analysis. *Scientific Reports, 6*, 22896.

Lee, S. H., Lee, S. M., Lee, J. H., Lee, H. S., & Kang, S. G. (2020). Biological process for coproduction of hydrogen and thermophilic enzymes during CO fermentation. *Bioresource Technology, 305*, 123067.

Lee, S. H., Lee, S.-M., Kang, S. G., & Lee, H. S. (2021). Enhanced H_2 production by deletion of the Tfx family DNA-binding protein in the hyperthermophilic archaeon *Thermococcus onnurineus* NA1. *International Journal of Hydrogen Energy, 46*, 35189–35197.

Leibold, H., Hornung, A., & Seifert, H. (2008). HTHP syngas cleaning concept of two stage biomass gasification for FT synthesis. *Powder Technology, 180*, 265–270.

Levin, D. B., Pitt, L., & Love, M. (2004). Biohydrogen production: Prospects and limitations to practical application. *International Journal of Hydrogen Energy, 29*, 173–185.

Liew, F. E., Nogle, R., Abdalla, T., Rasor, B. J., Canter, C., Jensen, R. O., Wang, L., Strutz, J., Chirania, P., De Tissera, S., Mueller, A. P., Ruan, Z., Gao, A., Tran, L., Engle, N. L., et al. (2022). Carbon-negative production of acetone and isopropanol by gas fermentation at industrial pilot scale. *Nature Biotechnology, 40*, 335–344.

Lim, J. K., Kang, S. G., Lebedinsky, A. V., Lee, J.-H., & Lee, H. S. (2010). Identification of a novel class of membrane-bound [NiFe]-hydrogenases in *Thermococcus onnurineus* NA1 by in silico analysis. *Applied and Environmental Microbiology, 76*, 6286–6289.

Lim, J. K., Bae, S. S., Kim, T. W., Lee, J.-H., Lee, H. S., & Kang, S. G. (2012). Thermodynamics of formate-oxidizing metabolism and implications for H_2 production. *Applied and Environmental Microbiology, 78*, 7393–7397.

Lim, J. K., Mayer, F., Kang, S. G., & Müller, V. (2014). Energy conservation by oxidation of formate to carbon dioxide and hydrogen via a sodium ion current in a hyperthermophilic archaeon. *Proceedings of the National Academy of Sciences of the United States of America, 111*, 11497–11502.

Lu, C., Zhang, H., Zhang, Q., Chu, C. Y., Tahir, N., Ge, X., Jing, Y., Hu, J., Li, Y., Zhang, Y., & Zhang, T. (2020). An automated control system for pilot-scale biohydrogen production: Design, operation and validation. *International Journal of Hydrogen Energy, 45*, 3795–3806.

Ma, K., Weiss, R., & Adams, M. W. (2000). Characterization of hydrogenase II from the hyperthermophilic archaeon *Pyrococcus furiosus* and assessment of its role in sulfur reduction. *Journal of Bacteriology, 182*, 1864–1871.

Maeda, T., Sanchez-Torres, V., & Wood, T. K. (2008). Metabolic engineering to enhance bacterial hydrogen production. *Microbial Biotechnology, 1*, 30–39.

Manish, S., & Banerjee, R. (2008). Comparison of biohydrogen production processes. *International Journal of Hydrogen Energy, 33*, 279–286.

Matsumi, R., Manabe, K., Fukui, T., Atomi, H., & Imanaka, T. (2007). Disruption of a sugar transporter gene cluster in a hyperthermophilic archaeon using a host-marker system based on antibiotic resistance. *Journal of Bacteriology, 189*, 2683–2691.

McDowall, J. S., Murphy, B. J., Haumann, M., Palmer, T., Armstrong, F. A., & Sargent, F. (2014). Bacterial formate hydrogenlyase complex. *Proceedings of the National Academy of Sciences of the United States of America, 111*, 3948–3956.

McInerney, M. J., Sieber, J. R., & Stams, A. J. M. (2009). Syntrophy in anaerobic degradation. *Environmental Microbiology, 11*, 3054–3064.

Megía, P. J., Vizcaíno, A. J., Calles, J. A., & Carrero, A. (2021). Hydrogen production technologies: From fossil fuels toward renewable sources. *Mini Reviews of Energy Fuels, 35*, 16403–16415.

Moon, Y. J., Kwon, J., Yun, S. H., Lim, H. L., Kim, M. S., Kang, S. G., Lee, J. H., Choi, J. S., Kim, S. I., & Chung, Y. H. (2012). Proteome analyses of hydrogen-producing hyperthermophilic archaeon *Thermococcus onnurineus* NA1 in different one-carbon substrate culture conditions. *Molecular & Cellular Proteomics, 11*(M111), 015420.

Munasinghe, P. C., & Khanal, S. K. (2010). Biomass-derived syngas fermentation into biofuels: Opportunities and challenges. *Bioresource Technology, 101*, 5013–5022.

Nath, K., & Das, D. (2004). Improvement of fermentative hydrogen production: Various approaches. *Applied Microbiology and Biotechnology, 65*, 520–529.

Nikolaidis, P., & Poullikkas, A. (2017). A comparative overview of hydrogen production processes. *Renewable and Sustainable Energy Reviews, 67*, 597–611.

Oelgeschläger, E., & Rother, M. (2008). Carbon monoxide-dependent energy metabolism in anaerobic bacteria and archaea. *Archives of Microbiology, 190*, 257–269.

Rittmann, S., & Herwig, C. (2012). A comprehensive and quantitative review of dark fermentative biohydrogen production. *Microbial Cell Factories, 11*, 115.

Rittmann, S. K. R., Lee, H. S., Lim, J. K., Kim, T. W., Lee, J. H., & Kang, S. G. (2014). One-carbon substrate-based biohydrogen production: Microbes, mechanism, and productivity. *Biotechnology Advances, 33*, 165–177.

Roy, S., Vishnuvardhan, M., & Das, D. (2014). Improvement of hydrogen production by newly isolated *Thermoanaerobacterium thermosaccharolyticum* IIT BT-ST1. *International Journal of Hydrogen Energy, 39*, 7541–7552.

Sapra, R., Bagramyan, K., & Adams, M. W. W. (2003). A simple energy-conserving system: Proton reduction coupled to proton translocation. *Proceedings of the National Academy of Sciences of the United States of America, 100*, 7545–7550.

Sato, T., Fukui, T., Atomi, H., & Imanaka, T. (2003). Targeted gene disruption by homologous recombination in the hyperthermophilic archaeon *Thermococcus kodakaraensis* KOD1. *Journal of Bacteriology, 185*, 210–220.

Sawers, G. (1994). The hydrogenases and formate dehydrogenases of *Escherichia coli*. *Antonie Van Leeuwenhoek, 66*, 57–88.

Schicho, R. N., Ma, K., Adams, M. W., & Kelly, R. M. (1993). Bioenergetics of sulfurreduction in the hyperthermophilic archaeon *Pyrococcus furiosus*. *Journal of Bacteriology, 175*, 1823–1830.

Schuchmann, K., & Müller, V. (2012). A bacterial electron bifurcating hydrogenase. *The Journal of Biological Chemistry, 287*, 31165–31171.

Schuchmann, K., & Müller, V. (2013). Direct and reversible hydrogenation of CO_2 to formate by a bacterial carbon dioxide reductase. *Science, 342*, 1382–1385.

Seifert, A. H., Rittmann, S., Bernacchi, S., & Herwig, C. (2013). Method for assessing the impact of emission gasses on physiology and productivity in biological methanogenesis. *Bioresource Technology, 136*, 747–751.

Sharma, S., & Ghoshal, S. K. (2015). Hydrogen the future transportation fuel: From production to applications. *Renewable and Sustainable Energy Reviews, 43*, 1151–1158.

Show, K. Y., Lee, D. J., Tay, J. H., Lin, C. Y., & Chang, J. S. (2012). Biohydrogen production: Current perspectives and the way forward. *International Journal of Hydrogen Energy, 37*, 15616–15631.

Sokolova, T. G., Henstra, A.-M., Sipma, J., Parshina, S. N., Stams, A. J. M., & Lebedinsky, A. V. (2009). Diversity and ecophysiological features of thermophilic carboxydotrophic anaerobes. *FEMS Microbiology Ecology, 68*, 131–141.

Spath, P. L., & Dayton, D. C. (2003). *Preliminary screening - technical and economic assessment of synthesis gas to fuels and chemicals with emphasis on the potential for biomass-derived syngas*. National Renewable Energy Laboratory.

Steinhilper, R., Höff, G., Heider, J., & Murphy, B. J. (2022). Structure of the membrane-bound formate hydrogenlyase complex from *Escherichia coli*. *Nature Communications, 13*, 5395.

Takors, R., Kopf, M., Mampel, J., Bluemke, W., Blombach, B., Eikmanns, B., Bengelsdorf, F. R., Weuster-Botz, D., & Dürre, P. (2018). Using gas mixtures of CO, CO_2 and H_2 as microbial substrates: The do's and don'ts of successful technology transfer from laboratory to production scale. *Microbial Biotechnology, 11*, 606–625.

Techtmann, S. M., Colman, A. S., & Robb, F. T. (2009). "That which does not kill us only makes us stronger": The role of carbon monoxide in thermophilic microbial consortia. *Environmental Microbiology, 11*, 1027–1037.

Thauer, R. K., Kafer, B., Jungermann, K., & Zahringer, M. (1974). The reaction of the iron-sulfur protein hydrogenase with carbon monoxide. *European Journal of Biochemistry, 42*, 447–452.

Thauer, R. K., Jungermann, K., & Decker, K. (1977). Energy conservation in chemotrophic anaerobic bacteria. *Bacteriological Reviews, 41*, 100–180.

Uffen, R. L. (1976). Anaerobic growth of a *Rhodopseudomonas* species in the dark with carbon monoxide as sole carbon and energy substrate. *Proceedings of the National Academy of Sciences of the United States of America, 73*, 3298–3302.

Valentini, F., Marrocchi, A., & Vaccaro, L. (2022). Liquid organic hydrogen carriers (LOHCs) as H-source for bio-derived fuels and additives production. *Advanced Energy Materials, 12*, 2103362.

Verhagen, M. F. J. M., & Adams, M. W. (2001). Fe-only hydrogenase from *Thermotoga maritima*. *Methods in Enzymology, 331*, 216–226.

Wu, M., Ren, Q., Durkin, A. S., Daugherty, S. C., Brinkac, L. M., Dodson, R. J., Madupu, R., Sullivan, S. A., Kolonay, J. F., Haft, D. H., Nelson, W. C., Tallon, L. J., Jones, K. M., Ulrich, L. E., et al. (2005). Life in hot carbon monoxide: The complete genome sequence of *Carboxydothermus hydrogenoformans* Z-2901. *PLoS Genetics, 1*, e65.

Xu, J., Lupeng, L., Minjuan, X., Oger, P., Fengping, W., Jebbar, M., & Xiang, X. (2011). Complete genome sequence of the obligate piezophilic hyperthermophilic archaeon *Pyrococcus yayanosii* CH1. *Journal of Bacteriology, 193*, 4297–4298.

Yang, J.-I., Jung, H.-C., Oh, H.-M., Choi, B. G., Lee, H. S., & Kang, S. G. (2023). $NADP^+$ or CO_2 reduction by *frhAGB*-encoded hydrogenase through interaction with formate dehydrogenase 3 in the hyperthermophilic archaeon *Thermococcus onnurineus* NA1. *Applied and Environmental Microbiology, 89*, e0147423.

Yu, H., Wu, C. H., Schut, G. J., Haja, D. K., Zhao, G., Peters, J. W., Adams, M. W. W., & Li, H. (2018). Structure of an ancient respiratory system. *Cell, 173*, 1636–49.e16.

Yue, M., Lambert, H., Pahon, E., Roche, R., Jemei, S., & Hissel, D. (2021). Hydrogen energy systems: A critical review of technologies, applications, trends and challenges. *Renewable and Sustainable Energy Reviews, 146*, 111180.

Zivanovic, Y., Armengaud, J., Lagorce, A., Leplat, C., Guérin, P., Dutertre, M., Anthouard, V., Forterre, P., Wincker, P., & Confalonieri, F. (2009). Genome analysis and genome-wide proteomics of *Thermococcus gammatolerans*, the most radioresistant organism known amongst the Archaea. *Genome Biology, 10*, R70.

Chapter 16
Development of Chitin-Derived Biohydrogen Production Technology Using Hyperthermophilic Archaea

Tamotsu Kanai

Abstract A biohydrogen platform from swollen chitin was developed using the hyperthermophilic archaeon *Thermococcus kodakarensis*. Combining targeted genetic engineering with adaptive engineering, a strain (KC04ΔtM1) was obtained that efficiently converts swollen chitin to hydrogen (H_2) at 85 °C. This work offers an important step toward the future biohydrogen production using waste biomass.

Keywords Biohydrogen · *Thermococcus kodakarensis* · Chitin · Adaptive engineering · High-temperature fermentation

16.1 Introduction

Bioenergy production is the process to produce energy materials by utilizing the metabolic reactions of living organisms. This process is closely linked to the metabolic pathways of anaerobic microorganisms (Harirchi et al., 2022). Organisms living in aerobic environments, such as humans, can efficiently produce energy in vivo by transferring electrons from organic materials, such as carbohydrates, lipids, and proteins, to oxygen, which is used as a terminal electron acceptor. Since oxygen has a higher electron affinity than other commonly used terminal electron acceptors, the net energy produced by the reaction is relatively high. Anaerobes, on the other hand, can produce less energy because they cannot use oxygen. This means that much energy can be extracted from anaerobic metabolites, such as ethanol, butanol, methane, or hydrogen (H_2), by simply air oxidizing them.

Currently, hydrogen is attracting much attention as a next-generation energy carrier because it can be used in fuel cells to generate electricity with a high energy conversion rate. At present, however, most of hydrogen is produced by steam

T. Kanai (✉)
Biotechnology Research Center and Department of Biotechnology, Toyama Prefectural University, Imizu, Toyama, Japan
e-mail: kanai@pu-toyama.ac.jp

M. Ito et al. (eds.), *Extremophiles for a Sustainable Future*,
https://doi.org/10.1007/978-981-95-6970-0_16

reforming of natural gas ($CH_4 + 2H_2O \rightarrow 4H_2 + CO_2$). Therefore, in order to establish a circular society that uses hydrogen as an energy carrier, it is necessary to develop hydrogen production methods that use renewable energy sources.

Biohydrogen production is one solution to this problem (Cha et al., 2025). Facultative mesophilic anaerobes, such as *Escherichia coli*, are known to produce hydrogen as a byproduct under anaerobic growth conditions. On the other hand, hydrogen production using (hyper)thermophilic anaerobes has several advantages. First, high cultivation temperatures reduce the risk of contamination in the culture vessel when using waste biomass as a source for hydrogen production, allowing a stable hydrogen production. This is especially important since hydrogen is an excellent substrate for methane fermentation; hydrogen production drops significantly if methane fermentation occurs in the same tank. However, since thermophilic methanogens have only been confirmed in limited environments, such as deep-sea hydrothermal vents, their presence in waste biomass is highly unlikely. Therefore, a stable hydrogen production can be expected in high-temperature environments. Furthermore, natural polysaccharides, such as starch, become more soluble at high temperatures. This allows for cultivation under high substrate concentrations and simplifies the pretreatment process.

In this chapter, the high hydrogen production capability of *Thermococcus kodakarensis*, an anaerobic hyperthermophilic archaeon, is described with its successful breeding to assimilate on chitin and produce hydrogen using genetic recombination and adaptive evolution techniques (Aslam et al., 2017).

16.2 Chitin Degradation and Assimilation in *T. kodakarensis* KOD1

Chitin, a linear homopolymer of *N*-acetylglucosamine (GlcNAc) linked by β-1,4 bonds, is abundant in the exoskeletons of crustaceans and insects, as well as in the cell walls of mushrooms. Its annual production on Earth is estimated to be between 10^{10} and 10^{11} tons (Gooday, 1990), making it the second most abundant biomass after cellulose. Like cellulose, chitin is an "unutilized biomass" that is currently discarded. Therefore, it is necessary to develop effective ways to utilize it for establishing a circular economy.

T. kodakarensis KOD1 is a hyperthermophilic archaeon that was isolated from a solfatara near the coast of Kodakara Island in Kagoshima, Japan (Morikawa et al., 1994; Atomi et al., 2004). Its optimal growth temperature is 85 °C. It grows under anaerobic conditions by assimilating various organic materials, such as proteins, amino acids, starch, and pyruvate. Continuous hydrogen production was achieved by continuous cultivation of this archaeon (Kanai et al., 2005). Investigating the hydrogen production rate during stable growth in continuous culture revealed a hydrogen production rate of 56.9 mmol H_2 g-dcw^{-1} h^{-1} per cell, demonstrating that *T. kodakarensis* possesses a relatively high hydrogen production capacity.

Membrane-bound [NiFe]-hydrogenase (MBH) is responsible for hydrogen production, while the soluble cytosolic hydrogenase is responsible for hydrogen assimilation (recycling) (Kanai et al., 2011).

Another notable feature of this archaeon is that *T. kodakarensis* is one of the few hyperthermophiles that can be genetically engineered (Sato et al., 2003, 2005). Practical genetic engineering systems using homologous recombination or shuttle vectors (Santangelo et al., 2008) have been developed. Using this system, a recombinant strain that can highly expresses MBH was developed. This strain exhibited the highest hydrogen production rate per cell ever reported in the literature (120.4 mmol H_2 g-dcw^{-1} h^{-1}) (Kanai et al., 2015).

T. kodakarensis is also unique in that it has a group of genes (TK1110, TK1754-TK1760, TK1764, and TK1765) involved in chitin degradation and assimilation on its genome (Fukui et al., 2005). Biochemical and genetic analyses of this archaeon have provided an overview of its chitin-degrading/assimilating metabolic pathways (Fig. 16.1). First, chitin is cleaved by an extracellular chitinase (ChiA, TK1765) to *N*, *N′*-diacetylchitobiose [$(GlcNAc)_2$], a dimeric form of GlcNAc (Tanaka et al., 1999, 2001). ChiA is a protein with two chitinase domains, each of which has an endo- or exo-type catalytic domain. The generated $(GlcNAc)_2$ is then transported into the cell by a specific ABC transporter (DppABCDF, TK1756-TK1760), which then cleaves it into two molecules of glucosamine (GlcN). This cleavage occurs through a concerted reaction involving diacetylchitobiose deacetylase (Dac, TK1764) and exo-β-D-glucosaminidase (GlmA, TK1754) (Tanaka et al., 2003, 2004). Glucosamine kinase (GlcNK, TK1110) then phosphorylates GlcN to form glucosamine 6-phosphate (GlcN6P) using ADP as a phosphate donor (Aslam et al., 2018). GlcN6P is then deaminized/isomerized by GlcN6P deaminase (GlmD, TK1755) to form fructose 6-phosphate (Tanaka et al., 2005), which is a major metabolite of the glycolytic pathway.

16.3 Molecular Breeding of Chitin Hyper-Degrading and Assimilating Strains of *T. kodakarensis*

Although *T. kodakarensis* KOD1, the wild-type strain, possesses a series of genes necessary for chitin degradation and assimilation, the strain has not been observed to grow on chitin in previous experiments. We hypothesized that this inability may be due to the low expression levels of these enzymes. Therefore, we set out to create a strain with high expression of the chitin-degrading and chitin-assimilating genes using high-expression promoters.

Using a host strain developed for genetic recombination (KU216) (Sato et al., 2005), we first generated a strain with high ChiA expression to enhance chitin degradation (KC01) (Fig. 16.2). In this strain, the *csg* promoter (P_{csg}), which strongly induces transcription, was placed upstream of *chiA*. Next, using the strain KC01, we constructed a daughter strain (KC04) in which two strong promoters ($P_{efl\alpha}$ and P_{gdh})

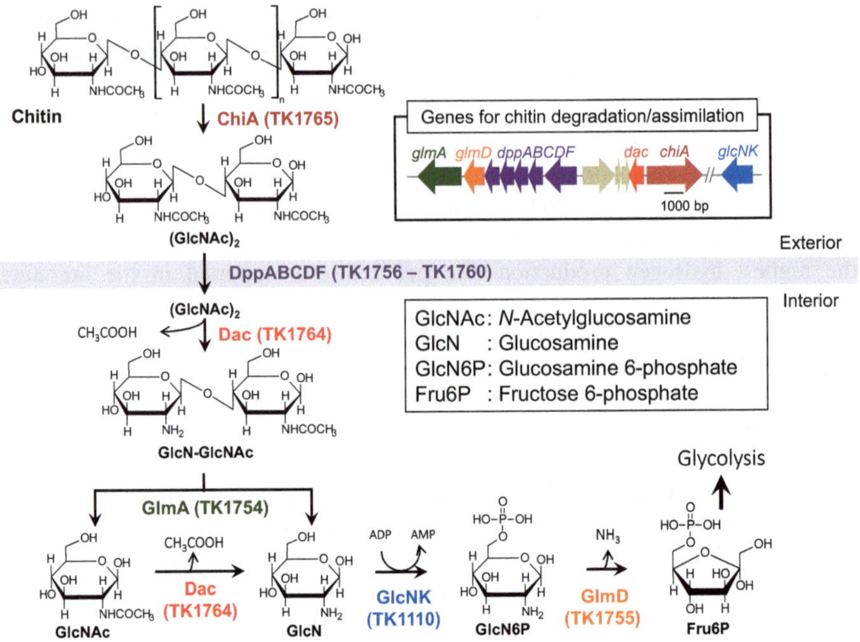

Fig. 16.1 Metabolic pathway of chitin degradation and assimilation in *T. kodakarensis*. Chitin is converted into *N*-, *N′*-diacetylchitobiose, $(GlcNAc)_2$, by extracellular chitinase (ChiA) encoded by the TK1765 gene. An ABC transporter (DppABCDF) encoded by the genes at locus TK1756 to TK1760 may allow $(GlcNAc)_2$ to enter into the cell, where deacetylation of the nonreducing end occurs by diacetylchitobiose deacetylase (Dac) encoded by the TK1764 gene. Exo-β-D-glucosaminidase (GlmA) encoded by the TK1754 gene hydrolyzes the product (GlcN-GlcNAc) to glucosamine (GlcN) and *N*-acetylglucosamine (GlcNAc). Dac further converts GlcNAc to GlcN with the release of one more acetate. In *T. kodakarensis*, there is no homolog for GlcN kinase (GlcNK) that produces glucosamine 6-phosphate (GlcN6P), but this reaction is performed by an ADP-dependent glucokinase encoded by the TK1110 gene (Aslam et al., 2018). GlcN6P is further deaminated, leaving ammonium, by GlcN6P deaminase (GlmD) encoded by the TK1755 gene. The fructose 6-phosphate (F6P) formed in this reaction further enters into the glycolytic pathway

were introduced separately upstream of the *dac* and *dppABCDF-glmDA* operons, respectively. Furthermore, a gene disruption strain of Tgr was generated from the KC04 strain to create the KC04Δt strain. Since Tgr is a transcriptional repressor of the glycolytic pathway (Kanai et al., 2007), this gene disruption was expected to increase the intracellular flow of glycolytic pathways. Therefore, at that time point, the KC04Δt strain was recognized as an ideal strain for chitin degradation and assimilation, with increased expression of all related enzymes.

These strains were incubated at 85 °C in a medium containing 0.4% (w/v) swollen chitin (Fig. 16.3). Comparing the chitin degradation of each strain revealed that all recombinant strains degraded more chitin than the host strain after 48 h of cultivation. However, even after 90 h of cultivation, the amount of chitin remaining in the culture medium did not change, and complete degradation was not achieved. By

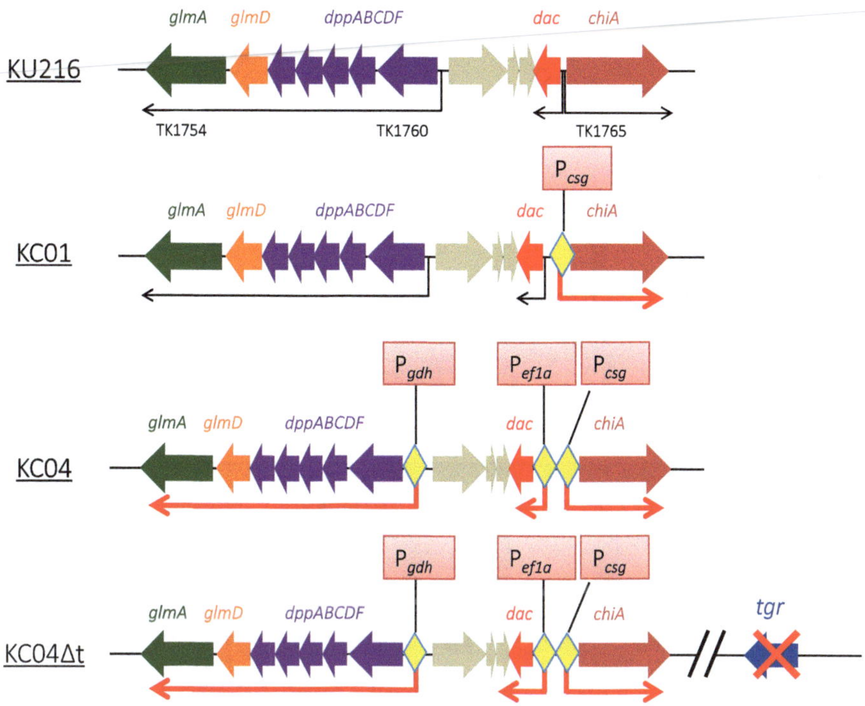

Fig. 16.2 Molecular breeding of *T. kodakarensis* KC04Δt strain. KU216 is a *T. kodakarensis* host strain expressing GlmA, GlmD, ABC transporter, Dac, and ChiA with their native promoters. In the strain KC01, the native promoter of *chiA* was replaced with a strong promoter, P_{csg}. KC01 was further modified to produce strain KC04 by placing promoters of an operon (*dppABCDF-glmDA*) and the *dac* gene with other strong promoters of P_{gdh} and $P_{ef1\text{-}\alpha}$, respectively. Strain KC04 was further modified to produce strain KC04Δt by deleting *tgr*, which is involved in transcriptional repression of glycolytic enzyme-encoding genes

contrast, browning of the culture medium appeared beginning at 72 h after cultivation, particularly in the strain KC04.

To confirm the presence of chitin degradation products, the supernatants of the culture medium were analyzed for these strains by thin-layer chromatography (TLC). The results showed the accumulation of $(GlcNAc)_2$ and GlcNAc in strain KC01 (Fig. 16.4). In addition to $(GlcNAc)_2$ and GlcNAc, the KC04 and KC04Δt strains showed accumulation of GlcN. Since GlcN is a monosaccharide with an amino group, its accumulation in a high-temperature environment was expected to cause the Maillard reaction, resulting in the browning of the medium.

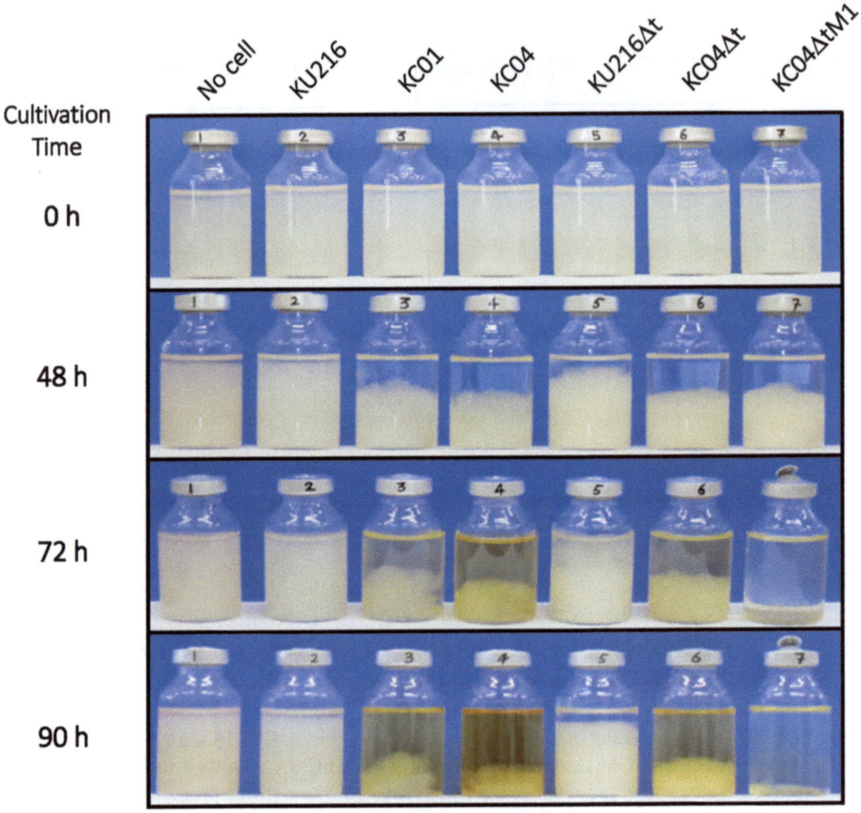

Fig. 16.3 Degradation of swollen chitin by engineered strains of *T. kodakarensis*. Cultivations were performed at 85 °C, and degradation of 0.4% (w/v) swollen chitin was observed at different time points until 90 h. An increased capacity to degrade swollen chitin was observed in strain KC04ΔtM1

16.4 Breeding of a Chitin-Hyper-Degrading and Chitin-Hyper-Assimilating Strain of *T. kodakarensis* Using the Adaptive Engineering Methods

Previous analyses have suggested that suppressing GlcN accumulation in *T. kodakarensis* is important to develop a strain that can degrade and assimilate chitin efficiently. However, it would be difficult in a non-model organism like *T. kodakarensis* to use metabolic engineering techniques to optimize the balance between GlcN production and consumption. Therefore, we employed an adaptive engineering (evolution) approach for the KC04Δt strain. This approach is based on the spontaneous evolution of living cells when they encounter severe growth defects. After repeated cultivation of a parent strain in a medium containing swollen chitin, adapted

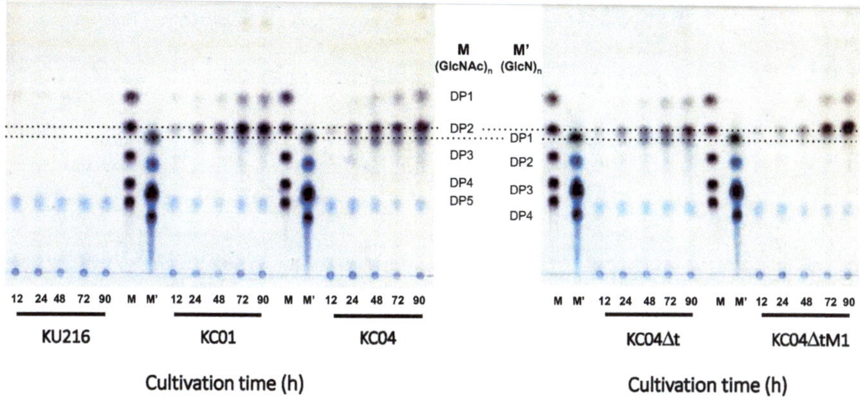

Fig. 16.4 TLC analysis of supernatants after cultivation of engineered *T. kodakarensis* strains. Chitin degradation products in culture supernatant were analyzed at different time points until 90 h. Lane M, standard GlcNAc oligomers ranging from DP1 to DP5 and standard GlcN oligomers ranging from DP1 to DP4, where DP indicates degree of polymerization. Culture supernatants of KC04ΔtM1 cells displayed increased accumulation of $(GlcNAc)_2$ with very little accumulation of GlcN

daughter strains with a high chitin-assimilating capability, thus showing a higher growth rate, appear, and they are enriched in the culture. After eight successive cultures of the KC04Δt strain in a medium containing swollen chitin, complete degradation of the chitin was achieved, suggesting the presence of evolved strains with improved chitin degradation and assimilation capabilities. One of the strains (KC04ΔtM1) was isolated from the culture by colony formation.

Culturing this strain in a medium containing swollen chitin resulted in nearly complete degradation of chitin after 72 h with little change in the color of the medium (Fig. 16.3). TLC analysis showed increased accumulation of $(GlcNAc)_2$, while no accumulation of GlcN was observed (Fig. 16.4). Furthermore, the accumulation of acetic acid and ammonium—metabolic byproducts of chitin deacetylation and deamination, respectively—was observed in the culture supernatant (data not shown).

Next, the hydrogen production capacity of the KC04ΔtM1 strain was examined (Fig. 16.5). In the host strain, hydrogen accumulation was constant with or without the addition of swollen chitin. On the other hand, the KC04ΔtM1 strain accumulated a significantly higher concentration of hydrogen (approximately ten-fold) in the medium containing swollen chitin than the host strain, while the amount of hydrogen accumulated in the medium without swollen chitin was similar to that of the host strain. These results suggest that nearly all of the hydrogen produced by the KC04ΔtM1 strain is produced through the degradation and assimilation of swollen chitin.

The genome of the strain KC04ΔtM1 was sequenced to investigate the cause of its high chitin assimilation capability. A 15-bp sequence (5′-ATCGAAAGGTTTATA-3′) containing a BRE/TATA sequence was found to be

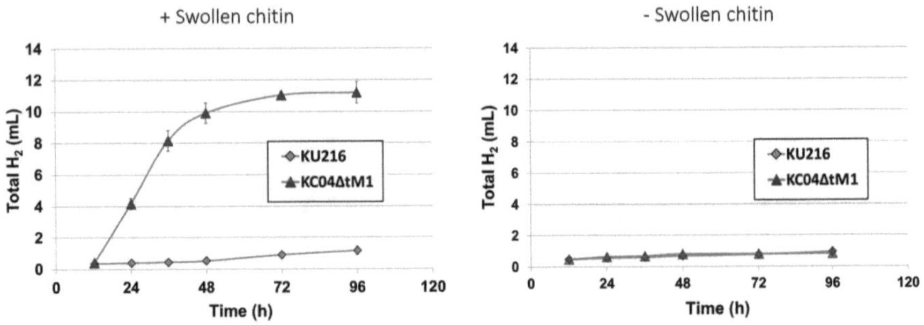

Fig. 16.5 Comparison of chitin-dependent hydrogen (H_2) production in KU216 and KC04ΔtM1 in medium with (**left**) or without (**right**) the addition of 0.4% (w/v) swollen chitin. KU216 is shown in white diamonds, while KC04ΔtM1 is shown in black triangles. Total hydrogen (mmol L^{-1}) represents the amount of molecular hydrogen present in the headspace of a culture bottle per mL of culture volume (media volume was 15 mL). Error bars represent standard deviations. The results show that strain KC04ΔtM1 exhibits hydrogen production dependent on swollen chitin

duplicated within the *gdh* promoter (P_{gdh}), which was introduced to overexpress the *dppABCDF-glmDA* operon. The expression levels of enzymes related to chitin degradation were determined by Western blot analysis. As a result, the levels of GlmD and GlmA, whose genes are under the control of P_{gdh}, were found to be decreased in the KC04ΔtM1 strain, while the levels of ChiA and Dac did not change (data not shown).

16.5 Discussion

The *T. kodakarensis* KC04ΔtM1 strain, which can produce high levels of hydrogen through the degradation and assimilation of chitin, was obtained through genetic engineering followed by adaptive engineering. While the genetically engineered KC04 and KC04Δt strains exhibited enhanced chitin degradation compared to the host strain, accumulation of GlcN in the culture medium promoted the Maillard reaction. The byproducts of the Maillard reaction are known to suppress cell growth (Helou et al., 2014). In contrast, the accumulation of GlcN did not occur in the KC04ΔtM1 strain, which suppressed the Maillard reaction. This led to efficient assimilation of chitin, resulting in growth and increased hydrogen production. The expected scenario for suppressing GlcN accumulation in strain KC04ΔtM1 is as follows. In this strain, the expression levels of GlmD and GlmA were significantly decreased. Therefore, the expression levels of DppABCDF are also expected to decrease, since the genes are encoded in the same operon. Since GlmA is an enzyme involved in the GlcN production, and DppABCDF is involved in supplying $(GlcNAc)_2$, which is eventually converted into GlcN, a decrease in the levels of these proteins would reduce the rate of GlcN production. The mutation is expected to balance the production and consumption rates of GlcN, contributing to reduced GlcN accumulation. Since the irreversible Maillard reaction causes the loss of carbon, nitrogen, and energy sources, the KC04ΔtM1 strain obtained by adaptive evolution can efficiently degrade and assimilate chitin by maintaining low GlcN accumulation.

For future applications of chitin-dependent biohydrogen production, it is essential to develop a system that can effectively utilize natural chitinous substances such as those found in crab and shrimp shells. While these materials contain chitin fibers, they are covered by proteins and calcium bicarbonate. Therefore, these compounds must be solubilized and/or broken down to obtain chitin fibers. As the genome of *T. kodakarensis* encodes numerous proteolytic enzymes (Fukui et al., 2005), further breeding of this strain through genetic and/or adaptive engineering could facilitate the development of processing methods for natural chitinous materials.

Acknowledgments I would like to express my sincere gratitude to Professor Haruyuki Atomi, Ms. Ayumi Horiuchi, Dr. Mehwish Aslam, and the other members of the Atomi Laboratory in the Department of Synthetic Chemistry and Biological Chemistry, Graduate School of Engineering, Kyoto University, Japan, who initiated this work.

References

Aslam, M., Horiuchi, A., Simons, J.-R., Jha, S., Yamada, M., Odani, T., Fujimoto, R., Yamamoto, Y., Gunji, R., Imanaka, T., Kanai, T., & Atomi, H. (2017). Engineering of a hyperthermophilic archaeon, *Thermococcus kodakarensis* for chitin-dependent hydrogen production. *Applied and Environmental Microbiology, 83*, e00280–e00217.

Aslam, M., Takahashi, N., Matsubara, K., Imanaka, T., Kanai, T., & Atomi, H. (2018). Identification of the glucosamine kinase in the chitinolytic pathway of *Thermococcus kodakarensis*. *Journal of Bioscience and Bioengineering, 125*, 320–326.

Atomi, H., Fukui, T., Kanai, T., Morikawa, M., & Imanaka, T. (2004). Description of *Thermococcus kodakaraensis* sp. nov., a well studied hyperthermophilic archaeon previously reported as *Pyrococcus* sp. KOD1. *Archaea, 1*, 263–267.

Cha, M., Park, M.-S., & Kim, S.-J. (2025). Biological routes for biohydrogen production: A clean and carbon-free fuel. *Biotechnology Journal, 20*, e70074.

Fukui, T., Atomi, H., Kanai, T., Matsumi, R., Fujiwara, S., & Imanaka, T. (2005). Complete genome sequence of the hyperthermophilic archaeon *Thermococcus kodakaraensis* KOD1 and comparison with *Pyrococcus* genomes. *Genome Research, 15*, 352–363.

Gooday, G. W. (1990). The ecology of chitin degradation. *Advances in Microbial Ecology, 11*, 387–430.

Harirchi, S., Wainaina, S., Sar, T., Nojoumi, S. A., Parchami, M., Parchami, M., Varjani, S., Khanal, S. K., Wong, J., Awasthi, M. K., & Taherzadeh, M. J. (2022). Microbiological insights into anaerobic digestion for biogas, hydrogen or volatile fatty acids (VFAs): A review. *Bioengineered, 13*, 6521–6557.

Helou, C., Marier, D., Jacolot, P., Abdennebi-Najar, L., Niquet-Léridon, C., Tessier, F. J., & Gadonna-Widehem, P. (2014). Microorganisms and Maillard reaction products: A review of the literature and recent findings. *Amino Acids, 46*, 267–277.

Kanai, T., Imanaka, H., Nakajima, A., Uwamori, K., Omori, Y., Fukui, T., Atomi, H., & Imanaka, T. (2005). Continuous hydrogen production by the hyperthermophilic archaeon, *Thermococcus kodakarensis* KOD1. *Journal of Biotechnology, 116*, 271–282.

Kanai, T., Akerboom, J., Takedomi, S., van de Werken, H. J. M., Blombach, F., van der Oost, J., Murakami, T., Atomi, H., & Imanaka, T. (2007). A global transcriptional regulator in *Thermococcus kodakaraensis* controls the expression levels of both glycolytic and gluconeogenic enzyme-encoding genes. *Journal of Biological Chemistry, 282*, 33659–33670.

Kanai, T., Matsuoka, R., Beppu, H., Nakajima, A., Okada, Y., Atomi, H., & Imanaka, T. (2011). Distinct physiological roles of the three [NiFe]-Hydrogenase orthologs in the hyperthermophilic archaeon *Thermococcus kodakarensis*. *Journal of Bacteriology, 193*, 3109–3116.

Kanai, T., Simons, J.-R., Tsukamoto, R., Nakajima, A., Omori, Y., Matsuoka, R., Beppu, H., Imanaka, T., & Atomi, H. (2015). Overproduction of the membrane-bound [NiFe]-hydrogenase in *Thermococcus kodakarensis* and its effect on hydrogen production. *Frontiers in Microbiology, 6*, 847.

Morikawa, M., Izawa, Y., Rashid, N., Hoaki, T., & Imanaka, T. (1994). Purification and characterization of a thermostable thiol protease from a newly isolated hyperthermophilic *Pyrococcus* sp. *Applied and Environmental Microbiology, 60*, 4559–4566.

Santangelo, T. J., Cubonovà, L., & Reeve, J. (2008). Shuttle vector expression in *Thermococcus kodakaraensis*: Contributions of *cis* elements to protein synthesis in a hyperthermophilic archaeon. *Applied and Environmental Microbiology, 74*, 3099–3104.

Sato, T., Fukui, T., Atomi, H., & Imanaka, T. (2003). Targeted gene disruption by homologous recombination in the hyperthermophilic archaeon *Thermococcus kodakaraensis* KOD1. *Journal of Bacteriology, 185*, 210–220.

Sato, T., Fukui, T., Atomi, H., & Imanaka, T. (2005). Improved and versatile transformation system allowing multiple genetic manipulations of the hyperthermophilic archaeon *Thermococcus kodakarensis*. *Applied and Environmental Microbiology, 71*, 3889–3899.

Tanaka, T., Fujiwara, S., Nishikori, S., Fukui, T., Takagi, M., & Imanaka, T. (1999). A unique chitinase with dual active sites and triple substrate binding sites from the hyperthermophilic archaeon *Pyrococcus kodakaraensis* KOD1. *Applied and Environmental Microbiology, 65*, 5338–5344.

Tanaka, T., Fukui, T., & Imanaka, T. (2001). Different cleavage specificities of the dual catalytic domains in chitinase from the hyperthermophilic archaeon *Thermococcus kodakaraensis* KOD1. *Journal of Biological Chemistry, 276*, 35629–35635.

Tanaka, T., Fukui, T., Atomi, H., & Imanaka, T. (2003). Characterization of an exo-β-D-glucosaminidase involved in a novel chitinolytic pathway from the hyperthermophilic archaeon *Thermococcus kodakaraensis* KOD1. *Journal of Bacteriology, 185*, 5175–5181.

Tanaka, T., Fukui, T., Fujiwara, S., Atomi, H., & Imanaka, T. (2004). Concerted action of diacetylchitobiose deacetylase and exo-β-D-glucosaminidase in a novel chitinolytic pathway in the hyperthermophilic archaeon *Thermococcus kodakaraensis* KOD1. *Journal of Biological Chemistry, 279*, 30021–30027.

Tanaka, T., Takahashi, F., Fukui, T., Fujiwara, S., Atomi, H., & Imanaka, T. (2005). Characterization of a novel glucosamine-6-phosphate deaminase from a hyperthermophilic archaeon. *Journal of Bacteriology, 187*, 7038–7044.

Part VI
Next-Generation Biotechnologies for a Sustainable Future

Chapter 17
Bioleaching and Mining Application

Satoshi Wakai

Abstract Extremophiles exhibit remarkable adaptations that enable survival in harsh environments, among which acidophiles have some industrial potential. Acidophiles thrive in low-pH environments and possess unique metabolic and enzymatic traits with high industrial relevance. Their ability to oxidize iron and sulfur compounds forms the basis of bioleaching, a technology applied to recover valuable metals from sulfide ores and increasingly from electronic waste. Heap leaching has been widely used for large-scale copper recovery, while tank leaching enables controlled processing of gold, nickel, and other metals. In recent years, bioleaching has expanded beyond natural ores to "urban mines," offering sustainable solutions for recycling critical elements such as cobalt, rare earths, and lithium from electronic waste. These developments demonstrate the versatility of acidophiles in supporting both primary and secondary resource recovery. Despite these advances, the application of bioleaching to marine mineral resources remains limited. Current acidophiles often fail to tolerate high salinity, underscoring the need to discover and characterize marine-adapted iron- and sulfur-oxidizers. Future progress will depend on integrating microbiology, biotechnology, and environmental considerations to expand biohydrometallurgy while minimizing ecological impact.

Keywords Extremophile · Acidophiles · Bioleaching · Metal · Sulfur oxidation

---

S. Wakai (✉)
Institute for Extra-cutting-edge Science and Technology Avant-garde Research (X-star), Japan Agency for Marine-Earth Science and Technology (JAMSTEC), Yokosuka, Japan

Graduate School of Engineering, Tokyo University of Agriculture and Technology (TUAT), Tokyo, Japan

Graduate School of Engineering, Kobe University, Kobe, Japan
e-mail: wakais@jamstec.go.jp

M. Ito et al. (eds.), *Extremophiles for a Sustainable Future*,
https://doi.org/10.1007/978-981-95-6970-0_17

17.1 Introduction

Life in extreme environments poses severe challenges for humans, who require protective equipment to survive under conditions such as high temperatures or high pressures. In contrast, diverse microorganisms are able to thrive in such habitats without external support, relying instead on intrinsic adaptive strategies (Cowan et al., 2024; Rekadwad et al., 2023; Wakai & Sambongi, 2018; Wakai, 2019). These adaptations allow survival across a wide range of physicochemical extremes, including high or low temperature, high pressure, salinity, acidity, and alkalinity, as well as in environments where multiple stresses occur simultaneously (Takai et al., 2008; Schleper et al., 1995b; Harrison et al., 2013). For example, deep-sea hydrothermal vents combine high temperature, elevated pressure, and high salinity, each of which demands specialized mechanisms of adaptation. While the adaptive strategies of extremophiles and their enzymes have been extensively reviewed elsewhere, the present chapter focuses on acidophilic microorganisms.

Acidophiles are of particular industrial interest due to their central role in bioleaching, a process that enables the recovery of valuable metals from sulfide minerals (Jones & Santini, 2023; Roberto & Schippers, 2022; Vera et al., 2022). Bioleaching has been widely applied in the extraction of copper from low-grade ores, with several large-scale industrial operations now in place. More recently, attention has shifted from conventional ore deposits to so-called "urban mines," where electronic waste and mobile devices represent new sources of valuable metals (Jaiswal et al., 2025; Vítová & Mezricky, 2024; Golzar-Ahmadi et al., 2024; Li et al., 2025). Unlike ore-based operations, which are constrained to mining regions, urban mining can be implemented in locations where electronic devices are concentrated, effectively expanding resource recovery into the sphere of daily human life. Given the global dependence on mobile technologies, demand for such recovery methods is expected to increase worldwide. Accordingly, this chapter highlights recent insights into the environmental adaptations of acidophilic microorganisms and their enzymes, the role of these organisms in bioleaching, and emerging technologies that extend metal recovery from natural ores to urban mines.

17.2 Extremophilic Environments and Microorganisms

17.2.1 Extreme Environments on the Earth

The Earth contains a very diverse range of extreme environments. Extreme environments on Earth include high-temperature sites such as deep-sea hydrothermal vents, hot springs, and mantles, where liquid water can exist at temperatures exceeding 400 °C under high pressure (Bischoff & Rosenbauer, 1988; Koschinsky et al., 2008). In contrast, low-temperature environments are represented by polar regions, alpine regions, and the deep sea, and liquid water is present at circa- or below-zero degrees Celsius. The deep sea also presents an extreme hydrostatic pressure environment, with the Challenger Deep reaching 110 MPa, while microbial growth has been observed at up to 140 MPa (Kato et al., 1997; Kusube et al., 2017). In terms of salinity, hypersaline lakes and salterns contain saturated NaCl conditions above 20%, and some extreme sites also feature high concentrations of $MgCl_2$, as in the Discovery Basin of the Mediterranean Sea (van der Wielen et al., 2005).

Extreme chemical conditions also occur in the form of strongly acidic and alkaline habitats. Extremely acidic environments, with pH values close to zero or below, are found in hot springs, solfataric fields, and acid mine drainage sites, while strongly alkaline waters, with pH above 12, occur in environments such as hyperalkaline lakes and ultradeep mines (Schleper et al., 1995a; Czop et al., 2011). Beyond single extremes, complex extreme environments combine multiple physicochemical stresses such as high temperature and pressure or heat and acidity, creating habitats where only polyextremophiles can persist (Takai et al., 2008; Schleper et al., 1995b; Harrison et al., 2013). These environments define the outer boundaries of the physicochemical conditions under which life can exist on Earth.

17.2.2 Acidophiles Living in Extreme Environments

Strongly acidic environments are widespread both in natural and anthropogenic settings. Natural acidic habitats include solfataric fields and acidic hot springs, where volcanic activity and sulfur oxidation generate pH values close to zero (Johnson & Hallberg, 2003; Brock, 1978). Acid mine drainage (AMD), one of the best-studied anthropogenic acidic environments, arises from the microbial oxidation of sulfide minerals, leading to waters with extremely low pH and high concentrations of dissolved metals such as Fe^{2+}/Fe^{3+}, Al^{3+}, and heavy metals (Nordstrom & Southam, 1997; Johnson & Hallberg, 2005). Such AMD systems are globally distributed at mining sites and represent ecologically significant examples of microbially driven geochemical alteration. In addition, acidic rivers, such as the Río Tinto in Spain (Fig. 17.1), provide large-scale natural analogues of AMD, where ferric ion and sulfur compounds maintain persistent acidity (González-Toril et al., 2003; Amils et al., 2007). These environments are not only harsh due to low pH but also because of the combined stress of elevated temperature, toxic metals, and limited nutrient availability, making them crucial natural laboratories for studying the limits of microbial life and biogeochemical cycling.

Acidophiles inhabiting these environments exhibit remarkable physiological and biochemical strategies that enable survival and growth under extreme acidity. Representative taxa include archaea such as *Picrophilus torridus*, which can grow optimally at pH values close to 0, and members of *Sulfolobus* and *Ferroplasma*, which are frequently associated with hot, sulfur-rich acidic springs (Schleper et al., 1995a, b; Schleper et al., 1996; Golyshina, 2011). Among bacteria, *Acidithiobacillus* species are of particular importance due to their ability to oxidize ferrous ions and reduced sulfur compounds, processes that not only sustain primary production in

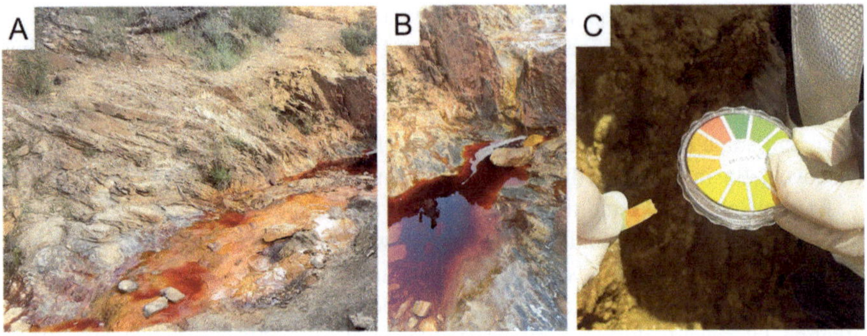

Fig. 17.1 Representative features of an extremely acidic river environment. (**a**) Flowing section of the river showing continuous discharge of acidic water. (**b**) Stagnant pool exhibiting intense reddish coloration due to high ferrous ion concentrations. (**c**) Field measurement of water pH using indicator test strips, confirming strongly acidic conditions (approximately pH 1~2)

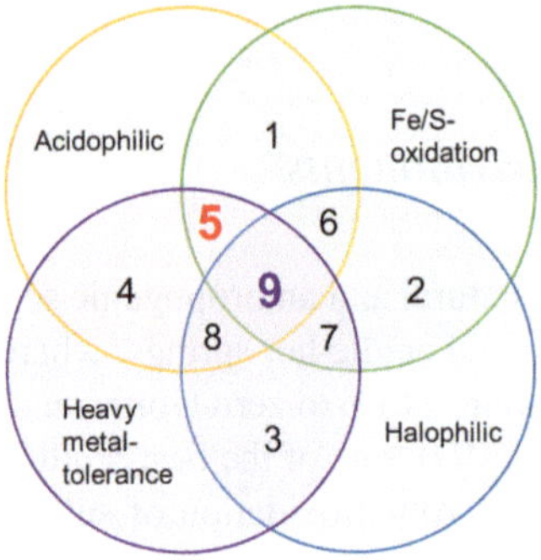

#	Category
1	Acidophilic Fe/S-oxidizing microbes
2	Halophilic Fe/S-oxidizing microbes
3	Halophile [heavy metal-tolerance]
4	Halophilic acidophile
5	Acidophilic Fe/S-oxidizing microbes [heavy metal-tolerance]
6	Halophilic, acidophilic Fe/S-oxidizing microbes
7	Halophilic Fe/S-oxidizing microbes [heavy metal-tolerance]
8	Halophilic acidophile [heavy metal-tolerance]
9	Halophilic, acidophilic Fe/S-oxidizing microbes [heavy metal-tolerance]

5 → Suitable category for bioleaching of onshore mineral resources

9 → Suitable category for bioleaching of offshore mineral resources

Fig. 17.2 Required properties for the bioleaching of onshore and offshore mineral resources

these ecosystems but also underpin industrial bioleaching operations (Johnson, 1998; Rawlings, 2002) (Fig. 17.2). Many acidophiles maintain intracellular near-neutral pH through impermeable membranes, active proton efflux systems, and specialized proteins that remain stable under acidic and metal-rich conditions (Baker-Austin & Dopson, 2007; Krulwich et al., 2011). These adaptations make acidophiles invaluable as model organisms for studying the molecular limits of life and as reservoirs of acid-stable enzymes with potential applications in biotechnology, from mineral processing to the development of acid-tolerant industrial biocatalysts.

17.2.3 Acidophilic Enzymes and Their Adaptation

Little is known regarding the adaptation mechanisms of acidophilic enzymes compared with those of thermophilic, piezophilic, and psychrophilic enzymes. Because acidophiles maintain neutral intracellular pH (Horikoshi, 1999; Sharma et al., 2012), intracellular enzymes do not need to adapt to acidic conditions. Conversely, extracellular enzymes must adapt to such pH extremes. Therefore, almost all studies have reported findings regarding enzyme function on extracellular substrates, large molecules such as starch, cellulose, and protein, and insoluble or nonpermeable substrates (Sarethy et al., 2011; Sharma et al., 2012).

The remarkable acid tolerance of enzymes produced by acidophiles arises from a combination of molecular adaptations that counteract protein denaturation under low-pH conditions (Huang et al., 2005; Kanao et al., 2010). A common strategy is the enrichment of acidic amino acid residues on the protein surface, which enhances solubility and minimizes disruptive protonation effects (Huang et al., 2005). In addition, acid-stable enzymes often display compact tertiary structures stabilized by dense hydrogen-bonding networks, increased numbers of salt bridges, and enhanced hydrophobic packing, all of which reduce the likelihood of denaturation in proton-rich environments (Feller, 2010; Reed et al., 2013). Some proteins also show modified isoelectric points that maintain favorable charge distributions, thereby supporting structural integrity and catalytic activity at low pH (Kumar & Nussinov, 2002). Furthermore, the presence of specialized molecular chaperones and protein-folding systems in acidophiles helps maintain correct conformation under stress conditions (Macario & Conway de Macario, 2007; Baker-Austin & Dopson, 2007). Collectively, these strategies not only safeguard enzymatic function in extreme acidity but also offer valuable design principles for protein engineering aimed at producing industrial biocatalysts capable of operating efficiently in harsh process environments.

Acidophiles produce a range of acid-stable enzymes that remain active under conditions where most conventional proteins denature. Examples include acidophilic β-glycosidases and proteases isolated from *Acidithiobacillus* species, which function efficiently at pH values below 3 (Rohwerder & Sand, 2007), and metallo-chaperones and oxidoreductases from *Sulfolobus* and *Ferroplasma*, which maintain structural integrity despite the presence of high proton concentrations and dissolved metals (Golyshina, 2011; Auernik & Kelly, 2010). The stability of these enzymes is supported by adaptive strategies such as an increased proportion of acidic amino acids on protein surfaces, enhanced intramolecular bonding networks, and chaperone systems that prevent misfolding in low-pH conditions (Dopson et al., 2005; Krulwich et al., 2011). From an industrial perspective, acid-stable enzymes are highly attractive for processes that require operation under acidic environments, including biomass hydrolysis in biofuel production, food and beverage processing, and the development of detergents and pharmaceuticals that must retain activity in gastric or other acidic conditions. Consequently, research on acidophilic microorganisms and their enzymes not only enhances our understanding of protein stability at the molecular level but also drives innovation in biotechnological applications where resilience to extreme acidity is essential.

17.3 Bioleaching

Heap leaching and tank leaching are both methods of bioleaching used to recover metals, but they differ significantly in reactor configuration, processing time, cost structure, and ore preparation (Li et al., 2025; Roberto & Schippers, 2022; Shiers et al., 2016; Panda et al., 2015; Schippers et al., 2014; Zhou et al., 2009) (Table 17.1). In heap leaching, crushed ore is piled onto an impermeable pad (heap), and leaching solution (often acidic, possibly with added microbes) is percolated through the heap; solids generally remain static, and leaching takes place over weeks to months, the process being relatively simple and low in capital cost. Tank leaching, by contrast, involves finely ground ore suspended in large stirred or agitated reactors (slurry), which allows for more controlled conditions and faster leaching kinetics but higher energy input and greater infrastructure investment. The choice between them depends on ore grade, mineralogy, particle size, extraction speed needed, and economic constraints—with heap leaching being more suited for low-grade, large-volume ores where slower recovery is acceptable, and tank leaching for higher-grade ores or when time, yield, and process control are more critical.

Table 17.1 Comparison of heap leaching and tank leaching

Feature	Heap Leaching	Tank Leaching
Process configuration	Crushed ore is stacked on an impermeable pad; acidic solution percolates through the heap	Finely ground ore is suspended in stirred tanks with acidic solution and microorganisms
Ore particle size	Coarsely crushed (several cm)	Finely ground (hundreds of μm)
Leaching time	Weeks to months	Days to weeks
Process control	Strongly influenced by external factors (e.g., rainfall, temperature); difficult to regulate	Precise control of aeration, temperature, and mixing
Metal recovery efficiency	Generally low to moderate	High recovery rates are achievable
Suitable ore type	Low-grade, large-volume sulfide ores	Higher-grade ores or cases requiring faster recovery
Capital cost	Relatively low; simple infrastructure suitable for large-scale operation	High; requires large reactors and aeration systems
Energy consumption	Low	High
Typical applications	Copper recovery from low-grade ores (e.g., Chilean copper mines)	Intensive leaching of gold, copper, uranium

17.3.1 Traditional Bioleaching of Sulfide Ore—Heap Leaching

Heap leaching has been the most widely applied bioleaching technology in industrial practice, particularly for the recovery of copper from low-grade sulfide ores. Large-scale operations, such as those in Chile, demonstrate its economic importance and feasibility for extracting copper directly from ore that would otherwise be uneconomical to process by conventional methods. The efficiency of heap leaching relies on a combination of abiotic chemical reactions and biotic microbial activity: ferric ion and sulfuric acid chemically oxidize sulfide minerals, while acidophilic microorganisms, such as *Acidithiobacillus* spp. and *Leptospirillum* spp., regenerate ferric ion and sulfuric acid through the oxidation of ferrous ion and reduced sulfur compounds (Jones & Santini, 2023; Rawlings & Johnson, 2007; Vera et al., 2022). This interplay between chemical and microbial processes sustains leaching reactions over long periods, enabling continuous dissolution of metal sulfides within the ore heaps.

The mechanism of heap leaching is driven by a coupled cycle of abiotic chemical oxidation and microbial catalysis that enables sustained dissolution of sulfide minerals. In the case of copper sulfide ores such as chalcopyrite ($CuFeS_2$), ferric ion (Fe^{3+}) and protons (H^+) chemically attack the mineral lattice, releasing copper ions into solution while producing ferrous ion (Fe^{2+}) and elemental sulfur. A typical reaction (Eq. 17.1) can be expressed as:

$$\mathrm{CuFeS_2} + 4\,Fe^{3+} \rightarrow Cu^{2+} + 5\,Fe^{2+} + 2\,\mathrm{S}^0 \tag{17.1}$$

Left unchecked, this reaction would halt as ferric ion is depleted; however, acidophilic microorganisms such as *A. ferrooxidans* and *Leptospirillum* spp. regenerate Fe^{3+} by oxidizing Fe^{2+} under aerobic conditions (Eq. 17.2).

$$4\,Fe^{2+} + \mathrm{O_2} + 4\,\mathrm{H^+} \rightarrow 4\,Fe^{3+} + 2\,\mathrm{H_2O} \tag{17.2}$$

Similarly, reduced sulfur compounds and elemental sulfur are oxidized to sulfuric acid (Eq. 17.3) by sulfur-oxidizing acidophiles, maintaining the acidic conditions required for leaching.

$$2\,\mathrm{S}^0 + 3\,\mathrm{O_2} + 2\,\mathrm{H_2O} \rightarrow 2\,\mathrm{H_2}SO_4 \tag{17.3}$$

This integrated cycle ensures the continuous supply of oxidants and protons, thereby sustaining the dissolution of sulfide minerals over extended periods in the heap environment.

17.3.2 Metabolism of Model Microorganisms

A. ferrooxidans is a chemolithoautotroph that gains energy primarily by oxidizing ferrous ion to ferric ion, a metabolism central to its role in sulfide mineral dissolution during bioleaching (Valdés et al., 2008; Quatrini & Johnson, 2009). The electron acquired from the oxidation of the ferrous ion is transferred across the cell envelope through a periplasmic-membrane electron transfer chain that classically involves outer-membrane redox proteins, the periplasmic copper protein rusticyanin, periplasmic cytochrome, and inner-membrane cytochrome *c* oxidase that ultimately reduce O_2 at terminal oxidases (Fig. 17.3); this electron flow is coupled to proton translocation and ATP synthesis, enabling carbon fixation through the Calvin–Benson–Bassham cycle (Quatrini & Johnson, 2009; Valdés et al., 2008; Blake & Shute, 1994; Appia-Ayme et al., 1999; Campodonico et al., 2016). The *rus* operon (encoding rusticyanin and associated *c*-type cytochromes) and related gene clusters are inducible by ferrous ion and form a "respiratory supercomplex" adapted to function at very low pH and in the presence of high metal loads (Yarzábal et al., 2004; Castelle et al., 2008). This iron-oxidation pathway not only provides metabolic energy but also regenerates the ferric oxidant that chemically attacks sulfide minerals in heaps, linking microbial physiology directly to mineral dissolution kinetics.

Sulfur oxidation in *A. ferrooxidans* is organized as a multi-branched network capable of oxidizing a variety of reduced inorganic sulfur compounds (RISCs), including sulfide, elemental sulfur, thiosulfate, tetrathionate, and sulfite (Wang et al., 2019; Ibáñez et al., 2023) (Fig. 17.3). A pivotal advance in understanding this network was the identification and characterization of sulfide:quinone oxidoreductase (SQR) in *A. ferrooxidans* (Wakai et al., 2004, 2007), which demonstrated that sulfide can be directly oxidized at the membrane and linked to the quinone pool. This discovery was significant because it provided a molecular explanation for the initial step of sulfide oxidation in acidophiles, establishing SQR as a key entry point for electron flow from sulfur compounds into the respiratory chain. In bioleaching systems, SQR activity enables the efficient coupling of sulfide oxidation with energy conservation and acid generation, processes that are central to sustaining microbial activity under strongly acidic and metal-rich conditions. The recognition of SQR's centrality also provided a molecular marker to trace sulfur oxidation capabilities across acidophilic communities in bioleaching heaps.

Beyond the biochemical identification of SQR, the examination of transcriptional responses of *A. ferrooxidans* to sulfur availability revealed that genes associated with sulfur oxidation are strongly upregulated in the presence of elemental sulfur and related substrates (Wakai et al., 2007). This regulatory plasticity indicates that *A. ferrooxidans* dynamically reallocates metabolic resources depending on the sulfur speciation in its environment, coordinating the expression of enzymes such as tetrathionate hydrolase (TetH), thiosulfate dehydrogenase (TsdA), and other periplasmic and membrane-associated sulfur oxidases (Ibáñez et al., 2023; Kanao et al., 2007, 2023; Yu et al., 2014). Such sulfur-responsive regulation underscores the

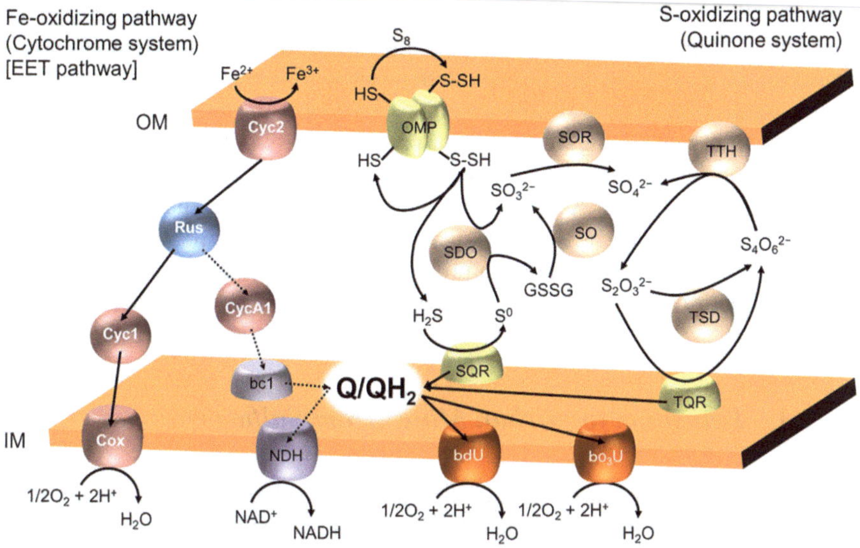

Fig. 17.3 Fe/S-oxidizing and EET pathway in *A. ferrooxidans*. The Fe-oxidizing pathway and extracellular electron transfer (EET) pathway consist of two cytochromes (Cyc2 and Cyc1), rusticyanin (Rus), and aa_3-type cytochrome *c* oxidase. This pathway connects reactions on the outer membrane (OM) to terminal oxidase on the inner membrane (IM). In contrast, the S-oxidizing pathway uses the quinone system, which has various enzymes (OMP thiol-bearing outer membrane protein, SDO sulfur dioxygenase, SOR sulfur oxygenase reductate, SO sulfur oxygenase, TSD thiosulfate dehydrogenase, TTH tetrathionate hydrolase) for oxidative conversions of the reduced sulfur compounds, substrate:quinone oxidoreductases (SQR sulfide:quinone oxidoreductase, TQR thiosulfate:quinone oxidoreductase), the quinone pool, and ubiquinol oxidases (bdU cytochrome *bd*-type ubiquinol oxidase, bo3U cytochrome bo_3-type ubiquinol oxidase). On the other hand, the uphill electron pathway for the reduction of NAD^+ to NADH consists of both the cytochrome and quinone systems, and dashed arrows represent this. This pathway includes one cytochrome (CycA1), the cytochrome bc_1 complex, the quinone pool, and NADH dehydrogenase (NDH)

ecological versatility of *A. ferrooxidans*, enabling it to thrive in heterogeneous leaching heaps where the distribution of RISCs changes spatially and temporally as mineral dissolution proceeds. In applied contexts, this inducible gene expression ensures that bioleaching consortia can adapt to shifting mineralogical and chemical conditions, maintaining robust oxidation activity over prolonged leaching cycles.

Beyond classical chemoautotrophy on soluble ferrous ion and RISCs, *A. ferrooxidans* has been shown in multiple studies to participate in extracellular electron transfer (EET) processes, which involve both donating electrons to and accepting electrons from solid surfaces or electrodes (Campodonico et al., 2016; Chabert et al., 2018; Ishii et al., 2015). Laboratory electrochemical experiments demonstrate that *A. ferrooxidans* can generate current when colonizing electrodes and that cell adhesion, outer-membrane redox proteins, and community signaling can modulate current output; conversely, under certain conditions *A. ferrooxidans* can grow autotrophically using electrons supplied from an electrode (electroautotrophy), implying direct uptake mechanisms or mediated transfer via soluble redox shuttles. In the bioleaching setting, these EET capabilities suggest additional routes by which microbes may accelerate mineral oxidation, either by direct electron withdrawal from a sulfide surface or by catalyzing redox cycling at the mineral–solution interface, potentially bypassing limitations of purely dissolved-phase Fe^{2+}/Fe^{3+} chemistry. Although mechanistic details (specific outer-membrane conduits, the role of conductive biofilms, and the relative importance of direct vs. mediated EET in natural heaps) remain under active investigation, integrating electrochemical perspectives into biomining models offers a promising framework to explain variability in leaching kinetics and to design bioelectrochemical strategies for enhanced recovery.

17.3.3 Transition to "Urban Mine" from Sulfide Ore

Tank leaching is a process in which finely ground sulfide ores are suspended in large, aerated, and agitated bioreactors. In contrast to heap leaching, where ore particles remain in a static pile, tank systems provide controlled mixing, temperature regulation, and efficient aeration, enabling higher reaction rates and shorter residence times (Brierley & Brierley, 2013; Zhang & Schippers, 2022). The ore slurry is continuously supplied with nutrients, oxygen, and acid, allowing operators to optimize physicochemical parameters and maintain stable microbial activity. Tank bioleaching has been applied industrially for the extraction of gold (via arsenopyrite and pyrite oxidation), copper, cobalt, and nickel, particularly when ores are of high value or refractory to conventional processing (Rawlings, 2002; Watling, 2006; Wang et al., 2023). Due to its high capital and operational costs, tank leaching is generally limited to ores with significant economic return, but it offers superior process control and metal recovery efficiency compared to heap leaching.

The microbial community in tank leaching systems is dominated by acidophilic chemolithotrophs, which catalyze the oxidative dissolution of sulfide minerals. *A. ferrooxidans* and *L. ferrooxidans* play key roles through ferrous ion oxidation, generating ferric ion as a powerful oxidant for sulfide dissolution (Zhang & Schippers, 2022). In parallel, *A. thiooxidans* and related sulfur-oxidizing bacteria contribute by oxidizing elemental sulfur and other reduced sulfur intermediates to sulfuric acid, thereby maintaining the acidic conditions necessary for continuous metal solubilization (Watling, 2006; Wang et al., 2023). The synergy of iron- and sulfur-oxidizing microorganisms ensures both the regeneration of chemical oxidants (Fe^{3+}, H^+) and the removal of passivating sulfur layers, leading to efficient mineral breakdown (Rawlings & Johnson, 2007; Tian et al., 2022). Moreover, tank reactors support the growth of extremophiles such as *A. caldus* and certain archaea under elevated temperatures (40~50 °C) (Dopson & Lindstrom, 1999; Schippers, 2007; Wang et al., 2023), further expanding the range of conditions under which bioleaching can be sustained. This integration of microbial metabolism with engineered reactor design exemplifies the controlled and accelerated nature of tank bioleaching.

Tank bioleaching is particularly advantageous for processing ores of high economic value or refractory mineralogy, where conventional techniques are ineffective (Brierley, 2008; Zhang & Schippers, 2022). A classic example is the treatment of refractory gold ores, in which pyrite (FeS_2) and arsenopyrite (FeAsS) encapsulate gold particles (Rawlings, 2002; Watling, 2006). Microbial oxidation of these sulfide matrices by *A. ferrooxidans*, *Leptospirillum* spp., and thermophilic archaea liberates the gold, enabling subsequent cyanidation (Rawlings & Johnson, 2007). Similarly, nickel and cobalt recovery from lateritic and sulfide ores has been achieved through stirred tank bioleaching, where iron-oxidizing and sulfur-oxidizing microorganisms accelerate dissolution and maintain acidic conditions (Watling, 2006; Zhang & Schippers, 2022). The controlled environment of the stirred tank allows optimization of temperature, aeration, and pulp density, providing high recovery efficiencies within weeks rather than the months or years required by heap leaching (Brierley & Brierley, 2013; Wang et al., 2023).

In tank leaching systems, microbial consortia are carefully managed to ensure robust oxidation of both iron and sulfur (Rawlings, 2002; Zhang &Schippers, 2022). Thermotolerant species such as *A. caldus*, *Sulfobacillus thermosulfidooxidans*, and the archaea *Ferroplasma* and *Metallosphaera* spp. become increasingly important under elevated temperature regimes, especially in the context of nickel and cobalt extraction (Dopson & Lindstrom, 1999; Schippers, 2007; Wang et al., 2023). These microorganisms oxidize ferrous ion, sulfur, and sulfide intermediates, ensuring a constant supply of ferric ion and protons that drive mineral dissolution. Process intensification strategies include multi-stage reactors, controlled redox potential management, and gas-lift or air-sparged systems to maximize oxygen transfer, all of which reinforce the metabolic activity of the microbial community (Brierley, 2008).

Beyond primary ores, tank bioleaching has emerged as a promising technology for recovering valuable metals from secondary resources, often referred to as "urban mines" (Jegan Roy et al., 2021; Wang et al., 2023). These include electronic waste (e-waste), spent batteries, and industrial residues rich in copper, cobalt, nickel, and rare Earth elements (Anaya-Garzon et al., 2021; Kadivar et al., 2021; Tian et al., 2022; Adetunji et al., 2023; Zhang et al., 2023; Biswal & Balasubramanian, 2023; Lalropuia et al., 2024). Microorganisms play a dual role in these systems: they leach base metals directly from complex waste matrices and facilitate subsequent separation steps by maintaining acidic and oxidizing conditions. For example, *A. ferrooxidans* has been employed to extract cobalt and lithium from spent lithium-ion batteries (Jegan Roy et al., 2021), while sulfur-oxidizing consortia help dissolve copper from printed circuit boards. Compared to pyrometallurgical or chemical recycling, bioleaching offers a lower-energy, environmentally benign alternative, though scaling and heterogeneity of waste streams remain challenges. Nevertheless, tank-based bioleaching reactors provide a controlled environment suitable for treating finely milled secondary materials, pointing to future applications in sustainable resource recovery (Wang et al., 2023).

17.4 Future Insight and Challenge

The development of bioleaching technologies for terrestrial mineral deposits and urban mines has already reached a practical level. In contrast, the application of bioleaching to marine mineral resources remains at an early stage. Japan possesses the world's sixth-largest Exclusive Economic Zone (EEZ), within which a vast abundance of marine mineral resources has been identified. However, the development of recovery technologies in this domain is significantly lagging behind, largely due to not only economic challenges but also technical hurdles and the complex interplay of environmental conservation concerns. Resource development heavily dependent on technological advancement carries the risk of long-term negative impacts on human society through ecosystem disruption (Kaikkonen et al., 2021; Yao et al., 2025; Alam et al., 2025). Therefore, the pursuit of sustainable technologies that are environmentally responsible and beneficial for humanity in the long term is of critical importance.

One of the major technical challenges in advancing marine bioleaching is the incompatibility of currently employed acidophilic iron- and sulfur-oxidizing bacteria, such as *A. ferrooxidans*, with high concentrations of sea salts typically present in marine mineral deposits (Inaba et al., 2021). These microorganisms, while highly effective in terrestrial and urban mining contexts, exhibit reduced activity or even complete inhibition under saline conditions. Consequently, the exploration and characterization of marine-adapted acidophilic iron- and sulfur-oxidizers represent a crucial step toward enabling sustainable and efficient bioleaching of seafloor mineral resources (Fig. 17.2). Such efforts will not only expand the scope of biohydrometallurgy but also contribute to establishing environmentally conscious strategies for marine resource utilization.

Recent studies have begun to shed light on acidophilic microorganisms inhabiting marine environments, which may serve as promising candidates for future bioleaching technologies. For example, a marine acidophilic sulfur-oxidizing bacterium, *Acidithiobacillus* sp. strain SH, has been isolated from seawater and shown to require salts such as NaCl for efficient oxidation of elemental sulfur and tetrathionate, with its genome sequence revealing multiple pathways for sulfur metabolism (Kamimura et al., 2003, 2005, 2018; Sharmin et al., 2016). These findings highlight the possibility of leveraging halotolerant sulfur-oxidizers for processing saline minerals in oceanic deposits. In parallel, laboratory studies have tested the interactions between *A. ferrooxidans* and deep-sea polymetallic nodules, showing distinctive electrochemical behaviors during bioleaching reactions (Jin-xing et al., 2018). Although these reports remain at the exploratory stage, they demonstrate that both sulfur-oxidizing and iron-oxidizing acidophiles can potentially adapt to marine mineral matrices under high-salinity conditions.

Moving forward, the discovery and characterization of truly marine-adapted iron- and sulfur-oxidizing acidophiles will be crucial for bridging the current gap between terrestrial bioleaching practices and marine mineral resource development. Integrating genomic, electrochemical, and ecological approaches will help identify the key adaptations that allow survival and metabolic activity in salt-rich, high-pressure deep-sea environments. Such insights could ultimately enable the design of environmentally conscious and economically feasible bioleaching technologies tailored for Japan's EEZ and other marine mineral reserves worldwide.

17.5 Conclusions

Acidophilic microorganisms and their unique metabolic capabilities have provided a foundation for bioleaching technologies that are now widely applied in the recovery of valuable metals from both terrestrial ores and urban mines. Advances in understanding iron and sulfur oxidation pathways, as well as the development of acid-stable enzymes, have contributed to optimizing industrial applications ranging from heap and tank leaching of copper, gold, and nickel to emerging approaches for recycling critical elements from electronic waste. Looking forward, the extension of bioleaching into marine mineral resources remains a key frontier, particularly for nations such as Japan with extensive exclusive economic zones. Overcoming the challenge of microbial activity under high-salinity conditions, alongside ensuring environmental sustainability, will be central to realizing the next generation of biohydrometallurgical technologies. Ultimately, integrating microbiology, biotechnology, and ecological considerations will enable the sustainable exploitation of mineral resources while minimizing the environmental footprint of resource recovery.

References

Adetunji, A. I., Oberholster, P. J., & Erasmus, M. (2023). Bioleaching of metals from e-waste using microorganisms: A review. *Minerals, 13*(6), 828.

Alam, L., Pradhoshini, K. P., Flint, R. A., & Sumaila, U. R. (2025). Deep-sea mining and its risks for social-ecological systems: Insights from simulation-based analyses. *PLoS One, 20*, e0320888.

Amils, R., González-Toril, E., Fernández-Remolar, D., Gómez, F., Aguilera, Á., Rodríguez, N., Malki, M., García-Moyano, A., Fairén, A. G., de la Fuente, V., & Sanz, J. L. (2007). Extreme environments as Mars terrestrial analogs: The Rio Tinto case. *Planetary and Space Science, 55*, 370–381.

Anaya-Garzon, J., Hubau, A., Joulian, C., & Guezennec, A. G. (2021). Bioleaching of E-waste: Influence of printed circuit boards on the activity of acidophilic iron-oxidizing bacteria. *Frontiers in Microbiology, 12*, 669738.

Appia-Ayme, C., Guiliani, N., Ratouchniak, J., & Bonnefoy, V. (1999). Characterization of an operon encoding two c-type cytochromes, an aa(3)-type cytochrome oxidase, and rusticyanin in *Thiobacillus ferrooxidans* ATCC 33020. *Applied and Environmental Microbiology, 65*, 4781–4787.

Auernik, K. S., & Kelly, R. M. (2010). Physiological versatility of the extremely thermoacidophilic archaeon Metallosphaera sedula supported by transcriptomic analysis of heterotrophic, autotrophic, and mixotrophic growth. *Applied and Environmental Microbiology, 76*, 931–935.

Baker-Austin, C., & Dopson, M. (2007). Life in acid: pH homeostasis in acidophiles. *Trends in Microbiology, 15*, 165–171.
Bischoff, J. L., & Rosenbauer, J. R. (1988). Liquid-vapor relations in the critical region of the system NaCl-H_2O from 380 to 415°C: A refined determination of the critical point and two-phase boundary of seawater. *Geochimica et Cosmochimica Acta, 52*, 2121–2126.
Biswal, B. K., & Balasubramanian, R. (2023). Recovery of valuable metals from spent lithium-ion batteries using microbial agents for bioleaching: A review. *Frontiers in Microbiology, 14*, 1197081.
Blake, R. C., & Shute, E. A. (1994). Respiratory enzymes of *Thiobacillus ferrooxidans*. Kinetic properties of an acid-stable iron: Rusticyanin oxidoreductase. *Biochemistry, 33*, 9220–9228.
Brierley, J. A. (2008). A perspective on developments in biohydrometallurgy. *Hydrometallurgy, 94*(1–4), 2–8.
Brierley, C. L., & Brierley, J. A. (2013). Progress in bioleaching: Part B: Applications of microbial processes by the minerals industries. *Applied Microbiology and Biotechnology, 97*, 7543–7552.
Brock, T. D. (1978). A Sour world: Life and death at low pH. In *Thermophilic microorganisms and life at high temperatures*. Springer Series in Microbiology. Springer.
Campodonico, M. A., Vaisman, D., Castro, J. F., Razmilic, V., Mercado, F., Andrews, B. A., Feist, A. M., & Asenjo, J. A. (2016). Acidithiobacillus ferrooxidans's comprehensive model driven analysis of the electron transfer metabolism and synthetic strain design for biomining applications. *Metabolic Engineering Communications, 3*, 84–96.
Castelle, C., Guiral, M., Malarte, G., Ledgham, F., Leroy, G., Brugna, M., & Giudici-Orticoni, M. T. (2008). A new iron-oxidizing/O_2-reducing supercomplex spanning both inner and outer membranes, isolated from the extreme acidophile *Acidithiobacillus ferrooxidans*. *The Journal of Biological Chemistry, 283*, 25803–25811.
Chabert, N., Bonnefoy, V., & Achouak, W. (2018). Quorum sensing improves current output with *Acidithiobacillus ferrooxidans*. *Microbial Biotechnology, 11*, 136–140.
Cowan, D. A., Albers, S. V., Antranikian, G., Atomi, H., Averhoff, B., Basen, M., Driessen, A. J. M., Jebbar, M., Kelman, Z., Kerou, M., Littlechild, J., Müller, V., Schönheit, P., Siebers, B., & Vorgias, K. (2024). Extremophiles in a changing world. *Extremophiles, 28*, 26.
Czop, M., Motyka, J., Sracek, O., & Szuwarzyński, M. (2011). Geochemistry of the hyperalkaline Gorka pit lake (pH> 13) in the Chrzanow region, southern Poland. *Water, Air, & Soil Pollution, 214*, 423–434.
Dopson, M., Baker-Austin, C., Bond, P. L. (2005). Analysis of differential protein expression during growth states of Ferroplasma strains and insights into electron transport for iron oxidation. *Microbiology 151*, 4127–4137.
Dopson, M., & Lindstrom, E. B. (1999). Potential role of *Thiobacillus caldus* in arsenopyrite bioleaching. *Applied and Environmental Microbiology, 65*, 36–40.
Feller, G. (2010). Protein stability and enzyme activity at extreme biological temperatures. *Journal of Physics. Condensed Matter, 22*, 323101.
Golyshina, O. V. (2011). Environmental, biogeographic, and biochemical patterns of archaea of the family *Ferroplasmaceae*. *Applied and Environmental Microbiology, 77*, 5071–5078.
Golzar-Ahmadi, M., Bahaloo-Horeh, N., Pourhossein, F., Norouzi, F., Schoenberger, N., Hintersatz, C., Chakankar, M., Holuszko, M., & Kaksonen, A. H. (2024). Pathway to industrial application of heterotrophic organisms in critical metals recycling from e-waste. *Biotechnology Advances, 30*, 108438.
González-Toril, E., Llobet-Brossa, E., Casamayor, E. O., Amann, R., & Amils, R. (2003). Microbial ecology of an extreme acidic environment, the Tinto River. *Applied and Environmental Microbiology, 69*, 4853–4865.
Harrison, J. P., Gheeraert, N., Tsigelnitskiy, D., & Cockell, C. S. (2013). The limits for life under multiple extremes. *Trends in Microbiology, 21*, 204–212.
Horikoshi, K. (1999). Alkaliphiles: Some applications of their products for biotechnology. *Microbiology and Molecular Biology Reviews, 63*, 735–750.

Huang, Y., Krauss, G., Cottaz, S., Driguez, H., & Lipps, G. (2005). A highly acid-stable and thermostable endo-beta-glucanase from the thermoacidophilic archaeon *Sulfolobus solfataricus*. *The Biochemical Journal, 385*, 581–588.

Ibáñez, A., Garrido-Chamorro, S., Coque, J. J. R., & Barreiro, C. (2023). From genes to bioleaching: Unraveling sulfur metabolism in *Acidithiobacillus*. *Genus, 14*, 1772.

Inaba, Y., West, A. C., & Banta, S. (2021). Glutathione Synthetase overexpression in Acidithiobacillus ferrooxidans improves Halotolerance of iron oxidation. *Applied and Environmental Microbiology, 87*, e0151821.

Ishii, T., Kawaichi, S., Nakagawa, H., Hashimoto, K., & Nakamura, R. (2015). From chemolithoautotrophs to electrolithoautotrophs: CO_2 fixation by Fe(II)-oxidizing bacteria coupled with direct uptake of electrons from solid electron sources. *Frontiers in Microbiology, 6*, 994.

Jaiswal, A., Raj, S. I., Isiaka Adetunji, A., Negadi, L., Singh, S., Tumba, K., Bahadur, I., & Uddin, I. (2025). Bioleaching as an eco-friendly nano-factory for sustainable inorganic waste management: Current advancements, challenges, and opportunities. *ChemistryOpen, 14*, e202500104.

Jegan Roy, J., Srinivasan, M., & Cao, B. (2021). Bioleaching as an eco-friendly approach for metal recovery from spent NMC-based lithium-ion batteries at a high pulp density. *ACS Sustainable Chemistry & Engineering, 9*, 3060–3069.

Jin-xing, K., Ya-li, F., Hao-ran, L., Zhu-wei, D., Xiang-yi, D., & Hong-jun, W. (2018). *Acidithiobacillus ferrooxidans*: Performances on the low-grade polymetallic oxidized-manganese and sulfur-nickel ores redox system. *Metallurgical Research & Technology, 115*, 601.

Johnson, D. B. (1998). Biodiversity and ecology of acidophilic microorganisms. *FEMS Microbiology Ecology, 27*, 307–317.

Johnson, D. B. (2014). Biomining-biotechnologies for extracting and recovering metals from ores and waste materials. *Current Opinion in Biotechnology, 30*, 24–31.

Johnson, D. B., & Hallberg, K. B. (2003). The microbiology of acidic mine waters. *Research in Microbiology, 154*, 466–473.

Johnson, D. B., & Hallberg, K. B. (2005). Acid mine drainage remediation options: A review. *Science of the Total Environment, 338*, 3–14.

Jones, S., & Santini, J. M. (2023). Mechanisms of bioleaching: Iron and sulfur oxidation by acidophilic microorganisms. *Essays in Biochemistry, 67*, 685–699.

Kadivar, S., Pourhossein, F., & Mousavi, S. M. (2021). Recovery of valuable metals from spent mobile phone printed circuit boards using biochar in indirect bioleaching. *Journal of Environmental Management, 280*, 111642.

Kaikkonen, L., Helle, I., Kostamo, K., Kuikka, S., Törnroos, A., Nygård, H., Venesjärvi, R., & Uusitalo, L. (2021). Causal approach to determining the environmental risks of seabed mining. *Environmental Science & Technology, 55*, 8502–8513.

Kamimura, K., Higashino, E., Shima, S., & Kanao, T. (2005). Requirement of sodium chloride for the oxidation of elemental sulfur and tetrathionate by a marine acidophilic sulfur-oxidizing bacterium, Acidithiobacillus sp. strain SH. *Journal of Bioscience and Bioengineering, 100*(4), 423–428.

Kamimura, K., Higashino, E., Moriya, S., & Sugio, T. (2003). Marine acidophilic sulfur-oxidizing bacterium requiring salts for the oxidation of reduced inorganic sulfur compounds. *Extremophiles, 7*, 95–99.

Kamimura, K., Sharmin, S., Yoshino, E., Tokuhisa, M., & Kanao, T. (2018). Draft genome sequence of *Acidithiobacillus* sp. Strain SH, a marine acidophilic sulfur-oxidizing bacterium. *Genome Announcements, 6*, e01603–e01617.

Kanao, T., Kamimura, K., & Sugio, T. (2007). Identification of a gene encoding a tetrathionate hydrolase in *Acidithiobacillus ferrooxidans*. *Journal of Biotechnology, 132*, 16–22.

Kanao, T., Matsumoto, C., Shiraga, K., Yoshida, K., Takada, J., & Kamimura, K. (2010). Recombinant tetrathionate hydrolase from *Acidithiobacillus ferrooxidans* requires exposure to acidic conditions for proper folding. *FEMS Microbiology Letters, 309*, 43–47.

Kanao, T., Kunihisa, T., Ohgimoto, S., Ito, M., Murakami, C., Nakayama, H., Tamura, T., & Kamimura, K. (2023). Recombinant expression using the tetrathionate hydrolase promoter in *Acidithiobacillus ferrooxidans*. *Journal of Bioscience and Bioengineering, 135*, 176–181.
Kato, C., Li, L., Tamaoka, J., & Horikoshi, K. (1997). Molecular analyses of the sediment of the 11,000-m deep Mariana Trench. *Extremophiles, 1*, 117–123.
Koschinsky, A., Garbe-Schönberg, D., Sander, S., Schmidt, K., Gennerich, H. H., & Strauss, H. (2008). Hydrothermal venting at pressure-temperature conditions above the critical point of seawater, 5 S on the Mid-Atlantic Ridge. *Geology, 36*, 615–618.
Krulwich, T. A., Sachs, G., & Padan, E. (2011). Molecular aspects of bacterial pH sensing and homeostasis. *Nature Reviews. Microbiology, 9*, 330–343.
Kumar, S., & Nussinov, R. (2002). Close-range electrostatic interactions in proteins. *Chembiochem, 3*, 604–617.
Kusube, M., Kyaw, T. S., Tanikawa, K., Chastain, R. A., Hardy, K. M., Cameron, J., & Bartlett, D. H. (2017). *Colwellia marinimaniae* sp. nov., a hyperpiezophilic species isolated from an amphipod within the Challenger Deep, Mariana Trench. *International Journal of Systematic and Evolutionary Microbiology, 67*, 824–831.
Lalropuia, L., Kucera, J., Rassy, W. Y., Pakostova, E., Schild, D., Mandl, M., Kremser, K., & Guebitz, G. M. (2024). Metal recovery from spent lithium-ion batteries via two-step bioleaching using adapted chemolithotrophs from an acidic mine pit lake. *Frontiers in Microbiology, 15*, 1347072.
Li, C., Jin, J., Yuan, Z., Zhang, C., Wu, L., & Wang, C. (2025). A review on leaching of spent lithium battery cathode materials adopting deep eutectic solvents. *ChemistryOpen, 14*, e202400258.
Macario, A. J., & Conway de Macario, E. (2007). Chaperonopathies and chaperonotherapy. *FEBS Letters, 581*, 3681–3688.
Nordstrom, D. K., & Southam, G. (1997). Geomicrobiology of sulfide mineral oxidation. In *Geomicrobiology: Interactions between microbes and minerals* (Vol. 35, pp. 361–390).
Panda, S., Akcil, A., Pradhan, N., & Deveci, H. (2015). Current scenario of chalcopyrite bioleaching: A review on the recent advances to its heap-leach technology. *Bioresource Technology, 196*, 694–706.
Quatrini, R., & Johnson, D. B. (2009). Extending the models for iron and sulfur oxidation in *Acidithiobacillus ferrooxidans*. *BMC Genomics, 10*, 394.
Rawlings, D. E. (2002). Heavy metal mining using microbes. *Annual Review of Microbiology, 56*, 65–91.
Rawlings, D. E., & Johnson, D. B. (2007). The microbiology of biomining: Development and optimization of mineral-oxidizing microbial consortia. *Microbiology (Reading), 153*, 315–324.
Reed, C. J., Lewis, H., Trejo, E., Winston, V., & Evilia, C. (2013). Protein adaptations in archaeal extremophiles. *Archaea, 2013*, 373275.
Rekadwad, B. N., Li, W. J., Gonzalez, J. M., Punchappady Devasya, R., Ananthapadmanabha Bhagwath, A., Urana, R., & Parwez, K. (2023). Extremophiles: the species that evolve and survive under hostile conditions. *3 Biotech, 13*, 316.
Roberto, F. F., & Schippers, A. (2022). Progress in bioleaching: Part B, applications of microbial processes by the minerals industries. *Applied Microbiology and Biotechnology, 106*, 5913–5928.
Rohwerder, T., & Sand, W. (2007). Mechanisms and biochemical fundamentals of bacterial metal sulfide oxidation. In E. R. Donati & W. Sand (Eds.), *Microbial processing of metal sulfides* (pp. 35–58). Springer.
Sarethy, I. P., Saxena, Y., Kapoor, A., Sharma, M., Sharma, S. K., Gupta, V., & Gupta, S. (2011). Alkaliphilic bacteria: Applications in industrial biotechnology. *Journal of Industrial Microbiology & Biotechnology, 38*, 769–790.
Schippers, A., Hedrich, S., Vasters, J., Drobe, M., Sand, W., & Willscher, S. (2014). Biomining: Metal recovery from ores with microorganisms. *Advances in Biochemical Engineering/Biotechnology, 141*, 1–47.

Schleper, C., Pühler, G., Kühlmorgen, B., & Zillig, W. (1995a). Life at extremely low pH. *Nature, 375*, 741–742.

Schleper, C., Puehler, G., Holz, I., Gambacorta, A., Janekovic, D., Santarius, U., Klenk, H. P., & Zillig, W. (1995b). *Picrophilus* gen. nov., fam. nov.: A novel aerobic, heterotrophic, thermoacidophilic genus and family comprising archaea capable of growth around pH 0. *Journal of Bacteriology, 177*, 7050–7059.

Schleper, C., Pühler, G., Klenk, H. P., & Zillig, W. (1996). Picrophilus oshimae and Picrophilus torridus fam. nov., gen. nov., sp. nov., two species of hyperacidophilic, thermophilic, heterotrophic, aerobic archaea. *International Journal of Systematic and Evolutionary Microbiology, 46*, 814–816.

Schippers, A. (2007). Microorganisms involved in bioleaching and nucleic acid-based molecular methods for their identification and quantification. In *Microbial processing of metal sulfides* (pp. 3-33). Dordrecht: Springer Netherlands.

Sharma, A., Kawarabayasi, Y., & Satyanarayana, T. (2012). Acidophilic bacteria and archaea: Acid stable biocatalysts and their potential applications. *Extremophiles, 16*, 1–19.

Sharmin, S., Yoshino, E., Kanao, T., & Kamimura, K. (2016). Characterization of a novel thiosulfate dehydrogenase from a marine acidophilic sulfur-oxidizing bacterium, *Acidithiobacillus thiooxidans* strain SH. *Bioscience, Biotechnology, and Biochemistry, 80*, 273–278.

Shiers, D. W., Collinson, D. M., & Watling, H. R. (2016). Life in heaps: A review of microbial responses to variable acidity in sulfide mineral bioleaching heaps for metal extraction. *Research in Microbiology, 167*, 576–586.

Takai, K., Nakamura, K., Toki, T., Tsunogai, U., Miyazaki, M., Miyazaki, J., Hirayama, H., Nakagawa, S., Nunoura, T., & Horikoshi, K. (2008). Cell proliferation at 122 degrees C and isotopically heavy CH_4 production by a hyperthermophilic methanogen under high-pressure cultivation. *Proceedings of the National Academy of Sciences of the United States of America, 105*, 10949–10954.

Tian, Y., Hu, X., Song, X., & Yang, A. J. (2022). Bioleaching of rare-earth elements from phosphate rock using *Acidithiobacillus ferrooxidans*. *Letters in Applied Microbiology, 75*, 1111–1121.

Valdés, J., Quatrini, R., & Holmes, D. S. (2008). Reconstructing the iron oxidation pathway of *Acidithiobacillus ferrooxidans* from genomics and proteomics data. *BMC Genomics, 9*, 597.

van der Wielen, P. W., Bolhuis, H., Borin, S., Daffonchio, D., Corselli, C., Giuliano, L., D'Auria, G., de Lange, G. J., Huebner, A., Varnavas, S. P., Thomson, J., Tamburini, C., Marty, D., McGenity, T. J., Timmis, K. N., & BioDeep Scientific Party. (2005). The enigma of prokaryotic life in deep hypersaline anoxic basins. *Science, 307*, 121–123.

Vera, M., Schippers, A., Hedrich, S., & Sand, W. (2022). Progress in bioleaching: Fundamentals and mechanisms of microbial metal sulfide oxidation - part A. *Applied Microbiology and Biotechnology, 106*, 6933–6952.

Vítová, M., & Mezricky, D. (2024). Microbial recovery of rare earth elements from various waste sources: A mini review with emphasis on microalgae. *World Journal of Microbiology and Biotechnology, 40*, 189.

Wakai, S. (2019). Biochemical and thermodynamic analyses of energy conversion in extremophiles. *Bioscience, Biotechnology, and Biochemistry, 83*, 49–64.

Wakai, S., & Sambongi, Y. (2018). Extremophilic enzymes related to energy conversion. In M. Suzuki (Ed.), *Role of water in ATP hydrolysis energy transduction by protein machinery* (pp. 275–302). Springer.

Wakai, S., Kikumoto, M., Kanao, T., & Kamimura, K. (2004). Involvement of sulfide:Quinone oxidoreductase in sulfur oxidation of an acidophilic iron-oxidizing bacterium, *Acidithiobacillus ferrooxidans* NASF-1. *Bioscience, Biotechnology, and Biochemistry, 68*, 2519–2528.

Wakai, S., Tsujita, M., Kikumoto, M., Manchur, M. A., Kanao, T., & Kamimura, K. (2007). Purification and characterization of sulfide:Quinone oxidoreductase from an acidophilic iron-oxidizing bacterium, Acidithiobacillus ferrooxidans. *Bioscience, Biotechnology, and Biochemistry, 71*, 2735–2742.

Wang, R., Lin, J. Q., Liu, X. M., Pang, X., Zhang, C. J., Yang, C. L., Gao, X. Y., Lin, C. M., Li, Y. Q., Li, Y., Lin, J. Q., & Chen, L. X. (2019). Sulfur oxidation in the acidophilic autotrophic bacteria: Systems and regulation. *Frontiers in Microbiology, 9*, 3290.

Wang, Y. J., Li, L. L., Zhao, S., Chen, Y., & Yu, A. F. (2023). Bioleaching of metals from spent fluid catalytic cracking catalyst using adapted *Acidithiobacillus caldus*. *Environmental Science and Pollution Research International, 30*, 125689–125701.

Watling, H. R. (2006). The bioleaching of sulphide minerals with emphasis on copper sulphides — A review. *Hydrometallurgy, 84*, 81–108.

Yao, W., Tian, C., Teng, Y., Diao, F., Du, X., Gu, P., & Zhou, W. (2025). Development of deep-sea mining and its environmental impacts: A review. *Frontiers in Marine Science, 12*, 1598584.

Yarzábal, A., Appia-Ayme, C., Ratouchniak, J., & Bonnefoy, V. (2004). Regulation of the expression of the *Acidithiobacillus ferrooxidans* rus operon encoding two cytochromes c, a cytochrome oxidase and rusticyanin. *Microbiology (Reading), 150*, 2113–2123.

Yu, Y., Liu, X., Wang, H., Li, X., & Lin, J. (2014). Construction and characterization of tetH overexpression and knockout strains of *Acidithiobacillus ferrooxidans*. *Journal of Bacteriology, 196*, 2255–2264.

Zhang, R., & Schippers, A. (2022). Stirred-tank bioleaching of copper and cobalt from mine tailings in Chile. *Mining and Engineering, 180*, 107514.

Zhang, X., Shi, H., Tan, N., Zhu, M., Tan, W., Daramola, D., & Gu, T. (2023). Advances in bioleaching of waste lithium batteries under metal ion stress. *Bioresources and Bioprocessing, 10*, 19.

Zhou, H. B., Zeng, W. M., Yang, Z. F., Xie, Y. J., & Qiu, G. Z. (2009). Bioleaching of chalcopyrite concentrate by a moderately thermophilic culture in a stirred tank reactor. *Bioresource Technology, 100*, 515–520.

Chapter 18
Microbial Fuel Cell and Bioenergy Generation

Kengo Inoue

Abstract Microbial fuel cells (MFCs) represent a key technology in microbial electrochemistry, enabling the direct conversion of chemical energy from organic compounds into electrical energy through microbial metabolism. While their potential for wastewater treatment and energy recovery has been widely demonstrated, practical applications are often constrained by suboptimal environmental conditions, such as extreme temperature, pH, salinity, or nutrient limitations. Recent research has highlighted the crucial role of extremophiles in overcoming these challenges, as their unique physiological and molecular adaptations allow efficient extracellular electron transfer under harsh conditions. This chapter reviews the principles of MFC operation, the diversity of electron transfer mechanisms—including direct contact, conductive appendages, and electron shuttles—and recent advances in understanding electricity generation by thermophiles, psychrophiles, acidophiles, alkaliphiles, halophiles, and other extremophiles. The application of extremophiles in MFCs not only enhances system performance and broadens technological versatility but also opens avenues for biosensing, bioremediation, and the development of biodegradable conductive materials relevant to e-waste mitigation. Despite substantial progress, the molecular basis of extracellular electron transfer in many extremophiles remains largely unexplored. Future discoveries are expected to expand the scope of microbial electrochemistry and accelerate the integration of MFCs into sustainable energy and environmental technologies.

Keywords Microbial fuel cell · Bioelectrochemical systems · Extracellular electron transfer · Wastewater treatment

K. Inoue (✉)
Department of Biochemistry and Applied Biosciences, Faculty of Agriculture, University of Miyazaki, Miyazaki, Japan
e-mail: kinoue@cc.miyazaki-u.ac.jp

M. Ito et al. (eds.), *Extremophiles for a Sustainable Future*,
https://doi.org/10.1007/978-981-95-6970-0_18

18.1 Introduction

Microbial fuel cells (MFCs) have attracted increasing attention as a sustainable technology for both wastewater treatment and renewable energy generation. An MFC is a type of bioelectrochemical system that directly converts the chemical energy stored in organic matter into electricity through microbial metabolism. Since the first demonstration of electricity generation using pure cultures and wastewater, a wide range of electroactive microorganisms have been identified, revealing considerable diversity in extracellular electron transfer mechanisms. One of the remarkable findings in recent years is the involvement of extremophiles—microorganisms that thrive under extreme conditions of temperature, pH, salinity, pressure, or nutrient availability—in MFC research. Their robust metabolic traits and unique molecular machinery enable electron transfer under conditions that are often inhibitory to mesophilic microorganisms. This offers new opportunities for improving MFC performance and expanding their applicability in diverse environments. Extracellular electron transfer in microorganisms is most clearly exemplified in MFCs, but its implications extend beyond energy generation. Bioelectrochemical systems utilizing extracellular electron transfer have been proposed for biosensing applications, in which the detection of specific compounds can be transduced into electrical signals, as well as for bioremediation strategies that couple microbial metabolism with pollutant degradation. Furthermore, recent discussions have highlighted the potential role of biodegradable conductive biomaterials—such as conductive biofilms, microbial nanowires, and electron shuttles produced by extremophiles—in addressing the growing environmental burden of electronic waste (e-waste). Despite increasing evidence of extracellular electron transfer across diverse microbial taxa, the molecular mechanisms underlying electron transfer have been elucidated in only a limited number of model organisms, often through genetic and biochemical approaches. Continued exploration of extremophiles is therefore expected to provide novel insights into extracellular electron transfer and to accelerate the development of microbial electrochemical technologies for sustainable energy and environmental applications.

18.2 Principles of Microbial Fuel Cells

MFCs are bioelectrochemical devices that directly convert the chemical energy of organic compounds into electrical energy through microbial metabolism. The concept that microorganisms can transfer electrons to electrodes was first demonstrated by Potter in 1910 in the United Kingdom (Potter, 1910). For much of the twentieth century, this phenomenon remained largely overlooked. A major breakthrough occurred in 1999, when Kim et al. experimentally demonstrated that the iron-reducing bacterium *Shewanella putrefaciens* could transfer electrons to an electrode (Kim et al., 1999). Shortly thereafter, Bond et al. reported that *Geobacter sulfurreducens*, an iron-reducing bacterium that naturally transfers electrons to metal minerals for energy acquisition, could be utilized as an electricity-producing bacterium in MFCs, generating remarkably high current densities (Bond & Lovley, 2003). These findings positioned MFCs as a promising environmental energy technology and stimulated intensive research and development.

The main components of a typical MFC include the anode, cathode, electrolyte, ion-exchange membrane, and electroactive microorganisms. MFCs are generally constructed in either a two-chamber configuration or a single-chamber air-cathode configuration (Fig. 18.1). The anode functions as the electron acceptor for electroactive microorganisms and must support efficient electron transfer as well as an adequate supply of electron donors from the electrolyte. Carbon-based materials such as carbon felt, graphite, and carbon fibers are widely employed (Banerjee et al., 2022). Metallic electrodes, including silver, stainless steel, and aluminum, have also been investigated. In two-chamber MFCs (Fig. 18.1a), electron acceptors, such as oxygen, permanganate, and ferricyanide, can be added to the catholyte. More recently, "biocathodes" have been developed, in which electroactive microorganisms colonize the cathode surface and transfer electrons from the electrode to oxygen or other terminal electron acceptors (Huang et al., 2011). In single-chamber air-cathode MFCs (Fig. 18.1b), oxygen from the atmosphere serves as the electron acceptor via an air cathode (Wang et al., 2017). On the electrolyte-facing surface of such cathodes, platinum (Pt) is commonly applied as a catalyst for the reaction that produces water molecules from protons in the electrolyte, oxygen molecules from the air, and electrons transferred from the anode. However, Pt suffers from disadvantages, including high cost and performance deterioration due to biofouling during long-term operation. Consequently, alternative materials such as carbon, metals, metal oxides (e.g., manganese oxide), conductive polymers, and nitrogen-doped carbons are under active investigation. To maintain separation between anode and cathode, ion-exchange membranes (anion- or cation-exchange) are frequently employed (Leong et al., 2013), although membrane-less configurations have also been reported (Ramirez-Nava et al., 2021).

In the anode chamber, electroactive microorganisms transfer electrons derived from intracellular metabolism to the anode surface. While this process may appear unusual, it represents one of many microbial strategies for energy acquisition throughout Earth's evolutionary history. The oxygen-dependent respiratory pathway, dominant in humans and other eukaryotes, is evolutionarily recent. On early Earth, when atmospheric oxygen was scarce, microorganisms relied on anaerobic respiration, using iron and other minerals as terminal electron acceptors (Vargas et al., 1998). In MFCs, the electrode substitutes for iron, enabling microorganisms—particularly those closely related to iron-reducing bacteria—to respire by transferring electrons to the electrode. Thus, electrogenic bacteria effectively use the electrode as their terminal electron acceptor. To maximize MFC performance, strict anaerobic conditions must be maintained around the anode. If oxygen is present, microorganisms preferentially transfer electrons to oxygen rather than to the electrode, thereby reducing current output. From a thermodynamic standpoint, this

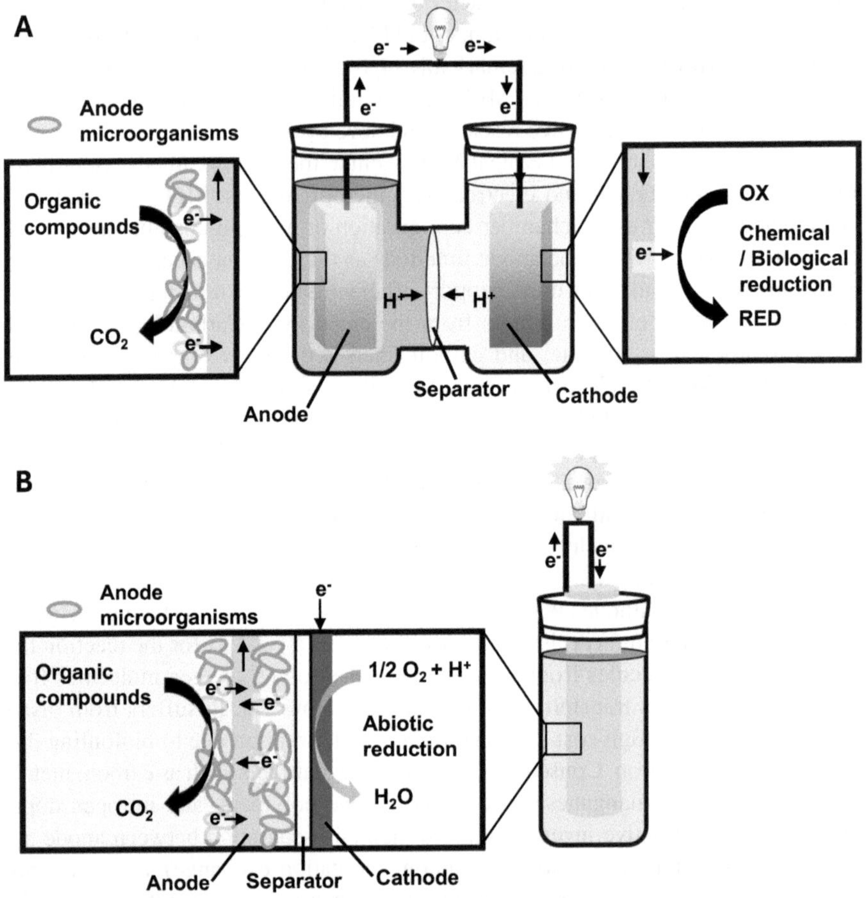

Fig. 18.1 Schematic diagram of a microbial fuel cell. (**a**) Two-chamber microbial fuel cell. The anode is immersed in an electrolyte containing organic compounds, where electricity-producing bacteria decompose the organic matter and transfer electrons to the electrode. The cathode is immersed in an electrolyte containing an electron acceptor, which is reduced either chemically or biologically (biocathode). OX and RED represent the oxidized and reduced forms of electroactive materials, respectively. The anode and cathode chambers are typically separated by an ion-exchange membrane. (**b**) Single-chamber microbial fuel cell. The anode functions essentially the same as in the two-chamber type, but the cathode is designed as an air cathode that uses oxygen in the air as the terminal electron acceptor. In the air cathode, protons in the electrolyte, electrons from the anode, and oxygen from the air react to produce water

is explained by the principle that electrons flow spontaneously from redox couples with lower potential to those with higher potential (Fig. 18.2).

In redox reactions, the electron donor possesses a lower redox potential, whereas the electron acceptor has a higher redox potential. The greater the potential difference between the donor and acceptor, the more energy (Gibbs free energy change)

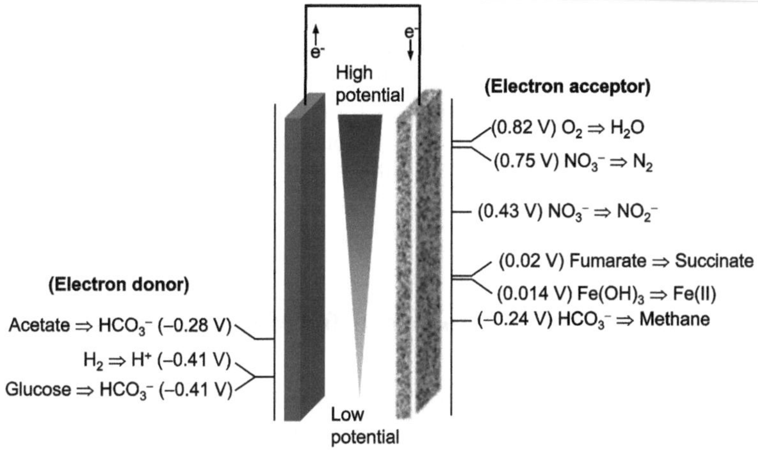

Fig. 18.2 Redox potentials of electron donors and acceptors (vs. standard hydrogen electrode)

is released per reaction. Under standard conditions, the standard electromotive force (emf) is defined as:

$$(\text{standard redox potential of the acceptor}) - (\text{standard redox potential of the donor}) = \text{standard emf.}$$

For example, when acetate serves as the electron donor and oxygen as the electron acceptor, the standard emf, using the values shown in Fig. 18.2, is calculated as $0.82-(-0.28) = 1.1$ V (vs. standard hydrogen electrode). From the standard emf, together with the number of electrons transferred and Faraday's constant, the standard Gibbs free energy of the reaction can be determined, thereby allowing the energy yield to be quantitatively evaluated. Consequently, in most cases where the same compound acts as the electron donor, microorganisms preferentially utilize oxygen as the electron acceptor because it provides a higher energy yield than the electrode.

The best-characterized electricity-producing microorganisms are iron-reducing species such as *Geobacter sulfurreducens* and *Shewanella oneidensis*. However, electricity generation is not restricted to iron-reducing bacteria; a broad diversity of microorganisms, including extremophiles, has been reported as capable of extracellular electron transfer (Logan et al., 2019; Lovley & Holmes, 2022). Examples include bacteria such as *Acetobacter aceti*, *Bacillus subtilis*, members of the genus *Clostridium*, *Klebsiella*, and *Pseudomonas*; archaea such as *Haloferax* and *Pyrococcus*; and yeasts such as *Saccharomyces* and *Candida*. The range of electron donors that can be metabolized differs among electricity-producing microorganisms, as do the optimal physicochemical conditions for electricity generation, including pH and temperature. Furthermore, the mechanisms of extracellular electron transfer are not uniform and are known to vary among microorganisms. Despite recent advances, the genes and enzymes involved in extracellular electron transfer remain incompletely elucidated. In complex environments where electrolytes contain a wide variety of organic compounds and the microbial community is diverse (e.g., sewage or livestock wastewater), microorganisms that do not directly transfer electrons to electrodes may nonetheless become dominant if they convert organic matter into substrates more readily utilized by electricity-producing microorganisms.

18.3 Extracellular Electron Transfer Mechanisms

Multiple factors influence the capacity of microorganisms to transfer electrons to electrodes. Based on current knowledge, including insights from genetic engineering approaches, three representative mechanisms have been identified: (i) direct cell–electrode contact, (ii) electrically conductive appendages, and (iii) electron shuttles (Fig. 18.3).

18.3.1 Direct Contact

In the direct-contact mechanism, microbial cells physically attach to the electrode surface to transfer electrons. Electricity-producing bacteria have been identified among both Gram-negative and Gram-positive species, although their cell surface structures differ substantially. Gram-negative bacteria possess a thin inner membrane and an outer membrane separated by a periplasm containing a thin peptidoglycan cell wall. In species such as *G. sulfurreducens* and *S. oneidensis*, *c*-type cytochromes embedded across the inner and outer membranes enable direct interaction with the electrode, thereby facilitating extracellular electron transfer (Gralnick & Bond, 2023). In contrast, Gram-positive bacteria lack an outer membrane and instead possess a thick peptidoglycan cell wall (Gomaa et al., 2022). Direct electron transfer has been demonstrated in *Thermincola* species, where multiheme *c*-type cytochromes form an electron relay from the inner membrane to surface reductases. In Gram-positive thermophiles, such as *Thermincola ferriacetica* and *T. potens*, multiheme cytochromes located within the cell wall are proposed to mediate electron transfer to surface-associated *c*-type cytochromes (Costa et al., 2019; Faustino et al., 2021). Furthermore, in the acidophilic *Ferrimicrobium acidophilum*, an alternative mechanism involving *a*-type cytochromes has been suggested for extracellular electron transfer (Blake et al., 2021).

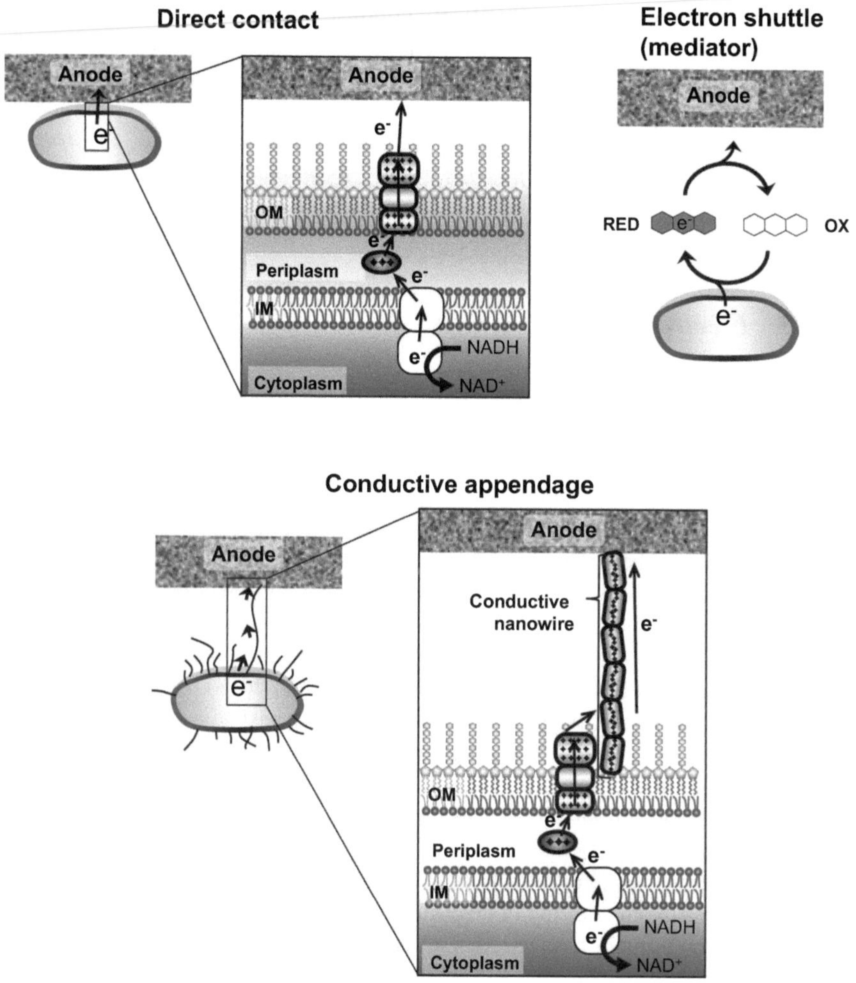

Fig. 18.3 Proposed model for extracellular electron transfer mechanisms to electrodes by Gram-negative bacteria

18.3.2 Electrically Conductive Appendage

Electrically conductive appendages in microorganisms were first reported in 2005 in *Geobacter sulfurreducens* (Reguera et al., 2005). Early studies suggested that these conductive nanowires were composed of PilA, the structural protein of type IV pili. However, subsequent investigations revealed that nanowires can also be formed from *c*-type cytochromes, including OmcS, OmcE, and OmcZ (Wang et al., 2019; Filman et al., 2019; Gu et al., 2023; Wang et al., 2022a, b). Although PilA lacks redox-active cofactors such as heme, the aromatic side chains of its amino acids are aligned along the pilus surface, allowing electron transfer via π–π stacking interactions (Lovley & Walker, 2019). By contrast, in the nanowires composed of *c*-type cytochromes, the heme-containing proteins are arranged linearly along the central axis, enabling electron conduction through the interior in a manner analogous to commercial copper wiring. Conductive appendages have also been identified in gram-positive bacteria, including *Paenibacillus dendritiformis*, *Clostridium ferrireducens*, *Enterococcus faecalis*, and *Lysinibacillus varians* (Gomaa et al., 2022).

18.3.3 Electron Shuttle

In the electron-shuttle mechanism, electricity-producing bacteria utilize redox-active compounds—such as flavins, phenazines, or humic substances—referred to as "electron shuttles" (also termed mediators) to transfer electrons to extracellular acceptors (Watanabe et al., 2009). In this process, microorganisms reduce the electron shuttle by transferring electrons to it; the reduced shuttle then delivers electrons to the electrode and becomes re-oxidized. The oxidized electron shuttle can subsequently receive electrons from the microorganism again, repeating the cycle. This strategy enables cells located at a distance from the electrode to participate in extracellular electron transfer and allows Gram-positive bacteria, which possess thick cell walls, to deliver electrons to electrodes (Light et al., 2018). *Listeria monocytogenes* employs a flavin-based extracellular electron transfer system involving an eight-gene locus, a specialized NADH dehydrogenase, and extracellular flavoproteins that use flavin shuttles to transfer electrons to iron or electrodes. This mechanism is widespread among diverse electricity-producing bacteria (Glasser et al., 2017; Gomaa et al., 2022). Electron shuttles may be synthesized endogenously by the bacteria themselves or derived from external sources, such as humic substances. Moreover, the artificial addition of electron shuttles has been shown to significantly enhance electron transfer to electrodes. Current research efforts are focused on the development of high-performance synthetic mediators with improved cell permeability and electron-transfer efficiency.

18.4 Extremophiles in Microbial Fuel Cells

A wide variety of organic wastes can be utilized as substrates in MFCs. In the anode electrolyte, extremophiles are particularly valuable when the organic waste exists under unusual or extreme conditions (Dopson et al., 2015). To date, most studies in bioelectrochemistry have been performed under mesophilic conditions, with near-neutral pH and moderate salinity, whereas investigations under extreme environments have remained relatively limited. In recent years, however, the scope of MFC research has expanded, and knowledge has gradually accumulated regarding electricity generation by extremophiles and their mechanisms of extracellular electron transfer. This section highlights studies on extremophiles in MFCs operating under extreme conditions, including thermophilic, psychrophilic, acidophilic, alkaliphilic, and halophilic environments. Among the extracellular electron transfer mechanisms described above (Fig. 18.3), direct electron transfer via cell-electrode contact and electron-shuttle-mediated transfer appear to represent the predominant strategies employed by Gram-positive extremophiles (Edel et al., 2022).

Thermophiles are microorganisms whose optimal growth occurs at elevated temperatures, typically above 45 °C and up to around 80 °C. Some thermophiles have even been found to grow in environments as hot as 122 °C (Takai et al., 2008). They possess specific molecular and physiological adaptations that maintain the stability of proteins, membranes, and nucleic acids under thermal stress (Lusk, 2019). Among studies on extremophiles in MFCs, relatively abundant knowledge has been accumulated on electricity generation and extracellular electron transfer mechanisms using thermophilic microorganisms. Microbial metabolism is generally more active at elevated temperatures, which can accelerate the degradation of organic compounds and enhance waste treatment, particularly for cellulosic biomass. However, electrolyte evaporation remains a significant technical challenge under high-temperature conditions (Carver et al., 2011). In thermophilic MFCs, Gram-positive *Thermincola* species have been most prominently identified (Mathis et al., 2008). Extracellular electron transfer via multiheme *c*-type cytochromes, similar to the well-studied mechanisms in *Geobacter sulfurreducens* and *Shewanella oneidensis*, has also been reported in *Thermincola ferriacetica* and *Thermincola potens* (Carlson et al., 2012). The multiheme *c*-type cytochrome CwcA (formerly TherJR_1122) is likely embedded tightly within the cell wall, suggesting that it forms an electron conduit analogous to the conductive nanowires composed of multiheme *c*-type cytochromes, such as OmcS, in *G. sulfurreducens*. Another extracellular electron transfer strategy in Gram-positive thermophiles involves lipoproteins and riboflavins, a mechanism first characterized in *Listeria monocytogenes* (Light et al., 2018). Genomic analyses suggest that this strategy may also be employed by thermophilic *Caldanaerobius* spp. and *Thermoanaerobacter* spp. Similarly, *Geobacillus* spp. have been shown to use related extracellular electron transfer mechanisms (Shrestha et al., 2020).

Psychrophiles (cryophiles) are microorganisms adapted to permanently cold environments, exhibiting optimal growth at 15 °C or lower. They are commonly found in polar regions, deep oceans, glaciers, and other constantly cold habitats, and possess specialized adaptations in their membranes, enzymes, and nucleic acids that allow cellular function under low-temperature conditions (De Maayer et al., 2014). In contrast to thermophilic MFCs, low temperatures generally exert a negative effect on decomposition of organic compounds and electricity generation (Larrosa-Guerrero et al., 2010). However, operation at low temperature is advantageous for achieving higher Coulombic efficiencies (Jadhav & Ghangrekar, 2009), making psychrophilic MFCs suitable for powering low-energy-consuming devices, such as environmental sensors. Microbial community analyses of anodes in low-temperature MFCs have revealed the presence of possible current-generating bacteria, including *Desulfuromonas* sp., *Geobacter* sp. *Arcobacter* sp., *Pseudomonas* sp., *Simplicispira psychrophila*, *Geobacter psychrophilus*, *Macellibacteroides fermentas*, and *Enterococcus lemanii* (Holmes et al., 2004a; Liu et al., 2012; Zhang et al., 2014; Dai et al., 2020). Furthermore, psychrophilic *Geopsychrobacter electrodiphilus* strains have been successfully isolated and demonstrated to produce current in pure-culture systems (Holmes et al., 2004b), suggesting that their extracellular electron transfer mechanisms are analogous to those characterized in *Geobacter sulfurreducens*.

Acidophiles are microorganisms that thrive in environments with very low pH, typically exhibiting optimal growth below pH 3. They are commonly found in acid mine drainage, volcanic soils, and acidic hot springs, and display physiological adaptations such as highly stable membrane structures, proton efflux systems, and acid-resistant proteins that maintain cellular homeostasis under extreme acidity (Sharma et al., 2012). In acidophilic MFCs, the cathode reaction benefits from the direct utilization of protons in the electrolyte, thereby preventing it from becoming rate-limiting (Erable et al., 2009). Generally, both cathode and anode potentials increase with decreasing pH (Dopson et al., 2015), indicating that acidic conditions are thermodynamically unfavorable for the anode (Fig. 18.2). Current generation at the anode in acidophilic MFCs has been reported for microorganisms such as *Acidiphilium cryptum*, *Acidiphilium* sp., and *Alicyclobacillus hesperidum* (Borole et al., 2008; Malki et al., 2008; Zhang et al., 2021). Members of the phylum Firmicutes have also been identified in acidophilic MFCs operating at pH 4–5 (Kim et al., 2014; Iigatani et al., 2019). At the cathode, the acidophilic bacterium *Acidithiobacillus ferrooxidans* has been demonstrated to function as a biocathode catalyst, facilitating reduction reactions and enabling electron transfer from the electrode to terminal electron acceptors such as oxygen (Carbajosa et al., 2010; García-Muñoz et al., 2011).

Alkaliphiles are microorganisms that thrive in environments with high pH, typically exhibiting growth above pH 9. They are commonly found in soda lakes, alkaline soils, and industrial alkaline wastewaters, and possess physiological adaptations such as specialized ion transport systems, stable cell wall structures, and enzymes that remain active under alkaline conditions. In MFCs, alkaline conditions are advantageous at the anode, particularly for organic waste utilization, because methanogens—competitors for electron donors—are depleted under high-pH conditions (Zhuang et al., 2010). In addition, the lower anode potential at alkaline pH results in an increased electromotive force of MFCs (Fig. 18.2). Current generation at the anode in alkaliphilic MFCs has been reported for microorganisms such as *Bacillus* sp., *Corynebacterium* sp., *Desulfuromonas acetexigens*, and *Pseudomonas alcaliphila* (Akiba et al., 1987; Liu et al., 2010; Zhang et al., 2011; Ketep et al., 2013). Moreover, *Anditalea andensis* and *Geoalkalibacter* spp. are capable of generating current not only under alkaliphilic but also halophilic conditions (Badalamenti et al., 2013; Shi et al., 2015).

Halophiles are microorganisms that grow optimally in environments with high salinity, typically at NaCl concentrations above 2 M. They are commonly found in salt lakes, salterns, and hypersaline soils, and exhibit physiological adaptations such as specialized ion pumps, compatible solute accumulation, and highly acidic proteins that maintain stability and functionality in high-salt conditions. In MFCs, high salinity can positively affect performance by increasing electrolyte conductivity, thereby decreasing internal resistance, as well as by inhibiting the growth of methanogens (Lefebvre et al., 2012; Guo et al., 2021). Halophilic MFCs have been extensively studied because of their high performance and wide technological applicability, particularly for saline wastewater treatment, leading to accumulation of substantial knowledge in this area (Grattieri & Minteer, 2018). Current generation at the anode has been reported for halophilic microorganisms such as *Bacillus selenitireducens*, *Geoalkalibacter subterraneus*, *Haloferax volcanii*, and *Natrialba magadii* (Miller & Oremland, 2008; Abrevaya et al., 2011; Carmona-Martínez et al., 2013). *Geoalkalibacter subterraneus, Desulfuromonas acetoxidans, Marinobacter,* and *Halomonas* spp. have also been identified in halophilic MFCs (Pierra et al., 2015; Zhao et al., 2024).

18.5 Conclusion

Research on MFCs has demonstrated that a wide range of microorganisms, including extremophiles, are capable of transferring electrons to electrodes. The extracellular electron transfer capacity of extremophiles offers significant potential to enhance MFC performance and expand the technological applicability of microbial electrochemical systems. While MFCs represent the most direct application of microbial extracellular electron transfer, additional applications are being actively explored, such as biosensors for detecting specific compounds via electrical signals and bioremediation strategies. Moreover, the rapid growth of global electronic waste (E-waste) has emerged as a major environmental challenge. Conductive materials derived from extremophiles—such as conductive biofilms, nanowires, and electron shuttles—are inherently biodegradable and may provide promising solutions for mitigating the e-waste problem. Although numerous microbial species have been identified as capable of extracellular electron exchange with electrodes or other organisms, only a limited number of molecular mechanisms have been elucidated through genetic and biochemical approaches. Nevertheless, insights obtained from extremophiles are expected to drive further advances in microbial electrochemistry and its applications.

References

Abrevaya, X. C., Sacco, N., Mauas, P. J. D., & Cortón, E. (2011). Archaea-based microbial fuel cell operating at high ionic strength conditions. *Extremophiles, 15*, 633–642.

Akiba, T., Bennetto, H. P., Stirling, J. L., & Tanaka, K. (1987). Electricity production from alkalophilic organisms. *Biotechnology Letters, 9*, 611–616.

Badalamenti, J. P., Krajmalnik-Brown, R., & Torres, C. I. (2013). Generation of high current densities by pure cultures of anode-respiring *Geoalkalibacter* spp. under alkaline and saline conditions in microbial electrochemical cells. *MBio, 4*, e00144–e00113.

Banerjee, A., Calay, R. K., & Mustafa, M. (2022). Review on material and design of anode for microbial fuel cell. *Energies, 15*, 2283.

Blake, R. C., Nautiyal, A., Smith, K. A., Walton, N. N., Pendleton, B., & Wang, Z. (2021). *Ferrimicrobium acidiphilum* exchanges electrons with a platinum electrode via a cytochrome with reduced absorbance maxima at 448 and 605 nm. *Frontiers in Microbiology, 12*, 705187.

Bond, D. R., & Lovley, D. R. (2003). Electricity production by *Geobacter sulfurreducens* attached to electrodes. *Applied and Environmental Microbiology, 69*, 1548–1555.

Borole, A. P., O'Neill, H., Tsouris, C., & Cesar, S. (2008). A microbial fuel cell operating at low pH using the acidophile *Acidiphilium cryptum*. *Biotechnology Letters, 30*, 1367–1372.

Carbajosa, S., Malki, M., Caillard, R., Lopez, M. F., Palomares, F. J., Martín-Gago, J. A., Rodríguez, N., Amils, R., Fernández, V. M., & de Lacey, A. (2010). Electrochemical growth of *Acidithiobacillus ferrooxidans* on a graphite electrode for obtaining a biocathode for direct electrocatalytic reduction of oxygen. *Biosensors & Bioelectronics, 26*, 877–880.

Carlson, H. K., Iavarone, A. T., Gorur, A., Yeo, B. S., Tran, R., Melnyk, R. A., Mathies, R. A., Auer, M., & Coates, J. D. (2012). Surface multiheme *c*-type cytochromes from *Thermincola potens* and implications for respiratory metal reduction by Gram-positive bacteria. *Proceedings of the National Academy of Sciences of the United States of America, 109*, 1702–1707.

Carmona-Martínez, A. A., Pierra, M., Trably, E., & Bernet, N. (2013). High current density via direct electron transfer by the halophilic anode respiring bacterium *Geoalkalibacter subterraneus*. *Physical Chemistry Chemical Physics, 15*, 19699–19707.

Carver, S. M., Vuoriranta, P., & Tuovinen, O. H. (2011). A thermophilic microbial fuel cell design. *Journal of Power Sources, 196*, 3757–3760.

Costa, N. L., Hermann, B., Fourmond, V., Faustino, M. M., Teixeira, M., Einsle, O., Paquete, C. M., & Louro, R. O. (2019). How thermophilic Gram-positive organisms perform extracellular electron transfer: Characterization of the cell surface terminal reductase OcwA. *MBio, 10*, e01210–e01219.

Dai, K., Sun, T., Yan, Y., Qian, D. K., Zhang, W., Zhang, F., & Zeng, R. J. (2020). Electricity production and microbial community in psychrophilic microbial fuel cells at 10°C. *Bioresource Technology, 313*, 123680.

De Maayer, P., Anderson, D., Cary, C., & Cowan, D. A. (2014). Some like it cold: Understanding the survival strategies of psychrophiles. *EMBO Reports, 15*, 508–517.

Dopson, M., Ni, G., & Sleutels, T. H. (2015). Possibilities for extremophilic microorganisms in microbial electrochemical systems. *FEMS Microbiology Reviews, 40*, 164–181.

Edel, M., Philipp, L. A., Lapp, J., Reiner, J., & Gescher, J. (2022). Electron transfer of extremophiles in bioelectrochemical systems. *Extremophiles, 26*, 31.

Erable, B., Etcheverry, L., & Bergel, A. (2009). Increased power from a two-chamber microbial fuel cell with a low-pH air-cathode compartment. *Electrochemistry Communications, 11*, 619–622.

Faustino, M. M., Fonseca, B. M., Costa, N. L., Lousa, D., Louro, R. O., & Paquete, C. M. (2021). Crossing the wall: Characterization of the multiheme cytochromes involved in the extracellular electron transfer pathway of *Thermincola ferriacetica*. *Microorganisms, 9*, 293.

Filman, D. J., Marino, S. F., Ward, J. E., Yang, L., Mester, Z., Bullitt, E., Lovley, D. R., & Strauss, M. (2019). Cryo-EM reveals the structural basis of long-range electron transport in a cytochrome-based bacterial nanowire. *Communications Biology, 2*, 219.

García-Muñoz, J., Amils, R., Fernández, V. M., de Lacey, A., & Malki, M. (2011). Electricity generation by microorganisms in the sediment-water interface of an extreme acidic microcosm. *International Microbiology, 14*, 73–81.

Glasser, N. R., Saunders, S. H., & Newman, D. K. (2017). The colorful world of extracellular electron shuttles. *Annual Review of Microbiology, 71*, 731–751.

Gomaa, O. M., Costa, N. L., & Paquete, C. M. (2022). Electron transfer in Gram-positive bacteria: Enhancement strategies for bioelectrochemical applications. *World Journal of Microbiology and Biotechnology, 38*, 83.

Gralnick, J. A., & Bond, D. R. (2023). Electron transfer beyond the outer membrane: Putting electrons to rest. *Annual Review of Microbiology, 77*, 517–539.

Grattieri, M., & Minteer, S. D. (2018). Microbial fuel cells in saline and hypersaline environments: Advancements, challenges and future perspectives. *Bioelectrochemistry, 120*, 127–137.

Gu, Y., Guberman-Pfeffer, M. J., Srikanth, V., Shen, C., Giska, F., Gupta, K., Londer, Y., Samatey, F. A., Batista, V. S., & Malvankar, N. S. (2023). Structure of *Geobacter* cytochrome OmcZ identifies mechanism of nanowire assembly and conductivity. *Nature Microbiology, 8*, 284–298.

Guo, F., Luo, H., Shi, Z., Wu, Y., & Liu, H. (2021). Substrate salinity: A critical factor regulating the performance of microbial fuel cells, a review. *Science of The Total Environment, 763*, 143021.

Holmes, D. E., Bond, D. R., O'Neil, R. A., Reimers, C. E., Tender, L. R., & Lovley, D. R. (2004a). Microbial communities associated with electrodes harvesting electricity from a variety of aquatic sediments. *Microbial Ecology, 48*, 178–190.

Holmes, D. E., Nicoll, J. S., Bond, D. R., & Lovley, D. R. (2004b). Potential role of a novel psychrotolerant member of the family *Geobacteraceae*, *Geopsychrobacter electrodiphilus* gen. nov., sp. nov., in electricity production by a marine sediment fuel cell. *Applied and Environmental Microbiology, 70*, 6023–6030.

Huang, L., Regan, J. M., & Quan, X. (2011). Electron transfer mechanisms, new applications, and performance of biocathode microbial fuel cells. *Bioresource Technology, 102*, 316–323.

Iigatani, R., Ito, T., Watanabe, F., Nagamine, M., Suzuki, Y., & Inoue, K. (2019). Electricity generation from sweet potato-shochu waste using microbial fuel cells. *Journal of Bioscience and Bioengineering, 128*, 56–63.

Jadhav, G. S., & Ghangrekar, M. M. (2009). Performance of microbial fuel cell subjected to variation in pH, temperature, external load and substrate concentration. *Bioresource Technology, 100*, 717–723.

Ketep, S. F., Bergel, A., Bertrand, M., Achouak, W., & Fourest, E. (2013). Lowering the applied potential during successive scratching/re-inoculation improves the performance of microbial anodes for microbial fuel cells. *Bioresource Technology, 127*, 448–455.

Kim, B. H., Kim, H. J., Hyun, M. S., & Park, D. H. (1999). Direct electrode reaction of Fe(III)-reducing bacterium, *Shewanella putrefaciens*. *Journal of Microbiology and Biotechnology, 9*, 127–131.

Kim, H., Kim, B., Kim, J., Lee, T., & Yu, J. (2014). Electricity generation and microbial community in microbial fuel cell using low-pH distillery wastewater at different external resistances. *Journal of Biotechnology, 186*, 175–180.

Larrosa-Guerrero, A., Scott, K., Head, I. M., Mateo, F., Ginesta, A., & Godinez, C. (2010). Effect of temperature on the performance of microbial fuel cells. *Fuel, 89*, 3985–3994.

Lefebvre, O., Tan, Z., Kharkwal, S., & Ng, H. Y. (2012). Effect of increasing anodic NaCl concentration on microbial fuel cell performance. *Bioresource Technology, 112*, 336–340.

Leong, J. X., Daud, W. R. W., Ghasemi, M., Liew, K. B., & Ismail, M. (2013). Ion exchange membranes as separators in microbial fuel cells for bioenergy conversion: A comprehensive review. *Renewable and Sustainable Energy Reviews, 28*, 575–587.

Light, S. H., Su, L., Rivera-Lugo, R., Cornejo, J. A., Louie, A., Iavarone, A., Ajo-Franklin, C. M., & Portnoy, D. A. (2018). A flavin-based extracellular electron transfer mechanism in diverse Gram-positive bacteria. *Nature, 562*, 140–144.

Liu, M., Yuan, Y., Zhang, L. X., Zhuang, L., Zhou, S. G., & Ni, J. R. (2010). Bioelectricity generation by a Gram-positive *Corynebacterium* sp. strain MFC03 under alkaline condition in microbial fuel cells. *Bioresource Technology, 101*, 1807–1811.

Liu, L., Tsyganova, O., Lee, D. J., Su, A., Chang, J. S., Wang, A., & Ren, N. (2012). Anodic biofilm in single-chamber microbial fuel cells cultivated under different temperatures. *International Journal of Hydrogen Energy, 38*, 15574–15579.

Logan, B. E., Rossi, R., Ragab, A., & Saikaly, P. E. (2019). Electroactive microorganisms in bioelectrochemical systems. *Nature Reviews. Microbiology, 17*, 307–319.

Lovley, D. R., & Holmes, D. E. (2022). Electromicrobiology: The ecophysiology of phylogenetically diverse electroactive microorganisms. *Nature Reviews. Microbiology, 20*, 5–19.

Lovley, D. R., & Walker, D. J. F. (2019). *Geobacter* protein nanowires. *Frontiers in Microbiology, 10*, 2078.

Lusk, B. G. (2019). Thermophiles; or, the modern prometheus: The importance of extreme microorganisms for understanding and applying extracellular electron transfer. *Frontiers in Microbiology, 10*, 818.

Malki, M., de Lacey, A. L., Rodríguez, N., Amils, R., & Fernandez, V. M. (2008). Preferential use of an anode as an electron acceptor by an acidophilic bacterium in the presence of oxygen. *Applied and Environmental Microbiology, 74*, 4472–4476.

Mathis, B. J., Marshall, C. W., Milliken, C. E., Makkar, R. S., Creager, S. E., & May, H. D. (2008). Electricity generation by thermophilic microorganisms from marine sediment. *Applied Microbiology and Biotechnology, 78*, 147–155.

Miller, L. G., & Oremland, R. S. (2008). Electricity generation by anaerobic bacteria and anoxic sediments from hypersaline soda lakes. *Extremophiles, 12*, 837–848.

Pierra, M., Carmona-Martínez, A. A., Trably, E., Godon, J. J., & Bernet, N. (2015). Specific and efficient electrochemical selection of *Geoalkalibacter subterraneus* and *Desulfuromonas acetoxidans* in high current-producing biofilms. *Bioelectrochemistry, 106*, 221–225.

Potter, M. C. (1910). On the difference of potential due to the vital activity of microorganisms. *Proceedings of the University of Durham Philosophical Society, 3*, 245–249.

Ramirez-Nava, J., Martínez-Castrejón, M., García-Mesino, R. L., López-Díaz, J. A., Talavera-Mendoza, O., Sarmiento-Villagrana, A., Rojano, F., & Hernández-Flores, G. (2021). The implications of membranes used as separators in microbial fuel cells. *Membranes, 11*, 738.

Reguera, G., McCarthy, D., Mehta, T., Nicoll, J. S., Touminen, M. T., & Lovley, D. R. (2005). Extracellular electron transfer via microbial nanowires. *Nature, 435*, 1098–1101.

Sharma, A., Kawarabayashi, Y., & Setyanarayana, T. (2012). Acidophilic bacteria and archaea: Acid stable biocatalysts and their potential applications. *Extremophiles, 16*, 1–19.

Shi, W., Wang, V. B., Zhao, C. E., Zhang, Q., Loo, S. C. J., Yang, L., & Xu, C. (2015). *Anditalea andensis* ANESC-S^T – An alkaliphilic halotolerant bacterium capable of electricity generation under alkaline-saline conditions. *PLoS One, 10*, e0132766.

Shrestha, N., Tripathi, A. K., Govil, T., Sani, R. K., Urgun-Demirtas, M., Kasthuri, V., & Gadhamshetty, V. (2020). Electricity from lignocellulosic substrates by thermophilic *Geobacillus* species. *Scientific Reports, 10*, 17047.

Takai, K., Nakamura, K., Toki, T., Tsunogai, U., Miyazaki, M., Miyazaki, J., Hirayama, H., Nakagawa, S., Nunoura, T., & Horikoshi, K. (2008). Cell proliferation at 122°C and isotopically heavy CH4 production by a hyperthermophilic methanogen under high-pressure cultivation. *Proceedings of the National Academy of Sciences of the United States of America, 105*, 10949–10954.

Vargas, M., Kashefi, K., Blunt-Harris, E. L., & Lovley, D. R. (1998). Microbiological evidence for Fe(III) reduction on early Earth. *Nature, 395*, 65–67.

Wang, Z., Mahadevan, G. D., Wu, Y., & Zhao, F. (2017). Progress of air-breathing cathode in microbial fuel cells. *Journal of Power Sources, 356*, 245–255.

Wang, F., Gu, Y., O'Brien, J. P., Yi, S. M., Yalcin, S. E., Srikanth, V., Shen, C., Vu, D., Ing, N. L., Hochbaum, A. I., Egelman, E. H., & Malvankar, N. S. (2019). Structure of microbial nanowires reveals stacked hemes that transport electrons over micrometers. *Cell, 177*, 361–369.

Wang, F., Chan, C. H., Suciu, V., Mustafa, K., Ammend, M., Si, D., Hochbaum, A. I., Egelman, E., & Bond, D. R. (2022a). Structure of *Geobacter* OmcZ filaments suggests extracellular cytochrome polymers evolved independently multiple times. *eLife, 11*, e81551.

Wang, F., Mustafa, K., Suciu, V., Joshi, K., Chan, C. H., Choi, S., Su, Z., Si, D., Hochbaum, A. I., Egelman, E., & Bond, D. R. (2022b). Cryo-EM structure of an extracellular *Geobacter* OmcE cytochrome filament reveals tetrahaem packing. *Nature Microbiology, 7*, 1291–1300.

Watanabe, K., Manefield, M., Lee, M., & Kouzuma, A. (2009). Electron shuttles in biotechnology. *Current Opinion in Biotechnology, 20*, 633–641.

Zhang, T., Zhang, L., Su, W., Gao, P., Li, D., He, X., & Zhang, Y. (2011). The direct electrocatalysis of phenazine-1-carboxylic acid excreted by *Pseudomonas alcaliphila* under alkaline condition in microbial fuel cells. *Bioresource Technology, 102*, 7099–7102.

Zhang, L., Shen, J., Wang, L., Ding, L., Xu, K., & Ren, H. (2014). Stable operation of microbial fuel cells at low temperatures (5–10°C) with light exposure and its anodic microbial analysis. *Bioprocess and Biosystems Engineering, 37*, 819–827.

Zhang, X., Wang, H., Xia, T., & Wang, X. (2021). Characterization of a new electrochemically active bacterium phylogenetically related to *Alicyclobacillus hesperidum* and its electrochemical performance in microbial fuel cell. *Biosensors & Bioelectronics, 175*, 112865.

Zhao, C., Jiao, T., Zhang, W., Zhang, W., Jia, M., Liu, S., Zhang, M., Han, F., Han, Y., Lei, J., Wang, X., & Zhou, W. (2024). Nutrients recovery by coupled bioreactor of heterotrophic ammonia assimilation and microbial fuel cell in saline wastewater. *Science of the Total Environment, 918*, 170697.

Zhuang, L., Zhou, S., Li, Y., & Yuan, Y. (2010). Enhanced performance of air-cathode two-chamber microbial fuel cells with high-pH anode and low-pH cathode. *Bioresource Technology, 101*, 3514–3519.

Chapter 19
Characteristics of Radiation-Resistant Bacteria and Their Biotechnological Applications

Issay Narumi

Abstract The most extensively studied radiation-resistant bacterium is *Deinococcus radiodurans*. Several radiation resistance mechanisms possessed by this bacterium have been elucidated at the molecular level. In contrast, microorganisms other than *D. radiodurans* exhibit resistance to ionizing radiation. Various radiation-resistant bacteria have been isolated from groups that are evolutionarily close to *Deinococcus* as well as those that are evolutionarily distant from it. In addition, some archaea are resistant to radiation. For applications in the biotechnology industry, research is being conducted to utilize radiation-resistant bacteria as hosts or harness the genes they possess, and further research and development are anticipated.

Keywords Radiation-resistant bacteria · Biotechnological applications · Radiation-resistance mechanisms · Genetic resources · Host-vector systems

19.1 Introduction

To a certain extent, all living organisms on Earth possess a certain level of resistance to ionizing radiation that exceeds the natural background radiation level (approximately 2 mGy/year). However, the degree of radiation resistance varies by more than ten thousandfold among different species. Ionizing radiation damages various biological macromolecules, such as DNA, proteins, and lipids, by directly imparting energy to them. Additionally, because cells contain a high proportion of water, reactive oxygen species are generated when ionizing radiation breaks down water both inside and outside the cell, causing indirect damage to these biological macromolecules. Among living organisms, certain radiation-resistant bacteria are particularly resistant to radiation. In this chapter, I discuss the sources and

I. Narumi (✉)
Faculty of Life Sciences, and Bio-Resilience Research Project (BRRP), Toyo University, Asaka, Saitama, Japan
e-mail: narumi@toyo.jp

M. Ito et al. (eds.), *Extremophiles for a Sustainable Future*,
https://doi.org/10.1007/978-981-95-6970-0_19

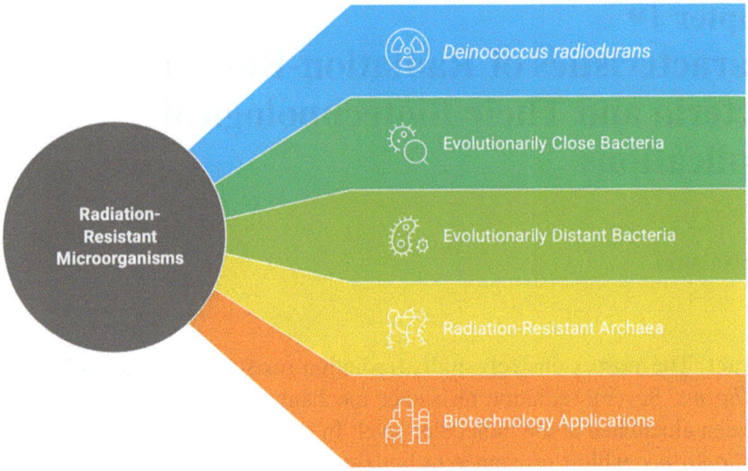

Fig. 19.1 Exploring radiation-resistant microorganisms

characteristics of several radiation-resistant bacteria, as well as their mechanisms of radiation resistance, which have been revealed thus far. Moreover, because a considerable number of archaeal species also exhibit radiation resistance, I will introduce some examples of these as well. Furthermore, I will discuss the industrial applications of radiation-resistant bacteria, as well as the use of enzymes and metabolic products produced by these bacteria (Fig. 19.1).

19.2 *Deinococcus* Species

The first report on the discovery of radiation-resistant bacteria was published in 1956 (Anderson et al., 1956). A bacterium discovered in canned beef from a gamma-ray sterilization test at an agricultural experiment station in Oregon, USA, was not completely sterilized even after exposure to a high dose of 60 kGy gamma rays. This bacterium multiplied when shaken in nutrient medium at 30 °C for 24 h, and when cultured on agar medium for three days, reddish-pink colonies were formed. The bacterium did not form spores and consistently exhibited vegetative cells (Fig. 19.2). In 1960, it was named *Micrococcus radiodurans* (Raj et al., 1960), and in the 1974 edition of the Guinness Book, it was listed as "the toughest microorganism in the world" (Moseley, 1983). Based on analyses of the cell wall and membrane lipid composition as well as 16S rRNA phylogenetic analysis, this radiation-resistant bacterium was reclassified in 1981 from the genus *Micrococcus* to a new genus, *Deinococcus*, within the newly established family *Deinococcaceae*, which is phylogenetically distant (Brooks & Murray, 1981). In 1997, a new order, *Deinococcales*, was established (Rainey et al., 1997), and a new class was created in 2001. As for the phylum, it was previously known as the *Deinococcus-Thermus*

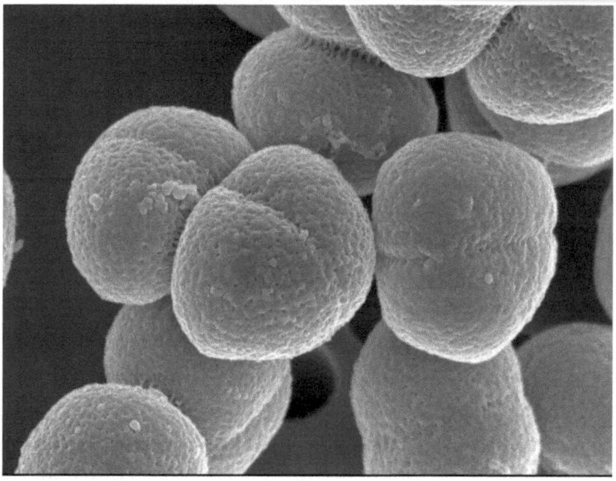

Fig. 19.2 Scanning electron microscope image of *D. radiodurans*

phylum (Weisburg et al., 1989), but in 2021, a new phylum, *Deinococcota*, was established (Oren & Garrity, 2021). The scientific name *Deinococcus radiodurans* means "strange coccus capable of withstanding radiation." The standard strain was referred to as R_I or R_1 in previous studies. However, since the second edition of Bergey's Manual of Systematic Bacteriology adopted the name R1 strain, it has been referred to as the R1 strain thereafter.

Since the discovery of *D. radiodurans*, >90 species of the genus *Deinococcus* have been isolated worldwide (Parte et al., 2020). The sources of isolation are highly diverse, including animal feces (llama, elephant), termite guts, freshwater fish (carp), seawater fish (cod, horse mackerel, Bombay duck), oysters, radium hot springs, mine aquifers, paper mill wastewater, Antarctic rocks, deserts, activated sludge, alpine soil, deep-sea sediments, hospital dust, the troposphere, and the stratosphere. Although not all species have been tested, bacteria of the genus *Deinococcus* not only exhibit resistance to ionizing radiation such as gamma rays but also show high resistance to physical and chemical agents that cause DNA damage, including ultraviolet light, desiccation, and DNA cross-linking agents. Because all sources of isolation are presumed to be locations with natural radiation dose levels, *Deinococcus* species are not explicitly found in areas with particularly high radiation levels in the current global environment. Despite the genus name, some *Deinococcus* species are rod-shaped.

Most bacteria in the genus *Deinococcus* are mesophiles, with an optimal growth temperature of approximately 30 °C. Strains with an optimal growth temperature of approximately 45 °C have also been isolated from terrestrial hot springs in Italy and Portugal and have been named *Deinococcus geothermalis* and *Deinococcus murrayi*, respectively (Ferreira et al., 1997). In 2003, a report was published on a *Deinococcus* species that could be cultured at 90 °C under anaerobic conditions and isolated from the surface of a hydrothermal vent in the Manus Basin, Papua New Guinea. Unfortunately, it was lost because it was unable to continue proliferating during subsequent subcultures (Kimura et al., 2003).

19.3 Radiation Resistance Mechanisms of *Deinococcus radiodurans*

D. radiodurans is not only resistant to electromagnetic radiation, such as gamma rays and X-rays, but also exhibits high resistance to particle radiation, including electrons, protons, and heavy charged particles (Moseley, 1983). Typical organisms, from ordinary microorganisms to humans, are generally more sensitive to radiation with higher energy transfer, such as heavy charged particles, than to radiation with lower energy transfer, such as gamma rays and electron beams. However, *D. radiodurans* exhibits equal resistance to all types of radiation, regardless of the kind, and maintains the same level of tolerance to any radiation quality (Kobayashi et al., 1995).

When *D. radiodurans* is exposed to high doses of radiation, its genomic DNA undergoes double-strand breaks and becomes fragmented, a process that is similar to that observed in other microorganisms. However, after a short period of cultivation, the broken DNA is repaired and the original genome is restored (Grimsley et al., 1991). In most organisms, double-strand breaks are the most challenging type of DNA damage to repair; in humans, even a few such breaks can be fatal. The desire to unravel the molecular mechanisms behind why *D. radiodurans* is proficient in repairing DNA double-strand breaks has motivated researchers to study radiation-resistant bacteria.

Kaplans and Zavarine examined the resistance to X-rays and ultraviolet radiation in eight types of bacteria with genomic GC content ranging from 34% to 67%. They found that X-ray resistance was inversely correlated with genomic GC content, whereas ultraviolet resistance was positively correlated with genomic GC content (Kaplan & Zavarine, 1962). However, *D. radiodurans*, which has a genomic GC content of 67%, exhibits high resistance to both X-rays and ultraviolet light. Therefore, it has long been considered that its resistance mechanisms are entirely different from those of typical bacteria (Moseley & Schein, 1964). Because *D. radiodurans* forms reddish-pink colonies, it was speculated that carotenoid pigments might be involved in X-ray resistance. However, this hypothesis was rejected because the X-ray resistance of white-colony mutants derived from *D. radiodurans* R1 was significantly higher than that of *Pseudomonas* species, a typical bacterium with similar genomic GC content (Moseley & Schein, 1964). Dean et al. were the first to report the presence of DNA double-strand break repair capability in *D. radiodurans* due to radiation-induced damage using a DNA viscosity measurement method (Dean et al., 1966). Kitayama and Matsuyama were the first to apply the sucrose density gradient centrifugation method to the genomic DNA of *D. radiodurans*. They have reported the capacity for DNA double-strand break repair (Kitayama & Matsuyama, 1968).

Based on the results of transformation experiments, it has been suggested that *D. radiodurans* possesses multiple genomes that are independently distributed during cell division (Tirgari & Moseley, 1980). *D. radiodurans* grows as tetrads under standard culture conditions. Furthermore, as each cell in the tetrad grows, septa develop from the cell wall toward the center of the cell, and just before cell division, the genome copy number in a tetrad increases to at least eight (Chou & Tan, 1991). Even when considering the tetrad growth form, it is known that in *D. radiodurans*, the genome copy number per cell ranges from four to ten (Hansen, 1978; Harsojo et al., 1981). In budding yeast, radiation resistance differs between haploid and diploid states, with diploid yeast having higher radiation resistance (Mortimer, 1958). However, no correlation has been observed between genome copy number and radiation resistance in *D. radiodurans* (Harsojo et al., 1981). Incidentally, *Azotobacter vinelandii*, which possesses a high genome copy number (40–80), is sensitive to radiation (Maldonado et al., 1994). Additionally, it has been found that *Thermus thermophilus*, a hyperthermophile belonging to the phylum *Deinococcota*, also possesses a polyploid genome (Ohtani et al., 2010), yet it is sensitive to radiation. Therefore, it is not possible to directly link genome copy number with recombinational repair ability; however, *D. radiodurans* may possess an as-yet-unknown mechanism for effectively utilizing duplicated genomic information. Moseley, who was the first to write a comprehensive review on *D. radiodurans*, concluded his review with the following words: "*M. radiodurans* probably has a nuclear structure in which a number of identical genomes (four or five) are arranged in such a way that recombinational exchange involving homologous regions of all copies may occur." (Moseley, 1983).

The genome of *D. radiodurans*, decoded in 1999, consists of four circular double-stranded DNA molecules (chromosome 1, chromosome 2, megaplasmid pMP1, and plasmid pCP1) (White et al., 1999). It has also been observed that in chromosome 2 and the megaplasmid that comprises the genome, both monomers and multimers are formed (Kikuchi et al., 1999). Genome analysis revealed that *D. radiodurans* possesses known DNA repair genes found in other organisms; however, it is possible that among its genes of unknown function, there may be DNA repair genes unique to the genus *Deinococcus*. Furthermore, since the development of gene manipulation systems using *D. radiodurans* plasmids had already begun in 1991 (Smith et al., 1991), techniques for functional analysis of known DNA repair protein homologs and genes of unknown function, such as gene disruption experiments, gene complementation experiments, and gene overexpression experiments, became readily available soon after the start of the post-genome era.

If there are multiple copies of the genome set in a single cell, even if a double-strand break occurs in some of them, intact homologous regions exist to serve as substrates for homologous recombination repair. Therefore, it is reasonable that a highly efficient homologous recombination repair mechanism plays an essential role in radiation resistance. Minton et al. proposed that in the genome of *D. radiodurans*, the four-stranded DNA intermediate (Holliday junction), which arises in the final stage of the homologous recombination repair pathway, is formed even under normal growth conditions, allowing for highly efficient homologous recombination repair (Minton & Daly, 1995). However, no evidence supporting this hypothesis was found when genomic DNA was directly observed using optimal mapping (Lin et al., 1999). Levin-Zaidman et al. proposed, based on transmission electron microscopy observations, that the genome of *D. radiodurans* is structured in a donut-like ring shape, which may be related to its radiation resistance (Levin-Zaidman et al.,

2003). On the other hand, Zimmerman & Battista argued that a ring-shaped genome structure is not necessary for radiation resistance and that it is the compact condensation of the genome within the cell that characterizes the *D. radiodurans* genome structure (Zimmerman & Battista, 2005). Based on these findings, Cox and Battista (2005) proposed the hypothesis that *D. radiodurans* possesses a mechanism that stabilizes DNA ends so that even when double-strand breaks occur, the two DNA ends do not dissociate (Cox & Battista, 2005).

Zahradka et al. proposed a model called extended synthesis-dependent strand annealing (ESDSA) (Zahradka et al., 2006). In this model, DNA polymerase and RecA are the primary enzymes responsible for DNA repair. The single-stranded ends of the DNA fragments invade partially overlapping double-stranded DNA, forming a D-loop structure and extending the DNA. Subsequently, the newly synthesized strand dissociates from the template, creating a long single-stranded DNA molecule that can then join complementary sequences to reconstruct the double-stranded DNA accurately.

Ishino and Narumi proposed a DNA double-strand break repair model called condensed nucleoid-dependent end-joining (CNDEJ), which is separate from ESDSA (Ishino & Narumi, 2015). Unlike eukaryotic non-homologous end joining (NHEJ), in which damaged DNA ends may drift apart during the repair process, this model describes a mechanism unique to radiation-resistant bacteria with condensed genomes. A protein found exclusively in the genus *Deinococcus*, discovered by Narumi et al. in 2004, plays a central role in CNDEJ. This protein preferentially binds to DNA with strand breaks, protects DNA ends from exonucleases, and promotes DNA ligation through the action of DNA ligase. Therefore, it was named pleiotropic protein promoting DNA repair (PprA) (Narumi et al., 2004). Besides DNA ligase, PprA also interacts with topoisomerase IB and DNA gyrase and is thought to facilitate efficient double-strand break repair while preventing the dissociation of double-strand break ends (Devigne et al., 2013; Kota et al., 2014a, b; Adachi et al., 2014).

A group of proteins induced by DNA damage has been identified in *D. radiodurans*. These include not only RecA (Narumi et al., 2001) and PprA (Ohba et al., 2005), but also DdrA, DdrB, DdrC, and DdrD (Tanaka et al., 2004). Although some functional analyses have been conducted to date (Lim et al., 2019), it remains to be clarified whether these Ddr proteins are involved in ESDSA or CNDEJ and how they interact with other DNA repair proteins.

A different induction mechanism, separate from the SOS response system, which is widely conserved in *Escherichia coli* and other bacteria, is involved in the induction of DNA repair genes in *D. radiodurans*. This was discovered by analyzing a radiation-sensitive mutant strain, KH840, of *D. radiodurans*. In this strain, a 1736 bp mobile genetic element (IS*8301*) (Islam et al., 2003) was inserted after the 307th nucleotide of the gene of unknown function located on the first chromosome (Hua et al., 2003). When the induction of PprA and RecA expression after gamma irradiation was examined in the strain KH840, neither was induced after radiation exposure in this strain. The protein encoded by this gene was named PprI, for "Inducer of PprA" (Hua et al., 2003). PprI has a protease domain at the N-terminus and activates the radiation response by degrading DdrO, a repressor of the radiation response mechanism (Ludanyi et al., 2014).

Proteome analysis of the wild-type strain of *D. radiodurans* and the *pprI* knockout strain revealed that the isoelectric point of a protein of approximately 10 kDa shifted slightly toward the acidic side. This protein has been identified as a homolog of cold shock proteins. Disruption of the gene encoding this protein resulted in a strain that showed gamma-ray sensitivity comparable to that of the *pprA* knockout strain. Additionally, it was found that the expression of PprA was not suppressed in the absence of irradiation. Based on these results, this protein was named PprM because it modulates PprA (Ohba et al., 2009). PprM possesses an RNA-binding domain, and it has been revealed that binding to RNA is essential for its functional expression (Li et al., 2017; Villa et al., 2021).

In addition to polyploidy, unique genome structures, and stress-induced DNA repair proteins, *D. radiodurans* also possesses passive defense mechanisms against DNA damage. It has long been established that elemental analysis using electron microscopy revealed a high concentration of Mn ions in the cytoplasm of *D. radiodurans* (Leibowitz et al., 1976). Daly et al. discovered that Mn ions are essential for radiation resistance (Daly et al., 2004). In *D. radiodurans* and other radiation-resistant bacteria, the intracellular Mn/Fe ratio tends to be high, and a relationship between defense against protein oxidation by carbonyl modifications and the presence of Mn ions has been reported (Daly et al., 2007). By replacing iron ions, Mn ions play a role in preventing the formation of hydroxyl radicals through the Fenton reaction (Imlay, 2014). Within the cells of *D. radiodurans*, Mn ions bind to low-molecular-weight metabolites, such as orthophosphates and peptides, which exist as manganese antioxidants (Sharma et al., 2017).

19.4 Other Radiation-Resistant Bacterial and Archaeal Species

In addition to the genus *Deinococcus*, the family *Deinococcaceae* also includes the genus *Deinobacterium*, from which the rod-shaped bacterium *Deinobacterium chartae*, with an optimal growth temperature of 37–45 °C, has been isolated from the biofilm of a paper mill wastewater facility in Finland (Ekman et al., 2011). Furthermore, within the order *Deinococcales*, there is a family called *Trueperaceae*, in addition to *Deinococcaceae*, from which the coccus bacterium *Truepera radiovictrix*, with an optimal growth temperature of 50 °C, has been isolated from a hot spring on São Miguel Island in the Azores (Albuquerque et al., 2005). Both bacteria have been reported to possess a similar degree of gamma-ray resistance as *D. radiodurans*, and their genome sequences have been decoded; however, no further analyses of their resistance mechanisms have been conducted. Additionally, for both bacteria, only one species from a single genus has been isolated, and much remains unknown regarding their habitats.

Rubrobacter radiotolerans is a gram-positive, aerobic, rod-shaped bacterium belonging to the phylum *Actinomycetota*, which was isolated from a biofilm in the riverbed bath of Misasa Onsen in Tottori Prefecture (Fig. 19.3) (Yoshinaka et al., 1973). Its D_{10} value, the gamma-ray dose at which 10% survival remains, is approximately 25 kGy, indicating a much higher level of radiation resistance than that of *D. radiodurans*. *R. radiotolerans* grows slowly on nutrient media, taking approximately 20 days to form visible colonies on the agar plates. Depending on the medium composition, bacterial clumps may form during liquid culture that are difficult to disrupt, making it challenging to obtain a uniformly concentrated cell suspension. Therefore, very few studies have used *R. radiotolerans* as a host, and it remains unknown whether it is naturally competent for transformation. *R. radiotolerans* produces a red carotenoid pigment. The primary pigment is bacterioruberin, a pigment produced by extreme halophiles that function as scavengers of reactive oxygen species (Saito et al., 1994). When the genomic DNA of *R. radiotolerans* was analyzed using sucrose density gradient centrifugation, the number of strand breaks after radiation exposure was found to be less than that observed in other bacteria, and the size of the fragmented DNA did not revert to its original state (Terato et al., 1999). Some studies have suggested that high intracellular accumulation levels of Mn ions and trehalose, which act as compatible solutes, are related to radiation resistance (Webb & DiRuggiero, 2012). Based on these findings, it is thought that the radiation resistance of *R. radiotolerans* is due to its high functionality in DNA damage protection mechanisms rather than in DNA repair mechanisms. *Kineococcus radiotolerans*, isolated from a high-level radioactive waste facility at the Savannah River Site in South Carolina, USA, also belongs to the phylum *Actinomycetota* (Phillips et al., 2002). The D_{10} value for gamma radiation in *K. radiotolerans* was approximately 7 kGy, which was nearly the same as that of *D. radiodurans*. Although the genome sequences of these microorganisms have been decoded, few studies have analyzed their resistance mechanisms.

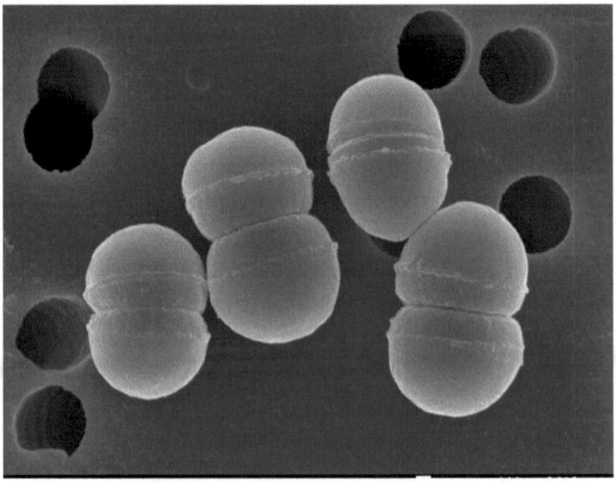

Fig. 19.3 Scanning electron microscope image of *R. radiotolerans*

Several terrestrial cyanobacteria of the genus *Chroococcidiopsis*, which have been isolated from deserts and high-salinity evaporation ponds and possess desiccation resistance, are known to exhibit resistance to ionizing radiation (Billi et al., 2000). The most highly resistant strains have a D_{10} value of approximately 5 kGy for gamma rays. Resistance to desiccation and radiation is correlated with the oxidative defense capability of proteins (Fagliarone et al., 2017). Because the compatible solute trehalose is thought to play an essential role in the desiccation resistance of *Chroococcidiopsis* species, trehalose is believed to be involved in its resistance to radiation (Li et al., 2025).

Thermococcus and *Pyrococcus*, which belong to the family *Thermococcaceae*, are anaerobic hyperthermophilic archaea (Zilling et al., 1983; Fiala & Stetter, 1986). The D_{10} values for gamma rays in *Thermococcus gammatolerans* and *Pyrococcus abyssi* were 6 and 3 kGy, respectively, indicating gamma-ray resistance comparable to that of *the Deinococcus* species (Jolivet et al., 2004). Exposure to gamma rays causes its genome to break, and DNA repair occurs during subsequent cultivation (DiRuggiero et al., 1997). Additionally, no proteomic changes induced by stress have been observed (Gérard et al., 2001). The intracellular Mn/Fe ratio in hyperthermophilic archaea is extremely low, and compatible solutes with high cytoplasmic concentrations have not yet been detected (Webb & DiRuggiero, 2012). Therefore, the mechanism underlying the radiation resistance of hyperthermophilic archaea remains unclear.

Strains of highly radiation-resistant archaea have also been reported among the extremely halophilic archaea that inhabit high-salinity environments, such as salt pans. *Halobacterium salinarum*, a member of the family *Halobacteriaceae*, is a highly halophilic archaeon with an optimal growth temperature of 50 °C. The D_{10} value for gamma radiation in *H. salinarum* strain NRC-1 was 5 kGy, and it is known to exhibit both high radiation resistance and strong desiccation tolerance (Kottemann et al., 2005). The intracellular Mn/Fe ratio in *H. salinarum* was exceptionally high, similar to that in *D. radiodurans* (Webb & DiRuggiero, 2012). Halide ions, which have high reactivity with hydroxyl radicals, are believed to be involved in the oxidative defense mechanisms of cellular macromolecules (Kish et al., 2009).

At present, a common feature shared by hyperthermophilic and highly halophilic archaea is that they possess polyploid genomes. *Thermococcus kodakarensis* has a polyploid genome, and the number of genome copies per cell varies from 7 to 19, depending on the growth stage (Spaans et al., 2015). *H. salinarum* has an average of approximately 25 genome copies during the logarithmic growth phase, which decreases to 15 copies during the early stationary phase (Breuert et al., 2006). The advantages of having polyploid genomes are thought to include a low mutation rate, high resistance to radiation, desiccation due to dependence on homologous recombination repair, and serving as a source of phosphate (Zerulla & Soppa, 2014).

19.5 Application to Bioremediation

If genes that confer the ability to remove persistent or hazardous substances or genes involved in the cellular uptake of elements can be introduced or highly expressed in radiation-resistant bacteria through genetic engineering techniques, it would be possible to remove persistent or hazardous substances contained in radioactive waste, as well as to remove radionuclides from soil or rivers contaminated with radioactive materials. This would be extremely useful as a bioremediation technology for removing radioactive substances.

Examples of bioremediation technologies using *D. radiodurans* include the following. By expressing the toluene dioxygenase gene derived from *Pseudomonas putida* on the chromosome of *D. radiodurans*, recalcitrant toluene was converted into substances that were more easily degradable (Lange et al., 1998). Hg(II) ions were reduced to less toxic and volatile metallic Hg by expressing an Hg(II) reduction gene derived from *E. coli* on the chromosome of *D. radiodurans* (Brim et al., 2000). Using *D. radiodurans* as a host, either a non-specific acid phosphatase gene derived from *Salmonella enterica* serovar Typhi or an alkaline phosphatase gene derived from *Sphingomonas* species was expressed on a plasmid, enabling the precipitation of uranium from uranyl nitrate solution (Appukuttan et al., 2006, 2011; Kulkarni et al., 2013). To enable bioremediation of cellulosic waste in radioactive environments, an endoglucanase gene derived from *Bacillus pumilus* was expressed on a plasmid using *D. radiodurans* as a host (Telang et al., 2014). Using *D. radiodurans* as a host, a synthetic gene encoding a phytochelatin analog or a metallothionein gene derived from *Synechococcus* species was expressed on a plasmid, thereby improving Cd accumulation capability (Chaturvedi & Archana, 2014). Using *D. radiodurans* as a host, Ni/Co transporter genes derived from either *Rhodopseudomonas palustris* or *Novosphingobium aromaticivorans* were expressed on a plasmid, resulting in improved removal efficiency of radioactive cobalt from contaminated water (Gogada et al., 2015). *D. geothermalis* has also been used as a host to express an *E. coli*-derived Hg(II) reduction gene on a plasmid, thereby reducing Hg(II) ions to the less toxic metallic mercury (Brim et al., 2003).

19.6 Utilization of Genetic Resources

Apart from their characteristic of radiation resistance, bacteria of the genus *Deinococcus* hold significant potential as valuable biological resources for bioremediation and various other industrial applications. Below are some examples. *Deinococcus* species produce deinoxanthin, an antioxidant with potent ROS-scavenging activity (Lemee et al., 1997). They degrade polycyclic aromatic hydrocarbons, which are urban air pollutants (Waight et al., 2007). They break down ethyl acetate, which is a marine pollutant (Kongpol et al., 2008). They produce polyamines, which are known to have significant effects on healthy lifespan (Hamana et al., 2009). They degrade dibutyl phthalate, a chemical that disrupts the endocrine system (Liao et al., 2010). They produce quorum-quenching enzymes that can be used to prevent food spoilage, prevent bacterial infections, and facilitate bioremediation (Kalia et al., 2011). They break down malachite green, the use of which is prohibited for aquatic animals (Lv et al., 2013). They degrade 17β-estradiol, which is another endocrine-disrupting chemical (Xiong et al., 2018). They alleviate heavy metal toxicity in rice plants (Dai et al., 2021). They reduce As(V) ions, which are highly toxic to the human body, thereby decreasing their toxicity (Zhang et al., 2022). The expression of beneficial traits such as the ability to decompose harmful substances or produce valuable materials often involves multiple genes. Introducing gene clusters from *Deinococcus* species involved in the expression of these beneficial traits into expression plasmids enables efficient generation of hosts with enhanced capacities to degrade harmful substances or produce functional materials. This not only benefits the field of bioremediation but also holds great promise for applications in the anti-aging cosmetics and health food manufacturing industries.

As mentioned above, it is believed that in the cells of *D. radiodurans*, Mn ions bind to low-molecular-weight metabolites, such as orthophosphate and peptides, existing as Mn antioxidants (Sharma et al., 2017). Interestingly, a synthetic Mn antioxidant complex, Mn-decapeptide-phosphate (MDP), can be used to protect against protein oxidative damage caused by radiation exposure during vaccine manufacturing (Tobin et al., 2020; Dollery et al., 2021).

The carotenoid deinoxanthin, produced by *D. radiodurans*, possesses high antioxidant activity, similar to other carotenoids (Lemee et al., 1997; Tian et al., 2007; Kuzucu, 2023). Jeong et al. (2021) reported that by placing the genes for 1-deoxy-D-xylulose-5-phosphate synthase and phytoene synthase (*dxs* and *crtB*), which are the rate-limiting enzymes in the carotenoid biosynthetic pathway of *D. radiodurans*, under the control of the *D. radiodurans groE* promoter and integrating them into plasmid pCP1, they were able to increase the expression level of deinoxanthin by 1.5-fold. The main surface layer protein of *D. radiodurans*, SlpA, forms a complex with deinoxanthin (S-layer deinoxanthin binding complex), which possesses antioxidant properties and is thermally stable (Farci et al., 2018). Interestingly, it has been reported that outer membrane vesicles released by *D. radiodurans* exhibit activity that alleviates oxidative stress in human epidermal keratinocyte cell lines. However, antioxidant activity against cultured cells is diminished in vesicles derived from an SlpA-deficient mutant of *D. radiodurans* (Han et al., 2023).

DNA repair-related proteins from radiation-resistant bacteria hold great potential as genetic engineering reagents. For example, the *D. radiodurans* PprA protein promotes the joining of DNA ends by DNA ligase, and this property has been utilized in the development of highly efficient DNA cloning kits (NIPPON GENE, 2025). Additionally, since the *D. radiodurans* PprA protein can recognize and bind to DNA break ends, this feature has been used to develop techniques for visualizing DNA breakage damage in cultured mammalian cells (Satoh et al., 2006).

19.7 Development of Host-Vector Systems

A host-vector system capable of stably expressing recombinant genes is required to realize the application of D. radiodurans in biotechnology. Plasmid vectors for expressing foreign genes in *D. radiodurans* and *D. geothermalis* have been limited to *D. radiodurans–E. coli* shuttle vectors, such as pI3 and its derivative plasmids, which were developed using the replicon of plasmid pUE10 from the *D. radiodurans* strain Sark and the *E. coli* vector (Masters & Minton, 1992; Meima & Lidstrom, 2000). It has been reported that pI3-series plasmids can also replicate in *Deinococcus deserti* (Dulermo et al., 2009). In recent years, systems utilizing conjugative plasmids have also been developed (Brumwell et al., 2022, 2023).

The plasmid vector pZT23 and its derivative plasmids, which enable gene expression in *Deinococcus grandis*, were developed based on the cryptic plasmid pUE30 derived from *Deinococcus radiopugnans* (Satoh et al., 2009). In the *D. grandis* strain TY3, it is possible to simultaneously maintain three types of shuttle plasmids (Sakai et al., 2004). When *D. grandis* spheroplasts are cultured in the presence of calcium, the periplasm enlarges to the extent that it becomes visible to the naked eye (Nishino et al., 2018, 2019; Morita et al., 2019; Nishino & Nishida, 2019). This enlarged periplasm may serve as a site for the accumulation of gene products.

The *groE* gene from *D. radiodurans* is commonly used as a promoter for constitutively high gene expression (Meima & Lidstrom, 2000). However, promoters of genes, such as *DR1261*, *rpmB*, and *dnaK*, can also be utilized to achieve high gene expression in *D. radiodurans* (Chen et al., 2019). For inducible gene expression, promoters located upstream of genes such as *pprA*, *ddrA*, *ddrB*, and *gyrB*, which have an operator sequence called the radiation/desiccation response motif (RDRM) nearby, can be utilized (Makarova et al., 2007). By placing the target gene downstream of these promoters, gene expression can be regulated by adding agents such as ultraviolet light, mitomycin C, methyl methanesulfonate, or ethidium bromide to the culture medium of the transformants (Narasimha & Basu, 2021).

19.8 Conclusions

Radiation-resistant bacteria include *Deinococcus*, *Deinobacterium*, *Truepera*, *Rubrobacter*, *Kineococcus*, and *Chroococcidiopsis*. Additionally, hyperthermophilic archaea, such as *Pyrococcus* and *Thermococcus*, and highly halophilic archaea, including *Halobacterium*, exhibit high levels of radiation resistance. These organisms have been isolated from a wide variety of sources, and there is not just one mechanism underlying radiation resistance. Because their diverse intracellular environments, biological defense mechanisms, and DNA repair systems do not share a common molecular foundation across species, radiation resistance is interpreted as the result of convergent evolution, with each group of organisms adapting to environmental pressures. Genetic resources of radiation-resistant bacteria provide valuable materials for promoting sustainable development. There is value in advancing research and development to utilize radiation-resistant bacteria as a useful biological resource.

References

Adachi, T., Hirayama, H., Shimizu, R., Satoh, K., Narumi, I., & Kuroki, R. (2014). Interaction of double-stranded DNA with polymerized PprA protein from *Deinococcus radiodurans*. *Protein Science, 23*, 1349–1358.

Albuquerque, L., Simões, C., Nobre, M. F., Pino, N. M., Battista, J. R., Silva, M. T., Rainey, F. A., & da Costa, M. S. (2005). *Truepera radiovictrix* gen. nov., sp. nov., a new radiation resistant species and the proposal of *Trueperaceae* fam. nov. *FEMS Microbiology Letters, 247*, 161–169.

Anderson, A. W., Nordan, H. C., Cain, R. F., Parrish, G., & Duggan, D. (1956). Studies on a radio-resistant micrococcus. I. Isolation, morphology, cultural characteristics and resistance to gamma radiation. *Food Technology, 10*, 575–578.

Appukuttan, D., Rao, A. S., & Apte, S. K. (2006). Engineering of *Deinococcus radiodurans* R1 for bioprecipitation of uranium from dilute nuclear waste. *Applied & Environmental Microbiology, 72*, 7873–7878.

Appukuttan, D., Seetharam, C., Padma, N., Rao, A. S., & Apte, S. K. (2011). PhoN-expressing, lyophilized, recombinant *Deinococcus radiodurans* cells for uranium bioprecipitation. *Journal of Biotechnology, 154*, 285–290.

Billi, D., Friedmann, E. I., Hofer, K. G., Caiola, M. G., & Ocampo-Friedmann, R. (2000). Ionizing-radiation resistance in the desiccation-tolerant cyanobacterium *Chroococcidiopsis*. *Applied & Environmental Microbiology, 66*, 1489–1492.

Breuert, S., Allers, T., Spohn, G., & Soppa, J. (2006). Regulated polyploidy in halophilic archaea. *PLoS One, 1*, e92.

Brim, H., McFarlan, S. C., Fredrickson, J. K., Minton, K. W., Zhai, M., Wackett, L. P., & Daly, M. J. (2000). Engineering *Deinococcus radiodurans* for metal remediation in radioactive mixed waste environments. *Nature Biotechnology, 18*, 85–90.

Brim, H., Venkateswaran, A., Kostandarithes, H. M., Fredrickson, J. K., & Daly, M. J. (2003). Engineering *Deinococcus geothermalis* for bioremediation of high-temperature radioactive waste environments. *Applied & Environmental Microbiology, 69*, 4575–4582.

Brooks, B. W., & Murray, R. G. E. (1981). Nomenclature for "*Micrococcus radiodurans*" and other radiation-resistant cocci: *Deinococcaceae* fam. nov. and Deinococcus gen. nov., including five species. *International Journal of Systematic Bacteriology, 31*, 353–360.

Brumwell, S. L., Van Belois, K. D., Giguere, D. J., Edgell, D. R., & Karas, B. J. (2022). Conjugation-based genome engineering in *Deinococcus radiodurans*. *ACS Synthetic Biology, 11*, 1068–1076.

Brumwell, S. L., Van Belois, K. D., Nucifora, D. P., & Karas, B. J. (2023). SLICER: A seamless gene deletion method for *Deinococcus radiodurans*. *Biodesign Research, 5*, 0009.

Chaturvedi, R., & Archana, G. (2014). Cytosolic expression of synthetic phytochelatin and bacterial metallothionein genes in *Deinococcus radiodurans* R1 for enhanced tolerance and bioaccumulation of cadmium. *Biometals, 27*, 471–482.

Chen, A., Sherman, M. W., Chu, C., Gonzalez, N., Patel, T., & Contreras, L. M. (2019). Discovery and characterization of native *Deinococcus radiodurans* promoters for tunable gene expression. *Applied & Environmental Microbiology, 85*, e01356–e01319.

Chou, F. I., & Tan, S. T. (1991). Salt-mediated multicell formation in *Deinococcus radiodurans*. *Journal of Bacteriology, 173*, 3184–3190.

Cox, M. M., & Battista, J. R. (2005). *Deinococcus radiodurans*—The consummate survivor. *Nature Review Microbiology, 3*, 882–892.

Dai, S., Chen, Q., Jiang, M., Wang, B., Xie, Z., Yu, N., Zhou, Y., Li, S., Wang, L., Hua, Y., & Tian, B. (2021). Colonized extremophile *Deinococcus radiodurans* alleviates toxicity of cadmium and lead by suppressing heavy metal accumulation and improving antioxidant system in rice. *Environmental Pollution, 284*, 117127.

Daly, M. J., Gaidamakova, E. K., Matrosova, V. Y., Vasilenko, A., Zhai, M., Venkateswaran, A., Hess, M., Omelchenko, M. V., Kostandarithes, H. M., Makarova, K. S., Wackett, L. P., Fredrickson, J. K., & Ghosal, D. (2004). Accumulation of Mn(II) in *Deinococcus radiodurans* facilitates gamma-radiation resistance. *Science, 306*, 1025–1028.

Daly, M. J., Gaidamakova, E. K., Matrosova, V. Y., Vasilenko, A., Zhai, M., Leapman, R. D., Lai, B., Ravel, B., Li, S. W., Kemner, K. M., & Fredrickson, J. K. (2007). Protein oxidation implicated as the primary determinant of bacterial radioresistance. *PLoS Biology, 5*, e92.

Dean, C. J., Feldschreiber, P., & Lett, J. T. (1966). Repair of X-ray damage to the deoxyribonucleic acid in *Micrococcus radiodurans*. *Nature, 209*, 49–52.

Devigne, A., Mersaoui, S., Bouthier-de-la-Tour, C., Sommer, S., & Servant, P. (2013). The PprA protein is required for accurate cell division of γ-irradiated *Deinococcus radiodurans* bacteria. *DNA Repair (Amst), 12*, 265–272.

DiRuggiero, J., Santangelo, N., Nackerdien, Z., Ravel, J., & Robb, F. T. (1997). Repair of extensive ionizing-radiation DNA damage at 95 degrees C in the hyperthermophilic archaeon *Pyrococcus furiosus*. *Journal of Bacteriology, 179*, 4643–4645.

Dollery, S. J., Zurawski, D. V., Gaidamakova, E. K., Matrosova, V. Y., Tobin, J. K., Wiggins, T. J., Bushnell, R. V., MacLeod, D. A., Alamneh, Y. A., Abu-Taleb, R., Escatte, M. G., Meeks, H. N., Daly, M. J., & Tobin, G. J. (2021). Radiation-inactivated *Acinetobacter baumannii* vaccine candidates. *Vaccines (Basel), 9*, 96.

Dulermo, R., Fochesato, S., Blanchard, L., & de Groot, A. (2009). Mutagenic lesion bypass and two functionally different RecA proteins in *Deinococcus deserti*. *Molecular Microbiology, 74*, 194–208.

Ekman, J. V., Raulio, M., Busse, H. J., Fewer, D. P., & Salkinoja-Salonen, M. (2011). *Deinobacterium chartae* gen. nov., sp. nov., an extremely radiation-resistant, biofilm-forming bacterium isolated from a Finnish paper mill. *International Journal of Systematic & Evolutionary Microbiology, 61*, 540–548.

Fagliarone, C., Mosca, C., Ubaldi, I., Verseux, C., Baqué, M., Wilmotte, A., & Billi, D. (2017). Avoidance of protein oxidation correlates with the desiccation and radiation resistance of hot and cold desert strains of the cyanobacterium *Chroococcidiopsis*. *Extremophiles, 21*, 981–991.

Farci, D., Slavov, C., & Piano, D. (2018). Coexisting properties of thermostability and ultraviolet radiation resistance in the main S-layer complex of *Deinococcus radiodurans*. *Photochemical & Photobiological Sciences, 17*, 81–88.

Ferreira, A. C., Nobre, M. F., Rainey, F. A., Silva, M. T., Wait, R., Burghardt, J., Chung, A. P., & da Costa, M. S. (1997). *Deinococcus geothermalis* sp. nov. and *Deinococcus murrayi* sp. nov., two extremely radiation-resistant and slightly thermophilic species from hot springs. *International Journal of Systematic Bacteriology, 47*, 939–947.

Fiala, G., & Stetter, K. (1986). *Pyrococcus furiosus* sp. nov. represents a novel genus of marine heterotrophic archaebacteria growing optimally at 100°C. *Archives of Microbiology, 145*, 56–61.

Gérard, E., Jolivet, E., Prieur, D., & Forterre, P. (2001). DNA protection mechanisms are not involved in the radioresistance of the hyperthermophilic archaea *Pyrococcus abyssi* and *P. Furiosus*. *Molecular Genetics & Genomics, 266*, 72–78.

Gogada, R., Singh, S. S., Lunavat, S. K., Pamarthi, M. M., Rodrigue, A., Vadivelu, B., Phanithi, P. B., Gopala, V., & Apte, S. K. (2015). Engineered *Deinococcus radiodurans* R1 with NiCoT genes for bioremoval of trace cobalt from spent decontamination solutions of nuclear power reactors. *Applied Microbiology & Biotechnology, 99*, 9203–9213.

Grimsley, J. K., Masters, C. I., Clark, E. P., & Minton, K. W. (1991). Analysis by pulsed-field gel electrophoresis of DNA double-strand breakage and repair in *Deinococcus radiodurans* and a radiosensitive mutant. *International Journal of Radiation Biology, 60*, 613–626.

Hamana, K., Hayashi, H., Niitsu, M., & Itoh, T. (2009). Polyamine analysis of thermophilic, acidophilic, alkaliphilic and radio-tolerant bacteria belonging to the domain bacteria and methanogens, thermophiles and extreme halophiles belonging to the domain archaea. *Journal of Japanese Society for Extremophiles, 8*, 59–68.

Han, J. M., Song, H. Y., Jung, J. H., Lim, S., Seo, H. S., Kim, W. S., Lim, S. T., & Byun, E. B. (2023). *Deinococcus radiodurans*-derived membrane vesicles protect HaCaT cells against H_2O_2-induced oxidative stress via modulation of MAPK and Nrf2/ARE pathways. *Biological Procedures Online, 25*, 17.

Hansen, M. T. (1978). Multiplicity of genome equivalents in the radiation-resistant bacterium *Micrococcus radiodurans*. *Journal of Bacteriology, 134*, 71–75.

Harsojo, Kitayama, S., & Matsuyama, A. (1981). Genome multiplicity and radiation resistance in *Micrococcus radiodurans*. *Journal of Biochemistry, 90*, 877–880.

Hua, Y., Narumi, I., Gao, G., Tian, B., Satoh, K., Kitayama, S., & Shen, B. (2003). PprI: A general switch responsible for extreme radioresistance of *Deinococcus radiodurans*. *Biochemical & Biophysical Research Communications, 306*, 354–360.

Imlay, J. A. (2014). The mismetallation of enzymes during oxidative stress. *Journal of Biological Chemistry, 289*, 28121–28128.

Ishino, Y., & Narumi, I. (2015). DNA repair in hyperthermophilic and hyperradioresistant microorganisms. *Current Opinion in Microbiology, 25*, 103–112.

Islam, M. S., Hua, Y., Ohba, H., Satoh, K., Kikuchi, M., Yanagisawa, T., & Narumi, I. (2003). Characterization and distribution of IS*8301* in the radioresistant bacterium *Deinococcus radiodurans*. *Genes & Genetic Systems, 78*, 319–327.

Jeong, S. W., Kim, J. H., Kim, J. W., Kim, C. Y., Kim, S. Y., & Choi, Y. J. (2021). Metabolic engineering of extremophilic bacterium *Deinococcus radiodurans* for the production of the novel carotenoid deinoxanthin. *Microorganisms, 9*, 44.

Jolivet, E., Corre, E., L'Haridon, S., Forterre, P., & Prieur, D. (2004). *Thermococcus marinus* sp. nov. and *Thermococcus radiotolerans* sp. nov., two hyperthermophilic archaea from deep-sea hydrothermal vents that resist ionizing radiation. *Extremophiles, 8*, 219–227.

Kalia, V. C., Raju, S. C., & Purohit, H. J. (2011). Genomic analysis reveals versatile organisms for quorum quenching enzymes: Acyl-homoserine lactone-acylase and -lactonase. *Open Microbiology Journal, 5*, 1–13.

Kaplan, H. S., & Zavarine, R. (1962). Correlation of bacterial radiosensitivity and DNA base composition. *Biochemical & Biophysical Research Communications, 8*, 432–436.

Kikuchi, M., Narumi, I., Kitayama, S., Watanabe, H., & Yamamoto, K. (1999). Genomic organization of the radioresistant bacterium *Deinococcus radiodurans*: Physical map and evidence for multiple replicons. *FEMS Microbiology Letters, 174*, 151–157.

Kimura, H., Asada, R., Masta, A., & Naganuma, T. (2003). Distribution of microorganisms in the subsurface of the Nanus Basin hydrothermal vent field in Papua New Guinea. *Applied & Environmental Microbiology, 69*, 644–648.

Kish, A., Kirkali, G., Robinson, C., Rosenblatt, R., Jaruga, P., Dizdaroglu, M., & DiRuggiero, J. (2009). Salt shield: Intracellular salts provide cellular protection against ionizing radiation in the halophilic archaeon, *Halobacterium salinarum* NRC-1. *Environmental Microbiology, 11*, 1066–1178.

Kitayama, S., & Matsuyama, A. (1968). Possibility of the repair of double-strand scissions in *Micrococcus radiodurans* DNA caused by gamma-rays. *Biochemical & Biophysical Research Communications, 33*, 418–422.

Kobayashi, Y., Shimizu, T., Tanaka, A., Kikuchi, M., Taucher-Scholz, G., & Watanabe, H. (1995). RBE/LET effects of heavy ions on inactivation in dry cells of *Deinococcus radiodurans*. *JAERI Review, 95-019*, 44–46.

Kongpol, A., Kato, J., & Vangnai, A. S. (2008). Isolation and characterization of *Deinococcus geothermalis* T27, a slightly thermophilic and organic solvent-tolerant bacterium able to survive in the presence of high concentrations of ethyl acetate. *FEMS Microbiology Letters, 286*, 227–235.

Kota, S., Charaka, V. K., & Misra, H. S. (2014a). PprA, a pleiotropic protein for radioresistance, works through DNA gyrase and shows cellular dynamics during postirradiation recovery in *Deinococcus radiodurans*. *Journal of Genetics, 93*, 349–354.

Kota, S., Charaka, V. K., Ringgaard, S., Waldor, M. K., & Misra, H. S. (2014b). PprA contributes to *Deinococcus radiodurans* resistance to nalidixic acid, genome maintenance after DNA damage and interacts with deinococcal topoisomerases. *PLoS One, 9*, e85288.

Kottemann, M., Kish, A., Iloanusi, C., Bjork, S., & DiRuggiero, J. (2005). Physiological responses of the halophilic archaeon *Halobacterium* sp. strain NRC1 to desiccation and gamma irradiation. *Extremophiles, 9*, 219–227.

Kulkarni, S., Ballal, A., & Apte, S. K. (2013). Bioprecipitation of uranium from alkaline waste solutions using recombinant *Deinococcus radiodurans*. *Journal of Hazardous Materials, 262*, 853–861.

Kuzucu, M. (2023). Extremophilic solutions: The role of deinoxanthin in counteracting UV-induced skin harm. *Current Issues in Molecular Biology, 45*, 8372–8394.

Lange, C. C., Wackett, L. P., Minton, K. W., & Daly, M. J. (1998). Engineering a recombinant *Deinococcus radiodurans* for organopollutant degradation in radioactive mixed waste environments. *Nature Biotechnology, 16*, 929–933.

Leibowitz, P. J., Schwartzberg, L. S., & Bruce, A. K. (1976). The *in vivo* association of manganese with the chromosome of *Micrococcus radiodurans*. *Photochemistry & Photobiology, 23*, 45–50.

Lemee, L., Peuchant, E., Clerc, M., Brunner, M., & Pfander, H. (1997). Deinoxanthin: A new carotenoid isolated from *Deinococcus radiodurans*. *Tetrahydron, 53*, 919–926.

Levin-Zaidman, S., Englander, J., Shimoni, E., Sharma, A. K., Minton, K. W., & Minsky, A. (2003). Ringlike structure of the *Deinococcus radiodurans* genome: A key to radioresistance? *Science, 299*, 254–256.

Li, W., Ma, Y., Yang, J., Xiao, F., Wang, W., & He, S. (2017). RNA-binding domain is necessary for PprM function in response to the extreme environmental stress in *Deinococcus radiodurans*. *Indian Journal of Microbiology, 57*, 492–498.

Li, C., Chen, Z., Chen, L., & Wang, G. (2025). The adaptation mechanism of desert soil cyanobacterium *Chroococcidiopsis* sp. to desiccation. *Plant Physiology & Biochemistry, 219*, 109414.

Liao, C. S., Chen, L. C., Chen, B. S., & Lin, S. H. (2010). Bioremediation of endocrine disruptor di-*n*-butyl phthalate ester by *Deinococcus radiodurans* and *Pseudomonas stutzeri*. *Chemosphere, 78*, 342–346.

Lim, S., Jung, J. H., Blanchard, L., & de Groot, A. (2019). Conservation and diversity of radiation and oxidative stress resistance mechanisms in *Deinococcus* species. *FEMS Microbiology Review, 43*, 19–52.

Lin, J., Qi, R., Aston, C., Jung, J., Anantharaman, T. S., Mishra, B., White, O., Daly, M. J., Minton, K. W., Venter, J. C., & Schwartz, D. C. (1999). Whole-genome shotgun optical mapping of *Deinococcus radiodurans*. *Science, 285*, 1558–1562.

Ludanyi, M., Blanchard, L., Dulermo, R., Brandelet, G., Bellanger, L., Pignol, D., Lemaire, D., & de Groot, A. (2014). Radiation response in *Deinococcus deserti*: IrrE is a metalloprotease that cleaves repressor protein DdrO. *Molecular Microbiology, 94*, 434–449.

Lv, G. Y., Cheng, J. H., Chen, X. Y., & Zhang, Z. F. (2013). Biological decolorization of malachite green by *Deinococcus radiodurans* R1. *Bioresource Technology, 144*, 275–280.

Makarova, K. S., Omelchenko, M. V., Gaidamakova, E. K., Matrosova, V. Y., Vasilenko, A., Zhai, M., Lapidus, A., Copeland, A., Kim, E., Land, M., Mavrommatis, K., Pitluck, S., Richardson, P. M., Detter, C., Brettin, T., Saunders, E., Lai, B., Ravel, B., Kemner, K. M., Wolf, Y. I., Sorokin, A., Gerasimova, A. V., Gelfand, M. S., Fredrickson, J. K., Koonin, E. V., & Daly, M. J. (2007). *Deinococcus geothermalis*: The pool of extreme radiation resistance genes shrinks. *PLoS One, 2*, e955.

Maldonado, R., Jiménez, J., & Casadesús, J. (1994). Changes of ploidy during the *Azotobacter vinelandii* growth cycle. *Journal of Bacteriology, 176*, 3911–3919.

Masters, C. I., & Minton, K. W. (1992). Promoter probe and shuttle plasmids for *Deinococcus radiodurans*. *Plasmid, 28*, 258–261.

Meima, R., & Lidstrom, M. E. (2000). Characterization of the minimal replicon of a cryptic *Deinococcus radiodurans* SARK plasmid and development of versatile *Escherichia coli-D. radiodurans* shuttle vectors. *Applied & Environmental Microbiology, 66*, 3856–3867.

Minton, K. W., & Daly, M. J. (1995). A model for repair of radiation-induced DNA double-strand breaks in the extreme radiophile *Deinococcus radiodurans*. *BioEssays, 17*, 457–464.

Morita, Y., Okumura, M., Narumi, I., & Nishida, H. (2019). Sensitivity of *Deinococcus grandis rodZ* deletion mutant to calcium ions results in enhanced spheroplast size. *AIMS Microbiology, 5*, 176–185.

Mortimer, R. K. (1958). Radiobiological and genetic studies on a polyploid series (haploid to hexaploid) of *Saccharomyces cerevisiae*. *Radiation Research, 9*, 312–326.

Moseley, B. E. B. (1983). Photobiology and radiobiology of *Micrococcus* (*Deinococcus*) radiodurans. *Photochemical & Photobiological Reviews, 7*, 223–274.

Moseley, B. E. B., & Schein, A. H. (1964). Radiation resistance and deoxyribonucleic acid base composition of *Micrococcus radiodurans*. *Nature, 203*, 1298–1299.

Narasimha, A., & Basu, B. (2021). New insights into the activation of radiation desiccation response regulon in *Deinococcus radiodurans*. *Journal of Biosciences, 46*, 10.

Narumi, I., Satoh, K., Kikuchi, M., Funayama, T., Yanagisawa, T., Kobayashi, Y., Watanabe, H., & Yamamoto, K. (2001). The LexA protein from *Deinococcus radiodurans* is not involved in RecA induction following γ irradiation. *Journal of Bacteriology, 183*, 6951–6956.

Narumi, I., Satoh, K., Cui, S., Funayama, T., Kitayama, S., & Watanabe, H. (2004). PprA: A novel protein from *Deinococcus radiodurans* that stimulates DNA ligation. *Molecular Microbiology, 54*, 278–285.

NIPPON GENE. (2025). *TA-Blunt Ligation Kit*. Retrieved September 19, 2025, from https://www.nippongene.com/siyaku/product/modifying-enzymes/ta-blunt-ligation/ta-blunt-ligation-kit.html

Nishino, K., & Nishida, H. (2019). Calcium ion induces outer membrane fusion of *Deinococcus grandis* spheroplasts to generate giant spheroplasts with multiple cytoplasms. *FEMS Microbiology Letters, 366*, fny282.

Nishino, K., Morita, Y., Takahashi, S., Okumura, M., Shiratani, S., Umemura, K., Narumi, I., Kondo, C., Ochiai, R., Oshima, T., & Nishida, H. (2018). Enlargement of *Deinococcus grandis* spheroplasts requires Mg^{2+} or Ca^{2+}. *Microbiology (Reading), 164*, 1361–1371.

Nishino, K., Tsuchikado, R., & Nishida, H. (2019). Sugar enhances outer membrane fusion in *Deinococcus grandis* spheroplasts to generate calcium ion-dependent extra-huge cells. *FEMS Microbiology Letters, 366*, fnz087.

Ohba, H., Satoh, K., Yanagisawa, T., & Narumi, I. (2005). The radiation responsive promoter of the *Deinococcus radiodurans pprA* gene. *Gene, 363*, 133–141.

Ohba, H., Satoh, K., Sghaier, H., Yanagisawa, T., & Narumi, I. (2009). Identification of PprM: A modulator of the PprI-dependent DNA damage response in *Deinococcus radiodurans*. *Extremophiles, 13*, 471–479.

Ohtani, N., Tomita, M., & Itaya, M. (2010). An extreme thermophile, *Thermus thermophilus*, is a polyploid bacterium. *Journal of Bacteriology, 192*, 5499–5505.

Oren, A., & Garrity, G. M. (2021). Valid publication of the names of forty-two phyla of prokaryotes. *International Journal of Systematic & Evolutionary Microbiology, 71*, 005056.

Parte, A. C., Sardà Carbasse, J., Meier-Kolthoff, J. P., Reimer, L. C., & Göker, M. (2020). List of prokaryotic names with standing in nomenclature (LPSN) moves to the DSMZ. *International Journal of Systematic and Evolutionary Microbiology, 70*, 5607–5612. Retrieved September 19, 2025, from https://lpsn.dsmz.de/

Phillips, R. W., Wiegel, J., Berry, C. J., Fliermans, C., Peacock, A. D., White, D. C., & Shimkets, L. J. (2002). *Kineococcus radiotolerans* sp. nov., a radiation-resistant, gram-positive bacterium. *International Journal of Systematic & Evolutionary Microbiology, 52*, 933–938.

Rainey, F. A., Nobre, M. F., Schumann, P., Stackebrandt, E., & da Costa, M. S. (1997). Phylogenetic diversity of the deinococci as determined by 16S ribosomal DNA sequence comparison. *International Journal of Systematic Bacteriology, 47*, 510–514.

Raj, H. D., Duryee, F. L., Deeney, A. M., Wang, C. H., Anderson, A. W., & Elliker, P. R. (1960). Utilization of carbohydrates and amino acids by *Micrococcus radiodurans*. *Canadian Journal of Microbiology, 6*, 289–298.

Saito, T., Terato, H., & Yamamoto, O. (1994). Pigment of *Rubrobacter radiotolerans*. *Archives of Microbiology, 162*, 414–421.

Sakai, M., Shimosaka, T., Katsumata, K., Yohda, M., & Narumi, I. (2004). Developing a new host-vector system for *Deinococcus grandis*. *Frontiers in Microbiology, 15*, 1387296.

Satoh, K., Wada, S., Kikuchi, M., Funayama, T., Narumi, I., & Kobayashi, Y. (2006). Method for detecting DNA strand breaks in mammalian cells using the *Deinococcus radiodurans* PprA protein. *Mutation Research, 596*, 36–42.

Satoh, K., Tu, Z., Ohba, H., & Narumi, I. (2009). Development of versatile shuttle vectors for *Deinococcus grandis*. *Plasmid, 62*, 1–9.

Sharma, A., Gaidamakova, E. K., Grichenko, O., Matrosova, V. Y., Hoeke, V., Klimenkova, P., Conze, I. H., Volpe, R. P., Tkavc, R., Gostinčar, C., Gunde-Cimerman, N., DiRuggiero, J., Shuryak, I., Ozarowski, A., Hoffman, B. M., & Daly, M. J. (2017). Across the tree of life, radiation resistance is governed by antioxidant Mn^{2+}, gauged by paramagnetic resonance. *Proceedings of the National Academy of Sciences of the United States of America, 114*, E9253.

Smith, M. D., Masters, C. I., Lennon, E., McNeil, L. B., & Minton, K. W. (1991). Gene expression in *Deinococcus radiodurans*. *Gene, 98*, 45–52.

Spaans, S. K., van der Oost, J., & Kengen, S. W. M. (2015). The chromosome copy number of the hyperthermophilic archaeon *Thermococcus kodakarensis* KOD1. *Extremophiles, 19*, 741–750.

Tanaka, M., Earl, A. E., Howell, H. A., Park, M. J., Eisen, J. A., Peterson, S. N., & Battista, J. R. (2004). Analysis of *Deinococcus radiodurans*'s transcriptional response to ionizing radiation and desiccation reveals novel proteins that contribute to extreme radioresistance. *Genetics, 168*, 21–33.

Telang, S., Patel, P., Sarangdhar, V., & Donde, S. (2014). Isolation and cloning of the endoglucanase gene from *Bacillus pumilus* and its expression in *Deinococcus radiodurans*. *3 Biotech, 4*, 57–65.

Terato, H., Kobayashi, M., Yamamoto, O., & Ide, H. (1999). DNA strand breaks induced by ionizing radiation on *Rubrobacter radiotolerans*, an extremely radioresistant bacterium. *Microbiological Research, 154*, 173–178.

Tian, B., Xu, Z., Sun, Z., Lin, J., & Hua, Y. (2007). Evaluation of the antioxidant effects of carotenoids from *Deinococcus radiodurans* through targeted mutagenesis, chemiluminescence, and DNA damage analyses. *Biochimica et Biophysica Acta, 1770*, 902–911.

Tirgari, S., & Moseley, B. E. B. (1980). Transformation in *Micrococcus radiodurans*: Measurement of various parameters and evidence for multiple, independently segregating genomes per cell. *Journal of General Microbiology, 119*, 287–296.

Tobin, G. J., Tobin, J. K., Gaidamakova, E. K., Wiggins, T. J., Bushnell, R. V., Lee, W. M., Matrosova, V. Y., Dollery, S. J., Meeks, H. N., Kouiavskaia, D., Chumakov, K., & Daly, M. J. (2020). A novel gamma radiation-inactivated Sabin-based polio vaccine. *PLoS One, 15*, e0228006.

Villa, J. K., Han, R., Tsai, C. H., Chen, A., Sweet, P., Franco, G., Vaezian, R., Tkavc, R., Daly, M. J., & Contreras, L. M. (2021). A small RNA regulates *pprM*, a modulator of pleiotropic proteins promoting DNA repair, in *Deinococcus radiodurans* under ionizing radiation. *Scientific Reports, 11*, 12949.

Waight, K., Pinyakong, O., & Luepromchai, E. (2007). Degradation of phenanthrene on plant leaves by phyllosphere bacteria. *Journal of General & Applied Microbiology, 53*, 265–272.

Webb, K. M., & DiRuggiero, J. (2012). Role of Mn^{2+} and compatible solutes in the radiation resistance of thermophilic bacteria and archaea. *Archaea, 2012*, 845756.

Weisburg, W. G., Giovannoni, S. J., & Woese, C. R. (1989). The *Deinococcus–Thermus* phylum and the effect of rRNA composition on phylogenetic tree construction. *Systematic & Applied Microbiology, 11*, 128–134.

White, O., Eisen, J. A., Heidelberg, J. F., Hickey, E. K., Peterson, J. D., Dodson, R. J., Haft, D. H., Gwinn, M. L., Nelson, W. C., Richardson, D. L., Moffat, K. S., Qin, H., Jiang, L., Pamphile, W., Crosby, M., Shen, M., Vamathevan, J. J., Lam, P., McDonald, L., Utterback, T., Zalewski, C., Makarova, K. S., Aravind, L., Daly, M. J., Minton, K. W., Fleischmann, R. D., Ketchum, K. A., Nelson, K. E., Salzberg, S., Smith, H. O., Venter, J. C., & Fraser, C. M. (1999). Genome sequence of the radioresistant bacterium *Deinococcus radiodurans* R1. *Science, 286*, 1571–1577.

Xiong, W., Peng, W., & Liang, R. (2018). Identification and genome analysis of *Deinococcus actinosclerus* SJTR1, a novel 17β-estradiol degradation bacterium. *3 Biotech, 8*, 433.

Yoshinaka, T., Yano, K., & Yamaguchi, H. (1973). Isolation of highly radioresistant bacterium, *Arthrobacter radiotolerans* nov. sp. *Agricultural & Biological Chemistry, 37*, 2269–2275.

Zahradka, K., Slade, D., Bailone, A., Sommer, S., Averbeck, D., Petranovic, M., Lindner, A. B., & Radman, M. (2006). Reassembly of shattered chromosomes in *Deinococcus radiodurans. Nature, 443*, 569–573.

Zerulla, K., & Soppa, J. (2014). Polyploidy in haloarchaea: Advantages for growth and survival. *Frontiers in Microbiology, 5*, 274.

Zhang, M., Lu, G., Xiao, T., Xiao, E., Sun, X., Yan, W., Liu, G., Wang, Q., Yan, G., Liu, H., & Sun, W. (2022). Characterization of arsenic-metabolizing bacteria in an alkaline soil. *Environmental Pollution, 312*, 120040.

Zilling, W., Holz, I., Janekovic, D., Schäfer, W., & Reiter, W. D. (1983). The archaebacterium *Thermococcus celer* represents, a novel genus within the thermophilic branch of the archaebacteria. *Systematic & Applied Microbiology, 4*, 88–94.

Zimmerman, J. M., & Battista, J. R. (2005). A ring-like nucleoid is not necessary for radioresistance in the *Deinococcaceae*. *BMC Microbiology, 5*, 17.

Chapter 20
The Future of AI and Biotechnology

Masahiro Ito

Abstract The convergence of AI and biotechnology is reshaping the design, construction, testing, and optimization of biological systems. This chapter outlines the current integration of AI into the Design–Build–Test–Learn (DBTL) cycle, supported by biofoundries, standardized data formats, and automation. We examined self-driving laboratory concepts and case studies that demonstrated closed-loop AI-driven experimentation. Advances in protein structure prediction (AlphaFold, RoseTTAFold, and ESM) and generative sequence design (ProteinMPNN and diffusion models) have been reviewed for applications in enzyme engineering and drug discovery. AI-enabled genome editing and breeding, including improvements in guided RNA design, directed evolution, and precision fermentation, are also discussed. In bioprocess engineering, we cover AI-assisted media design and scale-up and process analytical technologies, soft sensors, and digital twins for real-time monitoring and optimization. This chapter also addresses ethical, safety, and governance issues such as biosecurity, dual-use risks, DNA synthesis screening standards, and the transparency of AI models. Through selected examples, we highlight how AI tools accelerate discovery, enhance reproducibility, and expand the scope of feasible biological design. This chapter concludes with perspectives on the remaining challenges, regulatory considerations, and the need for interdisciplinary training to realize the full potential of AI-driven biotechnology.

Keywords Artificial intelligence · Biofoundry · Protein design · Bioprocess optimization · Biosecurity

M. Ito (✉)
Faculty of Life Sciences, and Bio-Resilience Research Project (BRRP), Toyo University, Asaka, Saitama, Japan
e-mail: masahiro.ito@toyo.jp

M. Ito et al. (eds.), *Extremophiles for a Sustainable Future*,
https://doi.org/10.1007/978-981-95-6970-0_20

20.1 The Current Landscape of AI × Biotechnology (Integration of Design–Build–Test–Learn [DBTL] and Biofoundries)

In recent years, the biotechnology field has needed to accelerate and streamline the DBTL cycle. Traditionally, each stage (gene sequence design, experimental system construction, functional evaluation, and result analysis) is conducted manually or with only partial automation. However, advances in large-scale data analytics, AI technologies, and robotics make it possible to integrate these processes into unified, iterative operations through "biofoundries."

The enhancement of the DBTL cycle requires data standardization and interoperability. To enhance the long-term accessibility and interoperability of research outputs, internationally recognized Findable, Accessible, Interoperable, Reusable (FAIR) principles have been widely implemented. Simultaneously, the Synthetic Biology Open Language (SBOL), now in its latest version, SBOL3, provides a standardized format for exchanging design information (Wilkinson et al., 2016; McLaughlin et al., 2020). These standards enable seamless data sharing between researchers and institutions, forming a foundation that increases the portability of automated protocols and improves AI model training.

An emerging concept is the self-driving laboratory (SDL), in which AI is integrated with robotics to conduct DBTL operations with minimal human supervision (Tom et al., 2024). SDLs are expected to deliver orders-of-magnitude improvements in efficiency, particularly in extensive parameter optimization and the exploration of unknown conditions. However, their successful deployment depends on the careful design of performance metrics (KPIs) and appropriate calibration of human oversight and intervention (Volk & Abolhasani, 2024).

Concrete examples of SDLs in life sciences are steadily increasing. One such example is the fully autonomous protein engineering platform SAMPLE, which integrates automated sequence design, expression, and evaluation with AI-driven data analysis and feedback for subsequent cycles (Rapp et al., 2024). This case illustrates the potential of such systems to dramatically expand the speed and scale of bio-based manufacturing.

The integration of AI and biofoundries is expected to extend beyond academic research and industrial applications. In synthetic biology, metabolic engineering, and drug discovery, the adoption of SDLs and standardized DBTL frameworks enables the discovery and optimization of new molecules and processes with unprecedented speed and accuracy.

20.2 AI Models and Design Space: Structure Prediction, Generative Design, and Protein Language Models

Accurate prediction and design of protein structures are fundamental for drug discovery and industrial enzyme development. Recently, advances in AI have dramatically improved the precision and scope of these tasks. AlphaFold2, introduced in 2021, has reached near-atomic resolution for predicting single protein structures and is regarded as a transformative advance in structural biology (Jumper et al., 2021). In 2024, AlphaFold3 extended its capabilities to include complex biomolecular interactions such as protein–nucleic acid complexes and ligand-bound states.

The large protein language model ESMFold (ESM-2) enables atomic-resolution structure prediction directly from primary sequences without requiring multiple sequence alignments (Lin et al., 2023). This approach can be applied to sequences with limited evolutionary information, including artificial designs, thereby greatly expanding design flexibility.

In sequence design, deep learning-based generative models have advanced rapidly. ProteinMPNN reliably designs functional sequences guided by structural templates (Dauparas et al., 2022). RFdiffusion employs diffusion-based generative methods to construct novel folds and enable targeted modifications of active sites and binding pockets (Watson et al., 2023; Chu et al., 2024). These methods go beyond modifying existing structures and open the possibility of creating completely new protein folds.

Looking ahead, the integration of structure prediction, generative design, and language models into multimodal AI platforms will allow cross-domain exploration of the design space, further shortening the cycle from molecular design to experimental validation and driving a qualitative transformation in biotechnology R&D.

20.3 AI-Assisted Genome Editing and Breeding (CRISPR Design and Off-Target Safety)

The CRISPR–Cas system has rapidly become widespread as a genome-editing technology with applications ranging from basic research to crop improvement and medical use. However, its success largely depends on the accuracy of the guide RNA (gRNA) design, making the balance between editing efficiency and specificity a critical challenge.

Currently, the Rule Set 2/Azimuth model developed by the Broad Institute is extensively used as a standard for predicting gRNA activity (Doench et al., 2016). This model analyzes large experimental datasets statistically and, via machine learning (ML), scores Cas9 cleavage activity based on sequence features. These techniques have been incorporated into several design tools.

As an application of deep learning, DeepCRISPR was developed to simultaneously optimize on-target efficiency and off-target risk (Chuai et al., 2018). By integrating diverse features such as epigenomic data and secondary structure predictions, improved predictive accuracy was demonstrated.

Recently, additional tools and reviews have provided broader perspectives. Methods such as CRISPRoff, which enables tunable control of editing activity, have expanded the practical landscape of genome editing (Nuñez et al., 2021). Moreover, comprehensive comparisons of gRNA design algorithms, including the evaluation of on- and off-target scoring by CRISPR (Haeussler et al., 2016), and systematic overviews of prediction tools (Konstantakos et al., 2022), have helped establish best practices. These developments have contributed to the standardization of the workflow from design to validation. This knowledge directly enhances the safety of medical applications and breeding efficiency.

AI models are expected to support gRNA design, off-target analysis, and experimental planning in an integrated manner, thereby enabling higher precision and reduced risk of genome editing.

20.4 Smarter Fermentation and Cultivation with AI: Process Analytical Technology (PAT), Raman × ML, and Digital Twins

The optimization and quality assurance of fermentation and cultivation processes are critical factors that determine the competitiveness of industrial bioprocesses. Since 2004, the U.S. FDA's PAT initiative, together with the Quality by Design concept, has provided a key regulatory framework for monitoring and controlling bioprocesses (Food and Drug Administration, 2004). In recent years, under the discussion of Bioprocessing 4.0, the advanced utilization of real-time data and the adoption of digital twins have accelerated (Isoko et al., 2024).

A notable application of AI is in the combination of Raman spectroscopy with ML. Raman spectroscopy can nondestructively obtain real-time molecular vibration information from the culture medium. When coupled with ML models, it enables high-accuracy in-line estimation of parameters, such as glucose, metabolites, pH, and CO_2 concentration (Poth et al., 2022; Wang et al., 2023). This approach allows for immediate assessment and automated control of culture conditions, contributing to product quality stabilization and productivity improvement.

Advancements in digital twin technology have also been remarkable. A digital twin builds a virtual model that reflects the actual process used for the simulation and predictive control. Hybrid approaches combining ML with mechanistic models are gaining interest (Herrera-Ruiz et al., 2024; Shariatifar et al., 2025). These methods improve the prediction accuracy for scaled-up conditions and anomaly detection, leading to reduced development time and cost.

The widespread adoption of intelligent fermentation platforms that integrate PAT, Raman × ML, and digital twins is expected to enable consistent optimization and quality assurance from the laboratory to commercial-scale production.

20.5 Industrialization, Human Resources, and Ecosystems: Precision Fermentation and Automation Talent

Precision fermentation uses microorganisms or cells as platforms to produce high-value food ingredients, functional proteins, and alternative proteins. In recent years, the movement toward commercialization has accelerated (Eastham & Leman, 2024; Cho et al., 2025). As the market continues to expand, regulatory compliance and cost reductions are key factors for wider adoption (Boukid et al., 2023; Eastham & Leman, 2024; Knychala et al., 2024).

For process efficiency, the design of the medium composition and optimization of cultivation conditions are critical. Approaches using Bayesian optimization and active learning are gaining attention because they can efficiently narrow the search space and identify optimal conditions while reducing the number of experiments required (Cho et al., 2025; Gisperg et al., 2025). Combined with sensors and automation systems, these methods are expected to shorten development time while increasing productivity.

Furthermore, the spread of SDLs and digital production platforms drives the demand for talent at the intersection of robotics, data, and biology (Hysmith et al., 2024). Such roles require interdisciplinary expertise in molecular biology, fermentation engineering, and data science, along with skills to design and operate data-driven processes. Developing training programs through universities, industry, and public institutions is directly tied to the sustainable growth of the precision fermentation sector.

Therefore, overcoming the technical challenges of commercializing precision fermentation, along with building standardized development platforms and talent ecosystems, will be key to enhancing competitiveness in the global market.

20.6 Risks, Ethics, and Governance: Safe-by-Design and Safeguarding the Supply Chain

The rapid advancement of AI-assisted biotechnology is driving the need to strengthen the frameworks for safety, ethics, and governance. Risk management in the design, ordering, and distribution of synthetic nucleic acids has received increasing international attention, with recent reviews synthesizing best practices in sequence screening, genetic biocontainment, and governance options (Diggans & Leproust, 2019; Hoffmann et al., 2023).

Recent peer-reviewed analyses in *Applied Biosafety* describe and contextualize the updated U.S. screening approach for synthetic nucleic acids, detailing how the expanded scope and implementation pathways affect researchers, providers, and institutional biosafety programs (Sharkey et al., 2024). Complementing this, a sector-wide survey of gene-synthesis providers reported the current strengths and gaps in customer vetting and sequence screening workflow, evidence that informs practical safeguards for the supply chain (Kane & Parker, 2024). Together, these studies outline the operational levers for safe-by-design across providers, users, and benchtop synthesis ecosystems (Kane & Parker, 2024; Sharkey et al., 2024).

From a global perspective, peer-reviewed work has proposed common baselines for nucleic acid screening and discussed how shared definitions (e.g., "sequences of concern") and interoperable practices can enhance security and trust in the bio-economy (Wheeler et al., 2024). Parallel analyses of laboratory biosafety and biosecurity management underscore the institutional measures (risk assessment, documentation, and quality systems) that support the responsible handling of high-consequence materials (Tang et al., 2024). These strands collectively point toward internationally harmonized, evidence-based safeguards that remain adaptive as AI expands its biological design capabilities (Hoffmann et al., 2023; Wheeler et al., 2024).

These peer-reviewed insights function as essential safeguards in environments in which AI enables advanced design. Researchers and companies should build systems that align with the synthesized best practices, such as robust sequence and customer screening, transparent documentation, and institutional oversight, so that innovation proceeds with accountability and resilience across the synthetic nucleic acid supply chain (Diggans & Leproust, 2019; Hoffmann et al., 2023; Sharkey et al., 2024).

20.7 Extremophiles for a Sustainable Future: The Future of AI and Biotechnology

Extremophiles, which are microorganisms that thrive under extreme physical and chemical conditions, offer unique adaptations and metabolic pathways that are particularly valuable for sustainable technologies, including renewable energy, bioremediation, and biomaterials. Their utility lies in extremozymes and survival mechanisms that operate under harsh conditions, beyond the capacity of traditional organisms (Gallo & Aulitto, 2024).

AI technologies have enabled the exploration and exploitation of extremophiles at unprecedented speeds and scales. For example, AI can mine genomic and metagenomic datasets to discover genes encoding robust enzymes, predict structure–function relationships, design synthetic pathways for extremozymes, and optimize cultivation parameters, even for currently unculturable extremophile species. Such approaches effectively "democratize" access to extremophilic resources and accelerate sustainable innovation (Groff-Vindman et al., 2025).

Looking ahead, synthetic biology offers a pathway to integrate extremophilic traits into tailored chassis or synthetic constructs, making them more accessible for industrial use. AI-guided DBTL cycles, AI-powered design of chimeric proteins combining extremophile stability with desired functionality, and scalable fermentation platforms could transform research insights into real-world sustainable processes (Ye et al., 2023; Zheng et al., 2025).

The synergy of AI accelerating the design and discovery, and extremophiles providing inherently robust biological components, defines a future where biotechnology supports environmental sustainability at industrial and global scales (Anabtawi et al., 2025).

References

Anabtawi, H. M., Ikhlaq, A., Kumar, S., Rafique, S., & Aly Hassan, A. (2025). Addressing challenges for eco-friendly and sustainable wastewater treatment solutions using extremophile microorganisms. *Sustainability, 17*(6), 2339.

Boukid, F., Hassoun, A., Zouari, A., Tülbek, M. Ç., Mefleh, M., Aït-Kaddour, A., et al. (2023). Fermentation for designing innovative plant-based meat and dairy alternatives. *Food, 12*(5), 1005.

Cho, S., Jung, S. Y., Eun, H., & Lee, S. Y. (2025). Precision fermentation for producing food ingredients. *Current Opinion in Food Science, 61*, 101242. https://doi.org/10.1016/j.cofs.2024.101242

Chu, A. E., Kim, J., Cheng, L., El Nesr, G., Xu, M., Shuai, R. W., et al. (2024). An all-atom protein generative model. *Proceedings of the National Academy of Sciences of the United States of America, 121*(27), e2311500121. https://doi.org/10.1073/pnas.2311500121

Chuai, G., Ma, H., Yan, J., Chen, M., Hong, N., Xue, D., et al. (2018). DeepCRISPR: Optimized CRISPR guide RNA design by deep learning. *Genome Biology, 19*(1), 80. https://doi.org/10.1186/s13059-018-1459-4

Dauparas, J., Anishchenko, I., Bennett, N., Bai, H., Ragotte, R. J., Milles, L. F., et al. (2022). Robust deep learning–based protein sequence design using ProteinMPNN. *Science, 378*(6615), 49–56. https://doi.org/10.1126/science.add2187

Diggans, J., & Leproust, E. (2019). Next steps for access to safe, secure DNA synthesis. *Frontiers in Bioengineering and Biotechnology, 7*, 86. https://doi.org/10.3389/fbioe.2019.00086

Doench, J. G., Fusi, N., Sullender, M., Hegde, M., Vaimberg, E. W., Donovan, K. F., et al. (2016). Optimized sgRNA design to maximize activity and minimize off-target effects of CRISPR-Cas9. *Nature Biotechnology, 34*(2), 184–191. https://doi.org/10.1038/nbt.3437

Eastham, J. L., & Leman, A. R. (2024). Precision fermentation for food proteins: ingredient innovations, bioprocess considerations, and outlook — a mini-review. *Current Opinion in Food Science, 58*, 101194. https://doi.org/10.1016/j.cofs.2024.101194

Food and Drug Administration. (2004). *PAT—A framework for innovative pharmaceutical development, manufacturing, and quality assurance*. U.S. Food and Drug Administration.

Gallo, G., & Aulitto, M. (2024). Advances in extremophile research: Biotechnological applications through isolation and identification techniques. *Life (Basel), 14*(9). https://doi.org/10.3390/life14091205

Gisperg, F., Klausser, R., Elshazly, M., Kopp, J., Brichtová, E. P., & Spadiut, O. (2025). Bayesian optimization in bioprocess engineering—Where do we stand today? *Biotechnology and Bioengineering, 122*(6), 1313–1325. https://doi.org/10.1002/bit.28960

Groff-Vindman, C. S., Trump, B. D., Cummings, C. L., Smith, M., Titus, A. J., Oye, K., et al. (2025). The convergence of AI and synthetic biology: The looming deluge. *Npj Biomedical Innovations, 2*(1), 20. https://doi.org/10.1038/s44385-025-00021-1

Haeussler, M., Schonig, K., Eckert, H., Eschstruth, A., Mianne, J., Renaud, J. B., et al. (2016). Evaluation of off-target and on-target scoring algorithms and integration into the guide RNA selection tool CRISPOR. *Genome Biology, 17*(1), 148. https://doi.org/10.1186/s13059-016-1012-2

Herrera-Ruiz, J. F., Fontalvo, J., & Prado-Rubio, O. A. (2024). Advances on hybrid modelling for bioprocesses engineering: Insights into research trends and future directions from a bibliometric approach. *Results in Engineering, 24*, 103548. https://doi.org/10.1016/j.rineng.2024.103548

Hoffmann, S. A., Diggans, J., Densmore, D., Dai, J., Knight, T., Leproust, E., et al. (2023). Safety by design: Biosafety and biosecurity in the age of synthetic genomics. *iScience, 26*(3). https://doi.org/10.1016/j.isci.2023.106165

Hysmith, H., Foadian, E., Padhy, S. P., Kalinin, S. V., Moore, R. G., Ovchinnikova, O. S., et al. (2024). The future of self-driving laboratories: From human in the loop interactive AI to gamification. *Digital Discovery, 3*(4), 621–636. https://doi.org/10.1039/D4DD00040D

Isoko, K., Cordiner, J. L., Kis, Z., & Moghadam, P. Z. (2024). Bioprocessing 4.0: A pragmatic review and future perspectives. *Digital Discovery, 3*(9), 1662–1681. https://doi.org/10.1039/D4DD00127C

Jumper, J., Evans, R., Pritzel, A., Green, T., Figurnov, M., Ronneberger, O., et al. (2021). Highly accurate protein structure prediction with AlphaFold. *Nature, 596*(7873), 583–589. https://doi.org/10.1038/s41586-021-03819-2

Kane, A., & Parker, M. T. (2024). Screening state of play: The biosecurity practices of synthetic DNA providers. *Applied Biosafety, 29*(2), 85–95. https://doi.org/10.1089/apb.2023.0027

Knychala, M. M., Boing, L. A., Ienczak, J. L., Trichez, D. B., & Stambuk, B. U. (2024). Precision fermentation as an alternative to animal protein, a review. *Fermentation, 10*(6), 315.

Konstantakos, V., Nentidis, A., Krithara, A., & Paliouras, G. (2022). CRISPR-Cas9 gRNA efficiency prediction: An overview of predictive tools and the role of deep learning. *Nucleic Acids Research, 50*(7), 3616–3637. https://doi.org/10.1093/nar/gkac192

Lin, Z., Akin, H., Rao, R., Hie, B., Zhu, Z., Lu, W., et al. (2023). Evolutionary-scale prediction of atomic-level protein structure with a language model. *Science, 379*(6637), 1123–1130. https://doi.org/10.1126/science.ade2574

McLaughlin, J. A., Beal, J., Mısırlı, G., Grünberg, R., Bartley, B. A., Scott-Brown, J., et al. (2020). The synthetic biology open language (SBOL) version 3: Simplified data exchange for bioengineering. *Frontiers in Bioengineering and Biotechnology, 8*, 1009. https://doi.org/10.3389/fbioe.2020.01009

Nuñez, J. K., Chen, J., Pommier, G. C., Cogan, J. Z., Replogle, J. M., Adriaens, C., et al. (2021). Genome-wide programmable transcriptional memory by CRISPR-based epigenome editing. *Cell, 184*(9), 2503–2519.e2517. https://doi.org/10.1016/j.cell.2021.03.025

Poth, M., Magill, G., Filgertshofer, A., Popp, O., & Großkopf, T. (2022). Extensive evaluation of machine learning models and data preprocessings for Raman modeling in bioprocessing. *Journal of Raman Spectroscopy, 53*(9), 1580–1591. https://doi.org/10.1002/jrs.6402

Rapp, J. T., Bremer, B. J., & Romero, P. A. (2024). Self-driving laboratories to autonomously navigate the protein fitness landscape. *Nature Chemical Engineering, 1*(1), 97–107. https://doi.org/10.1038/s44286-023-00002-4

Shariatifar, M., Rizi, M. S., Sotudeh-Gharebagh, R., Zarghami, R., & Mostoufi, N. (2025). On digital twins in bioprocessing: Opportunities and limitations. *Process Biochemistry, 156*, 274–299. https://doi.org/10.1016/j.procbio.2025.05.023

Sharkey, C. M., Lekveishvili, M., de la Rosa, T., & Danskin, K. (2024). Enhancing gene synthesis security: An updated framework for synthetic nucleic acid screening and the responsible use of synthetic biological materials. *Applied Biosafety, 29*(2), 63–70. https://doi.org/10.1089/apb.2023.0036

Tang, Q., Yan, F., Yuan, L., Tang, Y., Chen, H., Sun, Y., et al. (2024). Enhancing laboratory biosafety management: A comprehensive strategy from theory to practice. *Frontiers in Public Health, 12*. https://doi.org/10.3389/fpubh.2024.1439051

Tom, G., Schmid, S. P., Baird, S. G., Cao, Y., Darvish, K., Hao, H., et al. (2024). Self-driving laboratories for chemistry and materials science. *Chemical Reviews, 124*(16), 9633–9732. https://doi.org/10.1021/acs.chemrev.4c00055

Volk, A. A., & Abolhasani, M. (2024). Performance metrics to unleash the power of self-driving labs in chemistry and materials science. *Nature Communications, 15*(1), 1378. https://doi.org/10.1038/s41467-024-45569-5

Wang, J., Chen, J., Studts, J., & Wang, G. (2023). In-line product quality monitoring during biopharmaceutical manufacturing using computational Raman spectroscopy. *MAbs, 15*(1), 2220149. https://doi.org/10.1080/19420862.2023.2220149

Watson, J. L., Juergens, D., Bennett, N. R., Trippe, B. L., Yim, J., Eisenach, H. E., et al. (2023). De novo design of protein structure and function with RFdiffusion. *Nature, 620*(7976), 1089–1100. https://doi.org/10.1038/s41586-023-06415-8

Wheeler, N. E., Carter, S. R., Alexanian, T., Isaac, C., Yassif, J., & Millet, P. (2024). Developing a common global baseline for nucleic acid synthesis screening. *Applied Biosafety, 29*(2), 71–78. https://doi.org/10.1089/apb.2023.0034

Wilkinson, M. D., Dumontier, M., Aalbersberg, I. J., Appleton, G., Axton, M., Baak, A., et al. (2016). The FAIR guiding principles for scientific data management and stewardship. *Scientific Data, 3*(1), 160018. https://doi.org/10.1038/sdata.2016.18

Ye, J.-W., Lin, Y.-N., Yi, X.-Q., Yu, Z.-X., Liu, X., & Chen, G.-Q. (2023). Synthetic biology of extremophiles: A new wave of biomanufacturing. *Trends in Biotechnology, 41*(3), 342–357. https://doi.org/10.1016/j.tibtech.2022.11.010

Zheng, S., Shao, M., Wang, W., & Chen, G.-Q. (2025). Next-generation biotechnology inspired by extremes. *EMBO Reports, 26*(5), 1191–1195. https://doi.org/10.1038/s44319-025-00389-6

[illegible]

Chapter 21
Future Perspectives for Sustainable Development

Noriyuki Doukyu

Abstract As noted in other chapters, extremophiles are organisms adapted to thrive in extreme conditions, such as high temperature, pH, and salinity, which are unsuitable for many other life forms. These extremophiles represent a sustainable, cost-effective, and efficient resource for producing valuable materials. They generate extremolytes, extremozymes, biosurfactants, and so on, which are stable and adaptable across various environments. Their products find applications in a wide range of industries, including sustainable agriculture, food, cosmetics, pharmaceuticals, bioremediation, biofuel production, and biorefineries. A significant advantage of extremophiles and their enzymes is their robustness and stability under extreme conditions, facilitating more efficient and consistent bioprocesses. The reaction conditions associated with extremophiles often reduce the risk of contamination and enhance reaction rates, potentially leading to increased productivity and lower operational costs. Some applications of extremophiles have been discussed in previous chapters. This chapter comprehensively enumerates additional applications of extremophiles and their products across various industries not covered in other chapters, while also exploring their future prospects.

Keywords Extremophiles · Extremolytes · Extremozymes · Applications of extremophiles

21.1 Introduction

Extremophiles are microorganisms classified based on their specific extreme habitats and exist in environments that exceed the typical physical or chemical parameters conducive to the growth of common organisms (Wang et al., 2024). Major

N. Doukyu (✉)
Faculty of Life Sciences, and Bio-Resilience Research Project (BRRP), Toyo University, Asaka, Saitama, Japan
e-mail: dokyu@toyo.jp

M. Ito et al. (eds.), *Extremophiles for a Sustainable Future*,
https://doi.org/10.1007/978-981-95-6970-0_21

categories of extremophiles include thermophiles, hyperthermophiles, psychrophiles, psychrotolerants, acidophiles, alkaliphiles, piezophiles, metallophiles, xerophiles, and radiophiles. Organisms that can survive in multiple extreme environments are referred to as polyextremophiles (Caro-Astorga et al., 2024). Extremophiles hold immense value in industrial biotechnology due to their unique environmental adaptability and high tolerance to extreme conditions (Kochhar et al., 2022; Wang et al., 2024). Their application areas encompass fermentation, agriculture, therapeutics, dairy, food processing, bioenergy, bioremediation, bioleaching, and detergents. Extremolytes are small organic molecules synthesized or absorbed by extremophiles, accumulating intracellularly (Lentzen & Schwarz, 2006). They protect the macromolecules and cellular structures of extremophiles by forming a protective layer of water. Extremozymes are enzymes isolated from extremophiles (Salas-Bruggink et al., 2024). They are highly stable at extreme temperatures and pH levels and are resistant to denaturing agents, detergents, and organic solvents. These characteristics are especially valuable in industries that rely on biocatalysis under harsh conditions. They can reduce energy demands and improve efficiency by facilitating chemical reactions. In addition, extremozymes may exhibit unique traits such as tolerance to organic solvents, chiral selectivity, or the ability to degrade persistent substrates. Enzymes with such properties remain of significant industrial interest.

21.2 Applications of Extremophiles

As described in previous chapters, extremophiles can thrive in extreme environments and produce various useful substances by leveraging their unique growth characteristics. This section discusses several applications, including the production of organic acids, biofuels, and lipids under extreme conditions, that were not addressed in earlier chapters. These applications are summarized in Table 21.1 and described below.

21.2.1 Lactic Acid

Lactic acid is a key organic acid widely used in the food, pharmaceutical, and biodegradable plastic industries. Traditionally, its production has relied on mesophilic lactic acid bacteria. Microbial production methods offer advantages such as high purity, enantiomeric selectivity, low cost, and minimal environmental impact. However, microorganisms that are not acid-tolerant experience growth inhibition due to decreased pH caused by the accumulation of acidic products (Wang et al., 2024). In addition, adding a neutralizing agent like $CaCO_3$ to maintain the proper pH generates insoluble waste, such as $CaSO_4$, and increases production costs. Utilizing acid-tolerant microorganisms for organic acid production presents economic benefits and simplifies operations. *Lactobacillus acidophilus* is often

Table 21.1 Applications of selected extremophiles

Products	Extremophile	Remarks	References
Lactic acid	Acidophiles	An acid-resistant *Pichia kudriavzevii* NG7 can grow at pH 2.0 and produce lactic acid.	Toivari et al. (2013)
	Alkaliphiles	An alkaliphilic *Enterococcus hirae* BoM 1-2 efficiently produced lactic acid at pH 9.0.	Wang et al. (2024)
	Alkaliphiles	An engineered alkaliphilic *Bacillus* sp. strain N16–5 produced lactic acid at pH 10.	Assavasirijinda et al. (2016)
Amino acids	Halophiles	Genetically modified *Halomonas bluephagenesis* produced L-threonine and L-lysine in an open system with high salt concentration.	Du et al. (2020)
Ethanol	Thermophiles	A thermophilic *Clostridium thermocellum* can produce ethanol from cellulose.	Olson et al. (2023)
Butanol	Thermophiles	*Thermoanaerobacterium saccharolyticum* can produce butanol under high-temperature and anaerobic conditions.	Bhandiwad et al. (2014)
Biodiesel	Halophiles	A halophilic green microalga *Dunaliella salina* produces biodiesel in high-salinity conditions.	Converti et al. (2009)
Carotenoids	Piezophiles	*Haloterigena turcumenica* produces primarily C50 carotenoids when cultured in halobacterial medium.	Squillaci et al. (2017)
Omega-3-polyunsaturated fatty acids	Halophiles	The deep-sea marine bacterium *Shewanella electrodiphila* MAR441T produces remarkable amounts of EPA.	Zhang and Burgess (2017)
Polyhydroxyalkanoates	Halophiles	*Haloferax mediterranei* produces polyhydroxyalkanoates at high salt concentrations.	Don et al. (2006)

employed as an organic acid–producing strain due to its ability to thrive in acidic conditions. *Pichia kudriavzevii* NG7 is an extremely acid-resistant yeast that can grow at pH 2.0 and is capable of producing lactic acid (Toivari et al., 2013). The genetically engineered NG7 strain has been reported to produce 136 g/L of lactic acid at pH 3.6. Conversely, the use of alkaliphilic bacteria for lactic acid fermentation has also been documented (Abdel-Rahman et al., 2019). Most industrially available lactic acid bacteria are mesophilic and/or neutrophilic, which poses a risk of contamination by unwanted microorganisms that can reduce production efficiency and affect final product quality. Alkaliphilic strains are advantageous not only because they minimize contamination due to their tolerance to high pH but also because they reduce the need for neutralizers in lactic acid fermentation. The alkaliphilic *Enterococcus hirae* BoM 1–2 has been found to be an efficient producer of lactic acid at pH 9.0 (Wang et al., 2024). In the batch fermentation process, strain BoM 1–2 achieved a productivity of 39.9 g/L/h. Furthermore, alkaline organisms have the potential to produce organic acids using NaOH as a neutralizer instead of

$CaCO_3$. An engineered alkaliphilic *Bacillus* sp. strain N16–5 produced lactic acid with a titer of 143.99 g/L (Assavasirijinda et al., 2016).

21.2.2 Amino Acids

Amino acids are valuable industrial products produced through fermentation and enzymatic processes. These amino acids find widespread use in the food, animal feed, cosmetic, and pharmaceutical industries (Dong et al., 2011). Microorganisms such as *Corynebacterium glutamicum* and *Escherichia coli* are commonly used for amino acid production, but there is a risk of contamination when culturing these organisms. To mitigate this risk, attempts to utilize halophiles for amino acid production have been reported (Du et al., 2020; Zhao et al., 2022). The ability of genetically modified *Halomonas bluephagenesis* to produce L-threonine and L-lysine in an open system has been demonstrated (Du et al., 2020).

21.2.3 Biofuels

Biofuels are renewable energy sources derived from organic materials such as crops, algae, and animal waste. They serve as alternatives to fossil fuels, helping to reduce greenhouse gas emissions and dependence on nonrenewable energy sources. Microorganisms are crucial in the sustainable production of biofuels. Through biological processes, certain microbes can convert renewable biomass into various fuels, including bioethanol, biobutanol, and lipids for biodiesel production. However, microbial biofuel production faces several challenges. Lignocellulosic materials commonly used for industrial biofuel production are difficult for most microorganisms to degrade and often release inhibitory byproducts. In addition, compounds such as butanol can be toxic to microbial cells, reducing product yields. Furthermore, while high temperatures are typically beneficial for achieving high product yields in biorefining processes, most microorganisms cannot tolerate such conditions. To address these challenges, methods have been developed to harness extremophiles as biofuel producers. Their ability to thrive in extreme environments reduces the need for costly pretreatment and complex process control, making production more cost-effective.

Clostridium thermocellum is a thermophilic, anaerobic, gram-positive, spore-forming bacterium known for its remarkable ability to degrade cellulose directly (Lynd Lee et al., 2002). It thrives at high temperatures, typically around 55–60 °C, making it highly suitable for industrial processes. One of its most significant features is its production of a large extracellular enzyme complex called the cellulosome, which efficiently breaks down plant cell walls (Hirano et al., 2016). Unlike many other organisms that require pretreated cellulose, *C. thermocellum* can directly utilize raw biomass, offering great potential for cost-effective biofuel production.

During fermentation, *C. thermocellum* can produce ethanol from cellulose (Olson et al., 2023). Its natural capacity to convert cellulose into ethanol makes it a prime candidate for consolidated bioprocessing (CBP), a strategy aimed at simplifying and lowering the cost of bioethanol production from plant materials (Olson et al., 2012). However, a limitation is its relatively low ethanol tolerance, which can restrict industrial scalability. Mutations in alcohol dehydrogenase have been reported to improve ethanol tolerance in *C. thermocellum* (Brown et al., 2011). The genetically engineered *C. thermoceram* produced ethanol at 75% of its theoretical maximum yield (Tian et al., 2016).

Bio-butanol is an important biofuel that offers advantages over ethanol, including higher energy content, lower volatility, and better blending properties with gasoline. Although butanol production by mesotrophic *Clostridium acetobutylicum* has been extensively studied, the use of thermophilic bacteria may provide advantages in industrial biotechnological processes, such as reduced contamination risk, higher reaction rates, and lower differential costs for heating and cooling (Bhandiwad et al., 2014). Among various microbial producers, thermophilic and anaerobic bacteria of the *Thermoanaerobacterium* species have gained attention for their ability to produce butanol under high-temperature and anaerobic conditions (Bhandiwad et al., 2014). In *Thermoanaerobacterium saccharolyticum*, the integration of the *n*-butanol pathway in a lactate-deficient strain resulted in the production of 1.05 g/L of *n*-butanol from 10 g/L of xylose, corresponding to 26% of the theoretical maximum (Bhandiwad et al., 2014). In the butanol production process, increasing *n*-butanol concentration raises the NADH/NAD ratio, inhibiting microbial cell growth and severely affecting product yield (Tian et al., 2019). Several studies have shown that the *adhE* gene can convert large amounts of NADH to NADPH during *n*-butanol production, thereby reducing cytotoxicity. Overexpression of the *adhE* gene in strains such as *Thermoanaerobacterium saccharolyticum* JW/SL-YS485, *Clostridium thermocellum*, and *Clostridium acetobutylicum* has been reported to increase tolerance and *n*-butanol production (Bhandiwad et al., 2014; Yang et al., 2022).

Biodiesel is a renewable, biodegradable fuel made from vegetable oils, animal fats, or recycled cooking grease. It can be used in diesel engines, reducing greenhouse gas emissions compared to petroleum diesel. Recent research focuses on utilizing extremophiles as novel lipid sources for biodiesel production. Microalgae harness light energy and carbon dioxide to produce biomass with higher photosynthetic efficiency than plants. *Dunaliella salina* is a unicellular, halophilic green microalga widely studied for its potential in sustainable biodiesel production (Converti et al., 2009). One of its most significant advantages is its ability to grow in high-salinity environments, which minimizes contamination risks and allows the use of non-arable land and saline water, thereby lowering production costs. This alga can accumulate fatty acids (such as γ-linolenic, arachidonic, eicosapentaenoic, and docosahexaenoic acids), carotenoids, phycobiliproteins, and vitamins such as biotin, vitamin C, and vitamin E (Converti et al., 2009). The saturated fatty acids produced by this microorganism are expected to be suitable for biodiesel production (Converti et al., 2009).

21.2.4 Carotenoids

Carotenoids are naturally occurring pigments found in plants, algae, fungi, and some bacteria. They have garnered significant attention due to their industrial applications and potential health benefits for humans (Kirti et al., 2014). Carotenoids serve important functions as antioxidants, photoprotectants, and cell membrane stabilizers. Their functionality depends on their molecular structure, including the number of conjugated double bonds, the type of functional group, and molecular size (Albrecht et al., 2000). Possessing a high number of conjugated double bonds, carotenoids act as radical scavengers, protecting against oxidative damage caused by reactive oxygen species (Miller et al., 1996). Many carotenoids are utilized in the food industry; for instance, zeaxanthin serves as a coloring agent in poultry and fish, while canthaxanthin enhances the color of farmed salmon (Kirti et al., 2014). The advantage of using halophilic archaea is that they can be cultivated in non-sterile environments, allowing for carotenoid extraction without the need for mechanical disruption. This leads to a cost-effective and easy-to-manage production process (Squillaci et al., 2017). *Haloterigena turcumenica* primarily synthesizes C_{50} carotenoids when cultured in halobacterial medium (Squillaci et al., 2017). These molecules exhibit antioxidant properties applicable in the food, cosmetic, and pharmaceutical industries. *Haloterrigena thermotolerans* K15 produced 21.43 μg/g of dry cells of C_{50} carotenoids, demonstrating a higher antioxidant capacity than commonly used chemical antioxidants like ascorbic acid and butylated hydroxytoluene (Kesbiç & Gültepe, 2022). C_{50} carotenoids from the genera *Haloferax*, *Halogeometricum*, and *Haloarcula*, isolated from Algerian salt lakes, have also shown excellent antimicrobial activity, particularly against human pathogenic bacteria (Sahli et al., 2021).

21.2.5 Omega-3 Polyunsaturated Fatty Acids

Omega-3 polyunsaturated fatty acids (PUFAs) are essential fats primarily found in fatty fish, flaxseed, and walnuts. The most biologically significant omega-3 s are eicosapentaenoic acid (EPA) and docosahexaenoic acid (DHA). These fatty acids are crucial for cardiovascular health, brain function, and regulating inflammation (Calder, 2015). Numerous studies have demonstrated that omega-3 s can lower triglyceride levels, reduce blood pressure, and decrease the risk of heart disease (Calder, 2015). Moreover, DHA is vital for brain development and function, especially in infants and older adults (Innis, 2007). Since the human body cannot efficiently synthesize omega-3 s, they must be obtained through diet or supplementation. The most potent sources of EPA and DHA are fish oil; however, supply has not kept pace with the increasing global demand. Concerns arise that the global decline in fish stocks will lead to a shortage of omega-3 fatty acid sources (Rajan & Letourneau, 2012). Consequently, the production of omega-3 fatty acids by marine microorganisms is expected to be a viable alternative. Bacteria-derived PUFAs have mainly been reported from gram-negative bacteria from two bacterial phyla: *Gammaproteobacteria* (e.g., *Shewanella*, *Moritella*, and *Colwellia*) and *Bacteroidetes* (e.g., *Flexibacter* and *Psychroserpens*) (Nichols & McMeekin, 2002). The deep-sea marine bacterium *Shewanella electrodiphila* MAR441^T produced 20 mg of EPA/g (dry cell weight), achieving the highest percentage of EPA content (18%) when grown on L-proline and $(NH_4)_2SO_4$ (Zhang & Burgess, 2017). The addition of cerulenin further increased EPA production to 30 mg EPA/g (dry cell weight).

21.2.6 Polyhydroxyalkanoates (PHAs)

Plastic pollution has become a serious threat to the global environment, making the introduction of bioplastics essential for sustainable development. Polyhydroxyalkanoates (PHAs) are polyesters produced by various prokaryotes (Obruča et al., 2022). These renewable, biodegradable materials are considered alternatives to petrochemical polymers. In addition, PHAs are biocompatible and appealing for a wide range of biomedical applications. However, current challenges in PHA production include high costs and limited commercial applications. Halophilic bacteria present several unique features that are promising for PHA synthesis, such as high salinity, which prevents microbial contamination; high intracellular osmolality that facilitates cell lysis for PHA recovery; and broad availability of low-cost substrates (Mitra et al., 2020). Among all *Haloarchaea* strains, *Haloferax mediterranei* is reported to be the preferred PHA producer to date. This strain exhibits several advantageous characteristics for PHA production, including a high growth rate, metabolic diversity, and genetic stability. It has also demonstrated the ability to synthesize PHA using various industrial and household wastes as carbon sources (Don et al., 2006).

21.2.7 Mining

The extraction of metals from minerals and wastes typically involves chemical processes at high temperatures, such as metallurgy and hydrometallurgy, which contribute to environmental pollution. A new alternative technology known as "biomining" is expected to address these issues (Aparici-Carratalá et al., 2023). Biomining utilizes living organisms to extract metals from ores and other solid materials, and it is applied in ore refining, precious metal recovery, and bioremediation. Current applications of biomining include treating mine waste containing iron, copper, zinc, and gold (Kundu & Kumar, 2014). Thermophiles and acidophiles, belonging to the genera *Acidianus*, *Ferroplasma*, *Sulfolobus*, and *Metallosphaera*, are commonly used in biomining. One species with significant bioleaching potential is *Sulfolobus metallicus*, which solubilizes the brassite ($CuFeS_2$) film that forms in some mining operations, thereby recovering copper and iron and exposing precious metals for use (Huber & Stetter, 1991; Aparici-Carratalá et al., 2023). In addition, the acidophilic archaeon *Ferroplasma acidiphilum* can oxidize iron ions, while the hyperthermophilic archaeon *Pyrococcus furiosus* can bioleach gold and other metals (Naitam & Kaushik, 2021).

21.3 Extremozymes

Biocatalysts isolated from microorganisms in extreme environments are called extremozymes. These enzymes possess various properties, such as thermostability, salt tolerance, and cold adaptability. Extremozymes operate in extreme conditions, opening new possibilities for biocatalysis, biotransformation, economic development, and novel research fields. Extremophiles, including thermophilic, thermotropic, acidophilic, and halophilic proteins, have been studied, and their applications are currently being developed (Table 21.2). Amylases, proteases, lipases, pullulanases, cellulases, chitinases, xylanases, pectinases, isomerases, esterases, and dehydrogenases from extremophiles hold great potential for biotechnological applications in food, agriculture, chemistry, and medicine.

21.3.1 Thermophilic Enzymes

Thermozymes are enzymes that remain stable and active at high temperatures. They are primarily derived from thermophiles, which thrive in extreme environments such as hot springs, hydrothermal vents, and compost heaps (Vieille et al., 1996). Thermozymes are of significant interest for industrial and biotechnological applications due to their robustness and efficiency under harsh conditions that typically denature enzymes from mesophilic organisms. A major advantage of thermozymes is their ability to maintain catalytic activity at elevated temperatures, which can enhance reaction rates, reduce contamination risks, and improve substrate solubility. Many enzymes from thermophiles have been characterized, including cellulases, amylases, pullulanases, xylanases, mannanases, pectinases, chitinases, proteases, lipases, esterases, and phytases (Dumorné et al., 2017). Thermozymes provide valuable insights into protein stability and evolution, aiding protein engineering efforts. Their structural features, such as increased ionic interactions, hydrophobic cores, and rigid conformations, contribute to their thermal stability (Li et al., 2005). In biofuel production, enzymes like cellulases and xylanases from thermophiles are utilized to break down lignocellulosic biomass into fermentable sugars, enhancing efficiency and reducing contamination risks during fermentation (Turner et al., 2007). In the food industry, thermostable amylases, proteases, and lipases

Table 21.2 Applications of selected extremozymes

Types	Enzymes	Applications	References
Thermophilic	Cellulases and xylanases	Degradation of lignocellulosic biomsass for biofuel production	Turner et al. (2007)
	Amylases, proteases, and lipases	Food industry	Vieille et al. (1996)
	DNA polymerases	Molecular biology (PCR)	Akram et al. (2023)
Acidophilic	Amylases and glucoamylases	Starch processing	Parashar and Satyanarayana (2018)
	Endo-β-glucanases and endo-β-1,4-mannanase	Food industry	Lalithambika et al. (2012)
Alkaliphilic	Lipases, proteases, amylases, and cellulases	Additives to laundry detergents	Fujinami and Fujisawa (2010)
	Proteases	Leather manufacturing	Mesbah (2022)
	Cyclodextrin glycosyltransferases (CGTases)	Production of cyclodextrins	Martins and Hatti-Kaul (2002)
Psychrophilic	Proteases, lipases, and amylases	Detergent industry	Kumar et al. (2021)
	Psychrophilic alkaline phosphatases, nucleases, uracil-DNA N-glycosylases, proteases, and DNA ligases	Molecular biology tools	Mangiagalli et al. (2020)
Halophilic	Proteases and amylases	Manufacture of granular detergents	Mokashe et al. (2018)
	Cellulases, xylanases, and laccases	Pretreating lignocellulose with organic solvents or ionic liquids	Marhuenda-Egea and Bonete (2002)

improve baking, brewing, and dairy processes by retaining enzymatic activity during heating steps (Vieille & Zeikus, 2001). In molecular biology, the use of thermostable DNA polymerases is crucial for PCR, which requires DNA amplification at high temperatures (Akram et al., 2023). These enzymes are derived from bacteria and archaea that thrive in hot environments, such as *Thermus aquaticus* (Taq polymerase), *Pyrococcus furiosus* (Pfu polymerase), and *Thermococcus kodakarensis* (KOD polymerase).

21.3.2 Acidophilic Enzymes

Extracellular enzymes from acidophiles are resistant to low pH due to the prevalence of acidic amino acids on the protein surface (González et al., 2024). These enzymes, which tolerate low pH, include glucanases, amylases, glucoamylases, xylanases, cellulases, proteases, and oxidases. They are useful in biofuel production, starch processing, coal desulfurization, valuable metal recovery, and feed ingredients. The production of oligosaccharide syrup from starch requires gelatinization of the starch slurry, liquefaction by α-amylase, and digestion by glucoamylase.

In starch processing, the pH of the starch slurry is 3–5, and the use of acidophilic amylase eliminates the need to neutralize the starch slurry prior to liquefaction. This approach reduces both the time and cost of oligosaccharide production from raw starch. Several reports have documented acidophilic amylases and glucoamylases from acidophilic bacteria (*Bacillus*, *Alicyclobacillus*) and archaea (*Thermoplasma*, *Sulfolobus*, *Picrophilus*) (Parashar & Satyanarayana, 2018).

Numerous polysaccharide-degrading enzymes have been isolated from acidophiles. *Sulfolobus solfataricus* has three endo-β-glucanases. Among these enzymes, Sso1949 has optimum conditions at pH 1.8 and 80 °C (Lalithambika et al., 2012). In addition, the endoglucanase CelA4 from acidophilic *Alicyclobacillus* sp. operates optimally at 65 °C and pH 2.6, is stable over a wide pH range (1.8–7.6), and is resistant to both acidic and mesophilic proteases (Bai et al., 2010). Endo-1,4-β-glucanase from *Sulfolobus shibatae* is active at 95–100 °C, with maximum activity at pH 3.0–5.0 (Boyce & Walsh, 2018). Endo-β-1,4-mannanase produced by *Alicyclobacillus acidocaldarius* exhibits significant transglycosylation activity with relatively low hydrolytic activity and works optimally at pH 5.5 and 65 °C (Zhang et al., 2008). Carboxylesterases were isolated from a microbial community inhabiting acid mine drainage (pH ~ 2) (Vidal et al., 2022). In this study, 16 esterases were identified from microorganisms belonging to the genera *Acidithrix*, *Acidimicrobium*, and *Acidiphilium*. Several of these enzymes have demonstrated efficient degradation of acrylic acid and terephthalic acid-like esters, which may be relevant to the degradation of plastics. Activities of β-galactosidase, cellulase, lipase, xylanase, and protease have been identified in cell-free media obtained by culturing *Alicyclobacillus* sp. at pH 3 and 5 (Ortiz-Cortés et al., 2021).

21.3.3 Alkaliphilic Enzymes

Alkaliphilic enzymes exhibit maximum activity at pH 9 or higher and are produced primarily by alkaliphilic bacteria. These enzymes are characterized structurally by an abundance of arginine and histidine residues and a scarcity of glutamate residues, resulting in an alkaline isoelectric point compared to neutral enzymes. Arginine residues form ionic bonds with aspartic acid residues, which are thought to be important for enzyme stability under alkaline conditions (Fujinami & Fujisawa, 2010). Alkaliphilic enzymes are commonly added to laundry detergents. These enzymes, including alkaliphilic lipases, proteases, amylases, and cellulases, remain stable in alkaline detergents typically used in washing machines, resist the denaturing effects of surfactants and chelating agents, and are generally stable during long-term storage and in the presence of bleaching agents (Fujinami & Fujisawa, 2010). Alkaliphilic enzymes play a significant role in modern leather manufacturing, providing an eco-friendly alternative to traditional chemical processes. Leather production involves several steps, including soaking, dehairing, bating, and tanning. Historically, many of these steps relied on harsh chemicals such as lime and sodium sulfide, which can harm the environment and pose health risks to workers (Mesbah, 2022). Alkaliphilic enzymes, particularly alkaline proteases and lipases, have emerged as effective substitutes that enhance efficiency and sustainability. Cyclodextrin glycosyltransferases (CGTases) are crucial for the industrial production of cyclodextrins, cyclic oligosaccharides composed of six, seven, or eight glucose units (α-, β-, and γ-cyclodextrins, respectively). These are produced through the enzymatic conversion of starch and are widely used in pharmaceuticals, food, cosmetics, and agriculture for their ability to form inclusion complexes with various hydrophobic molecules, enhancing solubility and stability. Alkaliphilic bacteria, such as *Bacillus* species (e.g., *Bacillus firmus*, *Bacillus circulans*, and *Bacillus* sp.), are prolific producers of CGTases that function optimally at alkaline pH values (pH 9–11) (Martins & Hatti-Kaul, 2002). Other enzymes, such as alkaline amylases, lipases, and cellulases, are used in the food industry, paper and pulp processing, and biofuel production.

21.3.4 Psychrophilic Enzymes

Psychrophilic enzymes are produced by psychrophilic organisms and exhibit optimal activities between −20 and 10 °C, where typical enzymes from mesophilic organisms are usually inactive. Structurally, psychrophilic enzymes differ from their mesophilic and thermophilic counterparts by having greater molecular flexibility. Adaptations in protein structure to increase flexibility include a decreased ratio of arginine to lysine, a reduction in proline, an increase in glycine residues within the loop, decreased compactness of the hydrophobic core, increased surface hydrophobicity, weakened protein interactions (such as hydrogen bonds, disulfide bridges, electrostatic interactions, and weak metal bonds), and reduced formation of secondary structures (Sarmiento et al., 2015). One of the most prominent applications of these enzymes is in the detergent industry (Kumar et al., 2021). Psychrophilic proteases, lipases, and amylases are utilized in cold-water laundry detergents, effectively breaking down protein, fat, and starch stains at low washing temperatures (Mesbah, 2022). In addition, psychrophilic alkaline phosphatases, nucleases, uracil-DNA N-glycosylases, proteases, and DNA ligases have been commercialized as molecular biology tools (Mangiagalli et al., 2020). Hydrophilic enzymes are particularly appealing in molecular biology applications due to their instability at temperatures above 20 °C, allowing for easy inactivation through moderate heat treatment. This feature facilitates precise control over reaction timing, which is especially beneficial in multi-step molecular biology protocols. Enzymes are also employed in traditional processes such as brewing, baking, and cheese production. Recently, efforts have been made to replace traditional high-temperature processes with low-temperature alternatives, which offer advantages such as reduced environmental impact, lower costs, decreased microbial contamination, and improved shelf life (Mangiagalli et al., 2020).

21.3.5 Halophilic Enzymes

Halophilic enzymes are specialized proteins produced by halophilic microorganisms, uniquely adapted to remain stable and functional under conditions that denature or inhibit most other enzymes. These enzymes show optimal activity at sodium concentrations of 1.5 M or higher. Adaptations to high salt concentrations include an acidic isoelectric point and an increased number of aspartic acid and glutamic acid residues on the enzyme surface (Oren, 2013). Acidic amino acids help maintain protein solubility even at high salt concentrations due to their excellent water-binding capacity. Many halophilic enzymes are also tolerant of organic solvents, as these solvents reduce water activity, similar to high salt content. The stability of halophilic enzymes under low-water conditions makes them suitable for various biotechnological applications, allowing their use without stringent sterile conditions in processes involving organic solvents or salt water.

The salt stability of these enzymes is crucial for detergents, as NaCl is a component in the production of granular detergents. Consequently, halophilic proteases and amylases are incorporated into detergents (Mokashe et al., 2018). Halophilic enzymes serve as biocatalysts in unconventional organic solvents (Mesbah, 2022). In biofuel production, halophilic cellulases, xylanases, and laccases can be applied in the pretreatment of lignocellulose using organic solvents or ionic liquids. Furthermore, halophilic enzymes can be integrated into reverse micelles to develop biosynthetic processes in non-aqueous media (Marhuenda-Egea & Bonete, 2002).

21.4 Extremolytes

Extremolytes are low molecular weight organic compounds synthesized by extremophiles (Becker & Wittmann, 2020). These natural substances act as compatible solutes or chemical chaperones, helping maintain cell function without interfering with normal biochemical processes. Extremolytes play a crucial protective role by stabilizing proteins and cellular membranes, preventing protein denaturation, reducing oxidative stress, and preserving hydration levels under salt and temperature stress or in the presence of cytotoxins (Lentzen & Schwarz, 2006). Some well-known extremolytes include ectoine, hydroxyectoine, proline, mannitol, glycine-betaine, and trehalose, commonly found in halophilic bacteria (Becker & Wittmann, 2020). In addition, hyperthermophilic microorganisms often contain heterosides such as glucosylglycerol, glucosylglycerate, mannosylglycerate, and mannosylglyceramide (Becker & Wittmann, 2020). These compounds exhibit protective effects on proteins, DNA, food, feed, human cells, and human tissue. In industries such as cosmetics, medicine, and food, which require the protection of biomolecules and biological structures, extremolytes are gaining attention as substances with significant potential.

21.5 Conclusions

Extremophiles can produce a variety of useful substances, including organic acids, amino acids, biofuels, and lipids. Traditionally, the production of these products using mesophilic bacteria has been carried out under controlled fermentation conditions. However, as the demand for sustainable and cost-effective production increases, extremophiles are attracting attention as new cellular factories for useful substances. For instance, thermophiles, halophiles, alkaliphiles, and acidophiles provide significant advantages by enabling fermentation at high temperatures, high salt concentrations, high pH, and low pH, respectively. Thermophiles, for example, allow fermentation at high temperatures, reducing the risk of contamination while increasing substrate solubility and decreasing cooling costs in industrial settings. Similarly, acidophilic and alkaliphilic microorganisms can operate at pH levels that diminish the need for strict pH regulation and buffers, minimizing chemical use, environmental impact, and downstream processing costs.

Extremophiles are a source of valuable extremozymes, which represent a powerful tool for the next generation of biotechnological innovation. With continued interdisciplinary research and technological advancement, their applications are likely to expand across numerous industries, contributing to greener and more resilient industrial processes in the era of sustainable development. Furthermore, extremophiles produce extremolytes, which offer vast potential as natural, multifunctional protectants across multiple industries. With ongoing research, technological innovation, and commercialization efforts, extremolytes are poised to become key enablers of sustainable solutions in medicine, industry, agriculture, and beyond.

Despite these prospects, challenges remain. Extremophiles are often less well-characterized than conventional microbes, and their genetic manipulation can be more difficult. Moreover, optimizing industrial-scale processes for extreme conditions requires specialized equipment and engineering considerations. Nonetheless, advances in omics technologies, bioinformatics, and adaptive laboratory evolution are accelerating progress in this field.

Extremophiles present a promising option for sustainable production. With continued scientific and technical progress, they have the potential to transform the industrial biotechnology landscape and contribute significantly to a circular bioeconomy.

References

Abdel-Rahman, M. A., Hassan, S. E., Azab, M. S., Mahin, A. A., & Gaber, M. A. (2019). High improvement in lactic acid productivity by new Alkaliphilic bacterium using repeated batch fermentation integrated with increased substrate concentration. *BioMed Research International, 2019*, 7212870.

Akram, F., Shah, F. I., Ibrar, R., Fatima, T., Haq, I. U., Naseem, W., Gul, M. A., Tehreem, L., & Haider, G. (2023). Bacterial thermophilic DNA polymerases: A focus on prominent biotechnological applications. *Analytical Biochemistry, 671*, 115150.

Albrecht, M., Takaichi, S., Steiger, S., Wang, Z.-Y., & Sandmann, G. (2000). Novel hydroxycarotenoids with improved antioxidative properties produced by gene combination in *Escherichia coli*. *Nature Biotechnology, 18*, 843–846.

Aparici-Carratalá, D., Esclapez, J., Bautista, V., Bonete, M. J., & Camacho, M. (2023). Archaea: Current and potential biotechnological applications. *Research in Microbiology, 174*, 104080.

Assavasirijinda, N., Ge, D., Yu, B., Xue, Y., & Ma, Y. (2016). Efficient fermentative production ofpolymer-grade D-lactate by an engineered alkaliphilic *Bacillus* sp. strain under nonsterileconditions. *Microbial Cell Factories, 15*, 3.

Bai, Y., Wang, J., Zhang, Z., Shi, P., Luo, H., Huang, H., Feng, Y., & Yao, B. (2010). Extremely acidic beta-1,4-glucanase, CelA4, from thermoacidophilic *Alicyclobacillus* sp. A4 with high protease resistance and potential as a pig feed additive. *Journal of Agricultural and Food Chemistry, 58*, 1970–1975.

Becker, J., & Wittmann, C. (2020). Microbial production of extremolytes — High-value active ingredients for nutrition, health care, and well-being. *Current Opinion in Biotechnology, 65*, 118–128.

Bhandiwad, A., Shaw, A. J., Guss, A., Guseva, A., Bahl, H., & Lynd, L. R. (2014). Metabolic engineering of *Thermoanaerobacterium saccharolyticum* for *n*-butanol production. *Metabolic Engineering, 21*, 17–25.

Boyce, A., & Walsh, G. (2018). Expression and characterisation of a thermophilic endo-1,4-β-glucanase from *Sulfolobus shibatae* of potential industrial application. *Molecular Biology Reports, 45*, 2201–2211.

Brown, S. D., Guss, A. M., Karpinets, T. V., Parks, J. M., Smolin, N., Yang, S., Land, M. L., Klingeman, D. M., Bhandiwad, A., Rodriguez, M., Jr., Raman, B., Shao, X., Mielenz, J. R., Smith, J. C., Keller, M., & Lynd, L. R. (2011). Mutant alcohol dehydrogenase leads to improved ethanol tolerance in *Clostridium thermocellum*. *Proceedings of the National Academy of Sciences of the United States of America, 108*, 13752–13757.

Calder, P. C. (2015). Marine omega-3 fatty acids and inflammatory processes: Effects, mechanisms and clinical relevance. *Biochimica et Biophysica Acta, 1851*, 469–484.

Caro-Astorga, J., Meyerowitz, J. T., Stork, D. A., Nattermann, U., Piszkiewicz, S., Vimercati, L., Schwendner, P., Hocher, A., Cockell, C., & DeBenedictis, E. (2024). Polyextremophile engineering: A review of organisms that push the limits of life. *Frontiers in Microbiology, 15*, 1341701.

Converti, A., Casazza, A. A., Ortiz, E. Y., Perego, P., & Del Borghi, M. (2009). Effect of temperature and nitrogen concentration on the growth and lipid content of *Nannochloropsis oculata* and *Chlorella vulgaris* for biodiesel production. *Chemical Engineering and Processing: Process Intensification, 48*, 1146–1151.

Don, T. M., Chen, C. W., & Chan, T. H. (2006). Preparation and characterization of poly(hydroxyalkanoate) from the fermentation of *Haloferax mediterranei*. *Journal of Biomaterials Science Polymer Edition, 17*, 1425–1438.

Dong, X., Quinn, P. J., & Wang, X. (2011). Metabolic engineering of *Escherichia coli* and *Corynebacterium glutamicum* for the production of L-threonine. *Biotechnology Advances, 29*, 11–23.

Du, H., Zhao, Y., Wu, F., Ouyang, P., Chen, J., Jiang, X., Ye, J., & Chen, G.-Q. (2020). Engineering *Halomonas bluephagenesis* for L-threonine production. *Metabolic Engineering, 60*, 119–127.

Dumorné, K., Córdova, D. C., Astorga-Eló, M., & Renganathan, P. (2017). Extremozymes: A potential source for industrial applications. *Journal of Microbiology and Biotechnology, 27*, 649–659.

Fujinami, S. & Fujisawa, M. (2010). Industrial applications of alkaliphiles and their enzymes–past, present and future. Environmental Technology, 31, 845–856.

González, E., Vera, F., Scott, F., Guerrero, C., Bolívar, J. M., Aroca, G., Muñoz, J., Ladero, M., & Santos, V. E. (2024). Acidophilic heterotrophs: Basic aspects and technological applications. *Frontiers in Microbiology, 15*, 1374800.

Hirano, K., Kurosaki, M., Nihei, S., Hasegawa, H., Shinoda, S., Haruki, M., & Hirano, N. (2016). Enzymatic diversity of the *Clostridium thermocellum* cellulosome is crucial for the degradation of crystalline cellulose and plant biomass. *Scientific Reports, 6*, 35709.

Huber, G., & Stetter, K. O. (1991). Sulfolobus metallicus, sp. nov., a novel strictly Chemolithoautotrophic thermophilic archaeal species of metal-mobilizers. *Systematic and Applied Microbiology, 14*, 372–378.

Innis, S. M. (2007). Dietary (n-3) fatty acids and brain development. *The Journal of Nutrition, 137*, 855–859.

Kesbiç, F. I., & Gültepe, N. (2022). C(50) carotenoids extracted from *Haloterrigena thermotolerans* strain K15: Antioxidant potential and identification. *Folia Microbiologica, 67*, 71–79.

Kirti, K., Amita, S., Priti, S., Mukesh Kumar, A., & Jyoti, S. (2014). Colorful world of microbes: Carotenoids and their applications. *Advances in Biology, 2014*, 837891.

Kochhar, N., Kavya, I. K., Shrivastava, S., Ghosh, A., Rawat, V. S., Sodhi, K. K., & Kumar, M. (2022). Perspectives on the microorganism of extreme environments and their applications. *Current Research in Microbial Sciences, 3*, 100134.

Kumar, A., Mukhia, S., & Kumar, R. (2021). Industrial applications of cold-adapted enzymes: Challenges, innovations and future perspective. 3. *Biotech, 11*, 426.

Kundu, K., & Kumar, A. (2014). Biochemical engineering parameters for hydrometallurgical processes: Steps towards a deeper understanding. *Journal of Mining, 2014*, 290275.

Lalithambika, S., Peterson, L., Dana, K., & Blum, P. (2012). Carbohydrate hydrolysis and transport in the extreme thermoacidophile *Sulfolobus solfataricus. Applied and Environmental Microbiology, 78*, 7931–7938.

Lentzen, G., & Schwarz, T. (2006). Extremolytes: Natural compounds from extremophiles for versatile applications. *Applied Microbiology and Biotechnology, 72*, 623–634.

Li, W. F., Zhou, X. X., & Lu, P. (2005). Structural features of thermozymes. *Biotechnology Advances, 23*, 271–281.

Lynd Lee, R., Weimer Paul, J., van Zyl Willem, H., & Pretorius Isak, S. (2002). Microbial cellulose utilization: Fundamentals and biotechnology. *Microbiology and Molecular Biology Reviews, 66*, 506–577.

Mangiagalli, M., Brocca, S., Orlando, M., & Lotti, M. (2020). The "cold revolution". Present and future applications of cold-active enzymes and ice-binding proteins. *New Biotechnology, 55*, 5–11.

Marhuenda-Egea, F. C., & Bonete, M. J. (2002). Extreme halophilic enzymes in organic solvents. *Current Opinion in Biotechnology, 13*, 385–389.

Martins, R. F., & Hatti-Kaul, R. (2002). A new cyclodextrin glycosyltransferase from an alkaliphilic *Bacillus agaradhaerens* isolate: Purification and characterisation. *Enzyme and Microbial Technology, 30*, 116–124.

Mesbah, N. M. (2022). Industrial biotechnology based on enzymes from extreme environments. *Frontiers in Bioengineering and Biotechnology, 10*, 870083.

Miller, N. J., Sampson, J., Candeias, L. P., Bramley, P. M., & Rice-Evans, C. A. (1996). Antioxidant activities of carotenes and xanthophylls. *FEBS Letters, 384*, 240–242.

Mitra, R., Xu, T., Xiang, H., & Han, J. (2020). Current developments on polyhydroxyalkanoates synthesis by using halophiles as a promising cell factory. *Microbial Cell Factories, 19*, 86.

Mokashe, N., Chaudhari, B., & Patil, U. (2018). Operative utility of salt-stable proteases of halophilic and halotolerant bacteria in the biotechnology sector. *International Journal of Biological Macromolecules, 117*, 493–522.

Naitam, M. G., & Kaushik, R. (2021). Archaea: An agro-ecological perspective. *Current Microbiology, 78*, 2510–2521.

Nichols, D. S., & McMeekin, T. A. (2002). Biomarker techniques to screen for bacteria that produce polyunsaturated fatty acids. *Journal of Microbiological Methods, 48*, 161–170.

Obruča, S., Dvořák, P., Sedláček, P., Koller, M., Sedlář, K., Pernicová, I., & Šafránek, D. (2022). Polyhydroxyalkanoates synthesis by halophiles and thermophiles: Towards sustainable production of microbial bioplastics. *Biotechnology Advances, 58*, 107906.

Olson, D. G., McBride, J. E., Joe Shaw, A., & Lynd, L. R. (2012). Recent progress in consolidated bioprocessing. *Current Opinion in Biotechnology, 23*, 396–405.

Olson, D. G., Maloney, M. I., Lanahan, A. A., Cervenka, N. D., Xia, Y., Pech-Canul, A., Hon, S., Tian, L., Ziegler, S. J., Bomble, Y. J., & Lynd, L. R. (2023). Ethanol tolerance in engineered strains of *Clostridium thermocellum. Biotechnology for Biofuels and Bioproducts, 16*, 137.

Oren, A. (2013). Life at high salt concentrations, intracellular KCl concentrations, and acidic proteomes. *Frontiers in Microbiology, 4*, 315.

Ortiz-Cortés, L. Y., Ventura-Canseco, L. M. C., Abud-Archila, M., Ruíz-Valdiviezo, V. M., Velázquez-Ríos, I. O., & Alvarez-Gutiérrez, P. E. (2021). Evaluation of temperature, pH and nutrient conditions in bacterial growth and extracellular hydrolytic activities of two *Alicyclobacillus* spp. strains. *Archives of Microbiology, 203*, 4557–4570.

Parashar, D., & Satyanarayana, T. (2018). An insight into ameliorating production, catalytic efficiency, Thermostability and starch Saccharification of acid-stable α-amylases from Acidophiles. *Frontiers in Bioengineering and Biotechnology, 6*, 125.

Rajan, S. R., & Letourneau, D. K. (2012). What risk assessments of genetically modified organisms can learn from institutional analyses of public health risks. *Journal of Biomedicine & Biotechnology, 2012*, 203093.

Sahli, K., Gomri, M. A., Esclapez, J., Gómez-Villegas, P., Bonete, M.-J., León, R., & Kharroub, K. (2021). Characterization and biological activities of carotenoids produced by three haloarchaeal strains isolated from Algerian salt lakes. *Archives of Microbiology, 204*, 6.

Salas-Bruggink, D. I. J., Martín, J. S.-S., Leiva, G., & Blamey, J. M. (2024). Extremozymes: Challenges and opportunities on the road to novel enzymes production. *Process Biochemistry, 143*, 323–336.

Sarmiento, F., Peralta, R., & Blamey, J. M. (2015). Cold and hot Extremozymes: Industrial relevance and current trends. *Frontiers in Bioengineering and Biotechnology, 3*, 148.

Squillaci, G., Parrella, R., Carbone, V., Minasi, P., La Cara, F., & Morana, A. (2017). Carotenoids from the extreme halophilic archaeon *Haloterrigena turkmenica*: Identification and antioxidant activity. *Extremophiles, 21*, 933–945.

Tian, L., Papanek, B., Olson, D. G., Rydzak, T., Holwerda, E. K., Zheng, T., Zhou, J., Maloney, M., Jiang, N., Giannone, R. J., Hettich, R. L., Guss, A. M., & Lynd, L. R. (2016). Simultaneous achievement of high ethanol yield and titer in *clostridium thermocellum. Biotechnology for Biofuels, 9*, 116.

Tian, L., Cervenka, N. D., Low, A. M., Olson, D. G., & Lynd, L. R. (2019). A mutation in the AdhE alcohol dehydrogenase of *clostridium thermocellum* increases tolerance to several primary alcohols, including isobutanol, *n*-butanol and ethanol. *Scientific Reports, 9*, 1736.

Toivari, M., Vehkomäki, M.-L., Nygård, Y., Penttilä, M., Ruohonen, L., & Wiebe, M. G. (2013). Low pH D-xylonate production with *Pichia kudriavzevii. Bioresource Technology, 133*, 555–562.

Turner, P., Mamo, G., & Karlsson, E. N. (2007). Potential and utilization of thermophiles and thermostable enzymes in biorefining. *Microbial Cell Factories, 6*, 9.

Vidal, P., Martínez-Martínez, M., Fernandez-Lopez, L., Roda, S., Méndez-García, C., Golyshina, O. V., Guallar, V., Peláez, A. I., & Ferrer, M. (2022). Metagenomic mining for esterases in the microbial community of los rueldos acid mine drainage formation. *Frontiers in Microbiology, 13*, 868839.

Vieille, C., & Zeikus, G. J. (2001). Hyperthermophilic enzymes: Sources, uses, and molecular mechanisms for Thermostability. *Microbiology and Molecular Biology Reviews, 65*, 1–43.

Vieille, C., Burdette, D. S., & Zeikus, J. G. (1996). Thermozymes. *Biotechnology Annual Review, 2*, 1–83.

Wang, Y., Qian, J., Shi, T., Wang, Y., Ding, Q., & Ye, C. (2024). Application of extremophile cell factories in industrial biotechnology. *Enzyme and Microbial Technology, 175*, 110407.

Yang, R., Yang, C., Ma, L., Zhao, Y., Guo, Z., Niu, J., Chu, Q., Ma, Y., & Li, B. (2022). Identification of purine biosynthesis as an NADH-sensing pathway to mediate energy stress. *Nature Communications, 13*, 7031.

Zhang, J., & Burgess, J. G. (2017). Enhanced eicosapentaenoic acid production by a new deep-sea marine bacterium Shewanella electrodiphila MAR441T. *PLoS One, 12*, e0188081.

Zhang, Y., Ju, J., Peng, H., Gao, F., Zhou, C., Zeng, Y., Xue, Y., Li, Y., Henrissat, B., Gao, G. F., & Ma, Y. (2008). Biochemical and structural characterization of the intracellular mannanase AaManA of *Alicyclobacillus acidocaldarius* reveals a novel glycoside hydrolase family belonging to clan GH-A. *The Journal of Biological Chemistry, 283*, 31551–31558.

Zhao, C., Zheng, T., Feng, Y., Wang, X., Zhang, L., Hu, Q., Chen, J., Wu, F., & Chen, G.-Q. (2022). Engineered *Halomonas* spp. for production of L-Lysine and cadaverine. *Bioresource Technology, 349*, 126865.

Index

M. Ito et al. (eds.), *Extremophiles for a Sustainable Future*,
https://doi.org/10.1007/978-981-95-6970-0

MIX
Papier aus verantwortungsvollen Quellen
Paper from responsible sources
FSC® C105338

If you have any concerns about our products,
you can contact us on
ProductSafety@springernature.com

In case Publisher is established outside the EU,
the EU authorized representative is:
Springer Nature Customer Service Center GmbH
Europaplatz 3, 69115 Heidelberg, Germany

Printed by Libri Plureos GmbH
in Hamburg, Germany